Canadian Mathematical Society
Ouvrages de mathématiques de la SMC

Editors-in-Chief
Rédacteurs-en-chef
Jonathan Borwein
Peter Borwein

T0211726

Marian Mureşan

A Concrete Approach
to Classical Analysis

Marian Mureşan
Faculty of Mathematics & Computer Science
Babeş-Bolyai University
Str. Mihail Kogalniceanu nr. 1
400084 Cluj-Napoca
Romania
mmarianus24@yahoo.com

ISBN   978-1-4419-2705-7   e-ISBN   978-0-387-78933-0
DOI 10.1007/978-0-387-78933-0

Mathematics Subject Classification (2000): 46-xx

springer.com

To the memory of my mother

Computo ergo sum
D. H. Bailey[1]

Ἐν οἶδα ὅτι δέν οἶδα
(The only thing I know is that I don't know anything)
Socrates[2]

Also, he made a molten sea of ten cubits from brim to brim,
round in compass, and five cubits the height thereof;
and a line of thirty cubits did compass it round about.
Old Testament, 1 Kings 7:23

We would be amiss, however, if we did not emphasize that the
extended precision calculation of π has substantial application
as a test of the "global integrity" of a supercomputer.
D. H. Bailey, J. M. Borwein,[3] *and P. B. Borwein*[4] *in* [10]

... the main goal of computation in pure mathematics is
arguably to yield *insight*.
[22]

---

[1] David H. Bailey, Chief Technologist, Computational Research Dept., Lawrence
Berkeley Laboratory.
[2] Socrates of Athens (Σωκρατης ο Αθηναιος) ∼470–∼399 (BC).
[3] Jonathan Michael Borwein, 1951–.
[4] Peter Benjamin Borwein, 1953–.

Contents

List of Figures

Preface

This book reflects the conclusions of the author to some simple questions: "What should an easy comprehensible introduction to classical mathematical analysis look like? Can we avoid the basic results on differential and integral calculus to jump into abstract results? Actually, which results are considered as basic? Is the book a bridge to some new topics of research?" The influence of functional analysis and Bourbakism has been clear for a long time. At the same time, numerical methods emerged from analysis. It is hard to imagine discrete mathematics without analysis. New and even unexpected tendencies appeared. It is enough to mention some of them, experimental mathematics and scientific computing. These two topics are illustrated by two remarkable books [22] and [17].

Our answer to all these questions consists of our somehow taking all these fields into account. We mean, on the solid ground of classical results (sets, functions, metric spaces, sequences, series, limits, continuity, differentiability, and integrability) that we have to introduce newer results. Why introduce new results? They forcefully appear every day. Moreover, new and incredible methods appear. We mention only two of them presently considered as belonging to "experimental mathematics," namely the fast computation of the π number based on BBP methods, Ramanujan methods. Other methods explore strange functions by computers, that is, the nowhere differentiable functions. The latter topic has been considered as one belonging to "pure mathematics." Presently it came down into the laboratory of mathematical experiments. This means that by experimental methods we catch a result and then prove it rigorously.

We are pressed to take into account some parts from mathematics and to neglect many others. The present book is focused on differential and integral calculus.

Mathematical analysis offers a solid ground to many achievements in applied and discrete mathematics. In spite of the fact that this book concerns part of what is customarily called mathematical analysis, we have tried to include useful and relevant examples, exercises, and results enlightening the reader on the power of mathematical analysis tools. In this respect the topics covered by our book are quite "concrete."

The strong interplay between so-called theoretical mathematics and scientific computing has been emphasized by D. H. Bailey as "To this day I live in two worlds, theoretical math and scientific computing. I'm trying to marry these two by applying advanced computing to problems in pure mathematics. Experimental mathematics is the outcome."

We continuously had in front of our eyes a generic student wishing to know more about mathematical analysis at the beginning of his or her student life. We tried to offer paths from the standard knowledge of a student to modern and exciting topics in this way showing that a student from the first or second year is able to understand certain research problems.

The book has been divided into ten chapters and covers topics on sets and numbers, linear and metric spaces, sequences and series of numbers and functions, limits and continuity, differential and integral calculus of functions of one or several variables, constants (mainly π, but not only) and algorithms for finding them, the $W-Z$ method of summation, and estimates of algorithms and of certain combinatorial problems. Many challenging exercises accompany the text. Most of them have been the subjects of different mathematical competitions during the last few years. In this respect we consider that there is an appropriate balance between what is traditionally called theory and exercises.

The topics of the last two chapters bring the student closer to topics belonging also to computer science. In this way it is shown that the frontier between "pure" mathematics and other related topics is more or less a matter of taste.

It is the proper moment and place to express our sincere gratitude to Professor Heiner Gonska of the University of Duisburg-Essen, Germany, giving us, among others, the opportunity of using all the facilities of his department and library.

Thanks are due to Professor Jonathan M. Borwein of Dalhousie University, Halifax, Nova Scotia, Canada, for his constant and warm friendship along the years and to Professor Karl Dilcher of Dalhousie University, Halifax, Nova Scotia, Canada for his firm support in the publication of this book. The author is also grateful to the editors of Springer-Verlag, New York, for very strong and constant support offered to us.

Above all I express my deep gratitude to my wife Viorica for her unbroken encouragement, strong moral support, and constant understanding during the many days of work on this book.

Cluj-Napoca, December 2007   *Marian Mureşan*
                                                      Babeş-Bolyai University

1

Sets and Numbers

The aim of this chapter is to introduce several basic notions and results concerning sets and numbers.

1.1 Sets

1.1.1 The concept of a set

The basic notion of set theory which was first introduced by Cantor[1] occurs constantly in our results. Hence it would be useful to discuss briefly some of the notions connected to it before studying the mathematical analysis.

We take the notion of a set as being already known. Roughly speaking, a *set* (*collection, class, family*) is any identifiable collection of objects of any sort.

We identify a set by stating what its *members* (*elements*) are. The theory of sets has been described axiomatically in terms of the notion "member of" ([71]).

We make no effort to built the complete theory of sets, but will appeal throughout to intuition and elementary logic. The so-called "naive" theory of sets is completely satisfactory for us ([63]).

We usually adhere to the following notational conventions. Elements of sets are denoted by small letters: a, b, c, ..., x, y, z, α, β, γ, ... Sets are denoted by capital Roman letters: A, B, C, ..., X, Y, ... Families of sets are denoted by capital script letters: $\mathcal{A}$, $\mathcal{B}$, $\mathcal{C}$, ...

A set is often defined by some property of its elements. We write $\{x \mid P(x)\}$ (where $P(x)$ is some proposition about x) to denote the set of all x such that $P(x)$ is true. Here $\mid$ is read "such that".

If the object x *is* an element of the set A, we write $x \in A$; and $x \notin A$ means that this x *is not* in A or that *it does not belong* to A.

---

[1] Georg Ferdinand Ludwig Philipp Cantor, 1845–1918.

We write $\emptyset$ for the *empty* (*void*) set. It has no member at all.

For any object x, $\{x\}$ denotes the set whose only member is x. Then $x \in \{x\}$, but $x \neq \{x\}$. Similarly, $\{x_1, x_2, \ldots, x_n\}$ is the set whose elements are precisely x_1, x_2, ..., x_n. We emphasize that the order of elements in a set is irrelevant and that $\{x, x\} = \{x\}$.

Examples 1.1. Some examples of sets are listed below.

(a) The set of natural numbers, $\mathbb{N} = \{0, 1, 2, 3, \ldots\}$.
(b) The set of nonzero natural numbers, $\mathbb{N}^* = \{1, 2, 3, \ldots\}$.
(c) The set of integers, $\mathbb{Z} = \{0, \pm 1, \pm 2, \pm 3, \ldots\}$.
(d) The set of rational numbers, $\mathbb{Q} = \{p/q \mid p, q \in \mathbb{Z}, \ q \neq 0\}$; p is the *numerator* and q is the *denominator* of the fraction p/q.
(e) The set of positive integers less than 7.
(f) The set of Romanian cities having more than five million inhabitants.
(g) The set S of vowels in English alphabet. S may be written as $S = \{a, e, i, o, u\}$ or $S = \{x \mid x$ is a vowel in English alphabet$\}$.   △

The first axiomatic approach of the natural number system was realized by Peano[2] in [105]. His system is based on five axioms and a succession function satisfying

(a) There is a natural number 1.
(b) Every natural number a has a successor, denoted $s(a)$.
(c) There is no natural number whose successor is 1.
(d) Distinct natural numbers have distinct successors; that is, $a = b$ if and only if $s(a) = s(b)$.
(e) If a property is satisfied by 1 and also by the successor of every natural number that possesses it, then it is satisfied by all natural numbers.

It is clear that from the above axioms we get the set $\{1, 2, \ldots\}$. Nowadays zero is considered a natural number because it results in a richer algebraic structure. The only differences are that instead of (a), (c), and (e) one considers

(a′) There is a natural number 0.
(c′) There is no natural number whose successor is 0.
(e′) If a property is satisfied by 0 and also by the successor of every natural number that possesses it, then it is satisfied by all natural numbers.

Let A and B be sets such that every element of A is an element of B. Then A is called a *subset* of B and we write $A \subset B$ or $B \supset A$. In such a case we also say B is a *superset* of A. If $A \subset B$ and $B \subset A$, we write $A = B$. $A \neq B$ denies $A = B$. If $A \subset B$ and $A \neq B$, we say A is a *proper subset* of B and is sometimes written as $A \subsetneq B$. We remark that under this idea of equality of sets, the empty set is unique; that is, if $\emptyset_1$ and $\emptyset_2$ are any two empty sets, we have $\emptyset_1 \subset \emptyset_2$ and $\emptyset_2 \subset \emptyset_1$.

---

[2] Giuseppe Peano, 1858–1932.

Let A be a set. By $\mathcal{P}(A)$ we denote the family of subsets of A. Thus $\mathcal{P}(\emptyset) = \{\emptyset, \{\emptyset\}\}$. For $A = \{1, 2\}$, we have $\mathcal{P}(A) = \{\emptyset, \{1\}, \{2\}, \{1, 2\}\}$.

It is clear that if A is not a subset of B, the following statement has to be true "There exists an element x such that $x \in A$ and $x \notin B$."

1.1.2 Operations on sets

Let A and B be two sets. The set of elements that belong to A or to B (or to both) is called the *union of A and B*, and denoted $A \cup B$; see Figure 1.1. We write

$$A \cup B = \{x \mid x \in A \text{ or } x \in B\}.$$

Let $\mathcal{A}$ be a family of sets. Then we write

$$\cup \mathcal{A} = \{x \mid x \in A \text{ for some } A \in \mathcal{A}\}.$$

Similarly, let $\{A_\alpha\}_{\alpha \in I}$ be a family of sets indexed by the index set I. We write

$$\cup_{\alpha \in I} A_\alpha = \{x \mid x \in A_\alpha \text{ for some } \alpha \in I\}.$$

Let A and B be two sets. The set of elements that belong to both A and B is called the *intersection of A and B*, and denoted $A \cap B$; see Figure 1.2. We write

$$A \cap B = \{x \mid x \in A \text{ and } x \in B\}.$$

Let $\mathcal{A}$ be a family of sets. Then we write

$$\cap \mathcal{A} = \{x \mid x \in A \text{ for all } A \in \mathcal{A}\}.$$

Similarly, let $\{A_\alpha\}_{\alpha \in I}$ be a family of sets indexed by the index set I. We write

$$\cap_{\alpha \in I} A_\alpha = \{x \mid x \in A_\alpha \text{ for all } \alpha \in I\}.$$

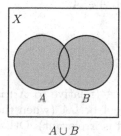

$A \cup B$

Fig. 1.1. Union

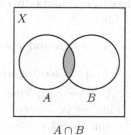

$A \cap B$

Fig. 1.2. Intersection

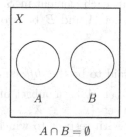

$A \cap B = \emptyset$

Fig. 1.3. Disjoint sets

Theorem 1.1. *Let* $A, B,$ *and* C *be sets. Then*

(a) $A \cup B = B \cup A$.

(a') $A \cap B = B \cap A$.

(b) $A \cup A = A$.

(b') $A \cap A = A$.

(c) $A \cup \emptyset = A$.

(c') $A \cap \emptyset = \emptyset$.

(d) $A \cup (B \cup C) = (A \cup B) \cup C$.

(d') $A \cap (B \cap C) = (A \cap B) \cap C$.

(e) $A \subset A \cup B$.

(e') $A \cap B \subset A$.

(f) $A \subset B \iff A \cup B = B$.

(f') $A \subset B \iff A \cap B = A$.

Thus the union and intersection are commutative, associative, and idempotent.

Theorem 1.2. *Let* $A, B,$ *and* C *be sets. Then*

(a) $\qquad A \cap (B \cup C) = (A \cap B) \cup (A \cap C)$ $\qquad$ distributive law;

(b) $\qquad A \cup (B \cap C) = (A \cup B) \cap (A \cup C)$ $\qquad$ distributive law.

Theorem 1.3. *Let* X *be a set and* $\{A_\alpha\}_{\alpha \in I}$ *a family of sets. Then*

(a) $X \cup (\cup_{\alpha \in I} A_\alpha) = \cup_{\alpha \in I}(X \cup A_\alpha)$.

(b) $X \cap (\cap_{\alpha \in I} A_\alpha) = \cap_{\alpha \in I}(X \cap A_\alpha)$.

(c) $X \cup (\cap_{\alpha \in I} A_\alpha) = \cap_{\alpha \in I}(X \cup A_\alpha)$.

(d) $X \cap (\cup_{\alpha \in I} A_\alpha) = \cup_{\alpha \in I}(X \cap A_\alpha)$.

We say that A and B are *disjoint*, provided $A \cap B = \emptyset$; see Figure 1.3. Let $\mathcal{A}$ be a family of sets such that any two distinct members of $\mathcal{A}$ are disjoint. Then family $\mathcal{A}$ is said to be *pairwise disjoint*. Thus an indexed family $\{A_\alpha\}_{\alpha \in I}$ is pairwise disjoint if $A_\alpha \cap A_\beta = \emptyset$ whenever $\alpha, \beta \in I$ and $\alpha \neq \beta$.

A family $\mathcal{A}$ of nonempty subsets of a set S is said to be a *partition* of S, provided the following two conditions are satisfied.

(i) $S = \cup_{A \in \mathcal{A}} A$.

(ii) $\mathcal{A}$ is a pairwise disjoint family.

Thus each element in S belongs to one and only one set $A \in \mathcal{A}$.

Let A and B be two sets. Then

$$A \setminus B = \{x \mid x \in A \text{ and } x \notin B\}$$

is said to be the *difference* of A and B; see Figure 1.4.

Let A be a subset of a set X. The *complement of* A (relative to X) is the set $\{x \mid x \in X, x \notin A\}$. This set is denoted by $\complement_X A$ or $\complement A$ (when the set with respect to which the complement is considered is irrelevant). Other notation is $X \setminus A$.

Proposition 1.1. *We are given two sets* X *and* A *so that* $A \subset X$. *Then* $\complement_X(\complement_X A) = A$.

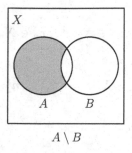

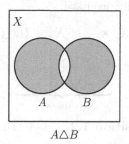

$A \setminus B$

$A \triangle B$

Fig. 1.4. Difference of sets   **Fig. 1.5.** Symmetric difference

Theorem 1.4. (de Morgan[3] laws) *One has*

(a) $\complement(A \cup B) = (\complement A) \cap (\complement B)$.

(b) $\complement(A \cap B) = (\complement A) \cup (\complement B)$.

(c) $\complement(\cup_{\alpha \in I} A_\alpha) = \cap_{\alpha \in I} \complement A_\alpha$.

(d) $\complement(\cap_{\alpha \in I} A_\alpha) = \cup_{\alpha \in I} \complement A_\alpha$.

Let A and B be two sets. The *symmetric difference of A and B* is the set $(A \setminus B) \cup (B \setminus A)$ and we write $A \triangle B$ for this set; see Figure 1.5. Note that $A \triangle B$ is the set consisting of those elements which are in exactly one of A or B. It can be defined as well by

$$A \triangle B = (A \cap \complement B) \cup (\complement A \cap B).$$

Sometimes it becomes significant to consider the order of the elements in a set. If we consider a pair (x_1, x_2) of elements in which we distinguish x_1 as the first element and x_2 as the second element, then (x_1, x_2) is called an *ordered pair*. Thus, two ordered pairs (x, y) and (u, v) are equal if and only if $x = u$ and $y = v$.

Let X and Y be nonempty sets. The *Cartesian*[4] *product of X and Y* (in this order) is the set $X \times Y$ of all ordered pairs (x, y) such that $x \in X$ and $y \in Y$. Hence, $X \times Y = \{(x, y) \mid x \in X, y \in Y\}$. Generally, $X \times Y \neq Y \times X$.

Remark. $(1, 2) \neq (2, 1)$ and $\{1, 2\} = \{2, 1\}$. $\triangle$

We usually write X^2 instead of $X \times X$.

The Cartesian product of three sets, say X, Y, and Z (in this order) is defined as

$$(X \times Y) \times Z = X \times (Y \times Z) = X \times Y \times Z = \{(x, y, z) \mid x \in X, y \in Y, z \in Z\}.$$

---

[3] August de Morgan, 1806–1871.

[4] Renatus Cartesius, the Latin name of René Du Perron Descartes, 1596–1650.

Generally, we have

$$X_1 \times X_2 \times \cdots \times X_n = \{(x_1, x_2, \ldots, x_n) \mid x_1 \in X_1, x_2 \in X_2, \ldots, x_n \in X_n\}.$$

Thus

$$X^n = \underbrace{X \times X \times \cdots \times X}_{n \text{ times}} = \{(x_1, x_2, \ldots, x_n) \mid x_i \in X, \ i = 1, 2, \ldots, n\}.$$

Proposition 1.2. *Let A, B, and C be sets. Then*

(a) $A \times (B \cup C) = (A \times B) \cup (A \times C)$.

(b) $A \times (B \cap C) = (A \times B) \cap (A \times C)$.

(c) *Suppose C and D are nonempty. Then $C \times D \subset A \times B \iff C \subset A$ and $D \subset B$.*

(d) $A \times B = \emptyset \iff A = \emptyset$ or $B = \emptyset$.

(e) $A \times B = B \times A \iff A = B$.

1.1.3 Relations and functions

A *(binary) relation R on two sets X and Y* is a subset of the Cartesian product of X and Y; that is, R is a relation on X and $Y \iff R \subset X \times Y$.

Let R be a relation on X and Y. The *domain of R* is the set

$$\mathrm{Dom}\, R = \{x \in X \mid (x, y) \in R \text{ for some } y \in Y\}.$$

The *range of R* is the set

$$\mathrm{Range}\, R = \{y \in Y \mid (x, y) \in R \text{ for some } x \in X\}.$$

The symbol R^{-1} denotes the *inverse of R*; that is, $R^{-1} = \{(y, x) \mid (x, y) \in R\}$.

Let R and Q be relations. The *composition (product)* of two relations R and Q is the relation

$$R \circ Q = \{(x, z) \mid \text{ for some } y, \ (x, y) \in Q \text{ and } (y, z) \in R\}.$$

The composition of R and Q may be empty. $R \circ Q \neq \emptyset \iff (\mathrm{Range}\, Q) \cap (\mathrm{Dom}\, R) \neq \emptyset$. Given R a relation on X and Y, $A \subset X$. The *image of A under R* is the set

$$R(A) = \{y \in Y \mid \exists x \in A \text{ such that } (x, y) \in R\}.$$

Proposition 1.3. *Let R, Q, and S be relations, and A and B be sets. Then*

(a) $(R^{-1})^{-1} = R$.

(b) $(R \circ Q)^{-1} = Q^{-1} \circ R^{-1}$.

(c) $R \circ (Q \circ S) = (R \circ Q) \circ S$.

(d) $(R \circ Q)(A) = R(Q(A))$.

(e) $R(A \cup B) = R(A) \cup R(B)$.

(f) $R(A \cap B) \subset R(A) \cap R(B)$.

An *equivalence relation* on a nonempty set X is a relation $\sim \subset X \times X$ such that for all x, y, and z in X the following conditions are satisfied.

(i) $x \sim x$ (reflexive).
(ii) $x \sim y$ implies $y \sim x$ (symmetric).
(iii) $x \sim y$ and $y \sim z$ imply $x \sim z$ (transitive).

Examples 1.2. (a) The usual "=" on $\mathbb{Q}$ is an equivalence relation on $\mathbb{Q}$.
(b) Let $\mathbb{Z}$ be the set of integers and settle down a natural number n. For every $a, b \in \mathbb{Z}$, we say "*a is congruent to b* modulo n" if $a - b = kn$ for some integer k. Here "congruence modulo n" is an equivalence relation on $\mathbb{Z}$. In notational form we write

$$a = b \quad (\text{mod } n) \Longleftrightarrow \exists \, k \in \mathbb{Z} \text{ such that } a - b = kn.$$

(c) Let $\mathbb{Z}$ be the set of integers and $x, y \in \mathbb{Z}$. Define $x \sim y$ if and only if $x - y$ is even. It is easy to check that $\sim$ is an equivalence relation on $\mathbb{Z}$. $\triangle$

Proposition 1.4. *Suppose $\sim$ is an equivalence relation on nonempty set X. Then it defines a partition of X. Conversely, each partition of X defines an equivalence on X.*

Proof. Define a nonempty subset S of X by

$$x, y \in S \Longleftrightarrow x \sim y.$$

If $x \in X$, then $S \neq \emptyset$. Obviously, the class of sets $\{S\}$ is a partition of X.
   Conversely, let $\{S\}$ be a partition of X. Define an equivalence relation on X by

$$x \sim y \Longleftrightarrow x, y \in S. \quad \square$$

The partition of X generated by an equivalence relation $\sim$ on X is denoted $X/\sim$ and is said to be the *equivalence classes* generated by $\sim$.
   Let P be a nonempty set. A *partial ordering* on P is a relation $\leq \subset P \times P$ such that for each x, y, and z in P one has

(i) $x \leq x$ (reflexive).
(ii) $x \leq y$ and $y \leq x$ imply $y = x$ (antisymmetric).
(iii) $x \leq y$ and $y \leq z$ imply $x \leq z$ (transitive).

Assume $\leq$ is a partial ordering on P. Then the pair $(P, \leq)$ is called a *partially ordered set.*

Examples 1.3. (a) Let X be a nonempty set and consider $A, B \subset X$. Define $A \leq B$ whenever $A \subset B$. Then " $\leq$ " is a partial ordering on the class of subsets of X.
(b) For $m, n \in \mathbb{N}$ define $m \leq n$ if there exists $k \in \mathbb{N}^*$ such that $m = kn$. Then " $\leq$ " is a partial ordering on $\mathbb{N}$. $\triangle$

If, moreover, the partial ordering relation $\leq$ satisfies

(iv) $x, y \in P$ implies $x \leq y$ or $y \leq x$,

then $\leq$ is called a *total ordering on* P. Assume $\leq$ is a total ordering on P. Then the pair $(P, \leq)$ is called a *totally ordered set*.

If $x \leq y$ and $x \neq y$, then we write $x < y$. The expression $x \geq y$ means $y \leq x$ and $x > y$ means $y < x$.

Law (iv) is sometimes stated "for arbitrary elements x and y in a totally ordered set exactly one of the relations $x < y$, $x = y$, $x > y$ holds." This law is called the *trichotomy law*.

Examples 1.4. (a) The usual $\leq$, meaning "less or equal", is a total ordering on $\mathbb{Q}$.
(b) Let A be a nonempty set. Then the relation $\subset$ on the class of all subsets of A is a partial ordering on it, as we already saw by (a) of Examples 1.3. We emphasize it is not a total ordering on A. $\triangle$

A subset Q of a partially ordered set P with order $\leq$ is said to be a *down-set* of P if whenever $x \in Q$, $y \in P$, and $y \leq x$, then $y \in Q$. Let R be an arbitrary subset of a partially ordered set P with order $\leq$. Then the smallest down-set containing R, denoted $\downarrow R$ is the set of all $x \in P$ for which there is a $y \in R$ such that $x \leq y$. If R is a singleton (i.e., $R = \{r\}$), then $\downarrow R = \downarrow r$ is said to be a *principal down-set* of P.

Suppose $\leq$ is a total ordering on P such that

(v) $\emptyset \neq A \subset P$ implies there exists an element $a \in A$ such that $a \leq x$ for each $x \in A$ (a is the *smallest element of* A);

then $\leq$ is called a *well ordering on* P.

Assume $\leq$ is a well ordering on P. Then the pair $(P, \leq)$ is called a *well-ordered set*.

Example 1.1. The set $\mathbb{N}$ of natural numbers with the usual ordering $\leq$ is a well-ordered set, whereas $\mathbb{Z}$ with the usual ordering $\leq$ is not a well-ordered set. $\triangle$

A binary relation $\preceq$ on a nonempty set X that is only transitive and reflexive is said to be a *quasi-order on* X. Then the *symmetric core* $\equiv$ of $\preceq$, defined by

$$x \equiv y \iff x \preceq y \text{ and } y \preceq x,$$

is an equivalence relation on X. Moreover, $\preceq$ defines a relation $\sqsubseteq$ on the quotient set $X/\equiv$ of equivalence classes $[x] = \{y \in X \mid x \equiv y\}$ as

$$[x] \sqsubseteq [y] \iff x \preceq y. \tag{1.1}$$

Note, the relation $\sqsubseteq$ is a partial order on $X/\equiv$. Conversely

Proposition 1.5. *Given an equivalence relation $\equiv$ on X together with a partial order $\sqsubseteq$ on the equivalence classes $X/\equiv = \{[x] \mid x \in X\}$, condition (1.1) defines a quasi-order $\preceq$ on X.*

Let P be a totally ordered set. For $x, y \in P$, we define $y = \max\{x, y\}$ if $x \leq y$, and $x = \max\{x, y\}$ if $y \leq x$. For a finite subset $\{x_1, \ldots, x_n\}$ (not all x_ks are necessarily distinct), we define $\max\{x_1, \ldots, x_n\} = \max\{\max\{x_1, \ldots, x_{n-1}\}, x_n\}$. Similarly, we define $\min\{x, y\}$. That is, $x = \min\{x, y\}$ whenever $x \leq y$, and $y = \min\{x, y\}$ whenever $y \leq x$. Also $\min\{x_1, \ldots, x_n\} = \min\{\min\{x_1, \ldots, x_{n-1}\}, x_n\}$.

Let $(P, \leq)$ be a partially ordered set and A be a nonempty subset of P. An element $x \in P$ is said to be

(i) A *lower bound of* A if $x \leq y$ for every $y \in A$. In this case we say that A is *bounded below*.

(ii) An *upper bound of* A if $y \leq x$ for every $y \in A$. In this case we say that A is *bounded above*.

(iii) The *greatest lower bound of* A or *infimum of* A if
   (iii$_1$) x is a lower bound of A,
   (iii$_2$) If $x < y$, then y is not a lower bound of A.

(iv) The *least upper bound of* A or *supremum of* A if
   (iv$_1$) x is an upper bound of A,
   (iv$_2$) If $y < x$, then y is not an upper bound of A.

A nonempty subset A of a partially ordered set is said to be *bounded* if it is bounded below and above. The empty set is bounded by definition. A is said to be *unbounded* if it is not bounded.

Remark 1.1. A nonempty subset A of a partially ordered set may have several lower and/or upper bounds whereas it has at most one infimum (denoted inf A) and at most one supremum (denoted sup A). $\triangle$

Example. Let A consist of all numbers $1/n$, where $n = 1, 2, \ldots$. Then by the usual $\leq$ on the set of rational numbers the set A is bounded, $\sup A = 1$, $\inf A = 0$, and $1 \in A$ whereas $0 \notin A$. $\triangle$

Let f be a relation and A be a set. As we already saw the image of A under f is the set $f(A) = \{y \mid (x, y) \in f$ for some $x \in A\}$. Note $f(A) \neq \emptyset \iff A \cap \text{Dom} f \neq \emptyset$. If $f(A) \subset B$, then this is interpreted as "f maps the set A into the set B." The *inverse image of* A *under* f is the set

$$f^{-1}(A) = \{x \mid (x, y) \in f \text{ for some } y \in A\} = \{x \mid f(x) \cap A \neq \emptyset\}.$$

A relation f is said to be *single-valued* if $(x, y) \in f$ and $(x, z) \in f$ imply $y = z$. In such a case we write $f(x) = y$. Otherwise it is also called a *multifunction* or *set-valued function* or even *correspondence*. A single-valued relation is said to be a *function* (*mapping, map, application, transformation, operator*). If f and f^{-1} are both single-valued, f is said to be a *bijective* function, Figure 1.8.

Theorem 1.5. *Let* X *and* Y *be nonempty sets and* $f \subset X \times Y$ *be a relation. Suppose that* $\{A_i\}_{i \in I}$ *is a family of subsets of* X *and* $\{B_j\}_{j \in J}$ *is a family of subsets of* Y. *Then*

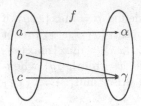 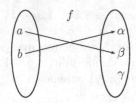 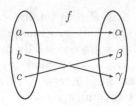

Fig. 1.6. Non-one-to-one   **Fig. 1.7.** Non-onto   **Fig. 1.8.** Bijective

(a) $f(\cup_{i \in I} A_i) = \cup_{i \in I} f(A_i)$.
(b) $f^{-1}(\cup_{j \in J} B_j) = \cup_{j \in J} f^{-1}(B_j)$.
(c) $f(\cap_{i \in I} A_i) \subset \cap_{i \in I} f(A_i)$.

The following statements are true if f is a function, but may fail for arbitrary relations.

(d) $f^{-1}(\cap_{j \in J} B_j) = \cap_{j \in J} f^{-1}(B_j)$.
(e) $f^{-1}(\complement_Y B) = \complement_X(f^{-1}(B))$, $B \subset Y$.
(f) $f(f^{-1}(B) \cap A) = B \cap f(A)$, $A \subset X$, $B \subset Y$.

Let f be a function such that $\operatorname{Dom} f = X$ and $\operatorname{Range} f \subset Y$. Then f is said to be a function *from* (on) X *into* (to) Y and we write $f : X \to Y$. If $\operatorname{Range} f = Y$, we say that f is *onto*, or *surjective*; that is, $f(X) = Y$. It means that for every $y \in Y$ there exists an $x \in X$ such that $y = f(x)$.

f is said to be *one-to-one* or *injective* if for any $x, t \in X$ with $x \neq t$, one has $f(x) \neq f(t)$. Equivalently, $f(x) = f(t)$ implies $x = t$. In other words, a function $f : X \to Y$ is said to be one-to-one if distinct elements in X have distinct images in Y, that is, if no two different elements in X have the same image. Figure 1.6 exhibits a surjective but not a one-to-one function, Figure 1.7 exhibits a one-to-one but not an onto function, and Figure 1.8 presents a bijective function.

Theorem 1.6. *Let X and Y be nonempty sets and $f : X \to Y$ be a function. Then f is bijective if and only if it is one-to-one and onto.*

Consider $f : X \to Y$ and $\emptyset \neq A \subset X$. A function $f_A : A \to Y$ defined as $f_A(x) = f(x)$, for all $x \in A$, is said to be the *restriction* of f to A.

Consider $f : A \to Y$ and a set $X \supset A$. A function $g : X \to Y$ satisfying $g(a) = f(a)$, for every $a \in A$, is said to be an *extension* of f to X.

Let X be a nonempty set. A *sequence* in X is a function having $\mathbb{N}^*$ or $\mathbb{N}$ or even an unbounded subset of it as its domain and X its range. Let x be such a function. We often write x_n instead of $x(n)$ for the value of x at n. The value x_n is called the nth *term* of the sequence. The sequence whose nth term is x_n is denoted by $(x_n)_{n=1}^{\infty}$, $(x_n)_{n \geq 1}$, $(x_n)_n$, or (x_n). A sequence (x_n) is said to be *in* a set X if $x_n \in X$ for each n.

1.2 Sets of numbers

A satisfactory discussion of the main concepts of analysis (e.g., convergence, continuity, differentiation, and integration) have to be based on an accurately defined number concept.

We do not, however, enter into any discussion of the axioms governing the arithmetic of the integers, but we take the rational number system as our starting point.

1.2.1 Two examples

It is well known that the rational number system is inadequate for many purposes. Maybe the oldest and the most frustrating case is the following. Consider an isosceles right triangle having the length of the catheti equal to 1. Can we express the length of the hypotenuse by a rational number? By the Pythagoras[5] theorem the square of the hypotenuse is 2.

Example 2.1. We start by showing that the equation

$$p^2 = 2 \tag{1.2}$$

is not satisfied by any rational p. For, suppose that (1.2) is satisfied. Then we can write $p = m/n$, where m and n are integers with $n \neq 0$. We may assume that m and n have no common divisor. Then (1.2) implies

$$m^2 = 2n^2. \tag{1.3}$$

This shows that m^2 is even. Hence m is even (if m is odd, m^2 is odd as well), and so m^2 is divisible by 4. It follows that the right-hand side of (1.3) is divisible by 4, so that n^2 is even, which implies that n is even.

Thus the assumption that (1.2) holds for a rational number leads us to the conclusion that both m and n are even, contrary to our choice on m and n. Hence (1.2) is impossible for rational p. So, the length of the hypotenuse to an isosceles right triangle with unitary catheti is nonrational. Several comments on this topic may be found at page 355.

We get the same conclusion following a different path. Relation (1.3) cannot have a nonzero solution in integers because the last nonzero digit of a square, written in base 3, is necessarily 1, whereas the last digit of twice a square is 2. $\triangle$

We examine this situation a little more closely.

Example 2.2. Let A be the set of all positive rationals p such that $p^2 < 2$, and let B be the set of all positive rationals p such that $p^2 > 2$. A and B are nonempty because $1 \in A$ and $2 \in B$. We show that A contains no largest element, and B contains no smallest element.

---

[5] Pythagoras of Samos ($\Pi\upsilon\vartheta\alpha\gamma\acute{o}\rho\alpha\varsigma$ o $\Sigma\acute{\alpha}\mu\iota o\varsigma$), $\sim$560–$\sim$480 (BC).

More explicitly, for every $p \in A$ we can find a rational $q \in A$ such that $p < q$, and for every $p \in B$ we can find a rational $q \in B$ such that $q < p$.

Suppose that $p \in A$. Then $p^2 < 2$. Choose a rational h such that $0 < h < 1$ and such that

$$h < \frac{2 - p^2}{2p + 1}.$$

Put $q = p + h$. Then $q > p$, and

$$q^2 = p^2 + (2p + h)h < p^2 + (2p + 1)h < p^2 + (2 - p^2) = 2,$$

so that q is in A. This proves the first part of our assertion.

Next, suppose that $p \in B$. Then $p^2 > 2$. Put

$$q = p - (p^2 - 2)/2p = p/2 + 1/p.$$

Then $0 < q < p$ and

$$q^2 = p^2 - (p^2 - 2) + \left(\frac{p^2 - 2}{2p}\right)^2 > p^2 - (p^2 - 2) = 2,$$

so that $q \in B$.   △

Remark. The purpose of the above discussion was to show that the rational number system has certain gaps, in spite of the fact that between any two distinct rationals there is another one (because $p < (p + q)/2 < q$). A deeper result is introduced by Corollary 2.11.   △

1.2.2 The real number system

There are several ways to introduce the real number set. We selected one of them assuming that it is easiest and shortest.

We say that a set X is the *real number set* provided on it there are defined two operations

$$X \times X \ni (x, y) \mapsto x + y \in X,$$
$$X \times X \ni (x, y) \mapsto xy \in X$$

called *addition* and *multiplication* as well as a binary relation $\leq$ called *ordering* and satisfying the following axioms (conditions, assumptions).

(R$_1$) $(x + y) + z = x + (y + z)$, $\forall x, y, z \in X$.
(R$_2$) There exists an element $0 \in X$, called *zero* or *null* such that $x + 0 = 0 + x = x$, $\forall x \in X$.
(R$_3$) For each $x \in X$ there exists an element $-x \in X$, called the *opposite* to x, such that $x + (-x) = (-x) + x = 0$.
(R$_4$) $x + y = y + x$, $\forall x, y \in X$.
(R$_5$) $(xy)z = x(yz)$, $\forall x, y, z \in X$.

(R_6) There exists an element $1 \in X \setminus \{0\}$, called *unity* or *identity*, such that $x \cdot 1 = 1 \cdot x = x$, $\forall x \in X$.

(R_7) For each element $x \in X \setminus \{0\}$, there exists an element $x^{-1} \in X$, called the *inverse* of x, such that $xx^{-1} = x^{-1}x = 1$.

(R_8) $xy = yx$, $\forall x, y \in X$.

(R_9) $x(y + z) = xy + xz$, $\forall x, y, z \in X$.

(R_{10}) $x \leq x$, $\forall x \in X$.

(R_{11}) For every $x, y \in X$, $x \leq y$ and $y \leq x$ imply $x = y$.

(R_{12}) For every $x, y, z \in X$, $x \leq y$ and $y \leq z$ imply $x \leq z$.

(R_{13}) For every $x, y \in X$ we have $x \leq y$ or $y \leq x$.

(R_{14}) For every $x, y, z \in X$, $x \leq y$ implies $x + z \leq y + z$.

(R_{15}) For every $x, y \in X$, $x \geq 0$ and $y \geq 0$, imply $xy \geq 0$.

(R_{16}) For every ordered pair (A, B) of nonempty subsets of X having the property that $x \leq y$ for every $x \in A$ and $y \in B$ there exists an element $z \in X$ such that

$$x \leq z \leq y, \quad \text{for every } x \in A \text{ and } y \in B.$$

Remarks. From (R_1)–(R_4) we have that $(X, +)$ is an *Abelian*[6] (*commutative*) group. So the null element and the opposite of an element are unique. Also $-0 = 0$ and $-(-x) = x$, for all $x \in X$. From (R_5)–(R_8) we have that $(X \setminus \{0\}, \cdot)$ is an Abelian group. Therefore the identity element and the inverse element of an element are unique. Also $1^{-1} = 1$ and $(x^{-1})^{-1} = x$, for all $x \in X \setminus \{0\}$. From (R_1)–(R_9) we have that $(X, +, \cdot)$ is a *field*; from (R_{10})–(R_{13}) it follows that $(X, \leq)$ is a *totally ordered* set; from (R_1)–(R_{15}) one has that $(X, +, \cdot, \leq)$ is a *totally ordered field*. (R_{14}) and (R_{15}) express the compatibility of the ordering relation with the algebraic operations. (R_{16}) has a special rôle that is made clear later.

We see at once that the set of rational numbers $\mathbb{Q}$ is a totally ordered field. At the same time the set of rational numbers does not fulfill Axiom (R_{16}), as we already saw by Example 2.2. $x^{-1} \in X \setminus \{0\}$ from (R_7) is denoted as $1/x$, too. Hence $y/x = yx^{-1}$. $\triangle$

For a while we ignore assumption (R_{16}).

Proposition 2.6. *There hold* $x \cdot 0 = 0 \cdot x = 0$.

Proof. We have

$$x \cdot 0 = x(0 + 0) = x \cdot 0 + x \cdot 0 \implies x \cdot 0 = x \cdot 0 - x \cdot 0 = 0.$$

Similarly, $0 \cdot x - 0$.   □

Remarks. From Proposition 2.6 it follows that 0 has no inverse because for each $x \in X$, $0 \cdot x = 0$ and $0 \neq 1 \in X \setminus \{0\}$. It also follows that the multiplication (defined on X and associative and commutative on $X \setminus \{0\}$)

---

[6] Niels Henrik Abel, 1802–1829.

is associative and commutative on X. Obviously, if $x = 0$ or $y = 0$, then $x \cdot y = 0$. $\triangle$

Proposition 2.7. *If $xy = 0$, then $x = 0$ or $y = 0$.*

Proof. Suppose $x \neq 0$. Then there exists x^{-1}. From one side $x^{-1}(xy) = (x^{-1}x)y = y$ and from the other side $x^{-1}(xy) = x^{-1}0 = 0$. Thus if $x \neq 0$, then $y = 0$. Similarly, by commutativity, if $y \neq 0$, then $x = 0$. $\square$

Proposition 2.8. *For every $x \in X$, $-x = (-1) \cdot x$.*

Proof. We have $x(1 + (-1)) = x \cdot 0 = 0$ and $x(1 + (-1)) = x \cdot 1 + x \cdot (-1) = x + x \cdot (-1)$. Then the conclusion follows. $\square$

Corollary 2.1. *One has $(-1)^2 = (-1)(-1) = 1$, $x(-y) = (-x)y = -(xy)$, for all $x, y \in X$.*

Proof. $(-1)^2 = (-1)(-1) = -(-1) = 1$. $x(-y) = x(-1)y = (-1)xy = -(xy)$ and $(-x)y = (-1)xy = -(xy)$. $\square$

Proposition 2.9. *Consider x, y, z, x_i, and y_i belonging to X, $i = 1, 2$. Then based on* (R_1)–(R_{15}) *one has*

(1) (a) $x_1 \leq x_2$ and $y_1 \leq y_2$ imply $x_1 + y_1 \leq x_2 + y_2$.

    (b) $x_1 < x_2$ and $y_1 \leq y_2$ imply $x_1 + y_1 < x_2 + y_2$.

(2) (a) $x > 0$ if and only if $x^{-1} > 0$.

    (b) $x \geq 0$ implies $-x \leq 0$.

    (c) $x > 0$ implies $-x < 0$.

(3) (a) $x \leq y$ and $z > 0$ imply $xz \leq yz$.

    (b) $x < y$ and $z > 0$ imply $xz < yz$.

    (c) $x \leq y$ and $z < 0$ imply $xz \geq yz$.

    (d) $x < y$ and $z < 0$ imply $xz > yz$.

(4)   If $xy > 0$, then $x \leq y$ if and only if $1/x \geq 1/y$.

(5) (a) $0 \leq x_1 \leq x_2$ and $0 \leq y_1 \leq y_2$ imply $x_1 y_1 \leq x_2 y_2$.

    (b) $0 < x_1 < x_2$ and $0 < y_1 \leq y_2$ imply $x_1 y_1 < x_2 y_2$.

    (c) $x_1 \leq x_2 \leq 0$ and $y_1 \leq y_2 \leq 0$ imply $x_1 y_1 \geq x_2 y_2$.

    (d) $x_1 < x_2 \leq 0$ and $y_1 \leq y_2 < 0$ imply $x_1 y_1 > x_2 y_2$.

The *absolute value function* is defined as

$$\text{for } x \in X, \quad |x| = \begin{cases} x, & x \geq 0, \\ -x, & x < 0. \end{cases}$$

So, $|\cdot| : X \to [0, \infty[$.

Proposition 2.10. *From* (R_1)–(R_{15}) *follow that for every* $x, y \in X$, *we have*

(1)   (a) $|x| \geq 0$.

  (b) $|x| = 0 \iff x = 0$.

  (c) $|x| = |-x|$.

(2)   (a) $|x + y| \leq |x| + |y|$.

  (b) $|x - y| \geq ||x| - |y||$.

(3)   (a) $|x| \leq a \iff -a \leq x \leq a$.

  (b) $|x| < a \iff -a < x < a$.

(4)   (a) $|xy| = |x| \cdot |y|$.

  (b) $\left|\dfrac{x}{y}\right| = \dfrac{|x|}{|y|}$.

  (c) $|x^n| = |x|^n$, $n \in \mathbb{N}^*$.

The *distance function* is defined as

$$\text{for } x, y \in X, \quad d(x, y) = |x - y|.$$

Thus $d : X \times X \to [0, \infty[$.

Proposition 2.11. *From Proposition 2.10 it follows that*

$$d(x, y) = 0 \text{ if and only if } x = y;$$
$$d(x, y) = d(y, x), \quad \forall x, y \in X;$$
$$d(x, y) \leq d(x, z) + d(z, y), \quad \forall x, y, z \in X.$$

The *signum function* is defined as

$$\text{for } x \in X, \quad \operatorname{sign} x = \begin{cases} 1, & x > 0, \\ 0, & x = 0, \\ -1, & x < 0. \end{cases}$$

Therefore $\operatorname{sign} : X \to \{-1, 0, 1\}$.

Warning. There exist several systems satisfying (R_1)–(R_{16}) axioms. But all are algebraically and order isomorphic, [66, theorem 5.34], [100, vol.1, § 2.9]. We choose one of them and call it *the set of real numbers*, and denoted it by $\mathbb{R} = (\mathbb{R}, +, \cdot, \leq)$. An element in $\mathbb{R}$ is said to be a *real number*. A real number x such that $0 \leq x$ is said to be *nonnegative*, whereas if $0 < x$, it is called a *positive* number. A real number x such $0 \geq x$ is said to be *nonpositive*, whereas if $0 > x$, it is called a *negative* number.

Proposition 2.12. *Number 1 is positive.*

Proof. Suppose that 1 is nonpositive; that is, $1 \leq 0$. Adding -1 to both sides we have $0 \leq -1$. Multiplying both sides by the nonnegative number -1 and using (R_{15}), we get $0 \leq (-1)(-1) \iff 0 \leq 1$. Now, 1 is simultaneously nonnegative and nonpositive, thus $0 = 1$. But this contradicts (R_6). Hence $1 > 0$. $\square$

It is clear that any set of real numbers (i.e., any subset of $\mathbb{R}$) having an infimum is nonempty and bounded below. The converse statement is also true.

Theorem 2.1. *Every nonempty and bounded below subset A of $\mathbb{R}$ has an infimum.*

Proof. Denote by A_0 the set of lower bounds of A. Because A is bounded below, $A_0 \neq \emptyset$. Remark that the ordered system (A_0, A) has the property that for every $x \in A_0$ and $y \in A$ it holds that $x \leq y$. From (R_{16}) it follows there exists a real number z such that

$$x \leq z \leq y, \text{ for every } x \in A_0 \text{ and } y \in A.$$

It results that number z is the greatest element in A_0, that is, an infimum of A. By Remark 1.1 we conclude that z is the infimum of A. $\square$

Corollary 2.2. *If A is a nonempty and bounded below subset of $\mathbb{R}$ and B is a nonempty subset of A, then*

$$\inf A \leq \inf B.$$

Theorem 2.2. *Every nonempty and bounded above subset A of $\mathbb{R}$ has a supremum.*

Corollary 2.3. *If A is a nonempty and bounded above subset of $\mathbb{R}$ and B is a nonempty subset of A, then*

$$\sup A \geq \sup B.$$

Remark. The proofs of the existence of an infimum and the existence of a supremum have used Axiom (R_{16}). At the same time it can be proved (and we see immediately) that (R_{16}) follows from any one of these theorems. Thus (R_{16}) is equivalent to any one of these theorems. Hence we may substitute (R_{16}) by one of these statements in order to get the same real number system. $\triangle$

Theorem 2.3. *Suppose X is a totally ordered field (i.e., it satisfies axioms (R_1)–(R_{15})) and, moreover, every nonempty and bounded above subset of it has a supremum. Then Axiom (R_{16}) is fulfilled.*

Proof. Consider an ordered pair (A, B) of nonempty subsets of X having the property that $x \leq y$ for any $x \in A$ and $y \in B$. Then A is nonempty and bounded above (by any element of B). It follows that there exists $z \in X$ such that

$$z = \sup A. \tag{1.4}$$

We have to show that $z \leq y$, for every $y \in B$. For, suppose there exists $y_0 \in B$ such that $y_0 < z$. Then y_0 is an upper bound of A strictly less then z, contradicting assumption (1.4). $\square$

A similar statement holds for the infimum.

Theorem 2.4. *Suppose X is a totally ordered field and, moreover, every nonempty and bounded below subset of it has an infimum. Then Axiom (R_{16}) is fulfilled.*

Theorem 2.5. (Archimedes'[7] principle) *For every two real numbers x and y such that $y > 0$ there exists a natural number n such that $x < ny$.*

Proof. Under the above-mentioned assumptions define

$$A = \{u \in \mathbb{R} \mid \exists n \in \mathbb{N}^*, \ u < ny\}.$$

and remark that $A \neq \emptyset$ (because at least $y \in A$). We show that $A = \mathbb{R}$. Suppose that $A \neq \mathbb{R}$ and denote $B = \mathbb{R} \setminus A$. Obviously, $B \neq \emptyset$.

Note that for every $u \in A$ and $v \in B$, $u < v$. Indeed, for every $u \in A$ there exists a natural n such that $u < ny$. Because $v \notin A$ and the real number set is a totally ordered set, it follows that $ny \leq v$. Then

$$u < ny \leq v \implies u < v.$$

Axiom (R_{16}) implies that for the ordered pair (A, B) there exists a real number z such that

$$u \leq z \leq v, \quad \forall u \in A, v \in B. \tag{1.5}$$

The real number $z - y$ belongs to A, because otherwise $z - y \in B$, and then by (1.5)

$$z \leq z - y \implies y \leq 0,$$

contradicting the hypothesis. Therefore $z - y \in A$. Then we can find a natural number n such that $z - y < ny$. We also have

$$z + y = (z - y) + 2y < (n + 2)y,$$

and it follows that $z + y \in A$. Then $z + y \leq z$, thus $y \leq 0$. The contradiction shows that $A = \mathbb{R}$ and the theorem is proved. $\square$

Remark. One can show that Axiom (R_{16}) is equivalent to the Archimedes principle. $\triangle$

---

[7] Archimedes of Siracusa ($'A\rho\chi\iota\mu\dot{\eta}\delta\eta\varsigma$ o $\Sigma\upsilon\rho\alpha\kappa o\dot{\upsilon}\sigma\iota o\varsigma$), 287–212 (BC).

Corollary 2.4. *Given a positive ε, there exists a natural number n so that $1/n < \varepsilon$.*

Corollary 2.5. *The set of natural numbers is unbounded.*

Theorem 2.6. *The supremum of a nonempty and bounded above set is unique.*

Proof. Let A be the set under discussion. Then by Theorem 2.2 we know that there exists $\sup A$. Suppose that $\sup A = a_1$ and $\sup A = a_2$ and $a_1 \neq a_2$. Then either $a_1 < a_2$ or $a_2 < a_1$. In both cases we get a contradiction.

We may argue equally well by Remark 1.1.   □

The following characterization of a supremum is useful.

Theorem 2.7. *A real number a is the supremum of a set $A \subset \mathbb{R}$ if and only if*

(i) *For every $x \in A$, $x \leq a$.*
(ii) *For every $\varepsilon > 0$ there is an element $y \in A$ such that $y > a - \varepsilon$.*

Proof. (i) says that a is an upper bound of A, and (ii) shows that there is no upper bound less then a.   □

Similar results hold in the case of an infimum.

Theorem 2.8. *The infimum of a nonempty and bounded below set is unique.*

Theorem 2.9. *A real number a is the infimum of a set $A \subset \mathbb{R}$ if and only if*

(i) *For every $x \in A$, $x \geq a$.*
(ii) *For every $\varepsilon > 0$ there is an element $y \in A$ such that $y < a + \varepsilon$.*

Theorem 2.10. *For every real $x > 0$ and every integer $n \geq 1$, there is one and only one real $y > 0$ such that $y^n = x$.*

Remark. This number y is written as $\sqrt[n]{x}$ or $x^{1/n}$ and it is called the *n*th *root* or *radical* (of index n) of the positive real number x. The second root is called the *square root,* and the third root is called the *cube root.*   △

Proof. If $n = 1$, y is precisely x. Suppose that $n \in \{2, 3, \ldots\}$.

That there is at most one such y is clear, because $0 < y_1 < y_2$ implies $y_1^n < y_2^n$.

Let E be the set consisting of all positive reals t such that $t^n < x$.

If $t = x/(1+x)$, then $0 < t < 1$; hence $t^n < t < x$, so E is not empty.

Put $t_0 = 1 + x$. Then $t_0 > 1$ implies $t_0^n > t_0 > x$, so that $t_0 \notin E$, and t_0 is an upper bound of E. Let $y = \sup E$ (which exists, by Theorem 2.2).

Suppose $y^n < x$. Choose h such that $0 < h < 1$ and

$$h < \frac{x - y^n}{(1+y)^n - y^n}.$$

We have

$$(y+h)^n = y^n + \binom{n}{1}y^{n-1}h + \cdots + h^n \leq y^n + h\left(\binom{n}{1}y^{n-1} + \cdots + 1\right)$$
$$= y^n + h((1+y)^n - y^n) < y^n + (x - y^n) = x.$$

Thus $y + h \in E$, contradicting the fact that y is an upper bound of E.

Suppose $y^n > x$. Choose k such that $0 < k < 1$, $k < y$, and

$$k < \frac{y^n - x}{(1+y)^n - y^n}.$$

Then, for $t \geq y - k$, we have

$$t^n \geq (y-k)^n = y^n - \binom{n}{1}y^{n-1}k + \binom{n}{2}y^{n-2}k^2 - \cdots + (-1)^n k^n$$
$$= y^n - k\left(\binom{n}{1}y^{n-1} - \binom{n}{2}y^{n-2}k + \cdots + (-1)^{n-1}k^{n-1}\right)$$
$$\geq y^n - k\left(\binom{n}{1}y^{n-1} + \binom{n}{2}y^{n-2} + \cdots + 1\right)$$
$$= y^n - k[(1+y)^n - y^n] > y^n + (x - y^n) = x.$$

Thus $y - k$ is an upper bound of E, contradicting the fact that $y = \sup E$.

It follows that $y^n = x$. $\square$

An *interval* A of the real number system is a subset of $\mathbb{R}$ so that for every $x, y \in A$ and $z \in \mathbb{R}$ satisfying $x \leq z \leq y$, we have $z \in A$. An interval bounded below and above is said to be *bounded*. Otherwise it is called *unbounded*. For any nonempty and bounded interval A, the nonnegative real number $l(A) = \sup A - \inf A$ is said to be the *length* of A.

We remark that for any real numbers a and b with $a \leq b$ the bounded intervals are listed in Figure 1.9.

$[a, b] = \{x \in \mathbb{R} \mid a \leq x \leq b\}$ closed interval

$[a, b[= \{x \in \mathbb{R} \mid a \leq x < b\}$ left closed right open interval

$]a, b] = \{x \in \mathbb{R} \mid a < x \leq b\}$ left open right closed interval

$]a, b[= \{x \in \mathbb{R} \mid a < x < b\}$ open interval

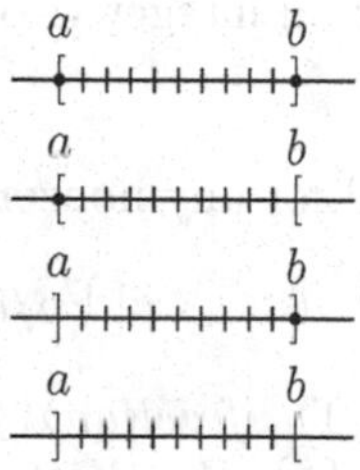

Fig. 1.9. Bounded intervals

1.2.3 Elements of algebra

Theorem 2.11. *For every real x there exists a unique integer k such that $k \leq x < k+1$.*

Let x be a real number. Its *floor* is the unique (Theorem 2.11) integer k satisfying

$$k \leq x < k+1,$$

and is denoted $\lfloor x \rfloor$ (read "the floor of x"). Equivalently, the floor of a real number x is the largest integer less than or equal to x. In some papers the floor of a real number is referred to as the *integer part* of it [30].

The *floor function* is defined by

$$\mathbb{R} \ni x \mapsto \lfloor x \rfloor \in \mathbb{Z}.$$

Let x be a real number. Its *ceiling* is the unique (Theorem 2.11) integer k satisfying

$$k - 1 < x \leq k,$$

and is denoted $\lceil x \rceil$ (read "the ceiling of x"). Equivalently, the ceiling of a real number x is the smallest integer greater than or equal to x. The *ceiling function* is defined by

$$\mathbb{R} \ni x \mapsto \lceil x \rceil \in \mathbb{Z}.$$

Note

$$\lfloor \sqrt{2} \rfloor = 1, \quad \lceil \sqrt{2} \rceil = 2, \quad \left\lfloor \frac{1}{2} \right\rfloor = 0, \quad \left\lceil \frac{1}{2} \right\rceil = 1, \quad \left\lceil -\frac{1}{2} \right\rceil = 0.$$

Obviously, for a real x,

$$x - 1 < \lfloor x \rfloor \leq x \leq \lceil x \rceil < x + 1,$$
$$x \in \mathbb{Z} \Longleftrightarrow \lfloor x \rfloor = \lceil x \rceil,$$
$$x \notin \mathbb{Z} \Longleftrightarrow \lceil x \rceil = \lfloor x \rfloor + 1.$$

For an integer n one has

$$\left\lfloor \frac{n}{2} \right\rfloor + \left\lceil \frac{n}{2} \right\rceil = n,$$

and for any nonzero integers a and b

$$\lfloor \lfloor n/a \rfloor / b \rfloor = \lfloor n/(ab) \rfloor \quad \text{and} \quad \lceil \lceil n/a \rceil / b \rceil = \lceil n/(ab) \rceil.$$

The *fractional part* of a real number x is defined as $x - \lfloor x \rfloor$ and is denoted by $\{x\}$. So, the *fractional part function* is defined by

$$\mathbb{R} \ni x \mapsto \{x\} \in [0, 1[.$$

Theorem 2.12. *For every two real numbers x and y such that $x < y$ there exists a rational lying between them; that is, $x < u < y$, for a certain $u \in \mathbb{Q}$.*

Proof. Based on Archimedes' principle (Theorem 2.5) for the positive real $y - x$ there exists a natural n such that $1 < n(y - x)$. Then

$$1/n < y - x. \tag{1.6}$$

From Theorem 2.11 it follows that there exists an integer m such that

$$m \le nx < m + 1. \tag{1.7}$$

Obviously, $u = (m+1)/n$ is a rational, and satisfies $x < u$. From the left-hand side of (1.7) as well as from (1.6) we infer that u also satisfies

$$u = \frac{m}{n} + \frac{1}{n} \le x + \frac{1}{n} < y. \quad \square$$

An *irrational* number is precisely a nonrational real number; that is, it belongs to $\mathbb{R} \setminus \mathbb{Q}$.

Corollary 2.6. *Given any two real numbers x and y such that $x < y$, there exists an irrational number v such that $x < v < y$.*

Proof. Choose any irrational number v_0 ($\sqrt{2}$, for example). Then $x - v_0 < y - v_0$. By Theorem 2.12, there exists a rational u such that $x - v_0 < u < y - v_0$; that is, $x < v_0 + u < y$. We remark that $v = v_0 + u$ is irrational, because otherwise it follows that v_0 itself is rational, and this is not the case. $\quad \square$

A real number is said to be an *algebraic number* if it is a root of a polynomial equation with integer coefficients.

A real number is said to be an *transcendental number* if it is not a root of any polynomial equation with integer coefficients.

Let $P(x)$ be a polynomial of degree n

$$P(x) = a_n x^n + \cdots + a_1 x + a_0,$$

with $a_0, \ldots, a_n \in \mathbb{Z}$, and $a_n \ne 0$, and $x \in \mathbb{R}$.

Theorem 2.13. *If an irreducible fraction p/q is a root of P, p divides a_0 and q divides a_n.*

Proof. Suppose p/q is a root of P; then $q^n P(p/q) = 0$, hence

$$a_0 q^n + a_1 q^{n-1} p + \cdots + a_{n-1} q p^{n-1} = -a_n p^n.$$

From this identity q divides $a_n p^n$, but p and q are relatively prime, so q divides a_n. Similarly one can show that p divides a_0. $\quad \square$

Corollary 2.7. *The real roots of the polynomial*

$$x^n + a_{n-1} x^{n-1} + \cdots + a_1 x + a_0,$$

are either integers or irrational numbers.

Corollary 2.8. *The real roots of the polynomial*

$$x^2 - m = 0,$$

where m is prime, are irrational numbers.

Let A and B be two sets. If there exists a bijective mapping from A onto B, we say that A and B have the same *cardinal number* or that A and B are *equivalent*, and we write $A \sim B$.

Theorem 2.14. *The relation $\sim$ defined above is an equivalence relation.*

Recall, for every positive integer n, $\mathbb{N}_n^*$ is the set whose elements are precisely the integers $1, 2, \ldots, n$. For a set A we say that

(a) A is *finite* if $A \sim \mathbb{N}_n^*$ for some n (the empty set is, by definition, finite). The number of elements of a nonempty finite set A is n provided $A \sim \mathbb{N}_n^*$. In this case we write $|A| = n$, and we read "the number of elements of the nonempty and finite set A is equal to n" or "the cardinality of A is n." By definition, $|\emptyset| = 0$.
(b) A is *infinite* if A is not finite.
(c) A is *countable* if $A \sim \mathbb{N}^*$. Obviously $\mathbb{N} \sim \mathbb{N}^*$. We write $|A| = \aleph_0$ and read "the cardinality of A is aleph zero." Aleph is the first letter of the Hebrew alphabet.
(d) A is *uncountable* if A is neither finite nor countable. We write $|A| \geq \aleph_1 > \aleph_0$.
(e) A is *at most countable* (or *denumerable*) if $A \sim \mathbb{N}^*$ or $A \sim \mathbb{N}_n^*$ for some $n \in \mathbb{N}^*$. We write $|A| \leq \aleph_0$.

Remarks 2.1. (a) For two finite sets A and B so that $B \subset A$ we have $A \sim B$ if and only if $A = B$. For infinite sets, however, this is not exactly so. Indeed, let M be the set of all even positive integers, $M = \{2, 4, 6, \ldots\}$. It is clear that M is a proper subset of $\mathbb{N}^*$. But $\mathbb{N}^* \sim M$, because $\mathbb{N}^* \ni n \mapsto 2n \in M$ is a bijection.
(b) The sets $\{1, -1, 2, -2, 3, -3, \ldots\}$ and $\{0, 1, -1, 2, -2, 3, -3, \ldots\}$ are equivalent. Indeed, the function

$$b_k = \begin{cases} 0, & k = 1, \\ a_{k-1}, & k > 1, \end{cases}$$

maps the kth rank term a_k of the first set to the $(k+1)$th rank term in the second set in a bijective way.

From (a) and (b) we conclude that the sets $\mathbb{N}^*$ and $\mathbb{Z}$ are equivalent. Therefore we write $|\mathbb{Z}| = \aleph_0$. $\triangle$

Theorem 2.15. *Every infinite subset of a countable set is countable.*

Theorem 2.16. *Let $\{A_n \mid n = 1, 2, \ldots\}$ be a countable family of countable sets, and put*

$$B = \cup_{n=1}^{\infty} A_n.$$

Then B is countable.

Proof. Let every set A_n be arranged in a sequence $(x_{n\,k})_k$, $n = 1, 2, \ldots$, and consider the infinite array in which the elements of A_n form the nth row, Figure 1.10.

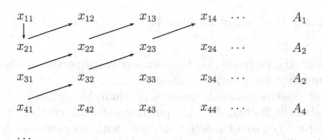

Fig. 1.10. Infinite array

The array contains all elements of B. As indicated by the arrows, these elements can be arranged in a sequence

$$x_{11}, \ x_{21}, \ x_{12}, \ x_{31}, \ x_{22}, \ x_{13}, \ \ldots . \tag{1.8}$$

If any two of the elements A_n have elements in common, these will appear more than once in (1.8). Hence there is a subset C of B such that $C \sim B$, which shows that B is at most countable (Theorem 2.15). Because $A_1 \subset B$, and A_1 is infinite, B is infinite, and thus countable. $\square$

Corollary 2.9. *Suppose A is at most countable and for every $\alpha \in A$, B_α is at most countable. Put*

$$C = \cup_{\alpha \in A} B_\alpha.$$

Then C is at most countable.

Proof. For C is equivalent to a subset of $\cup_\alpha B_\alpha$. $\square$

Theorem 2.17. *Let A be a countable set and let $B_n = A^n$, for some $n \in \mathbb{N}^*$. Then B_n is countable.*

Proof. That B_1 is countable is obvious, because $B_1 = A$. Suppose B_{n-1} is countable ($n = 2, 3, \ldots$). The elements of B_n are of the form

$$(b, a) \quad (b \in B_{n-1}, \ a \in A).$$

For every fixed b, the set of pairs (b, a) is equivalent to A, and hence, countable. Thus B_n is a countable union of countable sets. By Theorem 2.16, B_n is countable.   □

Corollary 2.10. (Cantor) *The set of all rational numbers is countable.*

Proof. We apply Theorem 2.17 with $n = 2$, noting that every rational r is of the form a/b, where a and b are integers and $b \neq 0$. The set of such pairs (a, b), and therefore the set of fractions a/b, are countable.   □

Therefore we write $|\mathbb{Q}| = \aleph_0$.

Corollary 2.11. *Each set of the form* $]a, b[\cap \mathbb{Q}$, *where* $a, b \in \mathbb{Q}$ *and* $a < b$, *is countable.*

Proof. Equivalently, we prove that between any two different rational numbers there are countably many rational numbers.

The set of positive rational numbers is countable (otherwise, the set of rational numbers is finite). To each positive rational number q we assign (uniquely) a pair (m, n) of positive integers with no common divisor such that $q = m/n$. The map $(m, n) \to (ma + nb)/(m + n)$ is one-to-one, $(ma + nb)/(m + n) \in \mathbb{Q}$, and $a < (ma + nb)/(m + n) < b$. Then to each positive rational number we assigned uniquely a rational number member of the open interval $]a, b[$.

On the other hand, to each rational $p \in]a, b[$ we can find a pair (m, n) of positive integers with no common divisors so that $p = (ma + nb)/(m + n)b$. m/n is a positive rational. Thus there is a bijection between the set of positive rational numbers and the set of rational numbers belonging to $]a, b[$.   □

Theorem 2.18. *Let A be the set of all sequences whose elements are the digits 0 and 1. Then A is uncountable.*

Proof. Let B be a countable subset of A, and let B consist of the sequences $s_1, s_2, \ldots$. We construct a sequence s as follows. If the nth digit in s_n is 1 we let the nth digit of s be 0, and vice versa. Then the sequence s differs from every member of B in at least one place; hence $s \notin B$. But clearly $s \in A$, so that B is a proper subset of A.

We have shown that every countable subset of A is a proper subset of A. It follows that A is uncountable (for otherwise A would be a proper subset of A, which is absurd).   □

Corollary 2.12. *The interval $[0, 1]$ is uncountable.*

Proof. Use the binary representation of the real numbers and apply Theorem 2.18.   □

Corollary 2.13. *Every interval $[a, b]$ $(a < b)$ is uncountable.*

Corollary 2.14. (Cantor) *The real number set is uncountable.*

Therefore we write $|\mathbb{R}| = \aleph_1$.

Theorem 2.19. (Cantor) *The algebraic number set is countable.*

Proof. For every polynomial $P(x) = a_n x^n + a_{n-1}x^{n-1} + \cdots + a_1 x + a_0$ of degree n and with integer coefficients, define its *height* $h(P)$ as the integer

$$n + |a_n| + |a_{n-1}| + \cdots + |a_1| + |a_0|.$$

Let U_m be the set of all the roots to all the polynomials P satisfying $h(P) \le m$. Obviously, for every integer m, the set U_m is finite. Because the set

$$\cup_{m=1}^{\infty} U_m$$

is countable, we conclude that the algebraic number set is countable.   □

Remark. Each number of the form $\sqrt{p}$, where p is prime, is irrational and algebraic.   △

Corollary 2.15. (Cantor) *The transcendental number set is uncountable.*

Proof. Suppose the transcendental number set is countable. Then the union of the algebraic number set and the transcendental number set is countable. This union precisely is the real number set. So, we contradict Corollary 2.19. The conclusion follows.   □

Remark. From Theorem 2.19 and Corollary 2.15 it follows that "almost all" real numbers are transcendental.   △

1.2.4 Elements of topology on $\mathbb{R}$

A set $A \subset \mathbb{R}$ is said to be *open* if for each $x \in A$ there is a positive ε such that $]x - \varepsilon, x + \varepsilon[\subset A$. Obviously, $\mathbb{R}$ is open. We consider by definition the empty set as being open. It follows immediately that

Theorem 2.20. *Let $\mathcal{O}$ denote the family of all open subsets of $\mathbb{R}$. Then*

(a) $\emptyset, \mathbb{R} \in \mathcal{O}$.
(b) *The union of any family of open sets is open.*
(c) *The intersection of any finite family of open sets is open.*

Remark. We may not relax the finiteness assumption in (c). For every $n \in \mathbb{N}^*$ the interval $]-1/n, 1/n[$ is an open set. The set $\{0\}$ containing just the origin is not an open set because it does not contain any open interval. Note that $\cap_{n \in \mathbb{N}^*}] - 1/n, 1/n[= \{0\}$.   △

Let A be a nonempty subset of $\mathbb{R}$ and $x \in A$. Then x is an *interior* point of A if there is an open set O with $x \in O \subset A$. The set of interior points of a set $A \subset \mathbb{R}$ is denoted by int A. If int $A = \emptyset$, we say that A has no interior point. This is the case for $A = \{1\}$ and $A = \{1/n \mid n = 1, 2, \dots\}$. Always int $A \subset A$.

Theorem 2.21. *Suppose $A, B \subset \mathbb{R}$. Then*

(a) int A *is the union of all open sets contained by A. Thus* int A *is open.*
(b) *A is open if and only if $A = $ int A.*
(c) *The interior of A is the largest open set (in respect to the inclusion of sets) contained in A.*
(d) *$A \subset B$ implies* int $(A) \subset$ int (B).
(e) int $(A \cap B) = $ int $(A) \cap$ int (B).
(f) int (int A) $=$ int A.
(g) int $\mathbb{R} = \mathbb{R}$.

Proof. (a) It follows from the definition of int A.
(b) Suppose that A is open. By definition, for every $x \in A$ there exists a positive ε such that $]x - \varepsilon, x + \varepsilon[\subset A$. This means that $x \in$ int A. Thus $A \subset$ int A. The other inclusion is trivial.
(c) We have to show that if int $A \subset B \subset A$ and B is open, then int $A = B$. For, suppose int $A \neq B$. Choose an $x \in B \setminus$ int A. Then there is a positive ε such that $]x - \varepsilon, x + \varepsilon[\subset B$ and $B \subset A$. Then $x \in$ int A. Thus we get a contradiction and the claim is proved.
(d) It is trivial.
(e) Suppose that $x \in$ int $(A) \cap$ int (B). Then there exist open sets O_1, O_2 such that $x \in O_1 \subset A$ and $x \in O_2 \subset B$. Thus $x \in O_1 \cap O_2 \subset A \cap B$, hence $x \in$ int $(A \cap B)$.

Suppose now that $x \in$ int $(A \cap B)$. We can find an open set O such that $x \in O \subset (A \cap B)$. Then $x \in O \subset A$ and $x \in O \subset B$. Hence $x \in$ int A and $x \in$ int B. We conclude $x \in$ int $(A) \cap$ int (B).
(f) int A is open by (a). The conclusion follows by (b).
(g) We noticed that $\mathbb{R}$ is open. The conclusion follows by (b). $\square$

Theorem 2.22. *Every open interval is an open set.*

A set $A \subset \mathbb{R}$ is said to be *closed* if $\complement_{\mathbb{R}} A$ is open. Then we have the following.

Theorem 2.23. *Let $\mathcal{C}$ denote the family of all closed subsets of $\mathbb{R}$. Then*

(a) $\emptyset, \mathbb{R} \in \mathcal{C}$.
(b) *The intersection of any family of closed sets is closed.*
(c) *The union of any finite family of closed sets is closed.*

Proof. (a) We have $\complement_{\mathbb{R}} \mathbb{R} = \emptyset$ and $\complement_{\mathbb{R}} \emptyset = \mathbb{R}$.
(b) It follows from (b) of Theorem 2.20 and Theorem 1.4.
(c) It follows from (c) of Theorem 2.20 and Theorem 1.4. $\square$

Remark. We may not relax the finiteness assumption in (c). For every $\mathbb{N}^*$ the interval $[1 - 1/n, 1]$ is a closed set. The set $]0, 1]$ is neither open nor closed. We note that $\cup_{n \in \mathbb{N}^*} [1 - 1/n, 1] =]0, 1]$. $\triangle$

Proposition 2.13. *Consider two sets $A, B \subset \mathbb{R}$ such that A is open and B is closed. Then $A \setminus B$ is open and $B \setminus A$ is closed.*

Proof. We write $A \setminus B = A \cap \complement_{\mathbb{R}} B$ and $B \setminus A = B \cap \complement_{\mathbb{R}} A$. $\quad\square$

For $A \subset \mathbb{R}$, the *closure* of A is the set $\operatorname{cl} A = \cap \{C \mid A \subset C \subset \mathbb{R}, C \text{ closed}\}$. We remark that $A \subset \operatorname{cl} A$ always.

The interior and the closure of a set are strongly tied.

Theorem 2.24. *Consider $A \subset \mathbb{R}$. Then*

$$\complement_{\mathbb{R}} (\operatorname{cl} A) = \operatorname{int} (\complement_{\mathbb{R}} A), \quad \complement_{\mathbb{R}} (\operatorname{int} A) = \operatorname{cl} (\complement_{\mathbb{R}} A).$$

Proof. We prove the first identity only. The proof of the second one runs similarly.

Set an arbitrary $x \in \complement_{\mathbb{R}} (\operatorname{cl} A)$.

$$x \notin \operatorname{cl} A \implies \exists C \text{ closed}, A \subset C, \text{ such that } x \notin C$$
$$\implies \exists O(= \complement_{\mathbb{R}} C) \text{ open}, O \subset \complement_{\mathbb{R}} A, x \in O \implies x \in \operatorname{int} (\complement_{\mathbb{R}} A).$$

Set an arbitrary $x \in \operatorname{int} (\complement_{\mathbb{R}} A)$. Then

$$\exists O \text{ open}, O \subset \complement_{\mathbb{R}} A, x \in O \implies \exists C(= \complement_{\mathbb{R}} O) \text{ closed}, A \subset C, x \notin C$$
$$\implies x \notin \operatorname{cl} A \implies x \in \complement_{\mathbb{R}} (\operatorname{cl} A). \quad \square$$

Corollary 2.16. *Suppose $A \subset \mathbb{R}$. Then $\operatorname{int} A = \complement_{\mathbb{R}} \operatorname{cl} (\complement_{\mathbb{R}} A)$ and $\operatorname{cl} A = \complement_{\mathbb{R}} (\operatorname{int} (\complement_{\mathbb{R}} A))$.*

Theorem 2.25. *Suppose $A, B \subset \mathbb{R}$. Then*

(a) $\operatorname{cl} A$ *is a closed set.*
(b) A *is closed if and only if $A = \operatorname{cl} A$.*
(c) $\operatorname{cl} A$ *is the smallest closed set (in respect to the inclusion of sets) containing A.*
(d) $A \subset B$ *implies $\operatorname{cl}(A) \subset \operatorname{cl}(B)$.*
(e) $\operatorname{cl}(A \cup B) = \operatorname{cl}(A) \cup \operatorname{cl}(B)$.
(f) $\operatorname{cl}(\operatorname{cl} A) = \operatorname{cl} A$.
(g) $\operatorname{cl} \mathbb{R} = \mathbb{R}$.

Proof. (a) It follows from the definition of the closure and (b) of Theorem 2.23.

(b) If A is closed, it appears as a member in the intersection defining the $\operatorname{cl} A$. Then $\operatorname{cl} A \subset A$. Thus $A = \operatorname{cl} A$.

If $A = \operatorname{cl} A$, A is closed by (a).

(c) We have to show that if $A \subset B \subset \operatorname{cl} A$ and B is closed, then $B = \operatorname{cl} A$. Because B is closed and $A \subset B$, B appears as a member in the intersection defining the $\operatorname{cl} A$. So, $B \supset \operatorname{cl} A$. Then $B = \operatorname{cl} A$, indeed.

(d) It is trivial.

(e) We apply Theorems 1.4 and 2.24. Then

$$\mathsf{C}_\mathbb{R}\mathrm{cl}\,(A \cup B) = \mathrm{int}\,\mathsf{C}_\mathbb{R}(A \cup B) = \mathrm{int}\,(\mathsf{C}_\mathbb{R}A \cap \mathsf{C}_\mathbb{R}B)$$
$$= \mathrm{int}\,(\mathsf{C}_\mathbb{R}A) \cap \mathrm{int}\,(\mathsf{C}_\mathbb{R}B) = (\mathsf{C}_\mathbb{R}\mathrm{cl}\,A) \cap (\mathsf{C}_\mathbb{R}\mathrm{cl}\,B) = \mathsf{C}_\mathbb{R}(\mathrm{cl}\,A \cup \mathrm{cl}\,B).$$

(f) By (a) $\mathrm{cl}\,A$ is a closed set. The conclusion follows by (b).
(g) The closure of any set is contained in $\mathbb{R}$. So, $\mathbb{R} \subset \mathrm{cl}\,\mathbb{R} \subset \mathbb{R}$.   □

Proposition 2.14. *Consider two open and disjoint sets $A, B \subset \mathbb{R}$. Then*

(a) *The closure of one set does not intersect the other; that is, $B \cap \mathrm{cl}\,A = \mathrm{cl}\,(B) \cap A = \emptyset$.*
(b) *$(\mathrm{int}\,\mathrm{cl}\,A) \cap (\mathrm{int}\,\mathrm{cl}\,B) = \emptyset$.*

Proof. (a) From the hypothesis we have $A \cap B = \emptyset$. Then $B \subset \mathsf{C}_\mathbb{R}A$ and $\mathrm{cl}\,B \subset \mathsf{C}_\mathbb{R}A$. Finally, $\mathrm{cl}\,B \cap A = \emptyset$.
(b) From (a) we have that $A \cap \mathrm{cl}\,B = \emptyset$. Then $A \cap (\mathrm{int}\,\mathrm{cl}\,B) = \emptyset$. Because $\mathrm{int}\,\mathrm{cl}\,B$ is open, we repeat the reasoning to this last open set and to A.   □

A *neighborhood* of a point $x \in \mathbb{R}$ is any set $A \subset \mathbb{R}$ containing an open interval O with $x \in O$ (i.e., $x \in O \subset A$). The system of all neighborhoods of a point $x \in \mathbb{R}$ is denoted by $\mathcal{V}(x)$. Obviously, $\mathbb{R} \in \mathcal{V}(x)$, for each $x \in \mathbb{R}$.

Proposition 2.15. *For a set $A \subset \mathbb{R}$ the following two sentences are equivalent.*

(a) *$x \in \mathrm{cl}\,A$.*
(b) *For every $V \in \mathcal{V}(x)$ one has $V \cap A \neq \emptyset$.*

Proof. We show that (a) implies (b). Suppose that for a point x and an open neighborhood $V \in \mathcal{V}(x)$, $V \cap A = \emptyset$. It means that $A \subset \mathsf{C}_\mathbb{R}V$ and $\mathsf{C}_\mathbb{R}V \in \mathcal{C}$. Then $\mathrm{cl}\,A \subset \mathsf{C}_\mathbb{R}V$ and $x \notin \mathrm{cl}\,A$, thus (a) does not hold.

We show that (b) implies (a). Suppose that $x \notin \mathrm{cl}\,A$. Then there exists a closed set C with $A \subset C$ and $x \notin C$. Set $V = \mathsf{C}_\mathbb{R}C$ and remark that V is open, $x \in V$, and $V \cap A = \emptyset$. But this contradicts (b).   □

Corollary 2.17. *Suppose V is an open set and $V \cap A = \emptyset$. Then $V \cap \mathrm{cl}\,A = \emptyset$.*

Proof. Suppose there is an element $x \in V \cap \mathrm{cl}\,A$. Then by Proposition 2.15 it follows that $V \cap A \neq \emptyset$.   □

For $x \in \mathbb{R}$ and $A \subset \mathbb{R}$ we say that x is a *limit point* or *accumulation point* of A if every $V \in \mathcal{V}(x)$ contains some point of A distinct from x; that is, $V \cap (A \setminus \{x\}) \neq \emptyset$. Denote by A' the set of limit (accumulation) points of the set A.

Theorem 2.26. *Consider $A, B, A_\alpha \subset \mathbb{R}$, $\alpha \in I$. Then*

(a) *$\mathrm{cl}\,A = A \cup A'$.*
(b) *$A \subset B$ implies $A' \subset B'$.*
(c) *$(A \cup B)' = A' \cup B'$.*
(d) *$\cup_{\alpha \in I} A_\alpha' = (\cup_{\alpha \in I} A_\alpha)'$.*

Corollary 2.18. *Consider $A \subset \mathbb{R}$. Then A is closed if and only if $A' \subset A$.*

Theorem 2.27. *Let A be a closed set of real numbers that is bounded above. Set $y = \sup A$. Then $y \in A$.*

Proof. Suppose $y \notin A$. For any $\varepsilon > 0$ there is a point $x \in A$ such that $y - \varepsilon < x < y$, Theorem 2.7, page 18. Thus, every neighborhood of y contains a point $x \in A$, and $x \neq y$, because $y \notin A$. It follows that y is a limit point of A which is not a point of A, so that A is not closed. This contradicts the hypothesis. $\square$

Corollary 2.19. *Let B be a closed set of real numbers that is bounded below. Set $y = \inf B$. Then $y \in B$.*

Proof. Set $A = -B = \{-b \mid b \in B\}$. Then A is closed and bounded above. Moreover, $y = -\sup A$. Then $y \in B$. $\square$

A point $x \in \mathbb{R}$ is said to be a *closure point* or *(adherent point)* of the set $A \subset \mathbb{R}$ if every neighborhood of x has a nonempty intersection with A.

Theorem 2.28. *Suppose $\emptyset \neq A \subset \mathbb{R}$. Then A is closed if and only if A coincides with the set of its closure points.*

Proof. Denote by B the set of closure points of A.

Suppose $A = B$. Then $\complement_{\mathbb{R}} A = \complement_{\mathbb{R}} B$ and for each $x \in \complement_{\mathbb{R}} A$ we can find a neighborhood V of x such that $V \cap A = \emptyset$; that is, $V \subset \complement_{\mathbb{R}} A$. Thus x is an interior point of $\complement_{\mathbb{R}} A$. The point x has been chosen arbitrary, so the set $\complement_{\mathbb{R}} A$ is open. Hence A is closed.

We have that $A \subset B$. Suppose that A is closed. Then $\complement_{\mathbb{R}} A$ is open. For any $x \notin A$, $\complement_{\mathbb{R}} A$ is a neighborhood of x not intersecting A. Then $x \notin B$. Thus we get that $B \subset A$, and finally $A = B$. $\square$

The *frontier* or *boundary* of a set $A \subset \mathbb{R}$ is the set $(\operatorname{cl} A) \cap (\operatorname{cl} \complement_{\mathbb{R}} A)$. We denote it by $\operatorname{fr} A$ and we note that it is closed. We remark that $\operatorname{fr} [0, 1] = \operatorname{fr} \,]0, 1[= \operatorname{fr} \{0, 1\} = \{0, 1\}$.

Theorem 2.29. *Suppose $A \subset \mathbb{R}$. Then A is open if and only if $A \cap \operatorname{fr} A = \emptyset$.*

Theorem 2.30. *Suppose $A, B \subset \mathbb{R}$. Then we have*

(a) $\operatorname{int} A = A \setminus \operatorname{fr} A$.
(b) $\operatorname{cl} A = A \cup \operatorname{fr} A$.
(c) $\operatorname{fr} (A \cup B) \subset \operatorname{fr} A \cup \operatorname{fr} B$.
(d) $\operatorname{fr} (A \cap B) \subset \operatorname{fr} A \cup \operatorname{fr} B$.

(e) $\operatorname{fr} (\mathbb{R} \setminus A) = \operatorname{fr} A$.
(f) $\mathbb{R} = \operatorname{int} A \cup \operatorname{fr} A \cup \operatorname{int} (\mathbb{R} \setminus A)$.
(g) $\operatorname{fr} (\operatorname{cl} A) = \operatorname{fr} A$.
(h) $\operatorname{fr} (\operatorname{int} A) \subset \operatorname{fr} A$.

(i) *A is open if and only if $\operatorname{fr} A = \operatorname{cl} A \setminus A$.*
(j) *A is closed if and only if $\operatorname{fr} A = A \setminus \operatorname{int} A$.*

A subset A of $\mathbb{R}$ is said to be *dense* in $\mathbb{R}$ provided $\operatorname{cl} A = \mathbb{R}$. Invoking Theorem 2.12, we find the following.

Corollary 2.20. $\mathbb{Q}$ *is dense in* $\mathbb{R}$.

Theorem 2.31. *Let* $(I_k)_{k \in \mathbb{N}}$ *be a nested sequence of nonempty closed and bounded intervals in* $\mathbb{R}$; *that is,*

$$I_{k+1} \subset I_k, \quad k \in \mathbb{N}. \tag{1.9}$$

Then

$$\cap_{k \in \mathbb{N}} I_k \neq \emptyset.$$

Proof. Denote $I_k = [a_k, b_k], \ k \in \mathbb{N}$. From (1.9) it follows that

$$a_k \leq a_{k+1} \leq b_{k+1} \leq b_k, \quad k \in \mathbb{N}. \tag{1.10}$$

Denote $A = \{x \mid x = a_k, \text{ for some } k \in \mathbb{N}\}$ and $B = \{y \mid y = b_k, \text{ for some } k \in \mathbb{N}\}$. Then for every $x \in A$ and every $y \in B$ we have $x \leq y$, because otherwise there exist $a_k \in A$ and $b_m \in B$ such that

$$b_m < a_k.$$

We have either $m < k$ or $k < m$. Suppose $m < k$. Then

$$b_m < a_k \leq b_k,$$

thus contradicting (1.10).

Axiom (R_{16}) supplies a real z such that

$$a_k \leq z \leq b_k, \quad k \in \mathbb{N}.$$

Then $z \in I_k$, for every $k \in \mathbb{N}$, and therefore $z \in \cap I_{k \in \mathbb{N}}$. $\square$

Remark. Theorem 2.31 is no longer true if all the closed intervals are unbounded. We can show it considering $I_k = [k, \infty[, \ k \in \mathbb{N}$. $\triangle$

1.2.5 The extended real number system

The extended real number set consists of the real number set to which two symbols, $+\infty \, (= \infty)$ and $-\infty$ have been adjoined, with the following properties.

(a) If x is real, $-\infty < x < +\infty$, and

$$x + \infty = \infty + x = +\infty, \quad x - \infty = -\infty + x = -\infty, \quad \frac{x}{+\infty} = \frac{x}{-\infty} = 0.$$

(b) If $x > 0$, $x(+\infty) = +\infty$, $x(-\infty) = -\infty$.
(c) If $x < 0$, $x(+\infty) = -\infty$, $x(-\infty) = +\infty$.

The extended real number system is denoted by $\overline{\mathbb{R}} = \mathbb{R} \cup \{+\infty\} \cup \{-\infty\}$ with the above-mentioned conventions.

Any element of $\mathbb{R}$ is called *finite* whereas $+\infty$ and $-\infty$ are called *infinities*.

Let A be a nonempty subset of the extended real number set. If A is not bounded above (i.e., for every real y there is an $x \in A$ such that $y < x$), we define $\sup A = +\infty$. Similarly, if A is not bounded below (i.e., for every real y there is an $x \in A$ such that $y > x$), we define $\inf A = -\infty$.

We define intervals involving infinities

$$[a, +\infty[= \{x \in \mathbb{R} \mid a \leq x\}. \qquad]a, +\infty[= \{x \in \mathbb{R} \mid a < x\}.$$
$$[a, +\infty] = \{x \in \overline{\mathbb{R}} \mid a \leq x \leq +\infty\}. \qquad]a, +\infty] = \{x \in \overline{\mathbb{R}} \mid a < x \leq +\infty\}.$$
$$]-\infty, a] = \{x \in \mathbb{R} \mid x \leq a\}. \qquad]-\infty, a[= \{x \in \mathbb{R} \mid x < a\}.$$
$$[-\infty, a] = \{x \in \overline{\mathbb{R}} \mid -\infty \leq x \leq a\}. \qquad [-\infty, a[= \{x \in \overline{\mathbb{R}} \mid -\infty \leq x < a\}.$$
$$[-\infty, \infty] = \overline{\mathbb{R}}. \qquad]-\infty, \infty[= \mathbb{R}.$$

By definition, an open neighborhood of $+\infty$ is an open interval $]a, +\infty[$, for some $a \in \mathbb{R}$. Similarly, an open neighborhood of $-\infty$ is an open interval $]-\infty, a[$, for some $a \in \mathbb{R}$.

1.2.6 The complex number system

Traditionally, there are two ways to introduce the set of *complex numbers*, denoted as $\mathbb{C}$. One way consists in defining a complex number as a pair of real numbers and afterwards two operations are defined, an addition and a multiplication. Thus we get that $\mathbb{C}$ together with the two operations is a field. This way is followed in [115].

More traditionally and more intuitively is considering a complex number as an algebraic entity of the form $a + ib$, where $i^2 = -1$ and $a, b \in \mathbb{R}$. Denote $z = a + ib$. Then a is its *real part* denoted $\operatorname{Re} z$, and b is its *imaginary part*, denoted $\operatorname{Im} z$. The *absolute value* of z is $|z| = \sqrt{a^2 + b^2}$.

A complex number $z = a + ib$ is null if and only if $a = b = 0$. A nonnull complex number $z = a + ib$ can be written as

$$z - a + ib = |z| \left(\frac{a}{|z|} + i \frac{b}{|z|} \right) = |z| (\cos \theta + i \sin \theta)$$

for a certain θ. Then the *Moivre*[8] *formula*

$$z = \cos \theta + i \sin \theta \implies z^n = \cos n\theta + i \sin n\theta$$

is valid.

The properties of complex numbers are introduced in many textbooks. We suggest one of them [115].

---

[8] Abraham de Moivre, 1667–1754.

1.3 Exercises

1.1. For arbitrary sets A, B, and C, show that

 (i) $(A \setminus B) \cap B = \emptyset$.
 (ii) $A \setminus (A \setminus B) = A \cap B$.
 (iii) $(A \setminus B) \cup (B \setminus A) = (A \cup B) \setminus (A \cap B)$.
 (iv) $(A \setminus B) \setminus C = A \setminus (B \cup C)$.
 (v) $(A \setminus B) \cap C = (A \cap C) \cap (C \setminus B) = (A \cap C) \setminus (B \cap C)$.

1.2. Show that the inclusion $A \setminus B \subset C$ holds if and only if $A \subset B \cup C$.

1.3. For arbitrary sets A, B, and C, show that

 (i) $(A \cup B) \setminus C = (A \setminus C) \cup (B \setminus C)$.
 (ii) $(A \cap B) \setminus C = (A \setminus C) \cap (B \setminus C)$.
 (iii) $(A \cup C) \setminus B \subset (A \setminus B) \cup C$.

1.4. Find the mutual relationships between sets X and Y (i.e., $X \subset Y$, $X = Y$, or $X \supset Y$) provided

 (i) $X = A \cup (B \setminus C)$, $\quad Y = (A \cup B) \setminus (A \cup C)$.
 (ii) $X = (A \cap B) \setminus C$, $\quad Y = (A \setminus C) \cap (B \setminus C)$.
 (iii) $X = A \setminus (B \cup C)$, $\quad Y = (A \setminus B) \cup (A \setminus C)$.
 (iv) $X = (A \times B) \cup (C \times B)$, $\quad Y = (A \cup C) \times B$.
 (v) $X = (A \times B) \cup (C \times D)$, $\quad Y = (A \times C) \cup (B \times D)$.
 (vi) $X = (A \cap B) \times (C \cap B)$, $\quad Y = (A \times C) \cap (B \times D)$.
 (vii) $X = (A \cup B) \times (C \cup B)$, $\quad Y = (A \times C) \cup (B \times D)$.

1.5. Let A_n, $n \in \mathbb{N}$, be a collection of sets.

 (i) Consider $B_n = \cup_{i=0}^{n} A_i$, $n \in \mathbb{N}$. Show that $\cup_{n=0}^{\infty} B_n = \cup_{n=0}^{\infty} A_n$.
 (ii) Consider $B_n = \cap_{i=0}^{n} A_i$, $n \in \mathbb{N}$. Show that $\cap_{n=0}^{\infty} B_n = \cap_{n=0}^{\infty} A_n$.

1.6. Let $A_{m,n}$, $m, n \in \mathbb{N}$, be a collection of sets. Which inclusion does it hold

 (i) $\cup_{m=0}^{\infty} (\cap_{n=0}^{\infty} A_{m,n}) \subset \cap_{n=0}^{\infty} (\cup_{m=0}^{\infty} A_{m,n})$;
 (ii) $\cup_{m=0}^{\infty} (\cap_{n=0}^{\infty} A_{m,n}) \supset \cap_{n=0}^{\infty} (\cup_{m=0}^{\infty} A_{m,n})$?

1.7. For any two sets A and B show that

 (i) $\complement_X \emptyset = X$ and $\complement_X X = \emptyset$.
 (ii) $\complement(A \setminus B) = (\complement A) \cup B$.
 (iii) $(A \cap \complement B) \cup (\complement A \cap B) = (A \cup B) \cap (\complement A \cup \complement B) = (A \cup B) \setminus (A \cap B)$.
 (iv) $A \subset B \implies \complement B \subset \complement A$.

1.8. There are given three sets A, B, and U with $A \subset U$ and $B \subset U$. Find the set $X \subset U$ satisfying

$$\complement(X \cup A) \cup (X \cup \complement A) = B.$$

1.9. Show that

(i) $\mathcal{P}(A) \cup \mathcal{P}(B) \subset \mathcal{P}(A \cup B)$.
(ii) $\mathcal{P}(A) \cup \mathcal{P}(B) = \mathcal{P}(A \cup B) \iff A \subset B$ or $B \subset A$.
(iii) $\mathcal{P}(A) \cap \mathcal{P}(B) = \mathcal{P}(A \cap B)$.
(iv) $\mathcal{P}(A \setminus B) \subset \mathcal{P}(A) \setminus \mathcal{P}(B)$.

1.10. Show whether each of the following functions is one-to-one and/or onto.

(i) Function $f : \mathbb{N} \to \mathbb{N}$, defined by $f(n) = 2n$, $n \in \mathbb{N}$.
(ii) Function $f : \mathbb{Q} \times \mathbb{Q} \to \mathbb{Q}$, defined by $f(p, q) = p$, $p, q \in \mathbb{Q}$.
(iii) Function $f : \mathbb{Q} \times \mathbb{Q} \to \mathbb{Q} \times \mathbb{Q}$, defined by $f(p, q) = (p, -q)$, $p, q \in \mathbb{Q}$.

1.11. Let A and B be sets and $f : A \to B$ be a mapping. Consider two mappings $f_* : \mathcal{P}(A) \to \mathcal{P}(B)$ and $f^* : \mathcal{P}(B) \to \mathcal{P}(A)$ defined by $f_*(M) = f(M)$, respectively, $f^*(N) = f^{-1}(N)$, where $M \subset A$ and $N \subset B$.

(a) Show that the following sentences are equivalent.
    (i) f is one-to-one.
    (ii) f_* is one-to-one.
    (iii) f^* is onto.
    (iv) $f(M \cap N) = f(M) \cap f(N)$, for every $M, N \in \mathcal{P}(A)$.
    (v) $f(\complement_A M) \subset \complement_B f(M)$, for every $M \in \mathcal{P}(A)$.
(b) Show that the following sentences are equivalent.
    (i) f is onto.
    (ii) f_* is onto.
    (iii) f^* is one-to-one.
    (iv) $\complement_B f(M) \subset f(\complement_A M)$, for every $M \in \mathcal{P}(A)$.
(c) Show that the following sentences are equivalent.
    (i) f is bijective.
    (ii) f_* is bijective.
    (iii) f^* is bijective.
    (iv) $f(\complement_A M) = \complement_B f(M)$, for every $M \in \mathcal{P}(A)$.

1.12. Let A and B be sets and $f : A \to B$ a mapping.

(a) Show that the following sentences are equivalent.
    (i) f is one-to-one.
    (ii) For any two functions $g, h : C \to A$ with $f \circ g = f \circ g$, it follows that $g = h$.
(b) Show that the following sentences are equivalent.
    (i) f is onto.
    (ii) For any two functions $g, h : B \to C$ with $g \circ f = h \circ f$, it follows that $g = h$.

1.13. Show that the following sets are countable.

(i) $\{2^n \mid n \in \mathbb{N}^*\}$.
(ii) The set of triangles in plane whose vertices have rational coordinates.

(iii) The set of points in the plane having rational coordinates.
(iv) The set of polynomials having rational coefficients.

1.14. Consider a set M having m elements and a set N having n elements, $m, n \geq 1$. Show that

(i) The number of functions $f : M \to N$ is equal to n^m.
(ii) If $m = n$, the number of bijective functions from M to N is equal to $m!$.
(iii) If $m \leq n$, the number of one-to-one functions from M to N is equal to $n(n-1)(n-2) \cdots (n-m+1)$.
(iv) If $m \geq n$, the number of onto functions from M to N is equal to

$$n^m - \binom{n}{1}(n-1)^m + \binom{n}{2}(n-2)^m - \binom{n}{3}(n-3)^m + \cdots + (-1)^{n-1}\binom{n}{n-1}.$$

1.15. Prove the following identities

$$\binom{n}{0} + \binom{n}{1} + \binom{n}{2} + \cdots + \binom{n}{n-1} + \binom{n}{n} = 2^n, \quad n \in \mathbb{N}^*,$$

$$\binom{n}{0} + \binom{n}{2} + \binom{n}{4} + \cdots = 2^{n-1}, \quad n \in \mathbb{N}^*,$$

$$\binom{n}{1} + \binom{n}{3} + \binom{n}{5} + \cdots = 2^{n-1}, \quad n \in \mathbb{N}^*.$$

1.16. Consider a set S containing n elements. The nth Bell[9] number B_n is defined as the number of partitions of S. If $S = \emptyset$, by definition $B_0 = 1$. Show that

$$B_1 = 1, \ B_2 = 2, \ B_3 = 5, \ B_4 = 15. \tag{1.11}$$

$$B_{n+1} = 1 + \sum_{k=1}^{n}\binom{n}{k}B_k = \sum_{k=0}^{n}\binom{n}{k}B_k. \tag{1.12}$$

1.17. Suppose A is a finite set so that $|A| = m$. Show that the number of solutions to

$$A = A_1 \cup A_2 \cup \cdots \cup A_k$$

is equal to $(2^k - 1)^m$.

1.18. Consider $A_1, A_2 \ldots, A_n$ finite sets. Show that

$$|A_1 \cup A_2 \cup \cdots \cup A_n| = \sum |A_i| - \sum_{i<j}|A_i \cap A_j| + \sum_{i<j<k}|A_i \cap A_j \cap A_k| - \cdots$$
$$+ (-1)^{n+1}|A_1 \cap A_2 \cap \cdots \cap A_n|,$$

$$|A_1 \cap A_2 \cap \cdots \cap A_n| = \sum |A_i| - \sum_{i<j}|A_i \cup A_j| + \sum_{i<j<k}|A_i \cup A_j \cup A_k| - \cdots$$
$$+ (-1)^{n+1}|A_1 \cup A_2 \cup \cdots \cup A_n|.$$

---

[9] Eric Temple Bell, 1883–1960.

1.19. Let x, y be real numbers with $y > 0$. Show that

$$\sum_{0 \le k < y} \left\lfloor x + \frac{k}{y} \right\rfloor = \lfloor xy + \lfloor x + 1 \rfloor \left(\lceil y \rceil - y \right) \rfloor.$$

1.20. Let $n \in \mathbb{N}$. Then show that

$$\left\lfloor \frac{n+1}{2} \right\rfloor + \left\lfloor \frac{n+2}{2^2} \right\rfloor + \cdots + \left\lfloor \frac{n+2^k}{2^{k+1}} \right\rfloor + \cdots = n. \tag{1.13}$$

1.21. Let n be a natural number greater than 1. Show that there is no system of three integers of the form $x = a - r$, $y = a$, and $z = a + r$, with $r > 0$, such that

$$x^n + y^n = z^n.$$

1.22. Let A be a finite set. Then show that $|\mathcal{P}(A)| = 2^{|A|}$.

1.23. (Square root inequality) For $x \ge 1$, show that

$$2\sqrt{x+1} - 2\sqrt{x} < \frac{1}{\sqrt{x}} < 2\sqrt{x} - 2\sqrt{x-1}. \tag{1.14}$$

1.24. (Schur [10] inequality) Let x, y, and z be nonnegative numbers. Show that for any $r > 0$, we have

$$x^r(x - y)(x - z) + y^r(y - z)(y - x) + z^r(z - x)(z - y) \ge 0.$$

1.25. (Bernoulli [11] inequality) Consider n real numbers $x_i \ge -1$, $i = 1, 2, \ldots,$ n, such that all of them have the same sign. Show that

$$(1 + x_1)(1 + x_2) \ldots (1 + x_n) \ge 1 + x_1 + x_2 + \cdots + x_n. \tag{1.15}$$

1.26. (Bernoulli inequality) For every $n \in \mathbb{N}^*$ and every $x \ge -1$, show that

$$(1 + x)^n \ge 1 + nx. \tag{1.16}$$

1.27. (i) For $n \ge \mathbb{N}^*$, show that

$$2 \le \left(1 + \frac{1}{n} \right)^n < 3.$$

(ii) Show that
$$\sqrt[n+1]{n+1} \le \sqrt[n]{n}, \quad \forall n \in \mathbb{N}^*, \quad n \ge 3.$$

1.28. (Mean inequality) Let $x_1, x_2, \ldots, x_m$ be positive reals. Show that the geometric mean is less than or equal to the arithmetic mean; that is,

$$\sqrt[m]{x_1 x_2 \ldots x_m} \le \frac{x_1 + x_2 + \cdots + x_m}{m}. \tag{1.17}$$

---

[10] Issai Schur, 1875–1941.
[11] Johann (I) Bernoulli, 1667–1748.

1.29. Let $x_1, x_2, \ldots, x_m$ be positive reals. Show that the harmonic mean is less than or equal to the geometric mean; that is,

$$\frac{m}{\dfrac{1}{x_1} + \dfrac{1}{x_2} + \cdots + \dfrac{1}{x_m}} \leq \sqrt[m]{x_1 x_2 \ldots x_m}. \tag{1.18}$$

1.30. (i) Show that in (1.17) and (1.18) we have equality if and only if $x_1 = \cdots = x_m$.

(ii) Let $x_1, x_2, \ldots, x_n$ be positive numbers. Show that

$$(x_1 + x_2 + \cdots + x_n)\left(\frac{1}{x_1} + \frac{1}{x_2} + \cdots + \frac{1}{x_n}\right) \geq n^2.$$

1.31. (Kantorovich [12] inequality) Suppose $x_1, x_2, \ldots, x_n \in [a, b]$, $0 < a \leq b$, and $t_i \geq 0$. Show that

$$\left(\sum_{i=1}^{n} t_i x_i\right)\left(\sum_{i=1}^{n} \frac{t_i}{x_i}\right) \leq \frac{(a+b)^2}{4ab}\left(\sum_{i=1}^{n} t_i\right)^2.$$

1.32. (i) Let $a_1, \ldots, a_n$ be positive numbers and $s = a_1 + \cdots + a_n$, where $n \geq 2$. Show that

$$\sum_{k=1}^{n} \frac{s}{s - a_k} \geq \frac{n^2}{n - 1}.$$

(ii) For every $n \in \mathbb{N}^*$, show that

$$n! \leq 2\left(\frac{n}{2}\right)^n. \tag{1.19}$$

1.33. Let $a_1, \ldots, a_n$ be positive numbers. Show that

(i)

$$\left(\frac{\sum a_1 a_2 \ldots a_{k-1}}{\binom{n}{k-1}}\right)^k \geq \left(\frac{\sum a_1 a_2 \ldots a_k}{\binom{n}{k}}\right)^{k-1}, \quad k \leq n.$$

Equality occurs if and only if $a_1 = a_2 = \cdots = a_n$.

(ii) (Maclaurin [13] formula) Show the inequality

$$\frac{a_1 + a_2 + \cdots + a_n}{n} \geq \left(\frac{\sum a_1 a_2 \ldots a_k}{\binom{n}{k}}\right)^{1/k}.$$

Note for $k = 1$, it reduces to an identity, whereas for $k = n$ it reduces to the mean inequality (1.17).

---

[12] Leonid Vitalyevich Kantorovich (Леонид Витальевич Канторович), 1912–1986.
[13] Colin Maclaurin, 1698–1746.

1.34. (i) (Lagrange [14] identity) Let a_i and b_i be real numbers, $i = 1, \ldots, m$. Show that

$$\left(\sum_{i=1}^{m} a_i^2 \right) \left(\sum_{i=1}^{m} b_i^2 \right) = \left(\sum_{i=1}^{m} a_i b_i \right)^2 + \sum_{1 \le i < j \le m} (a_i b_j - a_j b_i)^2. \qquad (1.20)$$

(ii) Let $n \ge 2$ be an integer and $t_1, t_2, \ldots, t_n$ be positive numbers. Show that

$$(t_1 + t_2 + \cdots + t_n) \left(\frac{1}{t_1} + \frac{1}{t_2} + \cdots + \frac{1}{t_n} \right) = n^2 + \sum_{1 \le i < j \le n} \left(\sqrt{\frac{t_i}{t_j}} - \sqrt{\frac{t_j}{t_i}} \right)^2.$$

(iii) Let $n \ge 3$ be an integer and $t_1, t_2, \ldots, t_n$ be positive numbers such that

$$n^2 + 1 > (t_1 + t_2 + \cdots + t_n) \left(\frac{1}{t_1} + \frac{1}{t_2} + \cdots + \frac{1}{t_n} \right).$$

Show that with any triple t_i, t_j, t_k, of numbers with $1 \le < i < j < k \le n$ can be the length of sides to a triangle.

1.35. (i) (Cauchy[15] inequality) Let a_i and b_i be real numbers, $i = 1, \ldots, m$. Show that

$$\left(\sum_{i=1}^{m} a_i^2 \right) \left(\sum_{i=1}^{m} b_i^2 \right) \ge \left(\sum_{i=1}^{m} a_i b_i \right)^2. \qquad (1.21)$$

(ii) Show that in (1.21) we have equality if and only if $a_i b_j - a_j b_i = 0$ for all $1 \le i < j \le m$.
(iii) Show that

$$\left| \sum_{1}^{n} a_i b_i \right| \le \sqrt{\sum_{1}^{n} |a_i|^2} \cdot \sqrt{\sum_{1}^{n} |b_i|^2}. \qquad (1.22)$$

(iv) Let a_i be real numbers, $i = 1, \ldots, m$. Show that

$$\frac{|a_1 + a_2 + \cdots + a_m|}{m} \le \sqrt{\frac{a_1^2 + a_2^2 + \cdots + a_m^2}{m}}.$$

(v) Let a, b and c be positive numbers. Show that

$$\frac{a^2}{b} + \frac{b^2}{c} + \frac{c^2}{a} \ge a + b + c + \frac{4(a-b)^2}{a+b+c}.$$

Now a converse to Cauchy inequality (1.21) is introduced.

---

[14] Joseph Louis Lagrange (baptized in the name of Giuseppe Lodovico Lagrangia), 1736–1813.
[15] Augustin Louis Cauchy, 1789–1857.

1.36. (Grinberg [16] inequality) Consider $0 < a_1 \le a_2 \le \cdots \le a_n$ and $0 < b_1 \le b_2 \le \cdots \le b_n$. Show that

$$\left(\sum_{k=1}^n a_k^2\right)\left(\sum_{k=1}^n b_k^2\right) < \left(\sum_{k=1}^n a_k b_k\right)^2 + \frac{1}{2}\left(\sum_{k=1}^n a_k\right)^2\left(\sum_{k=1}^n b_k\right)^2.$$

By the monotonicity of the two sequences we have

1.37. (i) (Chebyshev[17] inequality) If

$$a_1 < a_2 < \cdots < a_n \quad \text{and} \quad b_1 < b_2 < \cdots < b_n,$$

show that

$$n\sum_{k=1}^n a_k b_{n-k+1} < \left(\sum_{k=1}^n a_k\right)\left(\sum_{k=1}^n b_k\right) < n\sum_{k=1}^n a_k b_k.$$

(ii) (Chebyshev inequality) If

$$a_1 \le a_2 \le \cdots \le a_n \quad \text{and} \quad b_1 \le b_2 \le \cdots \le b_n,$$

show that

$$\left(\sum_{k=1}^n a_k\right)\left(\sum_{k=1}^n b_k\right) \le n\sum_{k=1}^n a_k b_k.$$

Show that the above inequality turns into an identity if and only if

$$a_1 = a_2 = \cdots = a_n \quad \text{or} \quad b_1 = b_2 = \cdots = b_n.$$

1.38. (Rearrangement inequality) Consider

$$a_1 \le a_2 \le \cdots \le a_n \quad \text{and} \quad b_1 \le b_2 \le \cdots \le b_n.$$

Suppose $\sigma : \{1, \ldots, n\} \to \{1, \ldots, n\}$ is a bijection. Show that

$$\sum_{k=1}^n a_k b_{n-k+1} \le \sum_{k=1}^n a_k b_{\sigma(k)} \le \sum_{k=1}^n a_k b_k.$$

1.39. (i) (Weierstrass[18] inequality) Show that for $0 \le a_1 \le a_2 \le \cdots \le a_n \le 1$,

$$\prod_{k=1}^n (1 - a_k) \ge 1 - \sum_{k=1}^n a_k. \tag{1.23}$$

---

[16] Darij Grinberg.

[17] Pafnuti Lvovich Chebyshev (Пафнутий Львович Чебышев), 1821–1894.

[18] Wilhelm Theodor Karl Weierstrass, 1815– 1897.

(ii) Let $a_1, \ldots, a_n$ be nonnegative numbers satisfying

$$a_1 + a_2 + \cdots + a_n \leq 1/2.$$

Then show

$$(1 - a_1) \cdots (1 - a_n) \geq 1/2.$$

1.40. Assume that $x_1, x_2, \ldots, x_n \geq -1$ and $\sum_{i=1}^n x_i^3 = 0$. Show that

$$\sum_{i=1}^n x_i \leq n/3.$$

1.41. Consider $0 \leq a_1, a_2, \ldots, a_n, b_1, b_2, \ldots, b_n$. Show that

$$\sqrt[n]{(a_1 + b_1)(a_2 + b_2) \cdots (a_n + b_n)} \geq \sqrt[n]{a_1 a_2 \cdots a_n} + \sqrt[n]{b_1 b_2 \cdots b_n}.$$

1.42. Suppose $n \in \mathbb{N}$. Show that

$$\binom{n}{0} - \binom{n}{2} + \binom{n}{4} - \binom{n}{6} + \cdots = 2^{n/2} \cos \frac{n\pi}{4},$$

$$\binom{n}{1} - \binom{n}{3} + \binom{n}{5} - \binom{n}{7} + \cdots = 2^{n/2} \sin \frac{n\pi}{4}.$$

1.43. Suppose $A = \{0, 1, 2, \ldots, n\}$. Let S_n be the set of triples (a_1, a_2, a_3), $a_1, a_2, a_3 \in A$, such that $|a_1 - a_2| = |a_2 - a_3|$. Find $|S_n|$.

1.44. Suppose $f : \mathbb{N} \to \mathbb{N}$ is onto, $g : \mathbb{N} \to \mathbb{N}$ is one-to-one and $f(n) \geq g(n)$, for all $n \in \mathbb{N}$. Prove that $f(n) = g(n)$, $\forall n \in \mathbb{N}$.

1.45. Find all bounded functions $f : \mathbb{Z} \to \mathbb{Z}$ such that $f(n+k) + f(k-n) = 2f(k)f(n)$, $\forall n, k \in \mathbb{Z}$.

1.46. Find all functions $f : \mathbb{N} \to \mathbb{N}$ such that $f(n+1) > f(f(n))$, $\forall n \in \mathbb{N}$.

1.47. Study the solvability of the exercise: let $\alpha, \beta \in \mathbb{R}$. Find all functions $f : [0, \infty[\to \mathbb{R}$ such that

$$f(x) \cdot f(y) = y^\alpha f\left(\frac{x}{2}\right) + x^\beta f\left(\frac{y}{2}\right), \quad \forall x, y \in [0, \infty[. \qquad (1.24)$$

1.48. Find all functions $f : \mathbb{R} \to \mathbb{R}$ satisfying for all $x, y \in \mathbb{R}$,

$$f(x - f(y)) = 1 - x - y. \qquad (1.25)$$

1.49. Consider $A =]0, 1]$. Show that $\text{int } A =]0, 1[$, $\text{cl } A = [0, 1]$, and $A' = [0, 1]$.

1.50. Consider $A =]0, 1] \cap \mathbb{Q}$. Show that $\inf A = 0$, $\sup A = 1$, $\text{int } A = \emptyset$, $\text{cl } A = [0, 1]$, $A' = [0, 1]$, and $\text{fr } A = [0, 1]$.

1.51. Show that

$$0 < |\sin x| < |x| < |\tan x|, \quad 0 < |x| < \frac{\pi}{2}.$$

1.4 References and comments

Many textbooks deal with basic concepts on sets and numbers. We mention only some of them here. An axiomatic introduction of natural numbers may be found in many textbooks, for example [110, Chapter 1] and [101].

For more details on relations see [92, §3.2], [93, §1.1], and [94, §2.3].

The second part of example 2.1 belongs to [54].

Example 2.2 may by found in several textbooks such as [48, vol.I, § 2.6], [99, p. 130], and in [115, p. 2].

Theorem 2.5 may be found in [95, Theorem 2.17, p. 23].

Theorem 2.10 belongs to [115, Theorem 1.37, p. 11].

Theorems 2.7 and 2.30 are in [47, §1.3].

Exercise 1.19 is from [30].

Many interesting papers appeared in connection with Exercise 1.20. See [136, 1969, pp. 268, 408, 456].

Exercise 1.21 may be found in [96].

The proofs of the inequalities in Exercise 1.33 may be found in [136, vol. 20(1969), 214–219].

Exercise (iii) in 1.34 is one of the exercises from IMO 2004.

More information on the rearrangement inequality as well as some of its applications may be found in [135] and [96].

The exercise (v) in 1.35 is an exercise from the Balkan Mathematical Olympiad, Iaşi, România, 2005.

Exercise (ii) in 1.39 appeared in [29, 1965–1966].

Exercise 1.40 is Exercise 3 of the second day of the Sixth International Mathematics Competition, 1999.

Exercise 1.41 is Exercise A2 in the 64th Putnam Competition 2003. Other solutions to this exercise may be found in [78] and [96, Exercise 2.3.46].

Exercises 1.44–1.48 are taken from [96].

2

Vector Spaces and Metric Spaces

This chapter is dedicated to introduce several basic notions and results concerning vector spaces and metric spaces.

2.1 Vector spaces

2.1.1 Finite-dimensional vector spaces

For every positive integer k let $\mathbb{R}^k$ be the set of all ordered k-tuples

$$x = (x_1, x_2, \ldots, x_k),$$

where $x_1, \ldots, x_k$ are real numbers, called *coordinates* of x. The elements of $\mathbb{R}^k$ are said to be *points* or *vectors*, especially when $k > 1$. If $y = (y_1, \ldots, y_k) \in \mathbb{R}^k$ and α is a real number, let

$$x + y = (x_1 + y_1, x_2 + y_2, \ldots, x_k + y_k), \qquad + : \mathbb{R}^k \times \mathbb{R}^k \to \mathbb{R}^k;$$
$$\alpha \cdot x = \alpha x = (\alpha x_1, \alpha x_2, \ldots, \alpha x_k), \qquad \cdot : \mathbb{R} \times \mathbb{R}^k \to \mathbb{R}^k.$$

These operations define addition of vectors and, respectively, multiplication of a real number (scalar) by a vector. It is easy to check the following sentences.

Theorem 1.1. (a) $(\mathbb{R}^k, +)$ *is a commutative group.*
(b) $\alpha(x + y) = \alpha x + \alpha y$, *for every* $\alpha \in \mathbb{R}$, $x, y \in \mathbb{R}^k$.
(c) $(\alpha + \beta)x = \alpha x + \beta x$, *for every* $\alpha, \beta \in \mathbb{R}$, $x \in \mathbb{R}^k$.
(d) $\alpha(\beta x) = (\alpha\beta)x$, *for every* $\alpha, \beta \in \mathbb{R}$, $x \in \mathbb{R}^k$.

These two operations make $\mathbb{R}^k$ into a *vector (linear) space over the field of the reals*. This vector space has *finite dimension;* more precisely, it is a *k-dimensional* vector space over the field of the reals. The $\mathbb{R}^k$ vector space is usually called the k-dimensional *Euclidean*[1] space. Why the real vector space

---
[1] Euclid of Alexandria ($E\upsilon\kappa\lambda\epsilon\iota\delta\eta\varsigma$ o $'A\lambda\epsilon\xi\alpha\nu\delta\rho\epsilon\upsilon\varsigma$), $\sim$325–$\sim$265 (BC).

$\mathbb{R}^k$ is said to be a k-dimensional one is clarified at page 46. The 0-dimensional vector space contains just one element, the scalar 0. The 1-dimensional vector space over the reals is $\mathbb{R}$.

The zero element 0 in $\mathbb{R}^k$ is said to be the *origin* of the space or its *null* vector. All coordinates of the null vector are equal to 0.

Remark. One may consider the $\mathbb{R}^k$ as a vector space over the field of rational numbers. Hereafter we consider only the $\mathbb{R}^k$ vector space over the field of reals. $\triangle$

Proposition 1.1. (Calculus rules in the linear space $\mathbb{R}^k$)

(a) $0 \cdot x = 0$, $\forall x \in \mathbb{R}^k$ (*the first 0 is the null scalar, and the second one is the null vector*).
(b) $\alpha \cdot 0 = 0$, $\forall \alpha \in \mathbb{R}$ (*the two 0 coincide being the null vector*).
(c) $(-1)x = (-x_1, \ldots, -x_n)$, $\forall x \in \mathbb{R}^k$.
(d) $\alpha(x_1 + \cdots + x_n) = \alpha x_1 + \cdots + \alpha x_n$, *for every* $\alpha \in \mathbb{R}$, *and* $x_i \in \mathbb{R}^k$, $i = 1, 2, \ldots, n$.
(e) $(\alpha_1 + \cdots + \alpha_n)x = \alpha_1 x + \cdots + \alpha_n x$, *for every* $\alpha_i \in \mathbb{R}$, $i = 1, 2, \ldots, n$, *and* $x \in \mathbb{R}^k$.

Define the *inner product* of $x, y \in \mathbb{R}^k$ by

$$\langle x, y \rangle = \sum_{i=1}^{k} x_i y_i,$$

thus $\langle \cdot, \cdot \rangle : \mathbb{R}^k \times \mathbb{R}^k \to \mathbb{R}$.

Theorem 1.2. *Consider the vector space* $\mathbb{R}^k$. *For every vector* $x, y, z \in \mathbb{R}^k$ *and scalar* $\alpha \in \mathbb{R}$ *the following properties hold.*

(a) $\langle x, x \rangle \geq 0$ *and* $\langle x, x \rangle = 0$ *if and only if* $x = 0$.
(b) $\langle x, y \rangle = \langle y, x \rangle$.
(c) $\langle x + y, z \rangle = \langle x, z \rangle + \langle y, z \rangle$.
(d) $\langle \alpha x, y \rangle = \alpha \langle x, y \rangle$.

Every mapping $f : \mathbb{R}^k \times \mathbb{R}^k \to \mathbb{R}$ satisfying (a)–(d) in the previous theorem is said to be an inner product on $\mathbb{R}^k$.

Suppose $a_1, a_2, \ldots, a_k$ are positive numbers. Define

$$\langle x, y \rangle = \sum_{i=1}^{k} a_i x_i y_i.$$

Then this is an inner product on $\mathbb{R}^k$ because it satisfies (a)–(d) in Theorem 1.2.

The *Euclidean norm* of an element $x \in \mathbb{R}^k$ is defined by

$$\|x\|_2 = \sqrt{\langle x, x \rangle} = \left(\sum_{i=1}^{k} x_i^2 \right)^{1/2}.$$

The *Minkowski*[2] *norm* or the l^1-*norm* of $x \in \mathbb{R}^k$ is defined by

$$\|x\|_1 = |x_1| + \cdots + |x_k|,$$

and the *uniform norm* of $x \in \mathbb{R}^k$ is defined by

$$\|x\|_\infty = \max\{|x_1|, \ldots, |x_k|\}.$$

For $1 \leq p < +\infty$ we define the l^p-*norm* of $x \in \mathbb{R}^k$ by

$$\|x\|_p = (|x_1|^p + \cdots + |x_k|^p)^{1/p}.$$

Therefore on $\mathbb{R}^k$ one can define several norms. In order to indicate which norm we are referring to we denote it by $\|\cdot\|_p$, $p \geq 1$, respectively, $\|\cdot\|_\infty$.

Theorem 1.3. *Let* $\|\cdot\|$ *be any of the norms defined above on* $\mathbb{R}^k$. *Suppose* $x, y, z \in \mathbb{R}^k$, *and* $\alpha \in \mathbb{R}$. *Then*
(a) $\|x\| \geq 0$.
(b) $\|x\| = 0 \iff x = 0$.
(c) $\|\alpha x\| = |\alpha| \|x\|$.
(d) $|\langle x, y \rangle| \leq \|x\|_2 \|y\|_2$.
(e) $\|x + y\| \leq \|x\| + \|y\|$ *(triangle inequality)*.
(f) $\|x - z\| \leq \|x - y\| + \|y - z\|$.

Proof. (a), (b), and (c) are trivial.
(d) If $y = 0$, we have equality. Suppose $y \neq 0$ and consider a real λ. Then

$$0 \leq \langle x + \lambda y, x + \lambda y \rangle = \langle x, x \rangle + 2\lambda \langle x, y \rangle + \lambda^2 \langle y, y \rangle.$$

Set $\lambda = -\langle x, y \rangle / \langle y, y \rangle$ and based on Theorem 1.2, we get

$$0 \leq \langle x, x \rangle - 2 \frac{\langle x, y \rangle^2}{\langle y, y \rangle} + \frac{\langle x, y \rangle^2 \langle y, y \rangle}{\langle y, y \rangle^2} = \frac{\langle x, x \rangle \langle y, y \rangle - \langle x, y \rangle^2}{\langle y, y \rangle}.$$

Then (d) follows. This inequality is the Cauchy inequality, Exercise 1.35 of Section 1.3.
(e) If the norm is an l_p-norm, $p \geq 1$, then our inequality is actually the Minkowski inequality; see Theorem 1.16 at page 56. If the norm under consideration is the uniform one, then a straightforward evaluation proves the inequality.
(f) It follows from (e).   □

Every function $\|\cdot\| : \mathbb{R}^k \to \mathbb{R}$ satisfying (a), (b), (c), and (e) in Theorem 1.3 is said to be a *norm* on $\mathbb{R}^k$. We already saw several norms on $\mathbb{R}^k$, namely $\|\cdot\|_p$, $p \geq 1$ and $\|\cdot\|_\infty$.

[2] Hermann Minkowski, 1864–1909.

2.1.2 Vector spaces

Some of the concepts introduced previously are considered in more general settings.

A nonempty set X is said to be a *vector space* over a field $\mathbb{K}$ ($\mathbb{R}$ or $\mathbb{C}$) provided

(a) $X = (X, +)$ is an Abelian group.
(b) A *scalar multiplication* is defined: to every $\lambda \in \mathbb{K}$ and $x \in X$ there is assigned an element $\lambda x \in X$ fulfilling the following four conditions.
  - (b_1) $\lambda(x + y) = \lambda x + \lambda y, \quad \forall x, y \in X, \; \lambda \in \mathbb{K}$.
  - (b_2) $(\lambda + \mu)x = \lambda x + \mu x, \quad \forall x \in X, \; \lambda, \mu \in \mathbb{K}$.
  - (b_3) $(\lambda\mu)x = \lambda(\mu x), \quad \forall x \in X, \; \lambda, \mu \in \mathbb{K}$.
  - (b_4) $1 \cdot x = x, \quad \forall x \in X$.

If $\mathbb{K} = \mathbb{R}$, the vector space is said to be a *real vector space*, whereas with $\mathbb{K} = \mathbb{C}$, the vector space is said to be a *complex vector space*. The elements in X are called *vectors* or *points*, and the elements in $\mathbb{K}$ are called *scalars*.

Proposition 1.2. *Let X be a vector space over a field $\mathbb{K}$. Then*

(a) $\lambda(x - y) = \lambda x - \lambda y$, *for every $x, y \in X$ and $\lambda \in \mathbb{K}$.*
(b) $(\lambda - \mu)x = \lambda x - \mu x$, *for every $x \in X$, $\lambda, \mu \in \mathbb{K}$.*
(c) $\lambda \cdot 0 = 0$, *for every $\lambda \in \mathbb{K}$.*
(d) $0 \cdot x = 0$, *for every $x \in X$.*
(e) $(-1)x = -x$, *for every $x \in X$.*
(f) *For $\lambda \in \mathbb{K}$ and $x \in X$, $\lambda x = 0 \implies \lambda = 0$ or $x = 0$.*

Proof. (a) $\lambda x = \lambda((x - y) + y) = \lambda(x - y) + \lambda y$.
(b) $\lambda x = ((\lambda - \mu) + \mu)x = (\lambda - \mu)x + \mu x$.
(c) $\lambda 0 = \lambda(x - x) = \lambda x - \lambda x = 0$.
(d) Take $\lambda = \mu$ in (b).
(e) $0 = 0x = (1 - 1)x = 1x + (-1)x = x + (-1)x$.
(f) It is trivial. □

Let X be a vector space over $\mathbb{K}$ and Y a nonempty subset of it. Y is said to be a *subspace* of X if Y is itself a vector space over $\mathbb{K}$ with respect to the operations of vector addition and scalar multiplication over X. Alternatively, Y is a subspace of X provided every $\alpha, \beta \in \mathbb{K}$ and $x, y \in Y$ imply $\alpha x + \beta y \in Y$. A *linear combination* of elements in Y is any element $x = \alpha_1 y_1 + \ldots + \alpha_n y_n$, with $\alpha_1, \ldots, \alpha_n \in \mathbb{K}$, and $y_1, \ldots, y_n \in Y$. The set of linear combinations of elements in Y is said to be the *linear hull* of Y and is denoted $\lin Y$.

Examples. (a) Let X be a vector space over a field $\mathbb{K}$. The the set $\{0\}$ consisting of the zero vector alone, and also the entire space X are subspaces of X.
(b) Let X be the vector space $\mathbb{R}^3$; then the set

$$Y = \{(x, y, 0) \mid x, y \in \mathbb{R}\}$$

is a subspace of X.

(c) Consider a homogeneous system of n linear equations in n unknowns with real coefficients

$$a_{1\,1}x_1 + a_{1\,2}x_2 + \cdots + a_{1\,n}x_n = 0$$
$$a_{2\,1}x_1 + a_{2\,2}x_2 + \cdots + a_{2\,n}x_n = 0$$
$$\cdots$$
$$a_{n\,1}x_1 + a_{n\,2}x_2 + \cdots + a_{n\,n}x_n = 0.$$

If we recognize each solution of the system as a point in $\mathbb{R}^n$, then the set Y of all solutions of the homogeneous system is a subspace of $\mathbb{R}^n$. $\triangle$

Proposition 1.3. *Let X be a vector space over a field $\mathbb{K}$ and $\emptyset \neq Y \subset X$. Then*

(a) *$Y \subset \mathrm{lin}\, Y$ and $\mathrm{lin}\, Y$ is a subspace of X.*
(b) *If Z is any other subspace of X containing Y, then $\mathrm{lin}\, Y \subset Z$.*

Proof. (a) For any $x \in Y$, $x = 1 \cdot x \in \mathrm{lin}\, Y$, thus $Y \subset \mathrm{lin}\, Y$. For any $\alpha, \beta \in \mathbb{K}$ and $x, y \in \mathrm{lin}\, Y$ we have

$$x = \alpha_1 x_1 + \alpha_2 x_2 + \cdots + \alpha_n x_n, \quad y = \beta_1 y_1 + \beta_2 y_2 + \cdots + \beta_m y_m,$$
$$\alpha x + \beta y = \gamma_1 z_1 + \gamma_2 z_2 + \cdots + \gamma_k z_k,$$

for some $\gamma_i \in \mathbb{K}$ and $z_i \in Y$ properly chosen.

(b) Choose $z \in \mathrm{lin}\, Y$. Then $z = \alpha_1 x_1 + \alpha_2 x_2 + \cdots + \alpha_n x_n$ for some scalars α_i and $x_i \in Y \subset Z$. Then $z \in Z$. $\square$

Remark. We conclude that $\mathrm{lin}\, Y$ is the smallest subspace (in respect to the inclusion of sets) of X containing Y. $\triangle$

A subset Y of a vector space X over a field $\mathbb{K}$ is said to be *linearly independent* provided for any $\alpha_1, \ldots, \alpha_n \in \mathbb{K}$, $x_1, \ldots, x_n \in Y$ such that $x_i \neq x_j$ for $i \neq j$,

$$\alpha_1 x_1 + \cdots + \alpha_n x_n = 0 \implies \alpha_1 = \cdots = \alpha_n = 0.$$

A characterization of this property is given by the following.

Proposition 1.4. *Let X be a vector space over a field $\mathbb{K}$, $Y \subset X$, Y nonempty. Then Y is linearly independent if and only if for any $x \in Y$, $x \notin \mathrm{lin}\,(Y \setminus \{x\})$.*

A subset Y of a vector space X over a field $\mathbb{K}$ is said to be *linearly dependent* if it is not linearly independent. Alternatively, there exist $\alpha_1, \ldots, \alpha_n \in \mathbb{K}$, not all of them equal to 0, and $x_1, \ldots, x_n \in Y$ such that $x_i \neq x_j$ for $i \neq j$, and

$$\alpha_1 x_1 + \cdots + \alpha_n x_n = 0.$$

A subset B of a vector space X over a field $\mathbb{K}$ is said to be a *basis* of X provided

(a) B is linearly independent.
(b) $\lim B = X$.

Theorem 1.4. *Any vector space nonidentical to its origin has a basis.*

Theorem 1.5. *Suppose that a vector space has a basis containing $n \in \mathbb{N}^*$ elements. Then any other basis of the vector space contains precisely $n \in \mathbb{N}^*$ elements.*

More generally,

Theorem 1.6. *Let B_0 and B_1 be two bases of a vector space over a field $\mathbb{K}$. Then $B_0 \sim B_1$; that is, they have the same cardinal.*

Thus, the cardinal of a basis is an invariant of the vector space. Then we define the *dimension* of a vector space X as the cardinal number

$$\dim X = \begin{cases} |B|, & X \neq \{0\} \\ 0, & X = \{0\} \end{cases}$$

and we say that the dimension of X is equal to $|B|$.

Examples. (a) Consider the vector space $\mathbb{R}$ over the field $\mathbb{R}$. Then $\dim \mathbb{R} = 1$ because $e_1 = 1$ is a basis of $\mathbb{R}$. We may consider as a basis for this vector space any nonzero real number. So, the basis of a vector space is not unique. Note $|\mathbb{R}| = \aleph_1$.
(b) Let X be the vector space $\mathbb{R}^k$, $k \in \mathbb{N}^*$, over the field of reals. We introduce its *canonical basis* by

$$e_1 = (1, 0, 0, \ldots, 0, 0), \quad e_2 = (0, 1, 0, \ldots, 0, 0), \quad e_3 = (0, 0, 1, \ldots, 0, 0), \ldots,$$
$$e_{k-1} = (0, 0, 0, \ldots, 1, 0), \quad e_k = (0, 0, 0, \ldots, 0, 1).$$

It is easy to check that the set $\{e_1, e_2, \ldots, e_k\}$ is indeed a basis of the vector space $\mathbb{R}^k$. Now we know that $\dim \mathbb{R}^k$ is precisely k. We emphasize that the vector space $\mathbb{R}^k$ over the reals has other bases as well. Indeed, it is easy to check that the following vectors

$$e_1 = (1, 0, 0, \ldots, 0, 0), \quad e_2 = (1, 1, 0, \ldots, 0, 0), \quad e_3 = (1, 1, 1, \ldots, 0, 0), \ldots,$$
$$e_{k-1} = (1, 1, 1, \ldots, 1, 0), \quad e_k = (1, 1, 1, \ldots, 1, 1).$$

form a basis to $\mathbb{R}^k$.
(c) Consider the vector space of polynomials with real coefficients over the field of reals. Then the dimension of this space is $\aleph_0$ because for every $n \in \mathbb{N}^*$ from

$$a_n x^n + \cdots + a_1 x + a_0 = 0,$$

where the right-hand side contains the constant null polynomial, it follows that all coefficients are zero. $\triangle$

Let X and Y be vector spaces over the same field $\mathbb{K}$. A mapping $f : X \to Y$ is said to be *additive* if

$$f(x + y) = f(x) + f(y), \quad \forall x, y \in X.$$

It follows immediately that $f(0) = 0$ and $f(-x) = -f(x)$, for all $x \in X$. A mapping $f : X \to Y$ is said to be *homogeneous* provided

$$f(\alpha x) = \alpha f(x), \quad \forall \alpha \in \mathbb{K}, \quad x \in X.$$

A mapping $f : X \to Y$ is said to be *linear* if it is additive and homogeneous; that is,

$$f(\alpha x + \beta y) = \alpha f(x) + \beta f(y), \quad \forall \alpha, \beta \in \mathbb{K}, \quad x, y \in X.$$

If $Y = \mathbb{K}$ and $f : X \to \mathbb{K}$, we say that function f is a *functional*.

Examples. (a) Consider two vector spaces over the same field $\mathbb{K}$. The mapping $f : X \to Y$ defined by $f(x) = 0$, for all $x \in X$, is linear.
(b) Let X be the vector space over a field $\mathbb{K}$. Then the *identity function* defined by $f(x) = x$, for all $x \in X$, is linear.
(c) Let us consider a field $\mathbb{K}$ and a matrix $A = (a_{ij})_{1 \le i \le m, 1 \le j \le n}$. All its entries are elements of $\mathbb{K}$; that is,

$$A = \begin{bmatrix} a_{11} & a_{12} & \cdots & a_{1n} \\ a_{21} & a_{22} & \cdots & a_{2n} \\ \cdots & & & \\ a_{m1} & a_{m2} & \cdots & a_{mn} \end{bmatrix}.$$

We define the function $f : \mathbb{K}^n \to \mathbb{K}^m$ by

$$x = (x_1, \ldots, x_n) \in \mathbb{K}^n, \quad f(x) = \left(\sum_{j=1}^{n} a_{1j} x_j, \sum_{j=1}^{n} a_{2j} x_j, \ldots, \sum_{j=1}^{n} a_{mj} x_j \right).$$

$$(2.1)$$

Function f thus defined is linear.
(d) Consider a field $\mathbb{K}$ and a linear mapping $f : \mathbb{K}^n \to \mathbb{K}^m$. Then f has the form given in (2.1), where the entries a_{ij} are supplied by

$$(a_{1j}, a_{2j}, \ldots, a_{mj}) = f(e_j), \quad 1 \le j \le n, \tag{2.2}$$

and $e_1, e_2, \ldots, e_n$ is the canonical basis of $\mathbb{K}^n$.
(e) From (c) and (d) it follows that every $m \times n$ matrix with entries in a field $\mathbb{K}$ defines a linear mapping from $\mathbb{K}^n$ to $\mathbb{K}^m$ by (2.1); conversely, every linear mapping from $\mathbb{K}^n$ to $\mathbb{K}^m$ is given by an $m \times n$ matrix with entries in $\mathbb{K}$ by (2.2). Thus we may identify a linear mapping $f : \mathbb{K}^n \to \mathbb{K}^m$ by the matrix generated through (2.2). $\triangle$

2.1.3 Normed spaces

A nonempty set X is said to be a *normed vector space* or a *normed space* over a field $\mathbb{K}$ of real or complex numbers provided we have the following.

(a) X is a vector space over the field $\mathbb{K}$.
(b) One can defined a function $X \ni x \to \|x\| \in \mathbb{R}$ such that for every $x, y \in X$ and $\lambda \in \mathbb{K}$ there hold
   (b$_1$) $\|x\| \geq 0$ and $\|x\| = 0 \Longleftrightarrow x = 0$.
   (b$_2$) $\|\lambda x\| = |\lambda| \cdot \|x\|$.
   (b$_3$) $\|x + y\| \leq \|x\| + \|y\|$ (triangle inequality).

In this case we write $(X, \|\cdot\|)$ or X if the norm precisely used is clear.

Remark. $(\mathbb{R}^k, \|\cdot\|_p)$, $p \geq 1$, and $(\mathbb{R}^k, \|\cdot\|_\infty)$ are normed spaces. $\triangle$

Let (x_n) be a sequence in a normed space X. It *converges* to $x \in X$ and we write $\lim x_n = x$ or $x_n \to x$ provided $\|x_n - x\| \to 0$ as $n \to \infty$. We say that (x_n) is *convergent*. Otherwise we say that the sequence (x_n) *diverges* or that it is *divergent*.

Let X be a normed space. A sequence (x_n) in X is said to be *Cauchy* or *fundamental* if for every $\varepsilon > 0$ there exists a rank $n_\varepsilon \in \mathbb{N}^*$ such that for any $n, m \in \mathbb{N}^*$, $n, m \geq n_\varepsilon$,

$$\|x_n - x_m\| < \varepsilon.$$

A normed space is said to be a *Banach*[3] *space* provided every Cauchy sequence is convergent.

Remarks. (a) A convergent sequence in a normed space is Cauchy. The converse statement is not always true.
(b) A convergent sequence (x_n) in a normed space is bounded; that is, there is a positive M so that $\|x_n\| \leq M$ for all terms. $\triangle$

Examples. (a) Consider the finite-dimensional Euclidean space $\mathbb{R}^n$. For $x = (x_1, \ldots, x_n) \in \mathbb{R}^n$ define (as we already did) the Euclidean norm

$$\|x\|_2 = \left(\sum_{k=1}^{n} x_k^2 \right)^{1/2}.$$

This space is a Banach space as shown by Theorem 1.8 at page 78.
(b) Let $C[a, b]$ be the vector space of real-valued continuous functions defined on $[a, b]$. For $x \in C[a, b]$ define

$$\|x\|_\infty = \max_{t \in [a,b]} |x(t)|. \tag{2.3}$$

This is a norm on the vector space $C[a, b]$. Indeed, first we remark by Theorem 2.10 at page 156, that the maximum is attained on a point in $[a, b]$. Then one

---

[3] Stefan Banach, 1892–1945.

can easily check Axioms (b$_1$) and (b$_2$). Now we check Axiom (b$_3$). For every $x, y \in C[a, b]$ we have

$$\|x + y\|_\infty = \max_{t \in [a,b]} |x(t) + y(t)| = |x(t_0) + y(t_0)| \leq |x(t_0)| + |y(t_0)|$$

$$\leq \|x\|_\infty + \|y\|_\infty,$$

for a point $t_0 \in [a, b]$ on which $[a, b] \ni t \to |x(t) + y(t)|$ attains its highest value. The norm defined by (2.3) on $C[a, b]$ is said to be the *uniform norm*. The space $(C[a, b], \|\cdot\|_\infty)$ is not only a normed space but also a Banach space, as follows from Theorem 2.1 at page 110 and from Theorem 8.2 at page 183.
(c) Let m be the vector space of bounded real sequences; that is, $x \in m$ provided $x = (x_1, x_2, \dots)$, $x_k \in \mathbb{R}$ for every $k \in \mathbb{N}^*$, and there is a nonnegative constant c such that $|x_k| \leq c$ for each term of the sequence x. For $x \in m$, define

$$\|x\| = \sup_k \{|x_k|\}.$$

Then the norm just defined on the vector space of bounded real sequences m determines a normed space. $\triangle$

Proposition 1.5. *In every normed space the following properties hold.*

(a) $\|(x_n + y_n) - (x + y)\| \leq \|x_n - x\| + \|y_n - y\|$.
(b) $\|\lambda_n x_n - \lambda x\| \leq |\lambda_n| \cdot \|x_n - x\| + \|x\| \cdot |\lambda_n - \lambda|$.
(c) $\big| \|x\| - \|y\| \big| \leq \|x - y\|$.

Corollary 1.1. *Consider X a normed space, two sequences (x_n) and (y_n) in X, and (λ_n) a sequence of scalars. Then*

(a) $x_n \to x$ and $y_n \to y$ imply $x_n + y_n \to x + y$.
(b) $x_n \to x$ and $\lambda_n \to \lambda$ imply $\lambda_n x_n \to \lambda x$.
(c) $x_n \to x$ implies $\|x_n\| \to \|x\|$.

2.1.4 Hilbert spaces

Let H be a vector space over $\mathbb{K}$, where $\mathbb{K} = \mathbb{C}$ or $\mathbb{K} = \mathbb{R}$. Define a mapping $\langle \cdot, \cdot \rangle : H \times H \to \mathbb{K}$, called the *inner product*, or *scalar product* satisfying for all $x, y, u, v \in H$ and $\lambda \in \mathbb{K}$,

(a) $\langle x, y \rangle = \overline{\langle y, x \rangle}$ (here the bar denotes the complex conjugation, thus $\langle x, x \rangle \in \mathbb{R}$); if $\mathbb{K} = \mathbb{R}$, this is just commutativity.
(b) $\langle v + u, y \rangle = \langle v, y \rangle + \langle u, y \rangle$.
(c) $\langle \lambda x, y \rangle = \lambda \langle x, y \rangle$.
(d) $\langle x, x \rangle \geq 0$, and $\langle x, x \rangle = 0 \iff x = 0$.

The pair $(H, \langle \cdot, \cdot \rangle)$, where H is a vector space and $\langle \cdot, \cdot \rangle$ is an inner product defined on it, is said to be an *inner product space* or a *pre-Hilbert*[4] *space*.

---

[4] David Hilbert, 1862–1943.

Theorem 1.7. (Cauchy–Buniakovski[5]–Schwarz[6]) *Let* $(H, \langle \cdot, \cdot \rangle)$ *be an inner product space. For every* $x, y \in H$ *it holds*

$$|\langle x, y \rangle|^2 \leq \langle x, x \rangle \langle y, y \rangle. \tag{2.4}$$

Proof. Suppose that $y \neq 0$, because otherwise inequality (2.4) is trivial.
Thus $y \neq 0$. Then $\langle y, y \rangle > 0$. For every $\lambda \in \mathbb{K}$ we have

$$0 \leq \langle x + \lambda y, x + \lambda y \rangle = \langle x, x \rangle + \lambda \langle y, x \rangle + \overline{\lambda} \langle x, y \rangle + \lambda \overline{\lambda} \langle y, y \rangle.$$

Now choose $\lambda = -\langle x, y \rangle / \langle y, y \rangle$. Then the above inequality supplies

$$0 \leq \langle x, x \rangle - \frac{\langle x, y \rangle}{\langle y, y \rangle} \langle y, x \rangle - \frac{\langle y, x \rangle}{\langle y, y \rangle} \langle x, y \rangle + \frac{\langle x, y \rangle \langle y, x \rangle}{\langle y, y \rangle^2} \langle y, y \rangle$$

$$= \langle x, x \rangle - \frac{\langle x, y \rangle}{\langle y, y \rangle} \langle y, x \rangle.$$

Thus (2.4) holds for all $y \neq 0$. Hence it holds for all y. $\square$

Corollary 1.2. *Let* $(H, \langle \cdot, \cdot \rangle)$ *be an inner product space. For every* $x, y \in H$ *it holds*

$$\langle x + y, x + y \rangle \leq \left(\sqrt{\langle x, x \rangle} + \sqrt{\langle y, y \rangle} \right)^2.$$

Proof. By (2.4) one has

$$\langle x + y, x + y \rangle = \langle x, x \rangle + 2\mathrm{Re}\, \langle x, y \rangle + \langle y, y \rangle$$

$$\leq \langle x, x \rangle + 2\sqrt{\langle x, x \rangle \langle y, y \rangle} + \langle y, y \rangle = \left(\sqrt{\langle x, x \rangle} + \sqrt{\langle y, y \rangle} \right)^2. \quad \square$$

We introduce the norm defined by an inner product as $\|x\| = \sqrt{\langle x, x \rangle}$. It is easy to check ($b_1$) and ($b_2$) from the definition of the norm, page 48. (b_3) follows by Corollary 1.2.

Remark. Thus an inner product space is a normed space. But not every normed space $(X, \| \cdot \|)$ originate in a vector space with an inner product $\langle \cdot, \cdot \rangle$ defined on it so that $\|x\| = \sqrt{\langle x, x \rangle}$, for all $x \in X$. $\triangle$

Proposition 1.6. *Let* $(H, \| \cdot \|)$ *be a pre-Hilbert space. Then every two elements* $x, y \in H$ *fulfill the identity*

$$\|x + y\|^2 + \|x - y\|^2 = 2(\|x\|^2 + \|y\|^2) \quad \text{(parallelogram law)}.$$

---

[5] Victor Yakovlevich Buniakovski (Виктор Яковлевич Буняковский), 1804–1889.
[6] Hermann Amaudus Schwarz, 1843–1921.

If a pre-Hilbert space is complete in respect to the norm defined by the inner product, it is called a *Hilbert space*.

The Cauchy–Buniakovski–Schwarz inequality is a basic tool for proving the triangle inequality in a pre-Hilbert space. The next result shows that the triangle inequality also holds in a much weaker space, that is, a space which is not a normed space.

Proposition 1.7. *Let V be a vector space over the real numbers, and let the operator $\|x\|$ be defined for all $x \in V$, satisfying the following conditions.*

$$\|x\| \geq 0, \quad \forall\, x \in V, \tag{2.5}$$

$$\|x + y\|^2 + \|x - y\|^2 = 2(\|x\|^2 + \|y\|^2), \quad \forall\, x, y \in V. \tag{2.6}$$

Then the triangle inequality holds.

Proof. From (2.6) we observe that $\|0\| = 0$ and $\|-x\| = \|x\|$, for each $x \in V$. Define an inner product $\langle \cdot, \cdot \rangle$ by

$$\langle x, y \rangle = \frac{1}{4} \left(\|x + y\|^2 + \|x - y\|^2 \right).$$

We have immediately $\langle x, x \rangle = \|x\|^2$ and $\langle x, y \rangle = \langle y, x \rangle$. Computing from (2.6) we obtain $\langle x + z, y \rangle = \langle x, y \rangle + \langle z, y \rangle$. Then it follows that $\langle rx, y \rangle = r \langle x, y \rangle$ for all $r \in \mathbb{Q}$. Hence, for every rational number r we have

$$0 \leq \langle rx + y, rx + y \rangle = r^2 \langle x, x \rangle + 2r \langle x, y \rangle + \langle y, y \rangle,$$

and thus this implies the Cauchy–Buniakovski–Schwarz inequality $|\langle x, y \rangle| \leq \|x\|\, \|y\|$. This in turn leads to the triangle inequality by Theorem 1.7.   □

Examples. (a) The simplest example of a Hilbert space is the following. The vector space is $\mathbb{R}$, the inner product is defined as $\langle x, y \rangle = xy$, for all $x, y \in \mathbb{R}$, and the completeness follows by (b) of Theorem 1.7 at page 77.
(b) A common Hilbert space is $(\mathbb{R}^k, \|\cdot\|_2)$ as follows from Theorem 1.8 at page 78.   △

2.1.5 Inequalities

The proofs of Hölder [7] inequalities are based on the W. H. Young [8] inequality. First we introduce the Young identity.

Theorem 1.8. *Let $f : [a, b] \to [c, d]$ be a bijective and increasing function. Then*

$$\int_a^b f(x)dx + \int_c^d f^{-1}(y)dy = bd - ac. \tag{2.7}$$

---

[7] Otto Ludwig Hölder, 1859–1937.
[8] William Henry Young, 1863–1942.

We introduce the integral form of Young inequality.

Theorem 1.9. *Let f be a continuous and strictly increasing function defined on $[0, \infty[$ such that $\lim_{u \to \infty} f(u) = \infty$ and $f(0) = 0$. Denote $g = f^{-1}$. For $x \in [0, \infty[$ define the following functions*

$$F(x) = \int_0^x f(u)du \ \text{ and } \ G(x) = \int_0^x g(v)dv. \tag{2.8}$$

Then $a, b \in [0, \infty[$ imply

$$ab \leq F(a) + G(b), \tag{2.9}$$

and equality holds if and only if $b = f(a)$.

See Figure 2.1.

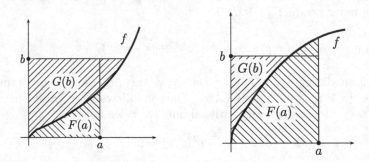

Fig. 2.1. Young inequality

Corollary 1.3. (Young) *Suppose $p > 1$ and α and β are nonnegative reals. Then*

$$\alpha\beta \leq \alpha^p/p + \beta^q/q, \ \text{ whenever } \ 1/p + 1/q = 1. \tag{2.10}$$

The equality holds if and only if $\beta = \alpha^{p-1}$.

Proof. First approach. For $u \in [0, \infty)$, define $f(u) = u^{p-1}$ and apply Theorem 1.9.

Second approach. Consider the function $f :]0, \infty[\to \mathbb{R}$ given by

$$f(x) = x^p/p + x^{-q}/q.$$

It has an absolute minimum at $x = 1$. The required inequality follows from $f(1) \leq f(\alpha^{1/q}\beta^{-1/p})$. $\square$

Let $a = (\alpha_1, \ldots, \alpha_n) \in \mathbb{R}^n$ or $\mathbb{C}^n$. If $r \neq 0$, define the *weighted mean with weight r of the finite sequence* a as

$$M_r(a) = \left(\sum_{k=1}^{n} |\alpha_k|^r\right)^{1/r} = \left(\sum |\alpha_k|^r\right)^{1/r}.$$

Proposition 1.8. *Suppose $p > 1$, $1/p + 1/q = 1$, and there are given two finite sequences $a = (\alpha_1, \ldots, \alpha_n)$ and $b = (\beta_1, \ldots, \beta_n)$ satisfying $M_p(a) = M_q(b) = 1$. Then*

$$M(ab) = M_1(ab) \le 1,$$

where $ab = (\alpha_1\beta_1, \ldots, \alpha_n\beta_n)$.

Proof. Choose $k \in \{1, \ldots, n\}$. We apply the Young inequality (2.10) to $|\alpha_k|$ and $|\beta_k|$. It follows that

$$|\alpha_k\beta_k| \le |\alpha_k|^p/p + |\beta_k|^q/q. \tag{2.11}$$

Summing up (2.11) for $k = 1, 2, \ldots, n$, we get

$$\sum |\alpha_k\beta_k| \le (1/p)\sum |\alpha_k|^p + (1/q)\sum |\beta_k|^q = 1/p + 1/q = 1. \quad \square$$

Theorem 1.10. (Hölder inequality for $p > 1$ and $q > 1$) *Suppose $p > 1$, $1/p + 1/q = 1$, $a = (\alpha_1, \ldots, \alpha_n)$, $b = (\beta_1, \ldots, \beta_n)$ are two finite sequences satisfying $M_p(a) > 0$ and $M_q(b) > 0$. Then*

$$M(ab) = \sum |\alpha_k\beta_k| \le M_p(a)M_q(b). \tag{2.12}$$

Proof. Define

$$\overline{\alpha}_k = \alpha_k/M_p(a), \qquad \overline{\beta}_k = \beta_k/M_q(b),$$
$$\overline{a} = (\overline{\alpha}_1, \ldots, \overline{\alpha}_n), \qquad \overline{b} = (\overline{\beta}_1, \ldots, \overline{\beta}_n).$$

We remark that $M_p(\overline{a}) = M_q(\overline{b}) = 1$, and therefore we can apply Proposition 1.8. Thus we find that $M(\overline{a}\overline{b}) \le 1$, that is, (2.12). $\square$

Theorem 1.11. *Inequality (2.12) turns into an equality if and only if the fraction $|\alpha_k|^p/|\beta_k|^q$ does not depend upon k (the fraction $0/0$ is excluded).*

Corollary 1.4. (Cauchy) *Set $p = 2$ and $q = 2$ in Theorem 1.10 and get Exercise 1.35 in Section 1.3.*

Corollary 1.5. (Cauchy–Buniakovski) *Consider $m \in \mathbb{N}^*$, $m \ge 2$, $a_j = (\alpha_{j1}, \ldots, \alpha_{jn})$, $j = 1, \ldots, m$, where all entries are nonnegative. Then*

$$\prod_{j=1}^{m}\left(\sum_{i=1}^{n}\alpha_{ji}^2\right)^{1/2} \ge \sum_{i=1}^{n}\prod_{j=1}^{m}\alpha_{ji}.$$

Proof. By induction in respect to m. $\square$

Corollary 1.6. *Suppose that $p_i \geq 0$ for all $i = 1, 2, \ldots, n$, and $p_1 + p_2 + \cdots + p_n = 1$. If $a_1, \ldots, a_n$ and $b_1, \ldots, b_n$ are nonnegative real numbers that satisfy the termwise bound $1 \leq a_i b_i$, for $i = 1, 2, \ldots, n$, then*

$$1 \leq \left(\sum_{i=1}^{n} p_i a_i \right) \left(\sum_{i=1}^{n} p_i b_i \right).$$

Proof. We have

$$1 = \left(\sum_{i=1}^{n} p_i \right)^2 \leq \left(\sum_{i=1}^{n} p_i \sqrt{a_i b_i} \right)^2 \leq \left(\sum_{i=1}^{n} p_i a_i \right) \left(\sum_{i=1}^{n} p_i b_i \right). \quad \square$$

Theorem 1.12. *Suppose $a, b \in \mathbb{R}$ and $p \geq 2$. Then*

$$|a + b|^p + |a - b|^p \leq 2^{p-1}(|a|^p + |b|^p). \tag{2.13}$$

Proof. Let $\alpha > 0, \beta > 0$, and $c = \sqrt{\alpha^2 + \beta^2}$. Then $0 < \min\{\alpha/c, \beta/c\} \leq \max\{\alpha/c, \beta/c\} < 1$. It follows that

$$(\alpha/c)^p + (\beta/c)^p \leq (\alpha/c)^2 + (\beta/c)^2 = 1.$$

Therefore $\alpha^p + \beta^p \leq c^p = (\alpha^2 + \beta^2)^{p/2}$. It follows that

$$|a + b|^p + |a - b|^p \leq (|a + b|^2 + |a - b|^2)^{p/2} = 2^{p/2}(|a|^2 + |b|^2)^{p/2}. \tag{2.14}$$

Because $2/p + (p - 2)/p = 1$, by the Hölder inequality we have

$$a^2 \cdot 1 + b^2 \cdot 1 \leq \left((a^2)^{p/2} + (b^2)^{p/2} \right)^{2/p} (1 + 1)^{(p-2)/p} = 2^{(p-2)/p} \left(|a|^p + |b^p| \right)^{2/p}.$$

This inequality together with (2.14) give (2.13). $\quad \square$

Theorem 1.13. (Hölder inequality for positive weights) *Consider $m \in \mathbb{N}^*$, $m \geq 2$, $a_j = (\alpha_{j1}, \ldots, \alpha_{jn})$, $j = 1, \ldots, m$, and $p_1, p_2, \ldots, p_m > 0$ so that $\sum 1/p_j = 1$. Suppose that $M_{p_j}(a_j) > 0$, $j = 1, \ldots, m$. Then*

$$\sum_{k=1}^{n} \left| \prod_{j=1}^{m} \alpha_{jk} \right| \leq \prod_{j=1}^{m} M_{p_j}(a_j). \tag{2.15}$$

Proof. If $m = 2$, Theorem 1.13 reduces to Theorem 1.10. Suppose that $m \geq 3$ and that we prove (2.15) by induction. Consider that (2.15) is true for $m - 1$ and we prove it for m. By Theorem 1.10 we have

$$\sum_{k=1}^{n} \left| \prod_{j=1}^{m} \alpha_{jk} \right| = \sum_{k=1}^{n} |\alpha_{1k}| \left| \prod_{j=2}^{m} \alpha_{jk} \right|$$

$$\leq M_{p_1}(a_1) \left[\sum_{k=1}^{n} \left(\prod_{j=2}^{m} |\alpha_{jk}| \right)^{p_1/(p_1-1)} \right]^{(p_1-1)/p_1}$$

$$= M_{p_1}(a_1) \left[\sum_{k=1}^{n} \prod_{j=2}^{m} |\alpha_{jk}|^{p_1/(p_1-1)} \right]^{(p_1-1)/p_1} . \tag{2.16}$$

We remark that

$$\frac{p_j(p_1-1)}{p_1} > 0, \; j = 2,\ldots,m, \; \text{and} \; \sum_{j=2}^{m} \frac{p_1}{p_j(p_1-1)} = \frac{p_1}{p_1-1} \sum_{j=2}^{m} \frac{1}{p_j} = 1.$$

Thus (2.16) is further evaluated as

$$\leq M_{p_1}(a_1) \left\{ \prod_{j=2}^{m} \left[\sum_{k=1}^{n} |\alpha_{jk}|^{(p_1/(p_1-1))(p_j(p_1-1)/p_1)} \right]^{p_1/(p_j(p_1-1))} \right\}^{(p_1-1)/p_1}$$

$$= M_{p_1}(a_1) \prod_{j=2}^{m} \left(\sum_{k=1}^{n} |\alpha_{jk}|^{p_j} \right)^{1/p_j} = \prod_{j=1}^{m} M_{p_j}(a_j). \quad \square$$

Theorem 1.14. (Hölder inequality for $0 < p < 1$) *Consider* $0 < p < 1$, $1/p + 1/q = 1$, *and two finite sequences of positive numbers* $a = (\alpha_1,\ldots,\alpha_n)$ *and* $b = (\beta_1,\ldots,\beta_n)$. *Then*

$$M(ab) \geq M_p(a)M_q(b).$$

Proof. Take $u = 1/p \; (> 1)$, $1/u + 1/v = 1$. Define $\gamma_k = \beta_k^{-1/u}$, $\delta_k = \beta_k^{1/u}\alpha_k^{1/u}$, $k = 1,\ldots,n$. So $\gamma_k^u = \beta_k^q$. By Theorem 1.10, we get

$$\sum_{k=1}^{n} \alpha_k^p = \sum_{k=1}^{n} \gamma_k \delta_k \leq \left(\sum \delta_k^u \right)^{1/u} \left(\sum \gamma_k^v \right)^{1/v}$$

$$= \left(\sum \alpha_k \beta_k \right)^p \left(\sum \beta_k^{-v/u} \right)^{1/v} = \left(\sum \alpha_k \beta_k \right)^p \left(\sum \beta_k^q \right)^{1/v} .$$

Hence

$$M_p(a)M_q(b) \leq M(ab). \quad \square$$

Theorem 1.15. (Hölder inequality for negative weights) *Consider* $m \in \mathbb{N}^*$, $m \geq 2$, *finite and nonzero sequences* $a_j = (\alpha_{j1},\ldots,\alpha_{jn})$, $j = 1,\ldots,m$, *and the weights* $p_1, p_2, \ldots, p_{m-1} < 0$ *and* $p_m \in \,]0,1[$ *satisfying* $\sum_{j=1}^{m} 1/p_j = 1$. *Then*

$$\sum_{k=1}^{n} \left| \prod_{j=1}^{m} \alpha_{jk} \right| \geq \prod_{j=1}^{m} M_{p_j}(a_j). \tag{2.17}$$

Proof. If $m = 2$, this theorem reduces to Theorem 1.14. Suppose that (2.17) holds for an $m \geq 2$. We show that it also holds for $m + 1$. Therefore consider $p_1, p_2, \ldots, p_m < 0$ and $p_{m+1} \in \mathbb{R}$ such that $\sum_{j=1}^{m+1} 1/p_j = 1$, and $a_j = (\alpha_{j1}, \ldots, \alpha_{jn})$ are nonzero for all $j = 1, \ldots, m + 1$. Thus $0 < p_{m+1} < 1$. Hence

$$\sum_{k=1}^{n} \left| \prod_{j=1}^{m+1} \alpha_{jk} \right| = \sum_{k=1}^{n} |\alpha_{1k}| \left| \prod_{j=2}^{m+1} \alpha_{jk} \right|$$

$$\geq M_{p_1}(a_1) \left[\sum_{k=1}^{n} \left(\prod_{j=2}^{m+1} |\alpha_{jk}| \right)^{p_1/(p_1-1)} \right]^{(p_1-1)/p_1}$$

$$= M_{p_1}(a_1) \left[\sum_{k=1}^{n} \prod_{j=2}^{m+1} |\alpha_{jk}|^{p_1/(p_1-1)} \right]^{(p_1-1)/p_1}. \tag{2.18}$$

We remark that

$$p_j(p_1 - 1)/p_1 < 0, \quad j = 2, \ldots, m, \quad p_{m+1}(p_1 - 1)/p_1 > 0, \quad \text{and}$$

$$\sum_{j=2}^{m+1} \frac{p_1}{p_j(p_1 - 1)} = \frac{p_1}{p_1 - 1} \sum_{j=2}^{m+1} \frac{1}{p_j} = 1.$$

Then (2.18) is further evaluated as

$$\geq M_{p_1}(a_1) \left\{ \prod_{j=2}^{m+1} \left[\sum_{k=1}^{n} |\alpha_{jk}|^{(p_1/(p_1-1))(p_j(p_1-1)/p_1)} \right]^{p_1/(p_j(p_1-1))} \right\}^{(p_1-1)/p_1}$$

$$= M_{p_1}(a_1) \prod_{j=2}^{m} \left(\sum_{k=1}^{n} |\alpha_{jk}|^{p_j} \right)^{1/p_j} = \prod_{j=1}^{m} M_{p_j}(a_j). \quad \square$$

Theorem 1.16. (Minkowski inequality for $p \geq 1$) *Suppose the assumptions of Theorem 1.10 are satisfied. Then*

$$M_p(a + b) \leq M_p(a) + M_p(b). \tag{2.19}$$

Proof. For $p = 1$ the above inequality follows from the inequality given in (2) (a) of Proposition 2.10.

Suppose that $p > 1$. We apply the Hölder inequality (2.12) for the following two pairs of finite sequences.

$$(a_k)_{k=1}^{n}, \ (|a_k + b_k|^{p-1})_{k=1}^{n} \quad \text{and} \quad (b_k)_{k=1}^{n}, \ (|a_k + b_k|^{p-1})_{k=1}^{n}.$$

Then we have the following estimates,

$$\sum_{k=1}^{n} |a_k + b_k|^p \le \sum_{k=1}^{n} |a_k| \cdot |a_k + b_k|^{p-1} + \sum_{k=1}^{n} |b_k| \cdot |a_k + b_k|^{p-1}$$

$$\le (M_p(a) + M_p(b)) \left(\sum_{k=1}^{n} |a_k + b_k|^p \right)^{1/q}.$$

Dividing both sides by $\left(\sum |a_k + b_k|^p \right)^{1/q}$, we get inequality (2.19). □

2.2 Metric spaces

A nonempty set X, whose elements are called *points*, is said to be a *metric space* if with every two points x and y of X there is associated a real number $\rho(x, y)$, called the *distance from x to y*, fulfilling

(a) $\rho(x, y) = 0$ if and only if $x = y$.
(b) $\rho(x, y) = \rho(y, x)$.
(c) $\rho(x, y) \le \rho(x, z) + \rho(z, y)$, for every $x, y, z \in X$.

Thus $\rho : X \times X \to \mathbb{R}$. Often the distance function ρ is called a *metric* on X. A pair (X, ρ), where X is a nonempty set and ρ is a metric on X, is said to be a *metric space*.

From (c) we have that the distance ρ is nonnegative because setting $x = y$, it follows that $0 \le \rho(x, z)$. From (b) we have that the distance function is symmetric; (c) is called the *triangle inequality*.

Examples. (a) Consider a nonempty set X and define the following function

$$\rho(x, y) = \begin{cases} 0, & x = y, \\ 1, & x \ne y. \end{cases} \tag{2.20}$$

One can immediately check that this function ρ fulfills (a), (b), and (c). Therefore it defines a metric on X. Thus we note that on every nonempty set one can define at least a metric.

(b) We recall the definition of the distance function given on $\mathbb{R}$ at page 15. It follows that $(\mathbb{R}, \rho)$, where $\rho(x, y) = |x - y|$, is a metric space. This metric is called the *Euclidean metric* or the *Euclidean distance* on $\mathbb{R}$.

(c) Consider the complex plane $\mathbb{C}$ and $z_1, z_2 \in \mathbb{C}$, $z_k = x_k + iy_k$, $x_k, y_k \in \mathbb{R}$, $k = 1, 2$, $i^2 = -1$. Define on $\mathbb{C}$ the following distance function

$$\rho(z_1, z_2) = |z_1 - z_2| = \sqrt{(x_1 - x_2)^2 + (y_1 - y_2)^2},$$

called the *Euclidean distance* on $\mathbb{C}$.

(d) Consider the plane $\mathbb{R}^2$ and define on it the following metrics, for $u_k = (x_k, y_k) \in \mathbb{R}^2$, $k = 1, 2$,

(d$_1$) $\rho_1(u_1, u_2) = |x_1 - x_2| + |y_1 - y_2|$.

(d$_2$) $\rho_2(u_1, u_2) = \sqrt{(x_1 - x_2)^2 + (y_1 - y_2)^2}$.

(d$_3$) $\rho_\infty(u_1, u_2) = \max\{|x_1 - x_2|, |y_1 - y_2|\}$.

Metric ρ_2 is said to be the *Euclidean metric* on $\mathbb{R}^2$, and ρ_∞ is said to be the *uniform metric*.

(e) From (d) it follows that on a given set one can define several metrics.

(f) Let $(X, \|\cdot\|)$ be a normed space. Taking $\rho(x, y) = \|x - y\|$, $x, y \in X$, we get the metric space (X, ρ) whose metric is generated by the norm. Thus each normed space is a metric space with the metric induced by the norm. $\triangle$

Warning. Let (X, ρ) be a metric space. All points and sets mentioned in this section are understood to be elements and subsets of X.

The *open ball* with center at x and radius $r > 0$ is the set $B(x, r)$ given by

$$B(x, r) = \{y \in X \mid \rho(x, y) < r\}.$$

Obviously, $x \in B(x, r)$ for every $x \in X$ and $r > 0$. $B[x, r]$, $r \geq 0$, is the *closed ball* centered at x and radius r.

Consider the open balls $B_i(0, 1)$ generated by the metrics $\rho_1, \rho_2, \rho_\infty$ on $\mathbb{R}^2$ defined earlier. The three balls are presented in Figure 2.2.

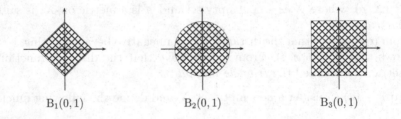

$B_1(0, 1)$ $B_2(0, 1)$ $B_3(0, 1)$

Fig. 2.2. Three open balls

Proposition 2.9. *Let p, q be two distinct points in a metric space (X, ρ). Then there exist two disjoint open balls centered in p and q.*

Proof. We have to find two positive numbers, say ε and δ, so that $B(p, \varepsilon) \cap B(q, \delta) = \emptyset$. Because p and q are distinct points, $\rho(p, q) = t > 0$. Set $\varepsilon = \delta = t/3$. Suppose $B(p, \varepsilon) \cap B(q, \delta) \neq \emptyset$. Then there exists a point that belongs to both open balls; that is, $x \in B(p, \varepsilon) \cap B(q, \delta)$. Thus

$$t = \rho(p, q) \leq \rho(p, x) + \rho(x, q) < 2t/3.$$

But this is impossible, hence our assumption that the two open balls have a common point is false. Thus the conclusion follows. $\square$

A set $A \subset X$ is said to be *open* if for each $x \in A$ there is a positive r such that $B(x, r) \subset A$. It follows that X is open. We consider the empty set as an open set. It follows immediately that

Theorem 2.1. *Let $\mathcal{O}$ denote the family of all open subsets of X. Then*

(a) $\emptyset, X \in \mathcal{O}$.
(b) *The union of any family of open sets is open.*
(c) *The intersection of any finite family of open sets is open.*

Proof. (a) The sentence follows from previous remarks.
(b) Suppose $A = \cup_{i \in I} A_i$, where all A_is are open. Then

$$x \in A \implies \exists i \in I, \ a \in A_i \implies \exists \, \mathrm{B}(x, r) \subset A_i \implies \mathrm{B}(x, r) \subset \cup_{i \in I} A_i.$$

(c) Suppose $A = \cap_{i=1}^{n} A_i$, where all A_is are open. Then

$$x \in A \implies a \in A_i, \ \forall i \implies \mathrm{B}(x, r_i) \subset A_i, \ \forall i$$
$$\implies \mathrm{B}(x, r) \subset \cap_{i=1}^{n} A_i, \text{ where } r = \min r_i > 0. \quad \square$$

Let A be a nonempty subset of X and $x \in A$. Then a point x is an *interior* point of A if there is an open set O with $x \in O \subset A$. The set of interior points of a set $A \subset X$ is denoted by $\mathrm{int}\, A$. If $\mathrm{int}\, A = \emptyset$, we say that A has no interior point. Always $\mathrm{int}\, A \subset A$.

Theorem 2.2. *Suppose $A, B \subset X$. Then*

(a) $\mathrm{int}\, A$ *is the union of all open sets contained by A. Thus $\mathrm{int}\, A$ is open.*
(b) A *is open if and only if $A = \mathrm{int}\, A$.*
(c) *The interior of A is the largest open set (in respect to the inclusion of sets) contained in A.*
(d) $A \subset B$ *implies $\mathrm{int}\, A \subset \mathrm{int}\, B$.*
(e) $\mathrm{int}\,(A \cap B) = \mathrm{int}\,(A) \cap \mathrm{int}\,(B)$.
(f) $\mathrm{int}\,(\mathrm{int}\, A) = \mathrm{int}\, A$.
(g) $\mathrm{int}\, X = X$.

Theorem 2.3. *Every open ball is an open set.*

Proof. Consider $A = \mathrm{B}(x, r)$ and let y be any point of A. Then there is a positive number h such that $\rho(x, y) = r - h$. For all points z such that $\rho(y, z) < h$, we have

$$\rho(x, z) \leq \rho(x, y) + \rho(y, z) < r - h + h = r.$$

Therefore $z \in A$. Thus y is an interior point of A. $\quad \square$

A set $A \subset X$ is said to be *closed* if $\complement_X A$ is open. Then we have the following.

Theorem 2.4. *Let $\mathcal{C}$ denote the family of all closed subsets of X. Then*

(a) $\emptyset, X \in \mathcal{C}$.
(b) *the union of any finite family of closed sets is closed.*
(c) *the intersection of any family of closed sets is closed.*

Proposition 2.10. *Consider two sets* $A, B \subset X$ *such that* A *is open and* B *is closed. Then* $A \setminus B$ *is open and* $B \setminus A$ *is closed.*

Proof. We only show that $A \setminus B$ is open. If $A \setminus B = \emptyset$, the conclusion follows. Now consider an arbitrary $x \in A \setminus B$. Then $x \in A$ and $x \in \complement B$; $\complement B$ is open. There exists an open ball $B(x, r) \subset A \cap \complement B = A \setminus B$.   □

For $A \subset X$, the *closure* of A is the set

$$\operatorname{cl} A = \cap \{C \mid A \subset C \subset X, \ C \text{ closed}\}.$$

$A \subset \operatorname{cl} A$ always.

The interior and the closure of a set are strongly connected.

Theorem 2.5. *Let* $A \subset X$. *Then*

$$\complement_X (\operatorname{cl} A) = \operatorname{int} (\complement_X A) \text{ and } \complement_X (\operatorname{int} A) = \operatorname{cl} (\complement_X A).$$

Corollary 2.7. *Suppose* $A \subset X$. *Then* $\operatorname{int} A = \complement_X \operatorname{cl} (\complement_X A)$.

Theorem 2.6. *Suppose* $A, B \subset X$. *Then*

(a) $\operatorname{cl} A$ *is a closed set.*
(b) A *is closed if and only if* $A = \operatorname{cl} A$.
(c) *The closure of* A *is the smallest closed set (in respect to the inclusion of sets) containing* A.
(d) $A \subset B$ *implies* $\operatorname{cl} A \subset \operatorname{cl} B$.
(e) $\operatorname{cl} (A \cup B) = \operatorname{cl} (A) \cup \operatorname{cl} (B)$.
(f) $\operatorname{cl} (\operatorname{cl} A) = \operatorname{cl} A$.
(g) $\operatorname{cl} X = X$.

Proposition 2.11. *Consider two open and disjoint sets* $A, B \subset X$. *Then*

(a) *The closure of one set does not intersect the other; that is,* $B \cap \operatorname{cl} A = \operatorname{cl} (B) \cap A = \emptyset$.
(b) $(\operatorname{int} \operatorname{cl} A) \cap (\operatorname{int} \operatorname{cl} B) = \emptyset$.

Proof. (a) $\complement A$ is closed. Because $A \cap B = \emptyset$, it follows that $B \subset \complement A$ and $\operatorname{cl} B \subset \operatorname{cl} \complement A = \complement A$. Then $A \cap \operatorname{cl} B \subset A \cap \complement A = \emptyset$.
(b) Applying (a) successively, we have that $A \cap B = \emptyset$ implies

$$\operatorname{int} \operatorname{cl} B \cap A = \emptyset \implies \operatorname{int} \operatorname{cl} B \cap \operatorname{cl} A = \emptyset \implies \operatorname{int} \operatorname{cl} B \cap \operatorname{int} \operatorname{cl} A = \emptyset. \quad □$$

A *neighborhood* of a point $x \in X$ is a set $A \subset X$ containing an open set O with $x \in O$; that is, $x \in O \subset A$. The system of all neighborhoods of a point $x \in X$ is denoted $\mathcal{V}(x)$. Obviously, $X \in \mathcal{V}(x)$, for each $x \in X$.

We have introduced the concept of neighborhood using the concept of open set. However, the concept of neighborhood can be introduced directly as follows [47]. Consider a nonempty set X. Then

(a) Each neighborhood of a point $x \in X$ contains x. The whole space is a neighborhood of all its points.
(b) A superset of a neighborhood of a point is a neighborhood of that point.
(c) The intersection of two neighborhoods of a point is a neighborhood of that point.
(d) Each neighborhood of a point x contains a neighborhood of x that is also a neighborhood of each of its points.

Application. Now we show that the notion of neighborhood can be used to show something not obviously belonging to this topic. Namely we show the infinitude of prime numbers.[9] The classical proof is by contradiction.

Consider $\mathbb{Z}$ the set of integers. For $a, b \in \mathbb{Z}$, $b > 0$, let

$$N_{a,b} = \{a + nb \mid n \in \mathbb{Z}\}.$$

Each $\mathbb{N}_{a,b}$ is a two-sided arithmetic progression. Note $a \in N_{a,b}$ for every b. N in the definition is supposed to remind us of neighborhoods. One might think of $N_{a,b}$ as those basic neighborhoods that are neighborhoods of each point (d). Add to the collection of all neighborhoods all supersets of $N_{a,b}$s. With this definition we only need to check property (c). Note

$$N_{a,b} \cap N_{a,c} = N_{a,\operatorname{lcm}(b,c)},$$

from which (c) follows.

Call a set O open if for every $a \in O$ there exists $b > 0$ so that $N_{a,b} \subset O$. We can check that the statements (a) and (b) hold as they are analogous for open sets. We also note that

(i) Any nonempty open set is infinite.
(ii) Besides being open, any set $N_{a,b}$ is also closed.

(i) follows from the definition and (ii) from $N_{a,b} = \mathbb{Z} \setminus \cup_{i=1,2,\dots,b-1} N_{a+i,b}$. Thus $N_{a,b}$ is closed as a complement of a union of open sets.

Except for ± 1 and 0, all integers have prime factors. Therefore each is contained in one or more $N_{0,p}$, where p is a prime. We have the identity

$$\mathbb{Z} \setminus \{\pm 1\} = \cup N_{0,p},$$

where the union is taken over the set of all primes. If the latter is finite, the right-hand side is closed as a union of a finite number of closed sets. Then $\{-1, 1\}$ is an open set as a complement to a closed set. This contradicts (i). $\triangle$

Proposition 2.12. *For a set $A \subset X$, the following two sentences are equivalent.*

(a) $x \in \operatorname{cl} A$.

---

[9] We follow a paper by Harry Fürstenberg from the Hebrew University of Jerusalem.

(b) *For every $V \in \mathcal{V}(x)$, one has $V \cap A \neq \emptyset$.*

Proof. Suppose $x \in \operatorname{cl} A$. In order to prove (b), suppose there exists an open $V \in \mathcal{V}(x)$ so that $V \cap A = \emptyset$. Then $A \subset \complement V$ and $\complement V$ is closed. So $\operatorname{cl} A \subset \complement V$. Hence $x \notin \operatorname{cl} A$, contradicting the hypothesis.

Suppose (a) does not hold; that is, $x \notin \operatorname{cl} A$. Then there exists a closed set F so that $A \subset F$ and $x \notin F$. But $V = \complement F$ is open, $x \in V$, and $V \cap A = \emptyset$. $\quad\square$

Corollary 2.8. *Suppose V is an open set and $V \cap A = \emptyset$. Then $V \cap \operatorname{cl} A = \emptyset$.*

For $x \in X$ and $A \subset X$ we say that x is a *limit point* or *accumulation point* of A if every $V \in \mathcal{V}(x)$ contains some point of A distinct from x; that is, $V \cap (A \setminus \{x\}) \neq \emptyset$. The set of limit points of a set A is denoted A'.

Theorem 2.7. *Consider $A, B, A_\alpha \subset X$, $\alpha \in I$. Then*

(a) $\operatorname{cl} A = A \cup A'$.
(b) *If $A \subset B$, $A' \subset B'$.*
(c) $(A \cup B)' = A' \cup B'$.
(d) $\cup_{\alpha \in I} A'_\alpha = (\cup_{\alpha \in I} A_\alpha)'$.

Corollary 2.9. *Consider $A \subset X$. Then A is closed if and only if $A' \subset A$.*

Theorem 2.8. *Let (X, ρ) be a metric space and let A' be the set of limit points of a set $A \subset X$. Then A' is closed.*

Proof. By Corollary 2.9, A' is closed if and only if $(A')' \subset A'$.

Set $y \in (A')'$. Then for every $\varepsilon > 0$ there is an $x \in A'$ such that $0 < \rho(x, y) < \varepsilon/2$. Because $x \in A'$ there is a $v \in A$ such that $\rho(x, v) < \rho(x, y)$. Hence $v \neq y$ and

$$0 < \rho(v, y) \leq \rho(x, v) + \rho(x, y) < \varepsilon.$$

Because $v \in A$, this shows that y is a limit point of A (i.e., $y \in A'$) and the proof is complete. $\quad\square$

Theorem 2.9. *If x is a limit point of a set A, then every neighborhood of x contains infinitely many points of A.*

Proof. Suppose there is a neighborhood V of x that contains only a finite number of points of A. Let $y_1, \ldots, y_n$ be those points of $V \cap A$, that are distinct from x, and put

$$r = \min_{1 \leq k \leq n} \rho(x, y_k).$$

Note that $r > 0$. The neighborhood $V = B(x, r)$ contains no point $y \in A$ such that $y \neq x$, so that x is not a limit point of A. This contradiction establishes the theorem. $\quad\square$

If $x \in A$ and x is not a limit point of A, x is said to be an *isolated* point of A. So, x is an isolated point of A if and only if it belongs to $A \setminus A'$.

A point $x \in X$ is said to be a *closure point* or (*adherent point*) of the set $A \subset X$ if every neighborhood of x has a nonempty intersection with A.

Theorem 2.10. *Suppose* $\emptyset \neq A \subset \mathbb{R}$. *Then* A *is closed if and only if* A *coincides with the set of its closure points.*

Suppose $A \subset X$. The *frontier* or *boundary* of the set A is the set $(\operatorname{cl} A) \cap (\operatorname{cl} \complement_X A)$. We denote it by $\operatorname{fr} A$ and we note that it is closed.

Theorem 2.11. *Suppose* $A \subset X$. *Then* A *is open if and only if* $A \cap \operatorname{fr} A = \emptyset$.

Theorem 2.12. *Suppose* $A, B \subset X$. *Then we have*

(a) $\operatorname{int} A = A \setminus \operatorname{fr} A$.   (e) $\operatorname{fr}(X \setminus A) = \operatorname{fr} A$.

(b) $\operatorname{cl} A = A \cup \operatorname{fr} A$.   (f) $X = \operatorname{int} A \cup \operatorname{fr} A \cup \operatorname{int}(X \setminus A)$.

(c) $\operatorname{fr}(A \cup B) \subset \operatorname{fr} A \cup \operatorname{fr} B$.   (g) $\operatorname{fr}(\operatorname{cl} A) = \operatorname{fr} A$.

(d) $\operatorname{fr}(A \cap B) \subset \operatorname{fr} A \cup \operatorname{fr} B$.   (h) $\operatorname{fr}(\operatorname{int} A) \subset \operatorname{fr} A$.

(i) A *is open if and only if* $\operatorname{fr} A = \operatorname{cl} A \setminus A$.

(j) A *is closed if and only if* $\operatorname{fr} A = A \setminus \operatorname{int} A$.

Proof. We only prove (a) and (c).

(a) $A \setminus \operatorname{fr} A = A \setminus (\operatorname{cl}(A) \cap \operatorname{cl}(X \setminus A)) = (A \setminus \operatorname{cl}(A)) \cup (A \setminus \operatorname{cl}(X \setminus A))$

$\qquad\qquad = A \setminus \operatorname{cl}(X \setminus A) = A \cap \operatorname{int} A = \operatorname{int} A.$

(c) $\operatorname{fr}(A \cup B) = (\operatorname{cl}(A \cup B)) \cap \operatorname{cl}(\complement(A \cup B))$

$\qquad\qquad = (\operatorname{cl}(A) \cup \operatorname{cl}(B)) \cap \operatorname{cl}(\complement(A) \cap \complement(B))$

$\qquad\qquad \subset (\operatorname{cl}(A) \cup \operatorname{cl}(B)) \cap (\operatorname{cl}(\complement A) \cap \operatorname{cl}(\complement B))$

$\qquad\qquad \subset (\operatorname{cl} A \cap \operatorname{cl} \complement(A)) \cup (\operatorname{cl} B \cap \operatorname{cl} \complement(B)) = \operatorname{fr}(A) \cup \operatorname{fr}(B).$ $\square$

A subset A of X is said to be *dense* in X provided $\operatorname{cl} A = X$. A is *bounded* if there is a real number m such that

$$\rho(x, y) \leq m, \text{ for every } x, y \in A.$$

Otherwise the set A is said to be *unbounded*. We introduce the *diameter* of a set A by

$$\operatorname{diam} A = \sup\{\rho(x, y) \mid x, y \in A\}.$$

The system of open sets corresponding to a metric space (X, ρ) is denoted by τ and is called the *topology* generated by the metric ρ. The pair (X, τ) is said to be a *topological space*. Sometimes we say that X is a topological space. This is the case when there is no chance of misunderstanding; that is, it is clear to what topology we are referring. $\tau \neq \emptyset$ always because $\emptyset, X \in \tau$.

Remark. All these topological notions (openness, closeness, etc.) are based ultimately on the notion of the metric. Thus changing the metric, the open (closed, etc.) sets, generally, change. For example, the family of open sets corresponding to the metric on $\mathbb{R}$ given by (2.20) coincides with the family of all subsets of $\mathbb{R}$. Obviously, this is not the case if we consider the Euclidean metric ρ on $\mathbb{R}$. For, the set containing precisely a point in $\mathbb{R}$ is open in the first case whereas it is not in the second case. $\triangle$

It may happen that different metrics on a set generate the same topology on that set.

Theorem 2.13. *Any two metrics on* $\mathbb{R}^k$ *induced by the norms* $\|\cdot\|_p$, $p \geq 1$, *or* $\|\cdot\|_\infty$ *generate the same open sets.*

Let (X, ρ_X) and (Y, ρ_Y) be two metric spaces. A mapping f of X into Y is said to be an *isometry* if $\rho_Y(f(x), f(y)) = \rho_X(x, y)$, for every pair of points $x, y \in X$. If there exists an isometry of X to Y, then X and Y are *isometric*. Metric spaces that are isometric are indistinguishable for many purposes, and are frequently identified.

Examples 2.1. We consider the following subsets of $\mathbb{R}^2$ (bijective to $\mathbb{C}$).

(a) The set of all complex z such that $|z| < 1$.
(b) The set of all complex z such that $|z| \leq 1$.
(c) A finite set.
(d) The set of all integers.
(e) The set $A = \{1/n \mid n = 1, 2, \dots\}$. A has a limit point (namely $x = 0$), but no point in A is a limit point of A; we stress the difference between a limit point and containing one.
(f) The set of all complex numbers.
(g) The interval $]0, 1[$.

We note that (d), (e), and (g) can also be regarded as subsets of $\mathbb{R}$. Some properties of these sets are tabulated below.

	Closed	Open	Bounded
(a)	No	Yes	Yes
(b)	Yes	No	Yes
(c)	Yes	No	Yes
(d)	Yes	No	No
(e)	No	No	No
(f)	Yes	Yes	No
(g)	No	No/Yes in $\mathbb{R}$	Yes. $\triangle$

Suppose $A \subset Y \subset X$, where (X, ρ) is a metric space. Recall we say that A is an open subset of X means that with each point $x \in A$ there is associated a positive number r such that $\rho(x, y) < r$, $y \in X$ imply that $y \in A$. Remark that $(Y, \rho|_{Y \times Y})$ is a metric space, too. We say that A is *open relative* to Y if with each $x \in A$ there is associated an $r > 0$ such that $y \in A$ whenever $\rho(x, y) < r$ and $y \in Y$.

Example. Example 2.1 (g) showed that a set may be open relative to Y without being an open subset of X. However, there is a simple relation between these concepts. $\triangle$

Theorem 2.14. *Suppose $Y \subset X$. A subset A of Y is open relative to Y if and only if $A = Y \cap G$ for some open subset G of X.*

Proof. Suppose A is open relative to Y. With each $x \in A$ there is associated a positive number r_x such that the conditions $\rho(x, y) < r_x$ and $y \in Y$ imply $y \in A$. Let V_x be the set of all $y \in X$ so that $\rho(x, y) < r_x$, and define

$$G = \cup_{x \in A} V_x.$$

Then G is an open subset of X.

Because $x \in V_x$ for all $x \in A$, it is clear that $A \subset G \cap Y$. By our choice of V_x we have $V_x \cap Y \subset A$ for $x \in A$, so that $G \cap Y \subset A$. Thus $A = G \cap Y$, and one half of the theorem is proved.

Conversely, if G is open in X and $A = G \cap Y$, for every $x \in A$ has a neighborhood $V_x \subset G$. Then $V_x \cap Y \subset A$, so that A is open relative to Y. $\square$

A metric space is said to be *separable* if it contains a countable dense set.

Theorem 2.15. $\mathbb{Q}$ *is a dense subset of* $\mathbb{R}$. $\mathbb{R}$ *is separable.*

Proof. The first assertion follows from Theorem 2.12, page 20. For the second part we also take into account Corollary 2.10, page 24. $\square$

2.3 Compact spaces

In this section we consider a metric space (X, ρ), although several results hold even in topological spaces. We restrict this presentation to the frame of metric spaces because it is generally enough to satisfy our purposes.

A *covering* of a set A in a metric space X is a family of subsets $\{G_\alpha\}$ of X such that $A \subset \cup_\alpha G_\alpha$. An *open covering* is a covering consisting of open subsets.

A subset K of a metric space X is said to be *compact* if every open covering of K contains a finite subcovering of it. More explicitly, the requirement is that if $\{C_\alpha\}$ is an open covering of K, then there are finitely many indices $\alpha_1, \ldots, \alpha_n$ such that

$$K \subset G_{\alpha_1} \cup \cdots \cup G_{\alpha_n}.$$

Example. Every finite set is compact. $\triangle$

Remark. We observed earlier that if $A \subset Y \subset X$, then A may be open relative to Y without being open relative to X. The property of being open thus depends on the space in which A is embedded. The same is true of the property of being closed. $\triangle$

Theorem 3.1. *Suppose $K \subset Y \subset X$. Then K is compact relative to X if and only if K is compact relative to Y.*

If K is a compact set of a metric space X and $K = X$, then we say that X is a *compact space*.

The compactness of a metric space can be characterized in several ways.

A family $\mathcal{F}$ of sets is said to have the *finite intersection property* if the intersection of the members of each finite subfamily of $\mathcal{F}$ is nonempty.

Theorem 3.2. *A metric space is compact if and only if every family of closed sets with the finite intersection property has a nonempty intersection.*

Proof. Suppose X is a compact space and there is given a family $\mathcal{F} = \{F_\alpha \mid \alpha \in I\}$ of closed sets such that the intersection of each finite subfamily is nonempty. Suppose that

$$\cap_\alpha F_\alpha = \emptyset. \tag{2.21}$$

Denote $G_\alpha = \complement_X F_\alpha$. Each G_α is open. By (2.21) we have that $\{G_\alpha\}$ is an open covering of the compact space X. Then there exists a finite subcovering of X; that is, there exists $\{G_{\alpha_i} \mid i = 1, 2, \ldots, n\}$ with

$$\cup_{i=1}^n G_{\alpha_i} = X.$$

It immediately follows that

$$\cap_{i=1}^n F_{\alpha_i} = \emptyset,$$

contradicting our hypothesis that the intersection of each finite subfamily is nonempty. So our assumption (2.21) is false and this part is proved.

Consider a family $\{G_\alpha \mid \alpha \in I\}$ of open sets that covers X, $X = \cup_\alpha G_\alpha$. Set $F_\alpha = \complement_X G_\alpha$ and remark that $\cap_\alpha F_\alpha = \emptyset$. Then there exists a finite subfamily $\{F_{\alpha_i} \mid i = 1, 2, \ldots, n\}$ with $\cap_{i=1}^n F_{\alpha_i} = \emptyset$. But this implies that X has a finite covering, namely $\{G_{\alpha_i} \mid i = 1, 2, \ldots, n\}$. Thus X is compact. $\square$

A metric space X is said to be *countably compact* if every countable open covering has a finite subcovering.

A sequence $(x_n)_{n \in \mathbb{N}}$ in a metric space (X, ρ) is said to *converge* if there is a point $x \in X$ with the property that for every $\varepsilon > 0$ there is a natural number n_ε such that $n \geq n_\varepsilon$ implies that $\rho(x_n, x) < \varepsilon$.

In this case we also say that (x_n) *converges* to x or that x is the *limit* of (x_n), and we write $x_n \to x$, or

$$\lim_{n \to \infty} x_n = \lim x_n = x.$$

If (x_n) does not converge, it is said to *diverge*.

A metric space X is said to be *sequentially compact* if every sequence has a convergent subsequence.

Theorem 3.3. *For a metric space X the following statements are equivalent.*

(a) X *is compact.*
(b) X *is countably compact.*
(c) X *is sequentially compact.*

Theorem 3.4. *Every compact subset of a metric space is closed and bounded.*

Proof. Let K be a compact subset of a metric space. First we prove that $\complement_X K$ is open.

If $K = X$, we are done. Suppose that $\complement_X K \neq \emptyset$. Choose $x \in X$, $x \notin K$. If $y \in K$, let V_y and W_y be neighborhoods of x, respectively, y of radius less than $\rho(x,y)/2$ (> 0, because $x \neq y$). Because K is compact, there are finitely many points $y_1, \ldots, y_n$ in K such that $K \subset W_{y_1} \cup \cdots \cup W_{y_n} = W$. If $V = V_{y_1} \cap \cdots \cap V_{y_n}$, then V is a neighborhood of x that does not intersect W. Hence $V \subset \complement_X K$, so that x is an interior point of $\complement_X K$. So, $\complement_X K$ is open and K is closed.

We show that K is bounded. Consider the family $\mathcal{B} = \{B(x,1) \mid x \in K\}$. Then $\mathcal{B}$ is an open covering and thus has a finite subcovering, say $\{B(x_1,1), \ldots, B(x_m,1)\}$. Let $k = \max\{\rho(x_i,x_j) \mid 1 \leq i,j \leq m\}$. Choose arbitrary $x,y \in K$. There exist $1 \leq i,j \leq m$, so that $x \in B(x_i,1)$ and $y \in B(x_j,1)$. Furthermore

$$\rho(x,y) \leq \rho(x,x_i) + \rho(x_i,x_j) + \rho(x_j,y) \leq 1 + k + 1.$$

Hence $\operatorname{diam} K \leq k + 2$, and thus K is bounded. $\square$

Remark. The converse of Theorem 3.4 generally does not hold. Under extra assumptions on the metric space it can be proved that a set is compact if and only if it is closed and bounded. This sentence is contained in Theorems 3.10 and 3.13. $\triangle$

Theorem 3.5. *If $\{K_\alpha\}$ is a family of compact subsets of a metric space X such that the intersection of every finite family of $\{K_\alpha\}$ is nonempty, then $\cap K_\alpha$ is nonempty.*

Proof. It follows from Theorems 3.2 and 3.4. $\square$

Theorem 3.6. *Every closed subset of a compact set is compact.*

Proof. Suppose $F \subset K \subset X$, F is closed (relative to X), and K is compact. Let $\{V_\alpha\}$ be an open covering of F. Then we have

$$F \subset K \subset \cup_\alpha V_\alpha \cup (X \setminus F).$$

Because K is compact, there is a finite subcovering by open sets of the form $\{V_{\alpha_i}\} \cup (X \setminus F)$. This finite subcovering of K is a covering of F. Because $X \setminus F$ does not cover F, there remain a finite number of V_α that cover F, namely $F \subset \cup_i V_{\alpha_i}$. We have thus shown that a finite number of members of $\{V_\alpha\}$ cover F. $\square$

Corollary 3.10. *If F is closed and K is compact in a metric space, then $K \cap F$ is compact.*

Proof. By Theorem 3.4, K is closed and by Theorem 2.4, $F \cap K$ is closed. Finally, by Theorem 3.6, $F \cap K$ is compact. $\square$

Theorem 3.7. *If A is an infinite subset of a compact set K, then it has a limit point in K.*

Proof. The result is a consequence of Theorem 3.3.

A direct proof is as follows. If no point of K is a limit point of A, then each $x \in K$ has a neighborhood V_x that contains at most one point of A (namely x, if $x \in A$). It is clear that no finite subfamily of $\{V_x\}$ can cover A, especially K because $A \subset K$. This contradicts the compactness of K. $\square$

Theorem 3.8. (Cantor) *If (K_n) is a sequence of nonempty nested $(K_{n+1} \subset K_n)$ and compact sets and if*

$$\lim_{n \to \infty} \mathrm{diam} K_n = 0,$$

then $\cap_1^\infty K_n$ consists in exactly one point.

Proof. Set $K = \cap_1^\infty K_n$. By Theorem 3.5, $K \neq \emptyset$. If K contains more than one element, then $\mathrm{diam} K > 0$. But for each n, $K_n \supset K$, so that $\mathrm{diam} K_n \geq \mathrm{diam} K$. This contradicts the assumption that $\mathrm{diam} K_n \to 0$. $\square$

Theorem 3.9. *Every closed and bounded interval is compact.*

Proof. Let $I = [a, b]$ be a closed and bounded interval. Denote $\delta = |a - b| = b - a$. Then $|x - y| \leq \delta$, for any $x, y \in I$. Suppose there exists an open covering $\{G_\alpha\}$ of I that contains no finite subcovering of I. Put $c = (a + b)/2$. The intervals $Q_1 = [a, c]$ and $Q_2 = [c, b]$ then determine two subintervals whose union is I. At least one of these sets, call it I_1, cannot be covered by any finite subfamily of $\{G_\alpha\}$ (otherwise I could be so covered). We next subdivide I_1 and continue the process. We obtain a sequence $\{I_n\}$ with the following properties.

(i) $I \supset I_1 \supset I_2 \supset \ldots$.
(ii) I_n is not covered by any finite subfamily of $\{G_\alpha\}$.
(iii) If $x, y \in I_n$, then $|x - y| \leq 2^{-n}\delta$.

By (i) and by Theorem 2.31 from page 30 there is a point $z \in \cap I_n$. For some α, $z \in G_\alpha$. Because G_α is open, there exists $r > 0$ such that $\mathrm{B}(z, r) \subset G_\alpha$. If n is so large that $2^{-n}\delta < r$, then (iii) implies that $I_n \subset G_\alpha$, which contradicts (ii). $\square$

Theorem 3.10. (Heine[10]–Borel[11]) *If a set $A \subset \mathbb{R}$ has one of the following properties, then it has the other two.*

---

[10] Heinrich Eduard Heine, 1821–1881.
[11] Emile Borel, 1871–1956.

(a) *A is closed and bounded.*
(b) *A is compact.*
(c) *Every infinite subset of A has a limit point in A.*

Proof. (a) $\Longrightarrow$ (b). There exists a closed and bounded interval I such that $A \subset I$. Then (b) follows by Theorems 3.9 and 3.4.
(b) $\Longrightarrow$ (c). This is Theorem 3.7.
(c) $\Longrightarrow$ (a). If not bounded, then A contains points x_n with $|x_n| > n$, $n = 1, 2, \ldots$. The set P consisting of these points x_n is infinite and clearly has no limit point in $\mathbb{R}$, hence has none in A. Thus (c) implies that A is bounded.

If A is not closed, then there is a point $x_0 \in \mathbb{R}$ that is a limit point of A but not a point of A. For $n = 1, 2, 3, \ldots$ there are points $x_n \in A$ such that $|x_n - x_0| < 1/n$. Let M be the set of these points x_n. Then M is infinite (otherwise, $|x_n - x_0|$ would have a constant positive value, for infinite many n), M has x_0 as a limit point, and M has no other limit point in $\mathbb{R}$. For if $y \in \mathbb{R}$, $y \neq x_0$, then

$$|x_n - y| \geq |x_0 - y| - |x_n - x_0| \geq |x_0 - y| - 1/n \geq |x_0 - y|/2$$

for all but finitely many n. This shows that y not a limit point of M (Theorem 2.9). Thus M has no limit point in A. Hence A is closed if (c) holds. $\square$

Remark. (b) $\Longleftrightarrow$ (c) in any metric space. (a) does not, in general, imply (b) and (c). $\triangle$

Theorem 3.11. (Weierstrass) *Every bounded infinite subset of $\mathbb{R}$ has a limit point in $\mathbb{R}$.*

Proof. Being bounded, the set A in question is a subset of an interval $[a, b] = I \subset \mathbb{R}$. By Theorem 3.9 I is compact, and so A has a limit point in I, by Theorem 3.7. $\square$

Theorems 3.9, 3.10, and 3.11 have correspondences in $\mathbb{R}^k$.

A *closed and bounded interval I* in $\mathbb{R}^k$ is defined as

$$I = [a_1, b_1] \times \cdots \times [a_k, b_k],$$

where each $[a_i, b_i]$ is a closed and bounded interval.

Theorem 3.12. *Every closed and bounded interval I in $\mathbb{R}^k$ is compact.*

Theorem 3.13. *Let E be a set in $\mathbb{R}^k$. Then the next statements are equivalent.*

(a) *E is closed and bounded.*
(b) *E is compact.*
(c) *Every infinite subset of E has a limit point in E.*

Theorem 3.14. (Weierstrass) *Every bounded infinite subset of $\mathbb{R}^k$ has a limit point in $\mathbb{R}^k$.*

2.4 Exercises

2.1. Show that $\mathbb{R}$ is a vector space over $\mathbb{Q}$, but $\mathbb{Q}$ is not a vector space over $\mathbb{R}$.

2.2. Show that the set of all polynomials with real coefficients endowed with the usual operations of addition and of multiplication by real scalars form a vector space over $\mathbb{R}$. Which is the null element?

2.3. Let n be a natural number. Show that the set of all polynomials of degree at most n with real coefficients endowed with the usual operations of addition and of multiplication by real scalars form a vector space over $\mathbb{R}$. Find its dimension.

2.4. Does the set of all polynomials of degree n with real coefficients endowed with the usual operations of addition and of multiplication by real scalars form a vector space? Prove it.

2.5. Show that the set c_0 of real sequences tending to zero endowed with the usual operations of addition and of multiplication by real scalars form a vector space.

2.6. Let c_0 be the set in the previous exercise. For $x \in c_0$, $x = (x_0, x_1, \ldots)$, define $\|x\| = \sup_{k \in \mathbb{N}} |x_k|$. Show that this a norm on c_0.

2.7. Let X be a real normed space and $x_1, x_2, \ldots, x_n$ be linear independent vectors in it. Show that there exists a positive ε such that for any system of vectors $y_1, y_2, \ldots, y_n$ with $\|x_i - y_i\| < \varepsilon$, $i = 1, 2, \ldots, n$ it follows that $\{y_1, y_2, \ldots, y_n\}$ is linear independent.

2.8. Let $(B_n)_n$ be a sequence of nested balls in a Banach space. For each n let x_n be the center of B_n. Show that the sequence $(x_n)_n$ is convergent.

2.9. Consider the set $A = [0, 1[\cup \{2\}$. Find $\operatorname{int} A$, $\operatorname{cl} A$, A', and $\operatorname{fr}(A)$.

2.10. Is the set $A = \{(\rho \cos \theta, \rho \sin \theta) \mid 0 \le \theta < \pi, \ 0 \le \rho < 1\} \cup \{(\rho \cos \theta, \rho \sin \theta) \mid 0 \le \theta < \pi, \ 0 \le \rho \le 1\}$ open or closed?

2.5 References and comments

Proposition 1.7 appeared as an exercise in [42].

Theorems 1.8, 1.9, and Corollary 1.3 can be seen in [66].

Theorem 1.13 coincides with [106, Part 2, Chapter 2, Exercise 81.3].

Theorem 1.15 has been published in [34].

Proposition 1.7 appears in [42].

The integral form of the Young inequality, Theorem 1.9, appears in [66, p. 189].

The Young inequality under the form given in Corollary 1.3 appears in [66, p. 190].

The Hölder inequality for positive weights appears in [106, Part 2, Chapter 2, Exercise 81.3].

The Hölder inequality for negative weights, Theorem 1.15 appears in [34].

The results in topology introduced here are classical. We followed mainly [47], [67], and [115].

In our presentation we considered the metric as a primary notion. However, it is possible and actually largely used, to consider the topology on a set as a primary notion [47].

3

Sequences and Series

The present chapter is devoted to introducing several results on numerical and functional sequences and series.

3.1 Numerical sequences

3.1.1 Convergent sequences

Recall that a sequence $(x_n)_{n \in \mathbb{N}}$ in a metric space (X, ρ) is said to converge if there is a point $x \in X$ with the property that for every $\varepsilon > 0$ there is a natural number n_ε such that $n \geq n_\varepsilon$ implies $\rho(x_n, x) < \varepsilon$.

In this case we also say that (x_n) converges to x or that x is the limit of (x_n), and we write $x_n \to x$, or

$$\lim_{n \to \infty} x_n = \lim x_n = x.$$

If (x_n) does not converge, it is said to diverge.

Remarks. (a) It might be well to point out that our definition of "convergent sequence" depends not only on $(x_n)_n$ but also on X; for instance, the sequence (x_n), $x_n = 1/n$ converges in $\mathbb{R}$ to 0, but fails to converge on the set $]0, \infty[$ (with $\rho(x, y) = |x - y|$).
(b) Suppose (x_n) is a sequence of real numbers. Then

$$\lim_{n \to \infty} x_n = a \iff \cap_{\varepsilon > 0} \cup_{m > 0} \cap_{n \geq m}]x_n - \varepsilon, x_n + \varepsilon[= \{a\}. \quad \triangle$$

A sequence $(x_n)_n$ in a metric space (X, ρ) is said to be *bounded* if its range $[x_n \mid n]$ is bounded (for bounded set, see page 63).

Theorem 1.1. *Let (x_n) be a sequence in a metric space (X, ρ).*

(a) *(x_n) converges to $x \in X$ if and only if every neighborhood of x contains all but finitely many of the terms of (x_n).*

(b) *If $x, y \in X$ and (x_n) converges to x and y, $x = y$.*
(c) *If (x_n) converges, its range is bounded.*
(d) *If $A \subset X$ and x is a limit point of A, then there is a sequence (x_n) in A such that $\lim x_n = x$.*
(e) *If $A \subset X$ and $x \in \operatorname{cl} A$, then there is a sequence (x_n) in A such that $\lim x_n = x$.*

Proof. (a) Suppose $\lim x_n = x$ and let $V \in \mathcal{V}(x)$. For some $\varepsilon > 0$, the conditions $\rho(y, x) < \varepsilon$, $y \in X$ imply $y \in V$. Corresponding to this ε, there exists n_ε such that $n \geq n_\varepsilon$ implies $\rho(x_n, x) < \varepsilon$. Thus $n \geq n_\varepsilon$ implies $x_n \in V$.

Conversely, suppose that every $V \in \mathcal{V}(x)$ contains all terms, but finitely many of them. Consider $V = \mathrm{B}(x, \varepsilon)$. Then there exists n_ε such that $x_n \in \mathrm{B}(x, \varepsilon)$ for all $n \geq n_\varepsilon$; that is, the sequence converges.
(b) Let $\varepsilon > 0$ be given. There exist positive integers n_ε and m_ε such that

$$[n \geq n_\varepsilon \implies \rho(x_n, x) < \varepsilon/2] \text{ and } [n \geq m_\varepsilon \implies \rho(x_n, y) < \varepsilon/2].$$

Hence if $n \geq \max\{n_\varepsilon, m_\varepsilon\}$, we have

$$0 \leq \rho(x, y) \leq \rho(x_n, x) + \rho(x_n, y) < \varepsilon.$$

ε has been chosen arbitrary, therefore we conclude that $\rho(x, y) = 0$.
(c) Suppose $\lim x_n = x$. There is a positive integer m such that $n > m$ implies $\rho(x_n, x) < 1$. Set

$$r = \max\{1, \rho(x_1, x), \rho(x_2, x), \ldots, \rho(x_m, x)\}.$$

Then $\rho(x_n, x) \leq r$, for $n = 1, 2, 3 \ldots$.
(d) For each positive integer n, there is a point $x_n \in A$ such that $\rho(x_n, x) < 1/n$. Given $\varepsilon > 0$, choose n_ε (which exists by the Archimedes' principle, Theorem 2.5 at page 17) so that $\varepsilon n_\varepsilon > 1$. If $n > n_\varepsilon$, it follows that $\rho(x_n, x) < \varepsilon$. Hence $x_n \to x$.
(e) We have that $\operatorname{cl} A = A \cup A'$. If $x \in A$, the constant sequence defined as $x_n = x$, for all n, supplies the conclusion. If $x \in A'$, the conclusion comes by (d).   □

Given a norm (page 43) on $\mathbb{R}^k$, we assign it a metric by

$$\rho(x, y) = \|x - y\|. \tag{3.1}$$

The next corollary follows from Theorem 2.13 at page 64.

Corollary 1.1. *A sequence converges in one of the metrics mentioned by Theorem 2.13 at page 64 if and only if it converges in any other of them.*

Below we introduce some arithmetic properties of limits of convergent real or complex sequences.

Theorem 1.2. *Suppose (x_n), (y_n) are real or complex sequences, and $\lim x_n = x$, $\lim y_n = y$. Then*

(a) $\lim(x_n + y_n) = x + y$.
(b) $\lim c \cdot x_n = cx$, $\lim(c + x_n) = c + x$, *for each real c.*
(c) $\lim x_n y_n = xy$.
(d) $\lim 1/x_n = 1/x$, *provided* $x_n \neq 0$ $(n = 1, 2, \dots)$ *and* $x \neq 0$.

Proof. (a) We can use (a) in Corollary 1.1 at page 49.
(b) We can make use Corollary 1.1 at page 49. Otherwise, the first claim follows from (c), and the second claim follows from (a).
(c) We can use Corollary 1.1 at page 49.
(d) We choose m such that $|x_n - x| < (1/2)|x|$ for $n > m$ and we see that

$$|x_n| > (1/2)|x|, \text{ for } n \geq m.$$

Given $\varepsilon > 0$, there is an integer $n_\varepsilon \geq m$ such that $n \geq n_\varepsilon$ implies

$$|x_n - x| < (1/2)|x|^2 \varepsilon.$$

Hence, for $n \geq n_\varepsilon$

$$\left| \frac{1}{x_n} - \frac{1}{x} \right| = \left| \frac{x_n - x}{x_n x} \right| < \frac{2}{|x|^2} |x_n - x| < \varepsilon. \quad \square$$

Theorem 1.3. (a) *Suppose* $x_n \in \mathbb{R}^k$, $n = 1, 2, \dots$, *and*

$$x_n = (\alpha_{1n}, \alpha_{2n}, \dots, \alpha_{kn}).$$

Then (x_n) *converges to* $x = (\alpha_1, \dots, \alpha_k)$ *if and only if*

$$\lim_{n \to \infty} \alpha_{jn} = \alpha_j, \ 1 \leq j \leq k. \tag{3.2}$$

(b) *Suppose* $(x_n)_n$, $(y_n)_n$ *are sequences in* $\mathbb{R}^k$, $(\beta_n)_n$ *is a sequence of real numbers, and* $x_n \to x$, $y_n \to y$, $\beta_n \to \beta$. *Then*

(i) $\lim(x_n + y_n) = x + y$
(ii) $\lim \langle x_n, y_n \rangle = \langle x, y \rangle$
(iii) $\lim \beta_n x_n = \beta x$.

We use the Euclidean metric. By Corollary 1.1, the conclusion is the same for all the metrics mentioned there.

Proof. (a) If $x_n \to x$, the inequalities

$$|\alpha_{jn} - \alpha_j| \leq \|x_n - x\|_2, \quad \forall j = 1, \dots, k$$

follow immediately from the definition of the Euclidean norm in $\mathbb{R}^k$. These imply that (3.2) holds.

Conversely, if (3.2) holds, then to each $\varepsilon > 0$ there corresponds a positive integer n_ε such that $n \geq n_\varepsilon$ implies

$$|\alpha_{jn} - \alpha_j| < \varepsilon/\sqrt{k}, \text{ whenever } 1 \leq j \leq k.$$

Hence, $n \geq n_\varepsilon$ implies

$$\|x_n - x\|_2 = (\sum_{j=1}^{k} |\alpha_{jn} - \alpha_j|^2)^{1/2} < \varepsilon,$$

so that $x_n \to x$. This proves (a).

(b) follows from (a) and from Theorem 1.2.   □

3.1.2 Subsequences

Given a sequence (x_n) in a metric space, consider a sequence $(n_k)_k$ of positive integers, such that $n_1 < n_2 < \ldots$. Then the sequence $(x_{n_k})_k$ is called a *subsequence* of $(x_n)_n$. If $(x_{n_k})_k$ *converges*, its limit is called a *subsequential limit* of $(x_n)_n$.

Remark. It is clear that (x_n) converges to x if and only if every subsequence of (x_n) converges to x.   △.

Theorem 1.4. *Every bounded sequence in $\mathbb{R}$ contains a convergent subsequence.*

Proof. Let E be the range of the bounded sequence (x_n); that is, $E = \{x_n\}$.

If E is finite, there is at least one point in E, say x, and a sequence $(n_k)_k$ with $n_1 < n_2 < \ldots$ such that

$$x_{n_1} = x_{n_2} = \cdots = x.$$

The subsequence $(x_{n_k})_k$ so obtained evidently converges.

If E is infinite, then E has a limit point $x \in \mathbb{R}$, Theorem 3.11 at page 69. Choose n_1 so that $|x_{n_1} - x| < 1$. Having chosen $n_1, \ldots, n_{i-1}$, we see by Theorem 2.9, page 62, that there is an integer $n_i > n_{i-1}$ such that $|x_{n_i} - x| < 1/i$. The sequence $(x_{n_i})_i$ thus obtained converges to x.   □

Theorem 1.5. *The subsequential limits of a sequence (x_n) in a metric space (X, ρ) form a closed set in X.*

Proof. Apply Theorem 2.8 at page 62.   □

Remark. How large could the set of subsequential limits of a sequence be? The two examples below show that the set under discussion could be large enough.   △

Example. The set of subsequential limits of the sequence

$$1, \underbrace{\frac{1}{2}, \frac{2}{2}, \frac{3}{2}}, \underbrace{\frac{1}{4}, \frac{2}{4}, \frac{3}{4}, \frac{4}{4}, \frac{5}{4}}, \cdots, \underbrace{\frac{1}{2^n}, \ldots, \frac{2^n + 1}{2^n}}, \cdots$$

is the closed interval $[0, 1]$, and the set of subsequential limits of the sequence

$$1, \underbrace{\frac{1}{2}, \frac{2}{2}, \frac{3}{2}}, \underbrace{\frac{1}{4}, \frac{2}{4}, \frac{3}{4}, \frac{4}{4}, \frac{5}{4}, \frac{6}{4}, \frac{7}{4}, \frac{8}{4}, \frac{9}{4}}, \ldots, \underbrace{\frac{1}{2^n}, \frac{2}{2^n}, \ldots, \frac{3^n}{2^n}}, \ldots$$

is the closed interval $[0, \infty[$. $\triangle$

3.1.3 Cauchy sequences

A sequence (x_n) in a metric space (X, ρ) is said to be a *Cauchy sequence* or *fundamental sequence* if for every $\varepsilon > 0$ there exists an integer n_ε such that $\rho(x_n, x_m) < \varepsilon$ provided $n \geq n_\varepsilon$, $m \geq n_\varepsilon$.

If (x_n) is a sequence in X and $E_n = \{x_n, x_{n+1}, \ldots\}$, (x_n) is a Cauchy sequence if and only if

$$\lim_{n \to \infty} \mathrm{diam} E_n = 0.$$

Theorem 1.6. *Let E be a set in a metric space X and $\mathrm{cl}\, E$ its closure. Then*

$$\mathrm{diam} E = \mathrm{diam}(\mathrm{cl}\, E).$$

Proof. Because $E \subset \mathrm{cl}\, E$, it is clear that

$$\mathrm{diam} E \leq \mathrm{diam}(\mathrm{cl}\, E).$$

Fix $\varepsilon > 0$ and choose $x \in \mathrm{cl}\, E$ and $y \in \mathrm{cl}\, E$. By the definition of $\mathrm{cl}\, E$, there are points $x', y' \in E$ such that $\rho(x, x') < \varepsilon$, $\rho(y, y') < \varepsilon$. Hence

$$\rho(x, y) \leq \rho(x, x') + \rho(x', y') + \rho(y', y) \leq 2\varepsilon + \rho(x', y') \leq 2\varepsilon + \mathrm{diam} E.$$

It follows that

$$\mathrm{diam}(\mathrm{cl}\, E) \leq 2\varepsilon + \mathrm{diam} E.$$

Because ε was arbitrary chosen, the theorem is established. $\square$

Theorem 1.7. (a) *Every convergent sequence in a metric space is a Cauchy sequence.*
(b) *Every Cauchy sequence in $\mathbb{R}$ converges.*

Proof. (a) If $\lim_{n \to \infty} x_n = x$ and $\varepsilon > 0$, there is an integer n_ε such that $\rho(x_n, x) < \varepsilon/2$ whenever $n \geq n_\varepsilon$. Hence for $n, m \geq n_\varepsilon$, we have

$$\rho(x_n, x_m) \leq \rho(x_n, x) + \rho(x_m, x) < \varepsilon,$$

so that (x_n) is a Cauchy sequence.
(b) Suppose (x_n) is a Cauchy sequence in $\mathbb{R}$. Set $E_n = \{x_n, x_{n+1}, \ldots\}$ and let $\mathrm{cl}\, E_n$ be the closure of E_n. By definition and Theorem 1.6 we see that

$$\lim_{n \to \infty} \mathrm{diam}(\mathrm{cl}\, E_n) = 0. \tag{3.3}$$

In particular, the sets $\mathrm{cl}\, E_n$ are bounded. They are also closed. Hence each $\mathrm{cl}\, E_n$ is compact. Also $\mathrm{cl}\, E_n \supset \mathrm{cl}\, E_{n+1}$. By Theorem 3.8 at page 68 there is a unique point $x \in \mathbb{R}$ that lies in every $\mathrm{cl}\, E_n$.

Let $\varepsilon > 0$ be given. By (3.3) there is an integer n_0 such that $\mathrm{diam}(\mathrm{cl}\, E_n) < \varepsilon$, if $n \geq n_0$. Because $x \in \mathrm{cl}\, E_n$, this means that $|y - x| < \varepsilon$ for all $y \in \mathrm{cl}\, E_n$, hence for all $y \in E_n$. In other words, if $n \geq n_0$, then $|x_n - x| < \varepsilon$. But this says precisely that $x_n \to x$, and thus the proof is finished. $\square$

A metric space (X, ρ) in which every Cauchy sequence converges is said to be *complete*.

Remark. $\mathbb{R}$ is a complete metric space, whereas $\mathbb{Q}$ is not. $\triangle$

Theorem 1.8. *Every space* $(\mathbb{R}^k, \| \cdot \|_p)$, $1 \leq p < \infty$, *and* $(\mathbb{R}^k, \| \cdot \|_\infty)$ *is complete.*

We consider the case of $(\mathbb{R}^k, \| \cdot \|_2)$. The other cases can be proved similarly. We prove the following claim. A sequence $(x_n)_n = (\alpha_{1n}, \ldots, \alpha_{kn})_n$ in $\mathbb{R}^k$ is $\| \cdot \|_2$-Cauchy if and only if every sequence $(\alpha_{j,n})_n$, $1 \leq j \leq k$, is a Cauchy sequence.

Proof of the claim. Suppose (x_n) is $\| \cdot \|_2$-Cauchy. Then there is an integer n_ε such that $n, m \geq n_\varepsilon$ imply

$$\|x_n - x_m\|_2 < \varepsilon.$$

It follows that

$$|\alpha_{jn} - \alpha_{jm}| \leq \|x_n - x_m\|_2 < \varepsilon, \; j = 1, \ldots, k.$$

Conversely, suppose that for every $\varepsilon > 0$ there is an integer n_ε such that $n, m \geq n_\varepsilon$ imply
$$|\alpha_{jn} - \alpha_{jm}| < \varepsilon/\sqrt{k}, \; j = 1, \ldots, k.$$
Hence, $n, m \geq n_\varepsilon$ imply
$$\|x_n - x_m\|_2 < \varepsilon.$$

Proof of Theorem 1.8. Suppose (x_n) is $\| \cdot \|_2$-Cauchy. Then every sequence $(\alpha_{jn})_n$, $1 \leq j \leq k$, is Cauchy, hence convergent to, say, $\alpha_j \in \mathbb{R}$. Based on (a) of Theorem 1.3, we conclude that (x_n) converges to $x = (\alpha_1, \cdots, \alpha_n)$. $\square$

Corollary 1.2. $\mathbb{C}$ *is a complete metric space.*

A characterization of completeness of a metric space is supplied by the next result.

Theorem 1.9. (Cantor) *A metric space is complete if and only if for every sequence* (F_n) *of nonempty, nested, and closed sets such that* $\lim \mathrm{diam} F_n = 0$ *one has* $\cap F_n \neq \emptyset$.

Theorem 1.10. *A closed subset of a complete metric space is complete.*

Proof. Let F be a closed subset of a complete metric space and let (x_n) be a Cauchy sequence in F. Then (x_n) converges to some $x \in X$. Because F is closed, $x \in F$. Because (x_n) has been chosen arbitrary, the conclusion follows.   □

Theorem 1.11. *A point x in a metric space (X, ρ) belongs to the closure of a set $A \subset X$ if and only if there exists a sequence in A that converges to x.*

Proof. In one sense the statement coincides with (e) in Theorem 1.1.

Suppose there is given a sequence (x_n) in A that converges to x. If $x \in A$, the conclusion follows. We suppose that $x \notin A$. Then each neighborhood $V \in \mathcal{V}(x)$ contains at least one term of the sequence; that is, there is a term $x_n \in A \cap V$. It means that $x \in A' \subset \operatorname{cl} A$.   □

Theorem 1.12. *A complete subset of a metric space is closed.*

Proof. Let F be a complete subset of a metric space (X, ρ). Let x be a limit point of F; that is, $x \in F'$. By (e) in Theorem 1.1 there is a sequence (x_n) in F that converges to x. By (a) in Theorem 1.7, the sequence (x_n) is Cauchy. Because F is complete, (x_n) converges to some point $y \in F$. By (b) in Theorem 1.1 we have that $x = y$. Thus we proved that $F' \subset F$. From here, by Corollary 2.9 at page 62, the conclusion follows.   □

A subset A of a metric space is said to be *precompact* if for each $\varepsilon > 0$ there is a finite F such that

$$A \subset \cup_{x \in F} B(x, \varepsilon).$$

Obviously, a compact set in a metric space is precompact.

A characterization of the precompact spaces is given by the following.

Theorem 1.13. *A metric space is precompact if and only if each sequence has a Cauchy subsequence.*

We introduce the relationship between compact spaces and complete spaces.

Theorem 1.14. *A metric space is compact if and only if it is complete and precompact.*

Theorem 1.15. (Baire[1]) *The intersection of a countable family of dense open sets in a complete metric space X is dense in X.*

---

[1] René-Louis Baire, 1874–1932.

3.1.4 Monotonic sequences

A sequence (x_n) of real numbers is said to be
  (a) *Monotonically increasing* if $x_n \leq x_{n+1}$, for all n.
  (b) *Monotonically decreasing* if $x_n \geq x_{n+1}$, for all n.
The class of *Monotonic sequences* consists of the increasing and decreasing sequences.
  A sequence (x_n) of real numbers is said to be
  (a) *Strictly increasing* if $x_n < x_{n+1}$, for all n.
  (b) *Strictly decreasing* if $x_n > x_{n+1}$, for all n.
The class of *strictly monotonic sequences* consists of the strictly increasing and strictly decreasing sequences.

Theorem 1.16. *Suppose $(x_n)_{n \in \mathbb{N}}$ is monotonic. Then (x_n) converges if and only if it is bounded.*

Proof. If (x_n) converges, it is bounded as we already saw by (c) in Theorem 1.1. No monotonicity is needed.
  Suppose $x_n \leq x_{n+1}$ for all $n \in \mathbb{N}$. Let E be the range of (x_n). Because (x_n) is bounded, let x be the least upper bound of E. Then $x_n \leq x$, for all $n \in \mathbb{N}$. For every $\varepsilon > 0$ there is a rank n_ε such that

$$x - \varepsilon < x_{n_\varepsilon} \leq x,$$

for otherwise $x - \varepsilon$ would be an upper bound of E. Because (x_n) increases, $n \geq n_\varepsilon$ therefore implies

$$x - \varepsilon < x_n \leq x < x + \varepsilon \iff |x_n - x| < \varepsilon,$$

which shows that (x_n) converges to x. $\square$

Theorem 1.17. *Suppose that beginning with a certain rank, the terms of a convergent sequence (x_n) satisfy the inequality $x_n \geq b$ ($x_n \leq b$). Then the limit a of the sequence (x_n) satisfies the inequality $a \geq b$ ($a \leq b$).*

Proof. Suppose that there exists $m \in \mathbb{N}$ such that for every $n \geq m$, $x_n \geq b$. We show that $a \geq b$.
  If $a < b$, denote $c = b - a$. Consider $\varepsilon = c/2$. Because a is the limit of the sequence (x_n), there is a rank $n_\varepsilon \in \mathbb{N}$ such that $|x_n - a| < \varepsilon$, for all $n \geq n_\varepsilon$; that is, $x_n < a + \varepsilon < b$ for all $n \geq n_\varepsilon$, contrary to our assumption. Hence $a \geq b$. $\square$

Corollary 1.3. *Suppose that beginning with a certain rank, the terms x_n and y_n of the convergent sequences (x_n) and (y_n) satisfy the inequality $x_n \leq y_n$. Then $\lim x_n \leq \lim y_n$.*

Proof. The sequence $(y_n - x_n)_n$ is convergent and has nonnegative terms. Thus its limit is nonnegative. Hence $\lim y_n - \lim x_n = \lim(y_n - x_n) \geq 0$. $\square$

Theorem 1.18. *Let (x_n) and (y_n) be two sequences. Suppose that (x_n) is convergent and that*

$$y_n - x_n \to 0, \quad \text{as } n \to \infty.$$

Then (y_n) converges and $\lim x_n = \lim y_n$.

Proof. Set $x = \lim x_n$. The conclusions follow from

$$|x - y_n| \le |x - x_n| + |x_n - y_n|. \quad \square$$

Corollary 1.4. *Suppose there are given three sequences (x_n), (a_n), and (y_n) satisfying $x_n \le a_n \le y_n$ from a certain rank. If the sequences (x_n) and (y_n) converge to the same limit, then (a_n) is convergent and its limit coincides with the limit of (x_n).*

Proof. It is obvious because $|a_n - x_n| \le |y_n - x_n|$. $\quad \square$

3.1.5 Upper limits and lower limits

Let (x_n) be a sequence of real numbers with the property that for every real m there is an integer n_m such that $n \ge n_m$ implies $x_n \ge m$. We then write

$$x_n \to \infty \quad \text{or} \quad \lim_{n \to \infty} x_n = \infty.$$

Similarly, if for every real m there is an integer n_m such that $n \ge n_m$ implies $x_n \le m$, we write

$$x_n \to -\infty \quad \text{or} \quad \lim_{n \to \infty} x_n = -\infty.$$

Let (x_n) be a sequence of real numbers. Let $x \in E \subset \overline{\mathbb{R}}$ be such that $x_{n_k} \to x$ for some subsequence $(x_{n_k})_k$. This set E contains all subsequential limits, plus possibly, the elements $+\infty$ and $-\infty$. We put

$$x^* = \sup E \quad \text{and} \quad x_* = \inf E.$$

The elements x^* and x_* are called the *upper*, respectively, the *lower* limits of (x_n). We use the notations

$$\limsup_{n \to \infty} x_n = x^*, \quad \liminf_{n \to \infty} x_n = x_*.$$

Theorem 1.19. *Let (x_n) be a sequence of real numbers. Let E and x^* be as defined earlier. Then x^* has the following two properties.*
(a) $x^* \in E$.
(b) *If $y > x^*$, there is an integer m such that $n \ge m$ implies $x_n < y$.*
  Moreover, x^* is the only number with the properties (a) *and* (b).

Proof. If $x^* = +\infty$, E is not bounded above. Hence (x_n) is not bounded above, and there is a subsequence (x_{n_k}) so that $x_{n_k} \to \infty$.

If x^* is real, then E is bounded above, and at least one subsequential limit exists, so that (a) follows from Theorem 1.5 and Theorem 2.27 at page 29.

If $x^* = -\infty$, E contains only one element, namely $-\infty$, and there is no subsequential limit; hence for any real m, $x_n > m$ for at most a finite number of values of n, so that $x_n \to -\infty$. This establishes (a) in all cases.

To prove (b) suppose there is a number $y > x^*$ such that $x_n \geq y$ for infinitely many values of n. In that case, there is a number $z \in E$ such that $z \geq y > x^*$, contradicting the definition of x^*.

Thus x^* satisfies (a) and (b).

To show the uniqueness, suppose there are two numbers, p and q, that satisfy (a) and (b) and suppose $p < q$. Choose x such that $p < x < q$. Because p satisfies (b), we have that $x_n < x$ for $n > m$. But then q cannot satisfy (a). $\square$

A similar statement holds for x_*.

Theorem 1.20. *Let (x_n) be a sequence of real numbers. Let E and x_* be as defined earlier. Then x_* has the following two properties.*
(a) $x_* \in E$.
(b) *If $y < x_*$, there is an integer m such that $n \geq m$ implies $x_n > y$.*
   Moreover, x_* is the only number with the properties (a) and (b).

Theorem 1.21. *A sequence (x_n) of real numbers converges if and only if*

$$\liminf_{n \to \infty} x_n = \limsup_{n \to \infty} x_n.$$

Proof. Suppose (x_n) is a convergent sequence. Then it is bounded and has a unique limit point. All subsequences converge precisely to that point, so the upper limit and the lower limit coincide.

Suppose $\liminf_{n \to \infty} x_n = \limsup_{n \to \infty} x_n$. Denote this common value by x. If $\liminf_{n \to \infty} x_n = \infty$, then by (b) in Theorem 1.20, $x_n \to \infty$, as $n \to \infty$. If $\limsup_{n \to \infty} x_n = -\infty$, then by (b) in Theorem 1.19, $x_n \to -\infty$, as $n \to \infty$. Now admit that $\liminf_{n \to \infty} x_n = \limsup_{n \to \infty} x_n \in \mathbb{R}$. By (b) in Theorem 1.19 and by (b) in Theorem 1.20 we have that for every $\varepsilon > 0$ there exists a natural n_ε such that for every natural $n \geq n_\varepsilon$, $x - \varepsilon < x_n < x + \varepsilon$. This means that $x_n \to x$, as $n \to \infty$. Hence (x_n) is convergent. $\square$

3.1.6 The big Oh and small oh notations

The process of selection among several algorithms regarding the same exercise is often based on easy implementation, elegance, time, and space allocation. Often we estimate the time consumption and based on it, we appreciate that one algorithm is faster than another.

Given are (a_n) and (b_n), two real sequences. We say that "(a_n) is big Oh or omicron of (b_n)" and write $a_n \in O(b_n)$, if there exist a natural n_0 and a positive constant C such that

$$|a_n| \leq C|b_n|, \quad \forall n \geq n_0.$$

In such a case we say that (a_n) does not grow faster than (b_n).

Obviously, $a_n \in O(a_n)$, $n \in O(n^2)$, whereas $n^2 \notin O(n)$. Often $a_n \in O(b_n)$ is written in the form $a_n = O(b_n)$.

The expression $n + O(n^2) = O(n^3)$ means that $n + O(n^2) \subset O(n^3)$.

Warning. We consider an equality sign "$=$" in an expression involving big Oh as "$\in$" or "$\subset$".

Remark 1.1. We saw earlier that $a_n = O(b_n)$ does not imply $b_n = O(a_n)$. Thus "$=$" in an expression involving big Oh is not symmetric. $\triangle$

Several properties concerning big Oh are introduced below.

Proposition 1.1. *The following properties hold.*

(a) *If c is a nonzero constant,*

$$cO(a_n) = O(a_n) \text{ and } O(ca_n) = O(a_n).$$

(b) *The big Oh is idempotent; that is,*

$$O(O(a_n)) = O(a_n).$$

(c) *The multiplication of big Oh expressions satisfy*

$$O(a_n)O(b_n) = O(a_n b_n) \text{ and } O(a_n b_n) = a_n O(b_n).$$

(d) *The big Oh expressions can be reduced:*

$$a_n = O(b_n) \implies O(a_n + b_n) = O(b_n).$$

(e) *For all $k \in \mathbb{N}^*$,*

$$(a_n + b_n)^k = O(a_n^k) + O(b_n^k).$$

(f) *If $a_n = O(b_n)$ and $c_n = O(b_n)$, then*

$$\alpha a_n + \beta c_n = O(b_n), \quad \forall \alpha, \beta \in \mathbb{R}.$$

(g)

$$a_n = b_n + O(c_n) \implies b_n = a_n + O(c_n).$$

(h) *The big Oh is transitive*

$$a_n = O(b_n) \text{ and } b_n = O(c_n) \implies a_n = O(c_n).$$

We introduce a necessary and sufficient condition that $a_n = O(b_n)$.

Lemma 3.1. *Given are* (a_n) *and* (b_n) *such that there exists* $n_0 \in \mathbb{N}$ *so that* $b_n \neq 0$ *for all* $n \geq n_0$, $n \in \mathbb{N}$. *Then*

$$\limsup_{n\to\infty} \left| \frac{a_n}{b_n} \right| < \infty \Longleftrightarrow a_n = O(b_n).$$

Proof. Suppose $\limsup_{n\to\infty} |a_n/b_n| = C \in [0, +\infty[$. Then for each $\varepsilon > 0$ there exists $n_0(= n_0(\varepsilon)) \in \mathbb{N}$ so that $|a_n/b_n| < C + \varepsilon$ for each $n \geq n_0$. Thus $a_n = O(b_n)$.

Suppose $a_n = O(b_n)$. Then there exists $n_0 \in \mathbb{N}$ such that $b_n \neq 0$ and $|a_n|/|b_n| \leq C$, for all $n \geq n_0$. So $\limsup_{n\to\infty} |a_n/b_n| \leq C$.   □

Remark. The assumption that $b_n \neq 0$ from an index n is only sufficient. The limit of the quotient $|a_n/b_n|$ might not exist. For, consider $a_n = (1 + (-1)^n)n^2$ and $b_n = n^2$. Then $\lim_{n\to\infty} |a_n/b_n|$ does not exist whereas $a_n = O(b_n)$.   △

Examples 1.1. Some examples are introduced below.

(a) $n/(n+1) = 1 + O(1/n)$ because $n/(n+1) = 1 - 1/(n+1) = 1 + O(1/n)$.
(b) For fixed k,

$$\binom{n}{k} = n^k/k! + O(n^{k-1}),$$

because

$$\binom{n}{k} = \frac{n^k - (k(k-1)/2)n^{k-1} + \cdots + (-1)^{k-1}(k-1)!}{k!} = \frac{n^k}{k!} + O(n^{k-1}).$$

(c) We have $\sum_{k=0}^{n} k! = n!(1 + O(n^{-1}))$. Indeed,

$$\sum_{k=0}^{n} k! = n! \left(1 + \frac{1}{n} + \frac{1}{n(n-1)} + \cdots + \sum_{k=0}^{n-3} \frac{k!}{n!} \right).$$ △

We introduce the small oh notation. Given are (a_n) and (b_n), two real sequences. We say that "(a_n) is small oh or omicron of (b_n)" and write $a_n \in o(b_n)$, if

$$\lim_{n\to\infty} \frac{a_n}{b_n} = 0.$$

In such a case we say that (b_n) grows faster than (a_n).

Obviously, $n \in o(n^2)$, and $n^2 \notin o(n)$. Often $a_n \in o(b_n)$ is written as $a_n = o(b_n)$.

The expression $n + o(n^2) = o(n^3)$ means that $n + o(n^2) \subset o(n^3)$.

Warning. We consider an equality sign "$=$" in an expression involving small oh as "$\in$" or "$\subset$".

Remark 1.2. We saw earlier that $a_n = o(b_n)$ does not imply $b_n = o(a_n)$. Thus " $=$ " in an expression involving small oh is not symmetric. $\triangle$

Below several properties concerning small oh are introduced.

Proposition 1.2. *The following properties hold.*

(a) $o(a_n) + o(a_n) = o(a_n)$.
(b) $o(a_n) \cdot o(b_n) = o(a_n b_n)$.
(c) $o(o(a_n)) = o(a_n)$.
(d) $o(a_n) \subset O(a_n)$.
(e) $a_n = o(b_n)$ *and* $b_n = o(c_n)$ *imply* $a_n = o(c_n)$.
(f) $n^\alpha = o(n^\beta)$ *if and only if* $\alpha < \beta$.
(g) *If* $0 < \varepsilon < 1 < c$, *then*

$$1 = o(\log \log n) = o(\log n) = o(n^\varepsilon) = o(n^c) = o(n^{\log n}) = o(n^n).$$

(h) *If* $a_n, b_n \notin \{0\}$, *then*

$$a_n = o(b_n) \quad \text{if and only if} \quad 1/b_n = o(1/a_n).$$

(i) $e^{a_n} = o(e^{b_n})$ *if and only if* $\lim_{n \to \infty}(a_n - b_n) = -\infty$.
(j) $1 = o(a_n) = o(b_n)$ *implies* $e^{a_n} = o(e^{b_n})$.

Given are (a_n) and (b_n), two real sequences. We say that "(a_n) is big omega of (b_n)" and write $a_n = \Omega(b_n)$ if $\liminf |a_n/b_n| > 0$; that is, there exists a positive constant C so that $b_n \geq C a_n$ from a given rank.

Given are (a_n) and (b_n), two real sequences. If (a_n) and (b_n) have the same growth, we denote $a_n = \Theta(b_n)$ if and only if $|a_n| \leq C|b_n|$ and $|b_n| \leq C|a_n|$ for some $C > 0$ and from a given rank. We say that the sequences are *similar*.

Two sequences (a_n) and (b_n) are said to be *equivalent* provided $\lim a_n/b_n = 1$. Then we write $a_n \sim b_n$. We also say that sequence (a_n) *grows asymptotically as* (b_n).

3.1.7 Stolz–Cesaro theorem and some of its consequences

Theorem 1.22. (Stolz [2]–Cesaro [3]) *Let* (a_n) *be a sequence of real numbers and* (b_n) *a strictly monotone and divergent sequence. Then*

$$\lim_{n \to \infty} \frac{a_{n+1} - a_n}{b_{n+1} - b_n} = l \quad (\in [-\infty, +\infty])$$

implies

$$\lim_{n \to \infty} \frac{a_n}{b_n} = l.$$

---

[2] Otto Stolz, 1842–1905.
[3] Ernesto Cesaro, 1856–1906.

Proof. Suppose that l is finite and (b_n) is strictly increasing. Choose a positive ε. For $\varepsilon/3$ we find a rank n_ε such that for every $n \geq n_\varepsilon$ we have $b_n > 0$ and

$$\left| \frac{a_{n+1} - a_n}{b_{n+1} - b_n} - l \right| < \varepsilon/3.$$

Thus

$$(b_{n+1} - b_n)(l - \varepsilon/3) < a_{n+1} - a_n < (b_{n+1} - b_n)(l + \varepsilon/3). \qquad (3.4)$$

Taking in (3.4), successively, $n = n_\varepsilon$, $n = n_\varepsilon + 1$, ..., $n = n_\varepsilon + p - 1$, and adding all these inequalities, we get successively

$$(b_{n_\varepsilon+p} - b_{n_\varepsilon})(l - \varepsilon/3) < a_{n_\varepsilon+p} - a_{n_\varepsilon} < (b_{n_\varepsilon+p} - b_n)(l + \varepsilon/3),$$

$$l - \frac{\varepsilon}{3} - \left(l - \frac{\varepsilon}{3}\right)\frac{b_{n_\varepsilon}}{b_{n_\varepsilon+p}} + \frac{a_{n_\varepsilon}}{b_{n_\varepsilon+p}} < \frac{a_{n_\varepsilon+p}}{b_{n_\varepsilon+p}} < l + \frac{\varepsilon}{3} - \left(l + \frac{\varepsilon}{3}\right)\frac{b_{n_\varepsilon}}{b_{n_\varepsilon+p}} + \frac{a_{n_\varepsilon}}{b_{n_\varepsilon+p}}. \qquad (3.5)$$

Note that the sequences $(b_{n_\varepsilon}/b_{n_\varepsilon+p})_p$ and $(a_{n_\varepsilon}/b_{n_\varepsilon+p})_p$ tend to 0. Then there exists a rank p_ε such that for every $p \geq p_\varepsilon$ there hold

$$\left| \frac{b_{n_\varepsilon}}{b_{n_\varepsilon+p}}\left(l \pm \frac{\varepsilon}{3}\right) \right| < \frac{\varepsilon}{3} \quad \text{and} \quad \left| \frac{a_{n_\varepsilon}}{b_{n_\varepsilon+p}} \right| < \frac{\varepsilon}{3}.$$

Hence, for every $n > n_\varepsilon + p_\varepsilon$, we finally have

$$\left| l - \frac{a_n}{b_n} \right| < \varepsilon.$$

Thus $a_n/b_n \to l$.

Suppose $l = +\infty$. We may assume that all b_ns are strictly positive. Choose a positive ε arbitrary large. There exists a rank n_ε such that for every $n \geq n_\varepsilon$ it holds

$$\frac{a_{n+1} - a_n}{b_{n+1} - b_n} > \varepsilon$$

or

$$a_{n+1} - a_n > \varepsilon(b_{n+1} - b_n).$$

Summing up these inequalities from n to $n + p$, we write

$$a_{n+p}/b_{n+p} > \varepsilon + (a_n - \varepsilon b_n)/b_{n+p}.$$

Now, there exists a rank p_ε such that for every $p > p_\varepsilon$ we have

$$|(a_n - \varepsilon b_n)/b_{n+p}| < \varepsilon/2.$$

Hence, for every $n > n_\varepsilon + p_\varepsilon$, we finally have

$$a_n/b_n > \varepsilon/2.$$

So, $a_n/b_n \to \infty$ and this case is proved.

The proof of the case $l = -\infty$ runs similarly. We omit it.   □

Remark. Generally, the converse statement of Theorem 1.22 does not hold. An example is supplied below by Remark 1.3.   △

However, a partial converse to Theorem 1.22 holds.

Theorem 1.23. *Let (a_n) be a sequence of real numbers and (b_n) a strictly monotone and divergent sequence. Suppose*

$$\lim_{n \to \infty} \frac{b_n}{b_{n+1}} = b \ (\in \mathbb{R} \setminus \{1\}).$$

Then

$$\lim_{n \to \infty} \frac{a_n}{b_n} = l \implies \lim_{n \to \infty} \frac{a_{n+1} - a_n}{b_{n+1} - b_n} = l.$$

Proof. Indeed,

$$\frac{a_{n+1} - a_n}{b_{n+1} - b_n} = \frac{\dfrac{a_{n+1}}{b_{n+1}} - \dfrac{a_n}{b_n} \dfrac{b_n}{b_{n+1}}}{1 - \dfrac{b_n}{b_{n+1}}} \to \frac{l - lb}{1 - b} = l. \quad \square$$

Corollary 1.5. *Suppose a sequence (x_n) is given. If*

$$\lim_{n \to \infty} x_n = x \ (\in [-\infty, \infty]), \quad then \quad \lim_{n \to \infty} \frac{x_1 + x_2 + x_3 + \cdots + x_n}{n} = x.$$

Proof. Take $b_n = n$ and $a_n = x_1 + x_2 + x_3 + \cdots + x_n$. Then

$$\frac{a_{n+1} - a_n}{b_{n+1} - b_n} = x_{n+1} \to x. \quad \square$$

Remark 1.3. Generally, the converse statement of Corollary 1.5 does not hold. Indeed, consider

$$a_n = (1 + (-1)^n)/2.$$

Obviously, this sequence does not converge. At the same time

$$\frac{a_1 + a_2 + \cdots + a_n}{n} = \begin{cases} 1/2, & n \text{ even} \\ (n-1)/2n, & n \text{ odd}. \end{cases}$$

Hence

$$\lim \frac{a_1 + a_2 + \cdots + a_n}{n} = \frac{1}{2}. \quad △$$

Corollary 1.6. *Suppose there is given a strictly positive sequence (a_n).*

(a) *If*

$$\lim_{n \to \infty} a_{n+1}/a_n = a \quad (a > 0),$$

then

$$\lim_{n \to \infty} \sqrt[n]{a_n} = a.$$

(b) *If*

$$\lim_{n\to\infty} a_n = a \quad (a > 0),$$

then

$$\lim_{n\to\infty} \sqrt[n]{a_1 a_2 \ldots a_n} = a.$$

Proof. (a) Take $x_n = \ln \sqrt[n]{a_n}$. Then $x_n = \ln a_n / n$. Now we apply Theorem 1.22 of Stolz–Cesaro and get $\lim x_n = \lim \ln(a_{n+1}/a_n) = \ln \lim(a_{n+1}/a_n) = \ln a$.

(b) Take $x_n = \ln \sqrt[n]{a_1 a_2 \ldots a_n}$ and reason as before. $\square$

Remark. Generally, the converses of (a) and (b) do not hold. In order to see directly that the converse of (a) is false it is enough to consider two distinct positive numbers p, q and the sequence (a_n) defined by $a_{2k-1} = p^k q^{k-1}$ and $a_{2k} = p^k q^k$, $k \in \mathbb{N}^*$. Then $a_n \to \sqrt{pq}$. At the same time the sequence (a_{n+1}/a_n) has no limit.

In order to see directly that the converse of (b) is false it is enough to consider as before two distinct positive numbers p, q and the sequence (a_n) defined by $a_{2k-1} = p$ and $a_{2k} = q$, $k \in \mathbb{N}^*$. Then $\sqrt[n]{a_1 a_2 \ldots a_n} \to \sqrt{pq}$ as $n \to \infty$ and the sequence (a_n) is divergent. $\triangle$

3.1.8 Certain combinatorial numbers

The aim of this subsection is to introduce certain combinatorial numbers. The sequences built up by these numbers have nice and useful properties.

Let σ be a permutation of $\mathbb{N}_n^*$. We say that σ has an increase (decrease) if there there is an i so that $\sigma(i) < \sigma(i+1)$ ($\sigma(i) > \sigma(i+1)$). The *Euler*[4] *number of the first kind*

$$\left\langle \begin{matrix} n \\ j \end{matrix} \right\rangle$$

is the number of permutations σ having j increases. Define

$$\left\langle \begin{matrix} 0 \\ 0 \end{matrix} \right\rangle = 1 \text{ and } \left\langle \begin{matrix} n \\ 0 \end{matrix} \right\rangle = 1.$$

We now introduce the *Stirling*[5] *numbers of the second kind*. Consider $n, k \in \mathbb{N}^*$ with $k \le n$. Then the Stirling number of the second kind

$$\left\{ \begin{matrix} n \\ k \end{matrix} \right\}$$

is defined as the number of partitions of a set of cardinality n into k nonempty sets. We define

---

[4] Leonhard Euler, 1707–1783.
[5] James Stirling, 1692–1770.

$$\left\{{0\atop 0}\right\} = 1, \quad \left\{{n\atop k}\right\} = 0 \text{ for } k > n, \quad \text{and} \quad \left\{{n\atop 0}\right\} = 0 \text{ for } n \geq 1.$$

Obviously, $\left\{{n\atop 1}\right\} = \left\{{n\atop n}\right\} = 1$, for $n \geq 1$.

Proposition 1.3. *The Stirling numbers of the second kind satisfy the recurrence*

$$\left\{{n\atop k}\right\} = k\left\{{n-1\atop k}\right\} + \left\{{n-1\atop k-1}\right\}, \quad n \geq 1. \tag{3.6}$$

Proof. Consider a set $A = \{a_1, \ldots, a_n\}$. Then there are $\left\{{n\atop k}\right\}$ classes of partitions of A, each partition having k subsets. Pick up a_n. There are two exclusive and exhaustive cases. Either a_n alone forms a subset and in this case there are $\left\{{n-1\atop k-1}\right\}$ partitions by the remaining $n-1$ elements or a_n belongs to a set containing at least two elements. In this case we can consider that we insert a_n to any one of the k subsets of partitions to $A \setminus \{a_n\}$. This can be done in $k\left\{{n-1\atop k}\right\}$ ways. Thus (3.6) is proved. $\quad\square$

Corollary 1.7. *One has*

$$\left\{{n\atop 2}\right\} = 2^{n-1} - 1, \quad n \in \mathbb{N}^*.$$

Proposition 1.4. *The Stirling numbers of the second kind satisfy the recurrences*

$$k\left\{{n\atop k}\right\} = \sum_{i=k-1}^{n-1} \binom{n}{i}\left\{{i\atop k-1}\right\}, \tag{3.7}$$

$$k\left\{{n\atop k}\right\} = \sum_{i=0}^{n-1} \binom{n}{i}\left\{{i\atop k-1}\right\}. \tag{3.8}$$

Proof. Consider an arbitrary partition of a set of cardinality n into k nonempty subsets. Neglect one of these subsets (which can be done in k ways). The remaining set is of cardinality i, $k-1 \leq i \leq n-1$, and using it one can get $\left\{{i\atop k-1}\right\}$ partitions.
  (3.8) follows from (3.7). $\quad\square$

Proposition 1.5. *The Stirling numbers of the second kind satisfy the recurrences*

$$\left\{{n\atop k}\right\} = \sum_{i=k-1}^{n-1} \binom{n-1}{i}\left\{{i\atop k-1}\right\}, \tag{3.9}$$

$$\left\{{n\atop k}\right\} = \sum_{i=0}^{n-1} \binom{n-1}{i}\left\{{i\atop k-1}\right\}. \tag{3.10}$$

Proof. Consider a set A of cardinality n. We eliminate the subset in a partition of A that contains a prescribed element in A. Consider that the subset is of cardinality $n-i$. The remaining set is of cardinality i, $k-1 \leq i \leq n-1$, and using it one gets $\left\{{i \atop k-1}\right\}$ partitions.

Identity (3.10) follows from (3.9). $\square$

Theorem 1.24. *For every positive integer n and real x there hold*

$$x^n = \sum_{k=1}^{n} \left\{{n \atop k}\right\} x(x-1)\cdots(x-k+1), \tag{3.11}$$

$$\left\{{n \atop k}\right\} = \frac{1}{k!}\sum_{j=1}^{k}(-1)^{k-j}\binom{k}{j}j^n. \tag{3.12}$$

Proof. Suppose x is a positive integer and consider two sets X and A satisfying $|X| = x$ and $|A| = n$. Then the number of mappings of A into X is x^n. Now suppose that k is the cardinality of the range to such a mapping. For k fixed there are $\left\{{n \atop k}\right\}$ ways in which elements of A are mapped onto the same elements of X. Thus partitions of A having k classes appear. There are $x(x-1)\cdots(x-k+1)$ images of these classes in a one-to-one way. Thus identity (3.11) is proved for positive integer x. Now if we consider both sides as polynomials in the real variable x, then the two polynomials (of finite degree) coincide on countably many points. Hence they coincide on every real x. Thus this identity is proved.

Identity (3.12) can be obtained in the following way.

$$\frac{1}{k!}\sum_{j=0}^{k}(-1)^{k-j}\binom{k}{j}j^n = \sum_{j=0}^{k}\frac{(-1)^{k-j}}{j!(k-j)!}\sum_{l=0}^{n}\left\{{n \atop l}\right\}j(j-1)\cdots(j-l+1)$$

$$= \sum_{j=0}^{k}\sum_{l=0}^{j}(-1)^{k-j}\left\{{n \atop l}\right\}\frac{1}{(k-j)!(j-l)!}$$

$$= \sum_{l=0}^{k}\left\{{n \atop l}\right\}\frac{1}{(k-l)!}\sum_{j=l}^{k}(-1)^{k-j}\binom{k-l}{k-j}$$

$$= \sum_{l=0}^{k}\left\{{n \atop l}\right\}\frac{1}{(k-l)!}(1-1)^{k-l} = \left\{{n \atop k}\right\}. \quad \square$$

We introduce the *Stirling numbers of the first kind*. Consider $n, k \in \mathbb{N}^*$ with $k \leq n$. Then the Stirling number of the first kind

$$\left[{n \atop k}\right]$$

is defined as the number of cycles of n elements formed by k subcycles. We define

$$\begin{bmatrix} 0 \\ 0 \end{bmatrix} = 1, \quad \begin{bmatrix} n \\ k \end{bmatrix} = 0 \text{ whenever } k > n, \text{ and } \begin{bmatrix} n \\ 0 \end{bmatrix} = 0 \text{ for } n \geq 1.$$

The cycle in Figure 3.1 is written as $[a, b, c, d]$ and we agree that

$$[a, b, c, d] = [b, c, d, a] = [c, d, a, b] = [d, a, b, c].$$

Fig. 3.1. The cycle $[a, b, c, d]$

We note that $[a, b, c, d] \neq [a, c, b, d]$. Clearly $\begin{bmatrix} 4 \\ 2 \end{bmatrix} = 11$ inasmuch as

$$\begin{array}{llll} [1, 2, 3] \, [4], & [1, 2, 4] \, [3], & [1, 3, 4] \, [2], & [2, 3, 4] \, [1] \\ [1, 3, 2] \, [4], & [1, 4, 2] \, [3], & [1, 4, 3] \, [2], & [2, 4, 3] \, [1] \\ [1, 2] \, [3, 4], & [1, 3] \, [2, 4], & [1, 4] \, [3, 2], & \end{array}$$

We have $\begin{bmatrix} n \\ n \end{bmatrix} = 1$ because the order of the subcycles is irrelevant. Also $\begin{bmatrix} n \\ 1 \end{bmatrix} = (n - 1)!$ because there are $n!$ permutations of n elements and each cycle might start with each element of it. A cycle containing one element corresponds bijectively to a set containing precisely that one element. A cycle containing two elements corresponds bijectively to a set containing precisely the two elements, because $\{a, b\} = \{b, a\}$ and $[a, b] = [b, a]$. Therefore we can write

$$\begin{bmatrix} n \\ n \end{bmatrix} = \begin{Bmatrix} n \\ n \end{Bmatrix} = 1, \quad \begin{bmatrix} n \\ n-1 \end{bmatrix} \geq \begin{Bmatrix} n \\ n-1 \end{Bmatrix} = \binom{n}{2},$$

$$\begin{bmatrix} n \\ k \end{bmatrix} \geq \begin{Bmatrix} n \\ k \end{Bmatrix}, \quad n, k \in \mathbb{N}, \ n \geq k.$$

Define the Fibonacci[6] sequence $(f_n)_{n \in \mathbb{N}}$ by

$$f_0 = 0, \quad f_1 = 1, \quad f_{n+2} = f_{n+1} + f_{n+1}, \quad \forall n \in \mathbb{N}. \tag{3.13}$$

We note all the terms are nonnegative and that for $n \geq 5$, $f_n \geq n$. Thus the sequence is monotonically increasing and unbounded. Some properties of the Fibonacci sequence or related to it are introduced below.

---

[6] Leonardo Fibonacci, 1180–1280.

Proposition 1.6.

(i) *The Cassini[7] identity is*

$$f_{n+1}^2 - f_n f_{n+2} = (-1)^n, \quad n \in \mathbb{N}. \tag{3.14}$$

(ii) *For $n \in \mathbb{N}^*$, f_n and f_{n+1} have no common divisor.*
(iii) *For each fixed $n \in \mathbb{N}^*$,*

$$f_{n+p} = f_{n-1} f_p + f_n f_{p+1}, \quad p \in \mathbb{N}. \tag{3.15}$$

(iv) *For $n, k \in \mathbb{N}^*$, f_n divides $f_{k \cdot n}$.*
(v) *The Binet[8] identity is*

$$f_n = \frac{1}{\sqrt{5}} \left(\left(\frac{1 + \sqrt{5}}{2} \right)^n - \left(\frac{1 - \sqrt{5}}{2} \right)^n \right), \quad n \in \mathbb{N}. \tag{3.16}$$

(vi) *The following identity holds*

$$f_0^2 + f_1^2 + \cdots + f_n^2 = f_n \cdot f_{n+1}, \quad n \in \mathbb{N}. \tag{3.17}$$

(vii) *It holds*

$$f_{m+2n} = \sum_{k=0}^n \binom{n}{k} f_{m+k}, \quad m, n \in \mathbb{N}. \tag{3.18}$$

(viii) *Define the sequence (u_n) by $u_n = f_{n+1}/f_n$, $n \in \mathbb{N}^*$. Then this sequence is convergent and nonmonotone.*

Proof. (i) We argue by induction. (3.14) holds for $n = 0$ and $n = 1$. Suppose that it holds for a fixed n and we prove it for $n + 1$. Then because $f_n = f_{n+2} - f_{n+1}$,

$$(-1)^n = f_{n+1}^2 - f_n f_{n+2} = f_{n+1}^2 - f_{n+2}^2 + f_{n+1} f_{n+2}$$
$$= f_{n+1}(f_{n+1} + f_{n+2}) - f_{n+2}^2 - (f_{n+2}^2 - f_{n+1} f_{n+3})$$
$$\implies f_{n+2}^2 - f_{n+1} f_{n+3} = (-1)^{n+1}.$$

(ii) Suppose there exists a $d \in \mathbb{N}^*$ which is a divisor of f_n and f_{n+1}. Then by (3.14), d is a divisor of 1. Then the conclusion follows.
(iii) It is clear that (3.15) holds for $p = 0$ and $p = 1$, and all $n \in \mathbb{N}^*$. Suppose there exists a natural p so that

$$f_{n+p} = f_{n-1} f_p + f_n f_{p+1} \text{ and } f_{n+p+1} = f_{n-1} f_{p+1} + f_n f_{p+2}, \quad n \in \mathbb{N}^*.$$

Then

---

[7] Giovanni Domenico Cassini, 1625–1712. Born near Nice, moved to Paris in 1669, and known since then under the name of Colbert.
[8] Jacques Philippe Marie Binet, 1786–1856.

$$f_{n+p+2} = f_{n+p+1} + f_{n+p} = f_{n+p}$$
$$= f_{n-1}f_p + f_nf_{p+1} + f_{n-1}f_{p+1} + f_nf_{p+2}$$
$$= f_{n-1}(f_p + f_{p+1}) + f_n(f_{p+1} + f_{p+2}) = f_{n-1}f_{p+2} + f_nf_{p+3}, \quad n \in \mathbb{N}^*.$$

(iv) We prove by induction on k. For $k = 1$, the claim is trivial. Suppose it holds for a $k \in \mathbb{N}^*$. We take $p = k \cdot n$ in (3.15). Then because

$$f_{(k+1)n} = f_{n+kn} = f_{n-1}f_{kn} + f_nf_{kn+1},$$

$f_{(k+1)n}$ is divisible by f_n.

(v) We write the characteristic equation for the second-order linear recurrence (3.13) and use the first terms f_0 and f_1.

(vi) For $n = 0$, (3.17) holds. Suppose there exists a natural n so that

$$f_0^2 + f_1^2 + \cdots + f_{n-1}^2 = f_{n-1} \cdot f_n.$$

Add f_n^2 to the both sides. Then

$$f_0^2 + f_1^2 + \cdots + f_n^2 = f_{n-1} \cdot f_n + f_n^2 = f_n(f_{n-1} + f_n) = f_n \cdot f_{n+1}.$$

(vii) We prove (3.18) by induction on n. For $n = 0$, we have $f_m = f_m$, which is obvious. Suppose there exists a natural n so that (3.18) holds and we prove it for $n + 1$. Then

$$\sum_{k=0}^{n+1} \binom{n+1}{k} f_{m+k} = f_m + \sum_{k=1}^{n} \left(\binom{n}{k-1} + \binom{n}{k} \right) f_{m+k} + f_{m+n+1}$$

$$= f_m + \sum_{k=1}^{n} \binom{n}{k-1} f_{m+k} + \sum_{k=1}^{n} \binom{n}{k} f_{m+k} + f_{m+n+1}$$

$$= \sum_{k=0}^{n} \binom{n}{k} f_{m+k+1} - f_{m+n+1} + \sum_{k=0}^{n} \binom{n}{k} f_{m+k} + f_{m+n+1}$$

$$= f_{m+2n+1} - f_{m+n+1} + f_{m+2n} + f_{m+n+1} = f_{m+2(n+1)}.$$

(viii) The terms of the sequence (u_n) are positive and greater then one for $n \geq 2$ because

$$u_{n+1} = f_{n+2}/f_{n+1} = (f_{n+1} + f_n)/f_{n+1} = 1 + 1/u_n.$$

Thus $1 < u_n < 2$, whenever $n \geq 3$. From

$$u_{n+1} - u_n = f_{n+2}/f_{n+1} - f_{n+1}/f_n = (-1)^{n+1}/(f_n f_{n+1})$$

it follows that the sequence (u_n) is nonmonotone and that for $n \geq 5$, $|u_{n+1} - u_n| < 1/(n(n+1))$. Then

$$|u_{n+k} - u_n| \le |u_{n+k} - u_{n+k-1}| + |u_{n+k-1} - u_{n+k-2}| + \cdots + |u_{n+1} - u_n|$$
$$< \frac{1}{(n+k)(n+k-1)} + \frac{1}{(n+k-1)(n+k-2)} + \cdots + \frac{1}{n(n+1)}$$
$$= 1/n - 1/(n+k) < 1/n.$$

Thus (u_n) is a Cauchy sequence, hence it converges. □

Remark. There is a geometric interpretation of (3.17). Consider squares of sides $f_0 = 1, f_1 = 1, f_2 = 2, f_3 = 3, f_4 = 5, f_5 = 8$, and $f_6 = 13$ and arrange them as shown in Figure 3.2. This arrangement is possible thanks to (3.17). Its left-hand side is a sum of areas of squares and the right-hand side is an area of a rectangle.

Figure (3.3) shows how one can draw a snail by arcs of circles. △

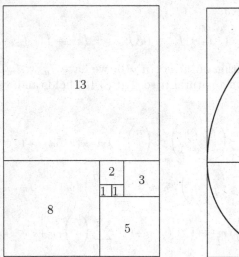
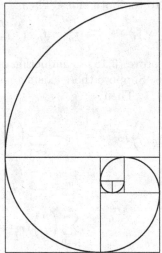

Fig. 3.2. Fibonacci numbers as length of sides

Fig. 3.3. The snail by arcs of circles

3.1.9 Unimodal, log-convex, and Pólya-frequency sequences

Let $a_0, a_1, a_2, \ldots$ be a sequence of nonnegative real numbers. It is *unimodal* if there is an index κ so that

$$a_0 \le a_1 \le \cdots \le a_\kappa \ge a_{\kappa+1} \ge \cdots .$$

It is said to be *concave* if

$$a_{k-1} + a_{k+1} \leq 2a_k, \quad \forall k \in \mathbb{N}^*.$$

It is said to be *convex* if

$$a_{k-1} + a_{k+1} \geq 2a_k, \quad \forall k \in \mathbb{N}^*.$$

It is said to be *log-concave* if

$$a_{k-1}a_{k+1} \leq a_k^2, \quad \forall k \in \mathbb{N}^*.$$

It is said to be *log-convex* if

$$a_{k-1}a_{k+1} \geq a_k^2, \quad \forall k \in \mathbb{N}^*.$$

It is said to be a *Pólya*[9] *frequency sequence* if for the matrix $(a_{i-j})_{i,j \geq 0}$ (where $a_k = 0$, if $k < 0$) all minors have nonnegative determinants. A finite sequence $a_0, a_1, \ldots, a_n$ is a Pólya frequency sequence if $a_0, a_1, \ldots, a_n, 0, 0, \ldots$ is a Pólya frequency sequence.

Theorem 1.25. *The sequence of binomial coefficient* $a_k = \binom{n}{k}$, $k = 0, 1, \ldots,$ n *is unimodal and* $\kappa = \lfloor n/2 \rfloor$.

Proof. The ratio of two consecutive terms is $\binom{n}{k} / \binom{n}{k-1} = (n-k+1)/k$. If $k \leq \lfloor n/2 \rfloor$, because $k < (n+1)/2$, it follows that

$$(n-k+1)/k > 1 \implies \binom{n}{k} / \binom{n}{k-1} > 1.$$

If $k > \lfloor n/2 \rfloor$, because $k - 1 \geq (n-1)/2$, it follows that

$$(n-k+1)/k \leq 1 \implies \binom{n}{k} / \binom{n}{k-1} \leq 1. \quad \square$$

Theorem 1.26. *For fixed* $n \in \mathbb{N}^*$ *the sequences* $\left(\binom{n}{k}\right)_{0 \leq k \leq n}$ *of binomial coefficients,* $\left(\langle\begin{smallmatrix}n\\k\end{smallmatrix}\rangle\right)_{0 \leq k \leq n}$ *of Euler numbers of the first kind,* $\left(\begin{bmatrix}n\\k\end{bmatrix}\right)_{0 \leq k \leq n}$ *of Stirling numbers of the first kind, and* $\left(\begin{Bmatrix}n\\k\end{Bmatrix}\right)_{0 \leq k \leq n}$ *of Stirling numbers of the second kind are log-concave.*

Proposition 1.7. *A log-concave sequence* $(a_n)_{n \geq 0}$ *is unimodal.*

Proof. Suppose $a_{m-1} > a_m < a_{m+1}$, for some $m \in \mathbb{N}^*$. Then $a_{m-1}a_{m+1} < a_m^2$, contradicting the hypothesis. $\square$

Remarks. (a) Unimodality does not imply log-concavity.
(b) A monotonic sequence is unimodal. $\triangle$

---

[9] György(=George) Pólya, 1887–1985.

Proposition 1.8. *A Pólya frequency sequence is log-concave.*

Proposition 1.9. *Suppose* (a_n) *has positive terms and define* $b_n = a_{n+1}/a_n$. *Then* (a_n) *is log-concave if and only if* (b_n) *is decreasing.*

Consider a finite sequence $a_0, a_1, \ldots, a_n$ and $A(x) = a_0 + \sum_{i=1}^{n} a_i x^i$ its generating polynomial. It follows in a sufficient condition for log-concavity.

Theorem 1.27. (Newton) *Given is a finite sequence* $a_0, a_1, \ldots, a_n$ *of nonnegative numbers. Suppose that its generating polynomial* A *has only real roots. Then*

$$a_k^2 \geq \left(\frac{k+1}{k}\right)\left(\frac{n-k+1}{n-k}\right) a_{k-1}a_{k+1},$$

and in particular, the sequence is log-concave.

A proof of this theorem based on elementary differential calculus is supplied at page 209.

Given two sequences (a_n) and (b_n), define their *ordinary convolution* (c_n) and *binomial convolution* (d_n) by

$$c_n = \sum_{k=0}^{n} a_k b_{n-k}, \text{ respectively, } d_n = \sum_{k=0}^{n} \binom{n}{k} a_k b_{n-k}, \quad n \in \mathbb{N}.$$

Proposition 1.10. *Suppose* (a_n) *and* (b_n) *have nonnegative terms and they are log-concave. Then their ordinary convolution* (c_n) *and binomial convolution* (d_n) *are log-concave.*

The Pólya frequency sequences are completely characterized by the next result.

Theorem 1.28. *A finite sequence* $a_0, a_1, \ldots, a_n$ *of nonnegative numbers is a Pólya frequency sequence if and only if its generating polynomial* $A(x) = a_0 + \sum_{i=1}^{n} a_i x^i$ *has only real roots.*

Theorem 1.29. *Let* $a_0, a_1, \ldots, a_{n-1}$ *be a Pólya frequency sequence and*

$$b_k = (a + c(k-1))a_{k-1} + (b + dk)a_k, \quad k = 0, 1, \ldots, n,$$

where $a_{-1} = a_n = 0$. *If* $ad \geq bc$ *and all* b_k *are nonnegative, then* $b_0, b_1, \ldots, b_n$ *is a Pólya frequency sequence.*

Consider a triangular array $\{a_{n,k}\}$ satisfying the recurrence

$$a_{n,k} = (rn + sk + t)a_{n-1,k-1} + (an + bk + c)a_{n-1,k}$$

for $n \geq 1$, $0 \leq k \leq n$, where $a_{n,k} = 0$ unless $0 \leq k \leq n$, and r, s, t, a, b, c are real numbers.

Theorem 1.30. *Let $\{a_{n,k}\}$ be a triangular array as defined earlier. Suppose that $rb \geq as$ and $(r+s+t)b \geq (a+c)s$. Then for each $n \geq 1$ the sequence $(a_{n,k})_k$ is a Pólya frequency sequence.*

We introduce some results on log-convex sequences.

Proposition 1.11. *If both (a_n) and (b_n) are log-convex, then so is the sequence $(a_n + b_n)$.*

Proof. By the Cauchy inequality and the log-convexity of (a_n) and (b_n) we have

$$(a_{k-1} + b_{k-1})(a_{k+1} + b_{k+1}) \geq (\sqrt{a_{n-1}a_{n+1}} + \sqrt{b_{n-1}b_{n+1}})^2 \geq (a_n + b_n)^2. \quad \square$$

Proposition 1.12. *Suppose (a_n) has positive terms and define $b_n = a_{n+1}/a_n$. Then (a_n) are log-convex if and only if (b_n) is increasing.*

Proposition 1.13. *Suppose (a_n) and (b_n) have nonnegative terms and are log-convex. Then their binomial convolution (d_n) is log-convex.*

Proof. We prove this result by induction. It is obvious that

$$d_0 d_2 = (a_0 b_0)(a_0 b_2 + 2a_1 b_1 + a_2 b_0) \geq (a_0 b_1 + a_1 b_0)^2 = d_1^2.$$

Suppose n is a positive integer at least 2. Then

$$d_n = \sum_{k=0}^{n-1} \binom{n-1}{k} a_k b_{n-k} + \sum_{k=0}^{n-1} \binom{n-1}{k} a_{k+1} b_{n-k-1}.$$

The first sum in the right-hand side is the binomial convolution of $(a_k)_{0 \leq k \leq n-1}$ with $(b_k)_{0 \leq k \leq n}$, thus log-convex. The second sum in the right-hand side is the binomial convolution of $(a_k)_{0 \leq k \leq n}$ with $(b_k)_{0 \leq k \leq n-1}$, thus log-convex. Hence by Proposition 1.11, their sum is also. $\quad \square$

Corollary 1.8. *Suppose (a_n) has nonnegative terms and is log-convex. Its binomial transform*

$$b_n = \sum_{k=0}^{n} \binom{n}{k} a_k$$

preserves the log-convexity.

Similar to Theorem 1.30 the next result holds.

Theorem 1.31. *Let $\{a_{n,k}\}$ be a triangle of nonnegative numbers. If $a_{n,k}$ is log-convex in n for each fixed k, then the linear transform $y_n = \sum_{k=0}^{n} a_{n,k} x_k$ preserves the log-convexity of sequences.*

The theorem states that if $\{a_{n,k}\}$ is a triangle of nonnegative numbers so that $a_{n,k}$ is log-convex in n for each fixed k and (x_k) is a log-convex sequence, then (y_k) is a log-convex sequence.

Proof. Denote $\Delta_n = y_{n-1}y_{n+1} - y_n^2$. Then Δ_n is a quadratic form in $n+2$ variables $x_0, x_1, \ldots, x_{n+1}$. For $0 \leq m \leq 2n$ and $0 \leq k \leq \lfloor m/2 \rfloor$, let a_k be the coefficient of the term $x_k x_{m-k}$ in Δ_n. Then

$$\Delta_n = \sum_{m=0}^{2n} a_k x_k x_{m-k},$$

where

$$a_k = \begin{cases} a_{n-1,k}\, a_{n+1,m-k} + a_{n+1,k}\, a_{n-1,m-k} - 2a_{n,k}\, a_{n,m-k}, & k < m/2 \\ a_{n-1,k}\, a_{n+1,k} - a_{n,k}^2, & k = m/2. \end{cases}$$

It is clear that $a_k \geq 0$ by the log-convexity of $a_{n,k}$ in n and the mean inequality. Thus $\Delta_n \geq 0$ and therefore (y_n) is log-convex. $\square$

3.1.10 Some special sequences

Proposition 1.14. *The following limits hold.*

(a) *If $p > 0$,* $\lim_{n \to \infty} \dfrac{1}{n^p} = 0$.

(b) *If $p > 0$,* $\lim_{n \to \infty} \sqrt[n]{p} = 1$.

(c) $\lim \sqrt[n]{n} = 1$.

(d) $\lim \sqrt[n]{n!} = \infty$.

Proof. (a), (b), (c), and (d) follow by Theorem 1.22 and Corollary 1.6. Other ways to approach the limits are introduced in the sequel.

One can prove using only the definition the limit in (a).

For (b) we can distinguish three cases. If $p = 1$, the result is obvious. Suppose that $p > 1$. Then for n large, $1 < \sqrt[n]{p} < \sqrt[n]{n}$, and the result follows by (a) in Corollary 1.6 and Corollary 1.4. If $p < 1$, we can use the previous case.

For (c) we can follow the next path. Obviously, $\sqrt[n]{n} = 1 + p_n$, with $p_n > 0$, for all $n \geq 2$. Then

$$n = (1 + p_n)^n = 1 + \binom{n}{1}p_n + \binom{n}{2}p_n^2 + \cdots + \binom{n}{n}p_n^n > \binom{n}{2}p_n^2.$$

Thus

$$0 < p_n < \sqrt{2/(n-1)} \longrightarrow 0, \quad \text{as } n \to \infty,$$

and the result follows.

For (d) we can use the inequality between the harmonic mean and geometric mean. Thus we can write

$$\frac{\ln n}{1/1 + 1/2 + \cdots + 1/n} \frac{n}{\ln n} \le \sqrt[n]{n!}.$$

In the left-hand side the first fraction tends to 1 by Corollary 1.11, and the second limit diverges by (c) in Exercise 3.16. Therefore the result follows. $\square$

Proposition 1.15. *Let (a_n) be the sequence defined as*

$$a_n = \left(1 + \frac{1}{n}\right)^n, \quad n \ge 1. \tag{3.19}$$

Then (a_n) is convergent.

Proof. First approach. From Exercise 1.27 at page 35 it follows that (a_n) is bounded.

We now study the monotonicity of the sequence (a_n).

$$a_{n+1} = 1 + 1 + \left(1 - \frac{1}{n+1}\right)\frac{1}{2!} + \cdots$$

$$+ \left(1 - \frac{1}{n+1}\right)\left(1 - \frac{2}{n+1}\right)\cdots\left(1 - \frac{n}{n+1}\right)\frac{1}{(n+1)!}$$

$$> 2 + \left(1 - \frac{1}{n+1}\right)\frac{1}{2!} + \cdots$$

$$+ \left(1 - \frac{1}{n+1}\right)\left(1 - \frac{2}{n+1}\right)\cdots\left(1 - \frac{n-1}{n+1}\right)\frac{1}{n!}$$

$$> 2 + \left(1 - \frac{1}{n}\right)\frac{1}{2!} + \cdots + \left(1 - \frac{1}{n}\right)\left(1 - \frac{2}{n}\right)\cdots\left(1 - \frac{n-1}{n}\right)\frac{1}{n!} = a_n.$$

Thus (a_n) is increasing. By Theorem 1.16, we conclude that (a_n) is convergent.

Second approach. We start with the identity

$$x^{n+1} - y^{n+1} = (x - y)\sum_{i=0}^{n} x^{n-i}y^i,$$

valid for any reals x and y and $n \in \mathbb{N}^*$. For $x > y > 0$, it results

$$(n+1)(x - y)y^n < x^{n+1} - y^{n+1} < (n+1)(x - y)x^n. \tag{3.20}$$

Substitute $x = 1 + 1/n$ and $y = 1 + 1/(n+1)$. Thus we get

$$x - y = 1/(n(n+1)), \quad x^{n+1} = x \cdot a_n, \quad y^{n+1} = a_{n+1}$$

$$x = \frac{n+1}{n} = \frac{x-1}{y-1} > y \implies y^2 < x + y - 1 = 1 + \frac{1}{n} + \frac{1}{n+1} \implies$$

$$y^{n+2} < \left(1 + \frac{1}{n} + \frac{1}{n+1}\right)y^n. \tag{3.21}$$

The right-hand side of (3.20) supplies

$$x \cdot a_n - a_{n+1} < \frac{1}{n} a_n \implies \left(1 + \frac{1}{n}\right) a_n - \frac{1}{n} a_n < a_{n+1} \implies a_n < a_{n+1}.$$

Thus (a_n) is increasing.

Denote $u_n = (1+1/n)^{n+1}$. Then $a_n < u_n$, for every $n \in \mathbb{N}^*$. From (3.21) it follows that

$$u_{n+1} = y^{n+2} < (1 + 1/n + 1/(n+1)) y^n = (1/n + y) y^n,$$

whereas from the left-hand side of (3.20) there results

$$\frac{1}{n} y^n < u_n - y^{n+1} \implies \left(\frac{1}{n} + y\right) y^n < u_n \implies u_{n+1} < \left(\frac{1}{n} + y\right) y^n < u_n.$$

Thus (u_n) is decreasing and

$$a_n < u_n < \cdots < u_5 = (1 + 1/5)^6 = 2,985984\ldots \implies a_n < 3, \quad \forall n \in \mathbb{N}^*.$$

Thus (a_n) is increasing and bounded above, hence it converges. $\square$

As usual, we denote by e the limit of the sequence (a_n); that is,

$$\lim_{n \to \infty} \left(1 + \frac{1}{n}\right)^n = e.$$

Actually, we saw that (a_n) increases, (u_n) decreases, and $a_n < u_n$, for each n. Moreover, $u_n - a_n \to 0$, as $n \to \infty$. Hence

$$\lim u_n = e \text{ and } a_n < e < u_n, \quad \forall n \in \mathbb{N}^*. \tag{3.22}$$

Corollary 1.9. *There hold the estimates*

$$\left(1 + \frac{1}{n}\right)^n < e < \left(1 + \frac{1}{n}\right)^{n+1}, \quad \forall n \in \mathbb{N}^*.$$

Proof. Just read the second part of (3.22). $\square$

The previous result is often used under the following form.

Corollary 1.10. *There hold the estimates*

$$n \ln\left(1 + \frac{1}{n}\right) < 1 < (n+1) \ln\left(1 + \frac{1}{n}\right), \quad \forall n \in \mathbb{N}^*.$$

A finer inequality than the one in the right hand-side of the previous corollary is given by Proposition 3.2 at page 390.

Below we introduce some properties of number e.

Proposition 1.16.

(i) *It holds that*

$$\lim_{n \to \infty} \left(1 + \frac{1}{1!} + \frac{1}{2!} + \frac{1}{3!} + \cdots + \frac{1}{n!} \right) = e.$$

(ii) *e can be written as*

$$e = 1 + \frac{1}{1!} + \frac{1}{2!} + \frac{1}{3!} + \cdots + \frac{1}{n!} + \frac{\theta_n}{n \cdot n!}, \tag{3.23}$$

where $\theta_n \in]0, 1[$.

(iii) *e can be approximated with an accuracy up to* 10^{-6}, *provided* $n \geq 8$.

(iv) *(Euler) e is a nonrational number; that is,* $e \in \mathbb{R} \setminus \mathbb{Q}$.

(v) $\lim \sqrt[n]{n!}/n = 1/e$.

(vi) *Suppose a sequence* (a_n) *is given so that* $\lim a_n = \infty$. *Then*

$$\lim_{n \to \infty} \left(1 + \frac{1}{a_n} \right)^{a_n} = e. \tag{3.24}$$

(vii) *Suppose a sequence* (a_n) *is given so that* $\lim a_n = -\infty$. *Then*

$$\lim_{n \to \infty} \left(1 + \frac{1}{a_n} \right)^{a_n} = e. \tag{3.25}$$

(viii) *Suppose a sequence* (b_n) *is given so that* $b_n \neq 0$, $n \in \mathbb{N}^*$, *and* $\lim b_n = 0$. *Then*

$$\lim_{n \to \infty} (1 + b_n)^{1/b_n} = e.$$

Proof. (i) We have $a_n < 1 + 1 + 1/2! + 1/3! + \cdots + 1/n!$, and the sequence (c_n) defined as

$$c_n = 1 + 1 + \frac{1}{2!} + \frac{1}{3!} + \cdots + \frac{1}{n!}, \quad n \in \mathbb{N}^*$$

is convergent and therefore the following inequality holds,

$$e = \lim_{n \to \infty} a_n \leq \lim_{n \to \infty} c_n. \tag{3.26}$$

On the other hand

$$a_n = \left(1 + \frac{1}{n} \right)^n = 1 + \binom{n}{1}\frac{1}{n} + \binom{n}{1}\frac{1}{n^2} + \cdots + \binom{n}{n}\frac{1}{n^n}$$

$$\geq 2 + \left(1 - \frac{1}{n} \right)\frac{1}{2!} + \cdots + \left(1 - \frac{1}{n} \right)\left(1 - \frac{2}{n} \right)\cdots\left(1 - \frac{k-1}{n} \right)\frac{1}{k!}, \quad \forall k \leq n$$

$$\implies \lim_{n \to \infty} a_n = e \geq 1 + 1 + \frac{1}{2!} + \cdots + \frac{1}{k!}, \quad \forall k \geq 2$$

$$\implies e \geq c_k \implies e \geq \lim_{k \to \infty} c_k. \tag{3.27}$$

Hence, from (3.26) and (3.27), we conclude

$$\lim_{k \to \infty} c_k = e.$$

(ii) By (3.42) we get

$$c_{n+m} - c_n = \frac{1}{(n+1)!} + \frac{1}{(n+2)!} + \cdots + \frac{1}{(n+m)!}$$

$$< \frac{1}{(n+1)!} \sum_{k=0}^{\infty} \frac{1}{(n+2)^k} = \frac{1}{(n+1)!} \frac{n+2}{n+1} < \frac{1}{n \cdot n!}.$$

For fixed n and passing $m \to +\infty$, follows

$$0 < e - c_n < 1/(n \cdot n!).$$

Denote

$$0 < \theta_n = (e - c_n) \cdot n \cdot n! < 1 \qquad\qquad (3.28)$$

and now (3.23) follows.
(iii) The inequality

$$0 < e - c_n < 1/(n \cdot n!) < 10^{-5}$$

is satisfied for any $n \geq 8$. So

$$e \cong 2 + \frac{1}{2!} + \cdots + \frac{1}{8!} \cong 2.71828.$$

(iv) Suppose that $e \in \mathbb{Q}$. Then we may write $e = m/n$, for some $m, n \in \mathbb{N}^*$.
Then

$$e = \frac{m}{n} = 1 + \frac{1}{1!} + \frac{1}{2!} + \frac{1}{3!} + \cdots + \frac{1}{n!} + \frac{\theta_n}{n \cdot n!}, \quad \theta_n \in]0,1[,$$

$$\implies (n)! \cdot m - n! \left(2 + \frac{1}{2!} + \frac{1}{3!} + \cdots + \frac{1}{n!} \right) = \frac{\theta_n}{n}.$$

The last equality is impossible because $\theta_n \in]0,1[$ (the left-hand side of it contains an integer number whereas its the right-hand side contains a noninteger number). Hence, clearly, $e \in \mathbb{R} \setminus \mathbb{Q}$.
(v) We have $\lim \sqrt[n]{n!}/n = \lim \sqrt[n]{n!/n^n}$. Denote $a_n = n!/n^n$ and apply (a) of Corollary (1.6). Then

$$\lim a_{n+1}/a_n = \lim n^{n+1}/(n+1)^{n+1} = 1/e.$$

A different approach is introduced by Proposition 1.19.
(vi) First suppose there is given a strictly increasing and divergent sequence of positive integers (n_k). Then the claim holds because the sequence (n_k) is a subsequence of the sequence of natural numbers.

Let (a_n) be a strictly increasing and divergent sequence of real numbers. Denote by $n_k = \lfloor a_k \rfloor$, the integer part of a_k. We may suppose that (n_k) is strictly increasing and divergent. Then (n_k) and $(n_k + 1)$ are strictly increasing and divergent sequences of positive integers. From

$$1 + 1/(n_k + 1) < 1 + 1/a_k \leq 1 + 1/n_k,$$

we get

$$\left(1 + \frac{1}{n_k + 1}\right)^{n_k} < \left(1 + \frac{1}{a_k}\right)^{a_k} \leq \left(1 + \frac{1}{n_k}\right)^{n_k + 1}.$$

The limits of extreme terms of the above inequality are equal to e, so the limit of the middle term is e, as well. Thus (3.24) is completely proved.

(vii) Denote $b_n = -a_n$. Then

$$\lim_{n \to \infty} \left(1 + \frac{1}{a_n}\right)^{a_n} = \lim_{n \to \infty} \left(1 - \frac{1}{b_n}\right)^{-b_n} = \lim_{n \to \infty} \left(\frac{b_n}{b_n - 1}\right)^{b_n}$$

$$= \lim_{n \to \infty} \left(1 + \frac{1}{b_n - 1}\right)^{b_n - 1} \left(1 + \frac{1}{b_n - 1}\right).$$

We apply (vi) and (3.25) is completely proved.

(viii) It follows from (vi) and (vii). $\square$

Proposition 1.17. θ_n *defined by* (ii) *in Proposition 1.16 satisfies that* $\theta_n \to 1$ *as* $n \to \infty$. *It holds*

$$\lim_{n \to \infty} n \cdot \sin(2\pi e n!) = 2\pi.$$

Proof. From (3.28) it follows that $\theta_n = n \cdot n!(e - c_n)$. Inasmuch as

$$e = 1 + \frac{1}{1!} + \frac{1}{2!} + \frac{1}{3!} + \cdots + \frac{1}{(n+1)!} + \frac{\theta_{n+1}}{(n+1) \cdot (n+1)!},$$

we have

$$\theta_n = nn! \left(\frac{1}{(n+1)!} \frac{\theta_{n+1}}{(n+1) \cdot (n+1)!}\right) = \frac{n}{n+1} + \frac{n\theta_{n+1}}{(n+1)!}.$$

The first fraction tends to 1 and because $\theta_{n+1} \in \,]0, 1[$, the second fraction tends to 0 as n tends to ∞. Thus $\theta_n \to 1$.

Because $2\pi n! c_n$ is a multiple of 2π, we have

$$\sin\left(2\pi n! c_n + \frac{2\pi\theta_n}{n}\right) = \sin\frac{2\pi\theta_n}{n}.$$

Thus

$$\lim_{n\to\infty} n\sin(2\pi en!) = \lim_{n\to\infty} n\sin\frac{2\pi\theta_n}{n} = \lim_{n\to\infty} \frac{\sin\dfrac{2\pi\theta_n}{n}}{\dfrac{2\pi\theta_n}{n}} 2\pi\theta_n = 2\pi.$$

We have used the the fact that $\lim(\sin\alpha_n)/\alpha_n = 1$, provided $\alpha_n \to 0$ as $n \to \infty$. □

Proposition 1.18. *The sequence*

$$a_n = 1 + \frac{1}{2} + \ldots \frac{1}{n} - \ln n, \quad n \geq 1,$$

is decreasing and bounded.

Proof. By Corollary 1.9 it follows that

$$(n+1)(\ln(n+1) - \ln n) > 1 > n(\ln(n+1) - \ln n)$$

$$\implies \frac{1}{n+1} < \ln(n+1) - \ln n < \frac{1}{n}. \tag{3.29}$$

Inequality (3.29) follows from the Lagrange mean value Theorem 2.3 at page 199, as well. Taking $n = 1, 2, \ldots, k$, we get

$$\sum_{n=1}^{k} \frac{1}{n+1} < \ln(k+1) < \sum_{n=1}^{k} \frac{1}{n}. \tag{3.30}$$

Remark that from (3.29) it follows that

$$a_{n+1} - a_n = 1/n + 1 - \ln(n+1) + \ln n < 0,$$

so (a_n) is decreasing.

From (3.30) we get

$$\ln k < \ln(k+1) < \sum_{n=1}^{k} 1/n.$$

Thus, (a_n) is bounded below by 0. Hence our sequence is convergent. Its limit is denoted by $\gamma = 0.5772156649\ldots$ and is called the *Euler–Mascheroni* [10] *constant.*

Remark 1.4 Denote

$$H_n = 1 + 1/2 + \cdots + 1/n, \quad n \in \mathbb{N}^*$$

and call them *harmonic numbers.*

Then Proposition 1.18 claims that

---

[10] Lorenzo Mascheroni, 1750–1800.

$$\lim_{n\to\infty}(H_n - \ln n) = \gamma.$$

Hence for large n, H_n is close to $\ln n + \gamma$. That is, H_n is as close as we want to $\ln n + \gamma$ provided n is large enough. We write

$$H_n \sim \ln n + \gamma, \ n \to \infty. \quad \triangle$$

Corollary 1.11. *One has*

$$\lim_{n\to\infty} \frac{1 + \dfrac{1}{2} + \cdots + \dfrac{1}{n}}{\ln n} = 1.$$

Proof. Pass to the limit in

$$\frac{1 + 1/2 + \cdots + 1/n - \ln n}{\ln n} + 1.$$

We may equally well use the Stolz–Cesaro theorem 1.22. $\quad\square$

Corollary 1.12. *It holds*

$$\lim_{n\to\infty}\left(\frac{1}{n+1} + \frac{1}{n+2} + \cdots + \frac{1}{kn}\right) = \ln k, \quad \forall k \in \{2, 3, \dots\}.$$

Proof. Pass to the limit in

$$\frac{1}{n+1} + \frac{1}{n+2} + \cdots + \frac{1}{kn}$$
$$= \left(1 + \frac{1}{2} + \cdots + \frac{1}{kn} - \ln kn\right) - \left(1 + \frac{1}{2} + \cdots + \frac{1}{n} - \ln n\right) + \ln k, \quad n \geq 1. \quad \square$$

Corollary 1.13. *Denote*

$$a_n = 1 - \frac{1}{2} + \frac{1}{3} - \frac{1}{4} + \cdots + (-1)^{n-1}\frac{1}{n}.$$

Then

$$\lim_{n\to\infty} a_n = \ln 2.$$

Proof. Recall the Catalan[11] identity

$$1 - \frac{1}{2} + \frac{1}{3} - \frac{1}{4} + \cdots - \frac{1}{2n} = \frac{1}{n+1} + \frac{1}{n+2} + \cdots + \frac{1}{2n}.$$

Then

$$a_{2n} = \frac{1}{n+1} + \frac{1}{n+2} + \cdots + \frac{1}{2n}$$
$$= 1 + \frac{1}{2} + \cdots + \frac{1}{2n} - \ln 2n - \left(1 + \frac{1}{2} + \cdots + \frac{1}{n} - \ln n\right) + \ln 2 \to \ln 2. \quad \square$$

---

[11] Eugéne Charles Catalan, 1814–1894.

Lemma 3.2. (Lalescu[12]) *Define*

$$a_n = \sqrt[n+1]{(n+1)!} - \sqrt[n]{n!}.$$

Then $a_n \to 1/e$ as $n \to \infty$.

The result follows from the next two propositions.

Proposition 1.19. *The sequence $x_n = n/\sqrt[n]{n!}$, $n > 3$, satisfies the inequalities*

$$e^{1-1/\sqrt{n}} < x_n < e^{1-1/n}. \tag{3.31}$$

Proof. We prove them by induction. For $n > 3$ the next inequality holds

$$x_n > e^{1-1/\sqrt{n}}. \tag{3.32}$$

Indeed, because $e < \sqrt{32/3}$ and $x_4 = \sqrt[4]{32/3}$, it follows that (3.32) holds for $n = 4$.

Suppose that (3.32) holds for some $n \geq 4$. Then

$$x_{n+1} = x_n^{n/(n+1)}\left(\frac{n+1}{n}\right)^{n/(n+1)} > \left(e^{1-1/\sqrt{n}}\right)^{n/(n+1)}\left(\left(1+\frac{1}{n}\right)^{n+1}\right)^{n/(n+1)^2}$$

$$> e^{(n-\sqrt{n})/(n+1)}e^{n/(n+1)^2} = e^{1-(1+(n+1)\sqrt{n})/(n+1)^2} \geq e^{1-1/\sqrt{n+1}}.$$

Thus

$$x_{n+1} > e^{1-1/\sqrt{n+1}},$$

and (3.32) is completely proved.

Now we suppose that for some $n \geq 4$, $x_n < e^{1-1/n}$. Then applying Corollary 1.9, we get

$$x_{n+1}^{n+1} = x_n^n \cdot \frac{(n+1)^n}{n^n} < \frac{(n+1)^n}{n^n} e^{n-1} < e^n \implies x_{n+1} < e^{1-1/(n+1)},$$

ending the proof in this way.   □

Proposition 1.20. *For every $\varepsilon > 0$ the next inequality is satisfied*

$$|a_n - 1/e| < \varepsilon$$

for every $n > n_\varepsilon$, where

$$n_\varepsilon = 1 + \lfloor 8 + 1/(2\varepsilon^2) + |8 - 1/(2\varepsilon^2)| \rfloor.$$

---

[12] Traian Lalescu, 1882–1929.

Proof. By (3.31) and Corollary 1.9, for $n \geq 16$, we get

$$a_n = \frac{n(\sqrt[n]{x_{n+1}} - 1)}{x_n} < \frac{n(e^{1/(n+1)} - 1)}{x_n} < \frac{n(e^{1/(n+1)} - 1)e^{1/\sqrt{n}}}{e}$$

$$< 1/e\, e^{1/\sqrt{n}} < 1/e + 1/\sqrt{n}.$$

At the same time

$$a_n = \frac{n(\sqrt[n]{x_{n+1}} - 1)}{x_n} > \frac{n(\sqrt[n]{x_{n+1}} - 1)}{e^{1-1/n}} = \frac{n}{e}\left(\sqrt[n]{x_{n+1}} - 1\right)e^{1/n}$$

$$> \frac{n}{e}\left(\sqrt[n]{x_{n+1}} - 1\right) > \frac{n}{e}\left(e^{(\sqrt{n+1}-1)/(n\sqrt{n+1})} - 1\right) > \frac{1}{e} - \frac{1}{\sqrt{n}}.$$

Thus, for $n \geq 16$, we have $|a_n - 1/e| < 1/\sqrt{n}$. Because

$$n_\varepsilon = 1 + [\max\{16, 1/\varepsilon^2\}] > \max\{16, 1/\varepsilon^2\},$$

for $n \geq n_\varepsilon$ we have $1/\sqrt{n} < \varepsilon$. Hence

$$|a_n - 1/e| < \varepsilon, \quad \forall n \geq n_\varepsilon. \quad \square$$

Proposition 1.21. *Given*

$$a_n = \sum_{k=0}^{n} 1\Big/\binom{n}{k}, \quad n \in \mathbb{N}. \tag{3.33}$$

Then

$$\lim_{n \to \infty} a_n = 2. \tag{3.34}$$

Proof. Claim 1. The sequence $(b_{n,k})_k = \left(\binom{n}{k}\right)_k$ of binomial coefficients is unimodal (Theorem 1.25) and because $\binom{n}{k+1} = (n-k)/(k+1)\binom{n}{k}$ we have that $(n-k)/(k+1) \geq 1 \iff k \leq (n-1)/2$.
Claim 2. For $n \geq 2$, one has

$$2 \leq a_n \leq 2 + \frac{2}{\binom{n}{1}} + \sum_{k=2}^{n-2} \frac{1}{\binom{n}{2}} = 2 + \frac{2}{n} + \frac{2(n-3)}{n(n-1)}. \tag{3.35}$$

Thus for n large enough, $2 \leq a_n \leq 3$.
Claim 3. For $n \geq 3$, one has

$$a_{n+1} \leq a_n. \tag{3.36}$$

Indeed, we have $1/\binom{n}{k} \geq 1/\binom{n+1}{k+1}$ and

$$\frac{1}{\binom{n}{1}} + \frac{1}{\binom{n}{2}} = \frac{n+1}{n(n-1)} \geq \frac{n^2+n+4}{n(n-1)(n+1)} = \frac{1}{\binom{n+1}{1}} + \frac{1}{\binom{n+1}{2}} + \frac{1}{\binom{n+1}{3}}.$$

Hence

$$a_n = 1 + \frac{1}{\binom{n}{1}} + \frac{1}{\binom{n}{2}} + \sum_{k=3}^{n} \frac{1}{\binom{n}{k}}$$

$$\geq \frac{1}{\binom{n+1}{1}} + \frac{1}{\binom{n+1}{2}} + \frac{1}{\binom{n+1}{3}} + \sum_{k=3}^{n} \frac{1}{\binom{n+1}{k+1}} = a_{n+1}.$$

Claim 4. For $n \geq 1$, one has

$$a_{n+1} = 1 + \frac{n+2}{2(n+1)} a_n. \tag{3.37}$$

Note that (3.36) follows from (3.37).

We have

$$(2n+2)a_{n+1} = 2n + 2 + 2\sum_{k=0}^{n} \frac{k!(n+1-k)!}{n!}$$

$$= 2n + 2 + \sum_{k=0}^{n} \frac{k!(n+1-k)!}{n!} + \sum_{m=0}^{n} \frac{(m+1)!(n-m)!}{n!}$$

$$= 2n + 2 + \sum_{k=0}^{n} \frac{k!(n+1-k)!}{n!} ((n+1-k) + (k+1)) = 2n + 2 + (n+2)a_n.$$

Thus (3.37) follows.

From Claims 2 and 3 we have that (a_n) is convergent. Denote $a = \lim a_n$. Passing to the limit in (3.37), we get $a = 2$.   □

3.2 Sequences of functions

In the previous section we have introduced several results concerning the convergence of numerical sequences. Here we discuss analogous exercises concerning the convergence of sequences of functions.

Consider a sequence $(f_n)_n = (f_0, f_1, \dots)$ of real-valued functions defined (all of them) on a set $A \subset \mathbb{R}$, $A \neq \emptyset$. We say that the *sequence (f_n) converges* on a point $t \in A$ provided the numerical sequence $(f_n(t))_n$ converges. It means that there exists a real number p such that for every $\varepsilon > 0$ there exists $n_\varepsilon > 0$ so that for all $n > n_\varepsilon$ one has $|f_n(t) - p| < \varepsilon$.

Suppose $f_n : A \to \mathbb{R}$, $n \in \mathbb{N}$ and for every $t \in A$, (f_n) converges on A. Then the set of limit points defines a function (we denote it as f) having A its domain of definition in the following way,

$$\lim_{n \to \infty} f_n(t) = f(t).$$

The limit is unique, so the function f is well defined. The function f is said to be the *pointwise limit* of the sequence (f_n). We denote this kind of convergence by $f_n \to f$ $(n \to \infty.)$ By the $\varepsilon - \delta$ language we can express the same concept as there exists a function $f : A \to \mathbb{R}$ such that for every $\varepsilon > 0$ and $t \in A$ there exists $n_{t,\varepsilon} \in \mathbb{N}$ so that for all $n > n_{t,\varepsilon}$, $n \in \mathbb{N}$, one has $|f_n(t) - f(t)| < \varepsilon$.

If in the above definition $n_{t,\varepsilon}$ does not depend on t, we get the *uniform convergence* of the sequence (f_n). The function f, its limit, is said to be the *uniform limit* of the sequence (f_n). We denote this kind of convergence by $f_n \xrightarrow{\;u\;} f$ $(n \to \infty)$.

Remark. The uniform convergence implies the pointwise convergence. The converse implication generally does not hold.   △

Examples 2.1. (a) Consider the sequence $f_n : [0,1[\to \mathbb{R}$ defined by $f_n(t) = t^n$, $n \in \mathbb{N}^*$. It can easily be settled that the pointwise limit of this sequence is the constant zero function on $[0,1[$, that is, $f : [0,1[\to \mathbb{R}$ defined by $f(t) = 0$.
(b) Consider the previous sequence of functions, but this time its terms are defined on the compact interval $[0,1]$. Then the pointwise limit of this sequence is

$$f(t) = \begin{cases} 0, & 0 \le t < 1 \\ 1, & t = 1. \end{cases}$$

(c) The convergence in (b) is only pointwise because from $0 < t^n < \varepsilon < 1$, we have the equivalent inequalities

$$n \ln t < \ln \varepsilon \Longleftrightarrow n > \ln(1/\varepsilon)/\ln(1/t). \tag{3.38}$$

If $t \uparrow 1$, then $\ln(1/t) \downarrow 0$. Thus the right-hand side in (3.38) is unbounded.
(d) Consider the sequence of functions

$$f_n(t) = \frac{t}{1 + n^2 t^2}, \quad t \in [0,1].$$

It is clear that its pointwise limit is the null function on $[0,1]$. Moreover, because

$$0 \le f_n(t) \le \frac{1}{2n} \cdot \frac{2nt}{1 + n^2 t^2} \le \frac{1}{2n},$$

we conclude that the sequence of functions (f_n) converges uniformly toward the null function on $[0,1]$.   △

Proposition 2.22. *Suppose* $f_n : A \to \mathbb{R}$, $\emptyset \ne A \subset \mathbb{R}$, *and for every* $t \in A$, $f(t) = \lim_{n\to\infty} f_n(t)$. *If* $a_n = \sup_{t\in A} |f_n(t) - f(t)|$ *and* $\lim_{n\to\infty} a_n = 0$, *the sequence* $(f_n)_n$ *converges uniformly on* A *to* f.

Proof. Because $\lim_{n\to\infty} a_n = 0$, for every ε there exists a natural number n_ε so that $|a_n| < \varepsilon$ for every $n > n_\varepsilon$. Thus $\sup_{t\in A} |f_n(t) - f(t)| < \varepsilon$ for every $n > n_\varepsilon$. This means that the sequence $(f_n)_n$ converges uniformly on A.   □

Similar to Theorem 1.7 at page 77 one has the next result.

Theorem 2.1. (General criterion of Cauchy) *Suppose $f_n : A \to \mathbb{R}$, $A \neq \emptyset$. The sequence $(f_n)_n$ converges uniformly on A if and only if for every $\varepsilon > 0$ there exists $n_\varepsilon \in \mathbb{N}$ such that for any $n, p \in \mathbb{N}$, $n \geq n_\varepsilon$, and $t \in A$,*

$$|f_{n+p}(t) - f_n(t)| < \varepsilon.$$

Proof. Suppose (f_n) converges uniformly on A to some function f. Set an arbitrary positive ε. We can find a natural n_ε such that for every natural $n \geq n_\varepsilon$ and every $t \in A$, it holds $|f_n(t) - f(t)| < \varepsilon/2$. Because $n + p \geq n_\varepsilon$ we have $|f_{n+p}(t) - f(t)| < \varepsilon/2$ for every $t \in A$. Thus

$$|f_{n+p}(t) - f_n(t)| \leq |f_{n+p}(t) - f(t)| + |f_n(t) - f(t)| < \varepsilon.$$

Suppose now that for every $\varepsilon > 0$ there exists $n_\varepsilon \in \mathbb{N}$ such that for any $n, p \in \mathbb{N}$, $n \geq n_\varepsilon$, and any $t \in A$, $|f_{n+p}(t) - f_n(t)| < \varepsilon/2$. We show that the sequence $(f_n)_n$ converges uniformly on A. Indeed, for every $t \in A$ the sequence $(f_n(t))_n$ is fundamental, hence convergent. We denote by $f(t)$ its limit. For p large enough, eventually depending on t and n, we have $|f_{n+p}(t) - f(t)| < \varepsilon/2$. Therefore

$$|f_n(t) - f(t)| < \varepsilon, \quad \forall n \geq n_\varepsilon, \quad t \in A.$$

This means precisely the sequence $(f_n)_n$ converges uniformly on A.   □

The above given definition of uniform convergence can be changed for more general spaces. Let E be a subset of a metric space and (f_n) a sequence of real-valued functions defined on E. The sequence (f_n) *converges uniformly on E to a function f* if for every ε there is a natural n_ε such that $n > n_\varepsilon$ implies

$$|f_n(x) - f(x)| < \varepsilon, \quad \forall x \in E.$$

3.3 Numerical series

All the results under consideration in the present section are complex-valued, unless the contrary is explicitly stated.

Given a sequence $(x_n)_{n \geq 1}$ (sometimes $(x_n)_{n \geq 0}$) we use the notation $\sum_{n=p}^{q} x_n$ $(p \leq q)$ to denote the finite sum $x_p + x_{p+1} + \cdots + x_q$. With $(x_n)_{n \geq 1}$ we associate a sequence (s_n) of *partial sums*, where $s_n = \sum_{k=1}^{n} x_k$. The pair $((x_n), (s_n))$ formed by a sequence and the sequence of partial sums associated with it is said to be a *numerical series* or a *series*. An older but still largely used notation is the symbolic expression

$$x_1 + x_2 + \cdots,$$

or, more concisely,

$$\sum_{n=1}^{\infty} x_n, \text{ or even } \sum x_n.$$

If (s_n) converges, say to s, then the series is said to *converge* and we write

$$\sum_{n=1}^{\infty} x_n = s.$$

The number s is said to be the *sum of the series*; but it should be clearly understood that s is the limit of the sequence of its partial sums, and not obtained simply by (indefinite) addition.

If (s_n) diverges, the series is said to *diverge*.

Given a convergent series $\sum a_n = s$ and let (s_n) be the sequence of partial sums. Then $r_n = s - s_{n-1} = \sum_{k=n}^{\infty} a_k$ is said to be the nth order *remainder*. A series $\sum a_n$ converges if and only if $r_n \to 0$.

Sometimes, for convenience of notation, we consider series of the form $\sum_{n=0}^{\infty} x_n$.

Theorem 3.1. (General criterion of Cauchy) $\sum x_n$ *converges if and only if for every $\varepsilon > 0$ there is a positive integer n_ε such that*

$$\left| \sum_{k=n}^{m} x_k \right| < \varepsilon \text{ if } m \geq n \geq n_\varepsilon. \tag{3.39}$$

Proof. Apply Theorem 1.7 at page 77 or Corollary 1.2 to (s_n).   □

In particular, for $m = n$, (3.39) becomes

$$|x_n| < \varepsilon, \quad n \geq n_\varepsilon,$$

that is, a necessary condition stated as follows.

Theorem 3.2. (Necessary condition) *If $\sum x_n$ converges, then $\lim x_n = 0$.*

Remark. The condition $x_n \to 0$ is not, however, sufficient ensuring convergence of $\sum x_n$. For instance, the series $\sum_1^{\infty} 1/n$ diverges as it results from Exercise 3.2 or it is made clear by Theorem 4.5.   △

We introduce some operations with convergent series.

Theorem 3.3. *Let $\sum a_n$ and $\sum b_n$ be two convergent series.*

(a) *If p is a real or a complex number, then the series $\sum p \cdot a_n$ and $\sum (a_n \pm b_n)$ are convergent and, moreover,*

$$\sum p \cdot a_n = p \sum a_n$$

and

$$\sum (a_n \pm b_n) = \sum a_n \pm \sum b_n. \tag{3.40}$$

(b) (Abel) *Moreover, denote $c_n = a_1 b_n + \cdots + a_n b_1$ and suppose $\sum c_n$ is convergent. Denote*
    (b₁) $\sum a_n = A$.

(b$_2$) $\sum b_n = B.$

(b$_3$) $\sum c_n = C.$

Then $C = AB.$

(c) (Mertens [13]) *Suppose*

(c$_1$) *The series* $\sum |a_n|$ *converges.*

(c$_2$) $\sum a_n = A.$

(c$_3$) $\sum b_n = B.$

(c$_4$) $c_n = \sum_{k=1}^{n} a_n b_{n+1-k}.$

Then

$$\sum c_n = AB.$$

(d) (Cauchy) *Suppose the series* $\sum |a_n|$ *and* $\sum |b_n|$ *converge. Then* $\sum |c_n|$, *with c_n in (c), converges.*

Proof. (a) Indeed,

$$p \sum a_n = p \lim_{n \to \infty} \sum_{k=1}^{n} a_k = \lim_{n \to \infty} \sum_{k=1}^{n} p\, a_k = \sum p\, a_n$$

and

$$\sum a_n \pm \sum b_n = \lim \sum^{n} a_k \pm \lim \sum^{n} b_k = \lim \left(\sum^{n} a_k \pm \sum^{n} b_k \right) = \sum (a_n \pm b_n).$$

(b) Denote

$$A_n = \sum_{k=1}^{n} a_k, \quad B_n = \sum_{k=1}^{n} b_k, \quad \text{and } C_n = \sum_{k=1}^{n} c_k.$$

We have

$$C_n = a_1 B_n + \cdots + a_n B_1,$$

from where

$$C_1 + \cdots + C_n = A_1 B_n + \cdots + A_n B_1, \quad \frac{C_1 + \cdots + C_n}{n} = \frac{A_1 B_n + \cdots + A_n B_1}{n}.$$

Apply (g) in Exercise 3.16 and note that the right-hand side (and implicitly the left-hand side, too) of the second equality tends to AB. By Corollary 1.5, the conclusion follows.

(c) From the necessary condition Theorem 3.2 it follows that $\lim a_n = \lim b_n = 0$. From (g) in Exercise 3.16 we have that the necessary condition is satisfied; that is $\lim c_n = 0$. Denote by A_n, B_n, and C_n the partial sums of the series $\sum a_n$, $\sum b_n$, respectively, $\sum c_n$. Let $d_n = B_n - B$. Then $\lim d_n = 0$. C_n can be written as

$$C_n = A_n B + a_1 d_n + a_2 d_{n-1} + \cdots + a_{n-1} d_2 + a_n d_1.$$

---

[13] Franz Mertens, 1840–1927.

Note that $A_n B \to AB$ and because $\lim a_n = \lim d_n = 0$, one has

$$\lim(a_1 d_n + a_2 d_{n-1} + \cdots + a_{n-1} d_2 + a_n d_1) = 0.$$

Hence $C_n \to AB$ and thus (c) is proved.
(d) Indeed,

$$|c_n| \le |a_1| \, |b_n| + \cdots + |a_n| \, |b_1|,$$

so the general term of $\sum |c_n|$ is less than or equal to the general term of the product of $\sum |a_n|$ and $\sum |b_n|$. This product series converges by (c). $\quad\Box$

Remarks. From (a) it follows that the set of convergent series is a vector space.

The assumption in (a) that both series are convergent is essential. Otherwise it could happen that the left-hand side of (3.40) exists whereas the right-hand side of it has no meaning. For,

$$0 = \sum(1 - 1) \text{ versus } \sum 1 - \sum 1. \quad \triangle$$

Similar to finite sums, we may group the terms of a convergent series in brackets (but no commutativity is allowed). For example,

$$a_1 + (a_2 + a_3) + (a_4 + a_5 + a_6 + a_7) + \ldots.$$

Theorem 3.4. *Let $\sum a_n$ be a convergent series. Then grouping its terms it results in a convergent series having the same sum.*

Proof. The sequence of partial sums of the transformed series is a subsequence of the convergent sequence of partial sums to the original series. $\quad\Box$

3.3.1 Series of nonnegative terms

We mean that $a_n \ge 0$, for every n.

Theorem 3.5. *A series of nonnegative terms converges if and only if its partial sums form a bounded sequence.*

Proof. The sequence of partial sums is increasing. It is convergent if and only if is bounded, according to Theorem 1.16. $\quad\Box$

Theorem 3.6. *Let $\sum a_n$ be a convergent series with nonnegative terms. Then the series $\sum b_n$ obtained from the former by rearranging (commuting) and renumbering its terms is also convergent having the same sum.*

Proof. Let s be the sum of the first series and s_n its partial sum of rank n. Denote by p_n the partial sum of rank n of the series $\sum b_n$. Fix a rank, let it be n, and consider p_n. Then

$$p_n = b_1 + \cdots + b_n = a_{k_1} + \cdots + a_{k_n}$$

and denote $N = \max\{k_1, \ldots, k_n\}$. Obviously, $p_n \leq s_N \leq s$. Thus we conclude that the series $\sum b_n$ converges and, if b is its sum, $b \leq s$. Reasoning viceversa, we get that $\sum a_n = \sum b_n$. $\square$

Theorem 3.7. (Comparison test) (a) *If $|x_n| \leq c_n$ for all $n \geq n_0$, where n_0 is some fixed integer, and if $\sum c_n$ converges, then $\sum x_n$ converges.*
(b) *If $x_n \geq y_n \geq 0$ for all $n \geq n_0$, where n_0 is some fixed integer, and if $\sum y_n$ diverges, then $\sum x_n$ diverges.*
(c) *Suppose $x_n, y_n > 0$ for all n large enough and*

$$\lim_{n \to \infty} \frac{x_n}{y_n} = k \in [0, \infty].$$

If $\sum y_n$ converges and $k < \infty$, then $\sum x_n$ converges. If $\sum y_n$ diverges and $k > 0$, then $\sum x_n$ diverges.
(d) *Suppose $x_n, y_n > 0$ and*

$$\frac{x_{n+1}}{x_n} \leq \frac{y_{n+1}}{y_n}, \quad \forall n \text{ large enough.} \tag{3.41}$$

If $\sum y_n$ converges, then $\sum x_n$ converges. If $\sum x_n$ diverges, then $\sum y_n$ diverges.

Proof. (a) Given $\varepsilon > 0$ there exists $n_\varepsilon \geq n_0$ such that $m \geq n \geq n_\varepsilon$ implies

$$\sum_{k=n}^{m} c_k \leq \varepsilon,$$

by the Cauchy criterion. Hence

$$\left| \sum_{k=n}^{m} x_k \right| \leq \sum_{k=n}^{m} |x_k| \leq \sum_{k=n}^{m} c_k \leq \varepsilon,$$

and (a) follows.
(b) Follows from (a), for if $\sum x_n$ converges, so must $\sum y_n$.
(c) Suppose $\sum y_n$ converges and $k < \infty$. Then for every $\varepsilon > 0$ there exists $n_\varepsilon \in \mathbb{N}$ such that

$$x_n / y_n < k + \varepsilon, \quad \forall n \geq n_\varepsilon \Longleftrightarrow x_n < (k + \varepsilon) y_n, \quad \forall n \geq n_\varepsilon.$$

The conclusion follows by (a).
    Suppose $\sum y_n$ diverges and $k > 0$. Then

$$\lim_{n \to \infty} \frac{y_n}{x_n} = \frac{1}{k} < \infty.$$

If $\sum x_n$ converges, then by the previous claim $\sum y_n$ converges as well. This contradicts the assumption. Hence $\sum x_n$ diverges.

(d) We may suppose that (3.41) holds from $n = 1$. Then

$$\frac{x_2}{x_1} \le \frac{y_2}{y_1}, \ \frac{x_3}{x_2} \le \frac{y_3}{y_2}, \ \dots, \ \frac{x_n}{x_{n-1}} \le \frac{y_n}{y_{n-1}}.$$

Hence

$$\frac{x_n}{x_1} \le \frac{y_n}{y_1} \iff x_n \le \frac{x_1}{y_1} y_n, \quad \forall n.$$

The conclusions follow by (a) and (b). $\square$

Examples. (a) The series

$$\sum_{n=1}^{\infty} \tan \frac{\pi}{4n}, \quad \sum_{n=1}^{\infty} \ln\left(1 + \frac{1}{2n}\right), \quad \sum_{n=1}^{\infty} \left(a^{1/n} - 1\right), \quad a > 1$$

diverge because

$$\lim_{n \to \infty} \frac{\tan \dfrac{\pi}{4n}}{\dfrac{\pi}{4n}} = 1, \quad \lim_{n \to \infty} \frac{\ln\left(1 + \dfrac{1}{2n}\right)}{\dfrac{1}{2n}} = 1, \quad \lim_{n \to \infty} \frac{(a^{1/n} - 1)}{\dfrac{1}{n}} = \ln a.$$

(b) The series

$$\sum_{n=1}^{\infty} \left(1 - \cos \frac{\pi}{n}\right), \quad \sum_{n=1}^{\infty} \left(e - \left(1 + \frac{1}{n}\right)^n\right)^2$$

converge because

$$\lim_{n \to \infty} \frac{1 - \cos(\pi/n)}{\pi^2/(2n^2)} = 1$$

and

$$\left(e - \left(1 + \frac{1}{n}\right)^n\right)^2 < \left(\left(1 + \frac{1}{n}\right)^{n+1} - \left(1 + \frac{1}{n}\right)^n\right)^2 = \left(1 + \frac{1}{n}\right)^{2n} \frac{1}{n^2} < \frac{e^2}{n^2},$$

$$\lim_{n \to \infty} \frac{\left(e - \left(1 + \frac{1}{n}\right)^n\right)^2}{\dfrac{1}{n^2}} \le e^2. \quad \triangle$$

Theorem 3.8. (Geometric series) *If* $-1 < x < 1$,

$$1 + \sum_{n=1}^{\infty} x^n = \frac{1}{1 - x}. \tag{3.42}$$

If $|x| \ge 1$, *the series diverges.*

Proof. If $x \neq 1$,

$$s_n = \sum_{k=0}^{n} x^k = \frac{1 - x^{n+1}}{1 - x}.$$

Then (3.42) follows letting $n \to \infty$. For $x = 1$, we get $1 + 1 + 1 + \ldots$, which obviously diverges.

For $x = -1$, we get $1 - 1 + 1 - 1 + \cdots + (-1)^n + \ldots$, which diverges.

If $|x| > 1$, the necessary condition is not satisfied.   □

Theorem 3.9. (Cauchy condensation test) *Suppose $x_1 \geq x_2 \geq x_3 \geq \cdots \geq 0$. Then $\sum_{n=1}^{\infty} x_n$ converges if and only if the series*

$$\sum_{k=0}^{\infty} 2^k x_{2^k} = x_1 + 2x_2 + 4x_4 + 8x_8 + \ldots$$

converges.

Proof. By Theorem 3.5 it suffices to consider the boundedness of the partial sums. Let

$$s_n = x_1 + x_2 + \cdots + x_n \text{ and } t_k = x_1 + 2x_2 + \cdots + 2^k x_{2^k}.$$

For $n < 2^k$,

$$s_n = x_1 + \cdots + x_n \overset{\text{add terms}}{\leq} x_1 + (x_2 + x_3) + \cdots + (x_{2^k} + \cdots + x_{2^{k+1}-1})$$
$$\leq x_1 + 2x_2 + \cdots + 2^k x_{2^k} = t_k,$$

so that

$$s_n \leq t_k. \tag{3.43}$$

On the other hand, for $n > 2^k$,

$$s_n = x_1 + \cdots + x_n \overset{\text{neglect terms}}{\geq} x_1 + x_2 + (x_3 + x_4) + \cdots + (x_{2^{k-1}+1} + \cdots + x_{2^k})$$
$$\geq (1/2)x_1 + x_2 + 2x_4 + \cdots + 2^{k-1} x_{2^k} = (1/2)t_k,$$

so that

$$2s_n \geq t_k. \tag{3.44}$$

By (3.43) and (3.44) we have that (s_n) and (t_k) are simultaneously either bounded or unbounded.   □

Theorem 3.10.

$$\sum_{n=1}^{\infty} \frac{1}{n^p} \text{ converges if and only if } p > 1.$$

Proof. If $p \leq 0$, its divergence follows by the necessary condition; that is, Theorem 3.2. If $p > 0$, the previous theorem is applicable, and we are led to the series

$$\sum_{k=0}^{\infty} 2^k \frac{1}{2^{kp}} = \sum_{k=0}^{\infty} 2^{(1-p)k}.$$

Now, $2^{1-p} < 1$ if and only if $1 - p < 0$, and the result follows by comparison with the geometric series (take $x = 2^{1-p}$). $\square$

Remark 3.1. The series

$$1 + 1/2 + 1/3 + 1/4 + \dots$$

is called the *harmonic* series and it diverges as we saw by Theorem 3.10. By Corollary 1.11 we deduce that the speed of divergence of the harmonic series agrees with the speed of divergence of the logarithmic function. $\triangle$

Theorem 3.11. (Logarithmic test) *Let $\sum a_n$ be a series with positive terms. The series converges if there exists a positive α such that*

$$\frac{\ln a_n^{-1}}{\ln n} \geq 1 + \alpha, \quad \forall n \text{ large enough.} \tag{3.45}$$

The series diverges if

$$\frac{\ln a_n^{-1}}{\ln n} \leq 1, \quad \forall n \text{ large enough.} \tag{3.46}$$

Proof. Suppose there exists a positive α such that (3.45) holds. Then

$$a_n \leq 1/n^{1+\alpha}$$

and the conclusion follows taking into account the comparison test and the convergence of the generalized harmonic series.

Suppose (3.46) holds. Then

$$a_n \geq 1/n$$

and the conclusion follows taking into account the comparison test and the divergence harmonic series. $\square$

Example 3.1. *The series with the general term $a_n = (\ln \ln n)^{-\ln n}$, $n \geq 3$, converges because by the logarithm test we have*

$$\frac{\ln a_n^{-1}}{\ln n} = \frac{\ln(\ln \ln n)^{\ln n}}{\ln n} = \ln \ln \ln n > 1 + 1, \quad \forall n \text{ large enough.} \quad \triangle$$

Corollary 3.14. *If $p > 1$,*

$$\sum_{n=2}^{\infty} \frac{1}{n(\ln n)^p} \tag{3.47}$$

converges. Otherwise, the series diverges.

Proof. The logarithmic function increases, hence $1/(n(\ln n)^p)$ decreases. We apply Theorem 3.9 to (3.47). This leads us to the series

$$\sum_{n=1}^{\infty} 2^k \frac{1}{2^k(\ln 2^k)^p} = \frac{1}{(\ln 2)^p} \sum_{k=1}^{\infty} \frac{1}{k^p}. \quad \square$$

Now we introduce the Carleman inequality.

Proposition 3.23. (Carleman[14])

(a) *For every sequence $(a_n)_{n \geq 1}$ with nonnegative terms such that $\sum_{n=1}^{\infty} a_n < \infty$, we have*

$$\sum_{n=1}^{\infty} (a_1 a_2 \cdots a_n)^{1/n} \leq e \sum_{n=1}^{\infty} a_n, \tag{3.48}$$

where the equality in (3.48) holds if and only if $a_n = 0$, $n = 1, 2, \ldots$.
(b) *e in (3.48) is the best constant satisfying it.*

Proof. (a) Put for $n \in \mathbb{N}^*$

$$c_n = (n+1)^n / n^{n-1}.$$

Note that $c_1 c_2 \cdots c_n = (n+1)^n$. Hence

$$(a_1 a_2 \cdots a_n)^{1/n} = \frac{(a_1 c_1 a_2 c_2 \cdots a_n c_n)^{1/n}}{n+1} \leq \frac{a_1 c_1 + a_2 c_2 + \cdots + a_n c_n}{n(n+1)}.$$

Consequently,

$$\sum_{n=1}^{\infty} (a_1 a_2 \cdots a_n)^{1/n} \leq \sum_{n=1}^{\infty} a_n c_n \left(\sum_{m=n}^{\infty} \frac{1}{m(m+1)} \right). \tag{3.49}$$

Because $\sum_{m=n}^{\infty} (1/m(m+1)) = 1/n$ we have

$$\sum_{n=1}^{\infty} a_n c_n \left(\sum_{m=n}^{\infty} \frac{1}{m(m+1)} \right) = \sum_{n=1}^{\infty} \frac{a_n c_n}{n} = \sum_{n=1}^{\infty} a_n((n+1)/n)^n < e \sum_{n=1}^{\infty} a_n.$$

Now invoking (3.49), we get the conclusion.
(b) Constant e in (3.48) is the best constant satisfying it means that if we substitute e by a smaller constant, we can find a sequence (a_n) for which inequality (3.48) is reversed. Choose an $\varepsilon \in \,]0, e[$. Set $a_n = n^{n-1}(n+1)^{-n}$ for $n = 1, 2, \ldots, N$ and $a_n = 2^{-n}$ for $n > N$, where N is settled later. Then

$$(a_1 a_2 \cdots a_n)^{1/n} = 1/(n+1) \tag{3.50}$$

for $n \leq N$. Choose k such that

---

[14] Torsten Carleman, 1892–1949.

$$(1+1/n)^n > e - \varepsilon/2 \text{ for } n > k. \qquad (3.51)$$

Choose $N > k$ from the condition

$$\sum_{n=1}^{k} a_n + \sum_{n=N+1}^{\infty} 2^{-n} \le \frac{\varepsilon}{(2e-\varepsilon)(e-\varepsilon)} \sum_{n=k+1}^{N} \frac{1}{n}. \qquad (3.52)$$

The last inequality is always possible because the harmonic series diverges. Using (3.50), (3.51), and (3.52) we have

$$\sum_{n=1}^{\infty} a_n = \sum_{n=1}^{k} a_n + \sum_{n=k+1}^{N} \frac{1}{n} \left(\frac{n}{n+1} \right)^n + \sum_{n=N+1}^{\infty} 2^{-n}$$

$$< \frac{\varepsilon}{(2e-\varepsilon)(e-\varepsilon)} \sum_{n=k+1}^{N} \frac{1}{n} + \left(e - \frac{\varepsilon}{2} \right)^{-1} \sum_{n=k+1}^{N} \frac{1}{n} = \frac{1}{e-\varepsilon} \sum_{n=k+1}^{N} \frac{1}{n}$$

$$\le \frac{1}{e-\varepsilon} \sum_{n=1}^{\infty} (a_1 a_2 \cdots a_n)^{1/n}. \quad \square$$

Theorem 3.12. *Suppose $(a_n)_{n\ge 1}$ is a sequence having positive terms. Then*

(a) *If*

$$\liminf_{n\to\infty} \frac{n \cdot a_n}{\sum_{k=1}^{n} a_k} > 2,$$

the series $\sum 1/a_n$ converges.

(b) *If*

$$\limsup_{n\to\infty} \frac{n \cdot a_n}{\sum_{k=1}^{n} a_k} < 2,$$

the series $\sum 1/a_n$ diverges.

Theorem 3.13. (Kummer[15]) *Suppose $a_n, u_n > 0$ for all n. Then*

(a) *If*

$$\liminf_{n\to\infty} \left(a_n \frac{u_n}{u_{n+1}} - a_{n+1} \right) > 0,$$

the series $\sum u_n$ converges.

(b) *If $\sum 1/a_n$ diverges and*

$$\limsup_{n\to\infty} \left(a_n \frac{u_n}{u_{n+1}} - a_{n+1} \right) < 0,$$

then $\sum u_n$ diverges.

---

[15] Ernest Eduard Kummer, 1810–1893.

Proof. (a) Denote

$$\liminf_{n\to\infty}\left(a_n\cdot\frac{u_n}{u_{n+1}}-a_{n+1}\right)=2\alpha>0.$$

Then there exists $n_0\in\mathbb{N}$ such that for all natural $n\geq n_0$,

$$a_n\cdot\frac{u_n}{u_{n+1}}-a_{n+1}>\alpha.$$

Therefore $\alpha u_{n+1}<a_n u_n-a_{n+1}u_{n+1}$ implies

$$\alpha(u_{n_0+1}+\cdots+u_n)=a_{n_0}u_{n_0}-a_{n+1}u_{n+1}<a_{n_0}u_{n_0}.$$

Thus the sequence of partial sums to $\sum u_n$ is bounded, so $\sum u_n$ is convergent.
(b) There exists $n_0\in\mathbb{N}$ such that for all natural $n\geq n_0$,

$$a_n\cdot\frac{u_n}{u_{n+1}}-a_{n+1}\leq 0.$$

Then

$$a_n u_n\leq a_{n+1}u_{n+1}\implies a_{n_0}u_{n_0}\leq a_{n_0+1}u_{n_0+1}\leq\cdots\leq a_n u_n\leq a_{n+1}u_{n+1}$$
$$\implies a_{n_0}u_{n_0}\leq a_{n+1}u_{n+1}\implies u_{n+1}\geq a_{n_0}u_{n_0}/a_{n+1}.$$

By the comparison test, the conclusion follows.   □
If $a_n=n$ in Theorem 3.13, the next result follows.

Theorem 3.14. (Raabe[16]–Duhamel[17]) *Suppose a sequence (u_n) of positive terms is given. Then*

(a) *If*

$$\liminf_{n\to\infty}n\left(\frac{u_n}{u_{n+1}}-1\right)>1,$$

the series $\sum u_n$ converges.
(b) *If*

$$\limsup_{n\to\infty}n\left(\frac{u_n}{u_{n+1}}-1\right)<1,$$

the series $\sum u_n$ diverges.

Example 3.2. The series

$$\sum_{n=1}^{\infty}\frac{1\cdot 3\cdots 5\cdots(2n-1)}{2\cdot 4\cdot 6\cdots 2n}\cdot\frac{1}{2n+1} \tag{3.53}$$

is convergent. Indeed, denote

---

[16] Josef Ludwig Raabe, 1801–1859.
[17] Jean Marie Constant Duhamel, 1797–1872.

$$u_n = \frac{1 \cdot 3 \cdot 5 \cdots (2n-1)}{2 \cdot 4 \cdot 6 \cdots 2n} \cdot \frac{1}{2n+1}$$

and apply Theorem 3.14 of Raabe–Duhamel to get

$$n\left(\frac{u_n}{u_{n+1}} - 1\right) = \frac{n(6n+5)}{4n^2 + 4n + 1} \xrightarrow{n \to \infty} \frac{3}{2} > 1.$$

Thus series (3.53) is convergent.   △

A useful test is the next one.

Theorem 3.15. (Gauss[18]) *Suppose we can write the fraction* u_n/u_{n+1} *as*

$$\frac{u_n}{u_{n+1}} = \alpha + \frac{\beta}{n} + \frac{\gamma_n}{n^{1+\lambda}}$$

where $\lambda > 0$ *and* (γ_n) *is a bounded sequence. If* $\alpha > 1$ *or* $\alpha = 1$ *and* $\beta > 1$, $\sum a_n$ *converges, whereas if* $\alpha \le 1$ *or* $\alpha = 1$ *and* $\beta \le 1$, *it diverges.*

Proof. Because

$$\lim_{n \to \infty} u_n/u_{n+1} = \alpha,$$

by the ratio test, the series $\sum u_n$ converges for $\alpha > 1$ and diverges for $\alpha < 1$. Suppose $\alpha = 1$. By the Raabe–Duhamel test, we have

$$n\left(u_n/u_{n+1} - 1\right) = \beta + \gamma_n/n^\lambda,$$

thus

$$\lim_{n \to \infty} n\left(u_n/u_{n+1} - 1\right) = \beta.$$

We conclude that $\sum u_n$ is convergent for $\beta > 1$ and is divergent for $\beta < 1$. Consider $\alpha = \beta = 1$. Then

$$u_n/u_{n+1} = 1 + 1/n + \gamma_n/n^{1+\lambda}.$$

Consider the divergent series

$$\sum_{n=2}^{\infty} \frac{1}{n \ln n}$$

and denote $a_n = n \ln n$. Then

$$\frac{u_n}{u_{n+1}} a_n - a_{n+1} = \left(1 + \frac{1}{n} + \frac{\gamma_n}{n^{1+\lambda}}\right) n \ln n - (n+1)\ln(n+1)$$

$$= (n+1)\ln n + \frac{\gamma_n}{n^\lambda} - (n+1)\ln(n+1) = \frac{\gamma_n}{n^\lambda} - \ln\left(1 + \frac{1}{n}\right)^{n+1} \xrightarrow{n \to \infty} -1.$$

Hence by Theorem 3.13 of Kummer, $\sum u_n$ diverges.   □

------
[18] Karl Friedrich Gauss, 1777–1855.

Example 3.3. We study the convergence of the series

$$\sum_{n=1}^{\infty} \left(\frac{1 \cdot 3 \cdot 5 \cdots (2n-1)}{2 \cdot 4 \cdot 6 \cdots 2n} \right)^{\lambda}. \tag{3.54}$$

Obviously, for $\lambda = 1$, the series diverges. We may suppose that $\lambda \neq 1$. We apply Theorem 3.15 of Gauss. Denote

$$u_n = \left(\frac{1 \cdot 3 \cdot 5 \cdots (2n-1)}{2 \cdot 4 \cdot 6 \cdots 2n} \right)^{\lambda}.$$

Then

$$\frac{u_n}{u_{n+1}} = \left(\frac{2n+2}{2n+1} \right)^{\lambda} \to 1 \quad \text{as} \quad n \to \infty.$$

Because $\alpha = 1$, we find β. Therefore

$$n \left(\frac{u_n}{u_{n+1}} - 1 \right) = n \frac{(1 + 1/n)^{\lambda} - (1 + 1/(2n))^{\lambda}}{(1 + 1/(2n))^{\lambda}}.$$

By (5.40) at page 211, we write

$$(1 + 1/n)^{\lambda} = 1 + \lambda/n + R_1(1/n), \quad (1 + 1/(2n))^{\lambda} = 1 + \lambda/(2n) + R_1(1/2n)$$

and

$$n \left(\frac{u_n}{u_{n+1}} - 1 \right) = n \frac{\lambda/(2n) + R_1(1/n) - R_1(1/2n)}{1 + \lambda/(2n) + R_1(1/2n)} \to \frac{\lambda}{2} \quad \text{as} \quad n \to \infty.$$

Thus $\beta = \lambda/2$. So, if $\lambda > 2$, by Theorem 3.14 of Raabe, series (3.54) converges and if $\lambda < 2$, the series diverges.

Suppose that $\lambda = 2$. Then

$$\frac{u_n}{u_{n+1}} = \left(\frac{2n+2}{2n+1} \right)^2,$$

$$\frac{u_n}{u_{n+1}} = 1 + \frac{1}{n} + \frac{\gamma_n}{n^2}, \quad \gamma_n = \frac{-n-1}{4n+4+1/n} \to -\frac{1}{4} \quad \text{as} \quad n \to \infty.$$

Hence, for $\lambda = 2$, series (3.54) diverges. $\triangle$

Let a be a real (or complex) number and n a nonnegative integer number. Then $(a)_n = a(a+1)\ldots(a+n-1)$ is the *rising factorial*.

Example 3.4. The numerical series

$$1 + \frac{(\alpha)_1(\beta)_1}{1!\,(\gamma)_1} + \frac{(\alpha)_2(\beta)_2}{2!\,(\gamma)_2} + \cdots + \frac{(\alpha)_n(\beta)_n}{n!\,(\gamma)_n} + \cdots = 1 + \sum_{k=1}^{\infty} \frac{(\alpha)_k(\beta)_k}{k!\,(\gamma)_k},$$

for some numbers α, β, and γ such that all the fractions are well defined, is said to be the *hypergeometric* series. It is a particular case (for $x = 1$) of (3.78) at page 138. Denote by a_n the general term of the hypergeometric series. Then

$$\frac{a_n}{a_{n+1}} = \frac{(n+1)(\gamma+n)}{(\alpha+n)(\beta+n)} = 1 + \frac{(\gamma+1-\alpha-\beta)n+\gamma-\beta\alpha}{(\alpha+n)(\beta+n)}$$

$$= 1 + \frac{\gamma+1-\alpha-\beta}{n} + \frac{\beta_n}{n^2},$$

where

$$\beta_n = \frac{(\gamma-\alpha\beta-\alpha-\beta)n^3 - \alpha\beta n^2}{n(\alpha+n)(\beta+n)}.$$

The sequence (β_n) converges to $\gamma - \alpha\beta - \alpha - \beta$, thus it is bounded. We apply the Gauss test theorem 3.15 and conclude that if $\gamma - \alpha - \beta > 0$, the hypergeometric series $\sum a_n$ converges, whereas if $\gamma-\alpha-\beta \le 0$, it diverges. $\triangle$

Theorem 3.16. (Bertrand[19]) *Suppose we are given a series $\sum u_n$ of positive terms. Denote*

$$\lim_{n\to\infty} \ln n \cdot \left(n\left(\frac{u_n}{u_{n+1}} - 1\right) - 1 \right) = \alpha.$$

Then if $\alpha > 1$, the series $\sum u_n$ converges, whereas for $\alpha < 1$, it diverges.

Proof. Apply Theorem 3.13 of Kummer, with $a_n = n\ln n$, $n \in \mathbb{N}^*$. Then

$$(u_n/u_{n+1})\, n\ln n - (n+1)\ln(n+1)$$

$$= \ln n \left(n\left(u_n/u_{n+1} - 1\right) - 1\right) - \ln(1+1/n)^{n+1} \to \alpha - 1 \text{ as } n \to \infty.$$

The conclusion follows. $\square$

3.3.2 The root and the ratio tests

The first test is the *root test* or the *Cauchy criterion*.

Theorem 3.17. (Cauchy) *Given $\sum x_n$, set $\alpha = \limsup_{n\to\infty} \sqrt[n]{|x_n|}$. Then*

(a) *If $\alpha < 1$, $\sum x_n$ converges.*
(b) *If $\alpha > 1$, $\sum x_n$ diverges.*
(c) *If $\alpha = 1$, the test gives no information.*

Proof. (a) If $\alpha < 1$, we can choose β so that $\alpha < \beta < 1$ and an integer m such that $\sqrt[n]{|x_n|} < \beta$, for all $n \ge m$. That is, $n \ge m$ implies $|x_n| < \beta^n$. Because $0 < \beta < 1$, $\sum \beta^n$ converges. Convergence of $\sum x_n$ follows from the comparison test, Theorem 3.7.

---

[19] Joseph Louis François Bertrand, 1822–1990.

(b) If $\alpha > 1$, then there is a sequence (n_k) such that $\sqrt[n_k]{|x_{n_k}|} \to \alpha$. Hence $|x_n| > 1$ for infinitely many values of n, so the condition $x_n \to 0$, necessary for convergence of $\sum x_n$, does not hold.

(c) Consider the series

$$\sum \frac{1}{n}, \quad \sum \frac{1}{n^2}. \tag{3.55}$$

For each of these series $\alpha = 1$, but the first diverges, whereas the second converges. $\square$

The next test is the *ratio test* or the *D'Alembert criterion*.

Theorem 3.18. (D'Alembert[20]) *The series* $\sum x_n$

(a) *Converges if*

$$\limsup_{n \to \infty} \left| \frac{x_{n+1}}{x_n} \right| < 1.$$

(b) *Diverges if*

$$\liminf_{n \to \infty} \left| \frac{x_{n+1}}{x_n} \right| > 1.$$

Proof. (a) We can find $\beta < 1$ and an integer m such that

$$|x_{n+1}/x_n| < \beta$$

for all $n \geq m$. In particular

$$|x_{m+1}| < \beta|x_m|, \quad |x_{m+2}| < \beta|x_{m+1}| < \beta^2|x_m|, \quad \ldots, |x_{m+p}| < \beta^p|x_m|.$$

That is,

$$|x_n| < \beta^{n-m}|x_m|$$

for all $n \geq m$, and (a) follows from the comparison test, because $\sum \beta^n$ converges.

(b) If $|x_{n+1}| \geq |x_n|$ for all $n \geq m$, it is easily seen that the necessary condition $x_n \to 0$ does not hold, and (b) follows. $\square$

Remark. We consider the series (3.55). In any case we have

$$\lim_{n \to \infty} x_{n+1}/x_n = 1,$$

but the first series diverges and the second converges. $\triangle$

Theorem 3.19. *For any sequence* (x_n) *of positive numbers*

$$\liminf_{n \to \infty} \frac{x_{n+1}}{x_n} \leq \liminf_{n \to \infty} \sqrt[n]{x_n} \leq \limsup_{n \to \infty} \sqrt[n]{x_n} \leq \limsup_{n \to \infty} \frac{x_{n+1}}{x_n}.$$

---

[20] Jean Baptiste Le Rond D'Alembert, 1717–1783.

3.3.3 Partial summation

Lemma 3.3. *Consider two sequences* (a_n), (b_n). *Put*

$$A_n = \sum_{k=0}^{n} a_k$$

if $n \geq 0$; *put* $A_{-1} = 0$. *Then, if* $0 \leq p \leq q$, *we have*

$$\sum_{n=p}^{q} a_n b_n = \sum_{n=p}^{q-1} A_n (b_n - b_{n+1}) + A_q b_q - A_{p-1} b_p. \tag{3.56}$$

The identity (3.56) is said to be the *partial summation formula*.

Theorem 3.20. (Abel) *Suppose there are given two sequences* $(a_n)_n$ *and* $(b_n)_n$. *Moreover,*

 (i) *The series* $\sum_{n=1}^{\infty} |b_n - b_{n+1}|$ *converges.*
 (ii) $\lim_{n \to \infty} b_n = 0$.
 (iii) *There exists a positive* α *such that for every* $n \in \mathbb{N}^*$ *and* $m \in \mathbb{N}^*$, $m \geq n$,
   we have $|a_n + a_{n+1} + \cdots + a_m| < \alpha$.

Then $\sum_{n=1}^{\infty} a_n b_n$ *converges.*

Proof. For every $n \in \mathbb{N}^*$ and $m \in \mathbb{N}^*$, $m \geq n$, we denote $\alpha_{n,m} = a_n + a_{n+1} + \cdots + a_m$. Then based on a variant of the partial summation formula (3.56) we have

$$
\begin{aligned}
a_n b_n + &\cdots + a_m b_m \\
&= \alpha_{n,n} b_n + (\alpha_{n,n+1} - \alpha_{n,n}) b_{n+1} + \cdots + (\alpha_{n,m} - \alpha_{n,m-1}) b_m \\
&= \alpha_{n,n}(b_n - b_{n+1}) + \cdots + \alpha_{n,m-1}(b_{m-1} - b_m) + \alpha_{n,m} b_m. \tag{3.57}
\end{aligned}
$$

We now choose a positive ε. For $\varepsilon/(2\alpha)$ we find an $n_\varepsilon \in \mathbb{N}^*$ that satisfies the following two requirements.

1. It is greater than or equal to the n_ε from the general criterion of Cauchy Theorem 3.1 applied to the convergent series $\sum_{n=1}^{\infty} |b_n - b_{n+1}|$.
2. For every $n \geq n_\varepsilon$, $|b_n| < \varepsilon/(2\alpha)$.

For every $m \geq n > n_\varepsilon$ we have

1. If $m = n$, $|a_n b_n| \leq \alpha |b_n| < \alpha(\varepsilon/(2\alpha)) < \varepsilon$.
2. If $m > n$, from (3.57) we get

$$\left| \sum_{k=n}^{m} a_k b_k \right| \leq \alpha(|b_n - b_{n+1}| + \cdots + |b_{m-1} - b_m|) + \alpha |b_m| < \alpha \frac{\varepsilon}{2\alpha} + \alpha \frac{\varepsilon}{2\alpha} = \varepsilon.$$

The conclusion follows by the general criterion of the Cauchy theorem 3.1.   □

Remark. If $b_n \downarrow 0$, the assumption that the series $\sum_{n=1}^{\infty} |b_n - b_{n+1}|$ converges in theorem 3.20 is useless. It may be neglected because the nth rank partial sum of the series $\sum_{n=1}^{\infty} |b_n - b_{n+1}|$ is equal to $b_1 - b_{n+1}$ which, in turn, tends to b_1.   △

Theorem 3.21. (Dirichlet[21]) *Suppose*

(i) *The partial sums* A_n *of* $\sum a_n$ *form a bounded sequence.*
(ii) $b_0 \geq b_1 \geq b_2 \geq \dots$.
(iii) $\lim_{n\to\infty} b_n = 0$.

Then $\sum a_n b_n$ *converges.*

Proof. It follows from the previous remark.   □

Theorem 3.22. (Leibniz [22]) *Suppose*

(i) $|c_1| \geq |c_2| \geq \dots$.
(ii) $c_{2m-1} \geq 0$, $c_{2m} \leq 0$, $m = 1, 2, \dots$.
(iii) $\lim_{n\to\infty} c_n = 0$.

Then $\sum c_n$ *converges.*

Proof. Apply Theorem 3.21 with $a_n = (-1)^{n+1}$, $b_n = |c_n|$.   □

Remark. The conclusion of the Leibniz theorem is not always true if the convergence of the sequence of absolute values is not monotonic. Consider the series

$$1 - \frac{1}{3} + \frac{1}{2} - \frac{1}{3^2} + \frac{1}{3} - \frac{1}{3^3} + \cdots + \frac{1}{n} - \frac{1}{3^n} + \dots . \quad △$$

Example 3.5. For $p > 0$ the series

$$1 - \frac{1}{2^p} + \frac{1}{3^p} - \frac{1}{4^p} + \cdots + (-1)^{n+1}\frac{1}{n^p} + \cdots$$

converges. For $p = 1$, based on Corollary 1.13, we can write the Leibnitz series

$$1 - \frac{1}{2} + \frac{1}{3} - \frac{1}{4} + \cdots + (-1)^{n+1}\frac{1}{n} + \cdots = \ln 2. \quad △ \qquad (3.58)$$

3.3.4 Absolutely and conditionally convergent series

Consider

$$\sum a_n. \qquad (3.59)$$

Series (3.59) with real or complex terms is said to be *absolutely convergent* if the series

$$\sum |a_n| \qquad (3.60)$$

converges.

[21] Peter Gustav Lejeune Dirichlet, 1805–1859.
[22] Gottfried Wilhelm Leibniz, 1646–1716.

Theorem 3.23. *The convergence of series* (3.60) *implies the convergence of series* (3.59).

Proof. We apply the Cauchy general criterion, Theorem 3.1. For arbitrary, but fixed, $\varepsilon > 0$ there exists a rank n_ε such that for every rank n and m, $n \geq n_\varepsilon$, and $n \geq m$ it holds

$$\left| \sum_{k=n}^{m} |a_k| \right| = \sum_{k=n}^{m} |a_k| < \varepsilon.$$

But using a well-known inequality, we get

$$\left| \sum_{k=n}^{m} a_k \right| \leq \sum_{k=n}^{m} |a_k| < \varepsilon.$$

Now, we apply once again Theorem 3.1.   □

Remarks. The converse statement to Theorem 3.23 is not true. It is enough to consider the harmonic series in Remark 3.1 and series (3.58).

It is obvious that a convergent series with nonnegative terms is absolutely convergent.   △

Series (3.59) is said to be *conditionally convergent* if the series converges whereas the corresponding series of the absolute values (3.60) diverges.

A very important property of a sum of a finite number of real (complex) summands is the commutative property; that is, a rearrangement of the terms does not affect their sum. Unfortunately, this is not the case for series.

Example. Consider the convergent series (3.58); that is,

$$1 - \frac{1}{2} + \frac{1}{3} - \frac{1}{4} + \cdots + (-1)^{n+1} \frac{1}{n} + \cdots.$$

We write it, based on Theorem 3.4, as

$$\left(1 - \frac{1}{2}\right) + \left(\frac{1}{3} - \frac{1}{4}\right) + \cdots + \left(\frac{1}{2k-1} - \frac{1}{2k-2}\right) + \cdots.$$

We notice, as we already saw, that its sum is $s = \ln 2$.

We rearrange (3.58) according to the rule: two negative terms after a positive one. We find

$$\left(1 - \frac{1}{2} - \frac{1}{4}\right) + \left(\frac{1}{3} - \frac{1}{6} - \frac{1}{8}\right) + \cdots + \left(\frac{1}{2k-1} - \frac{1}{4k-2} - \frac{1}{4k}\right) + \cdots. \quad (3.61)$$

Denote by S_n the nth order partial sum of (3.61). Then we have

$$S_{3n} = \sum_{k=1}^{n} \left(\frac{1}{2k-1} - \frac{1}{4k-2} - \frac{1}{4k} \right) = \sum_{k=1}^{n} \left(\frac{1}{4k-2} - \frac{1}{4k} \right)$$

$$= \frac{1}{2} \sum_{k=1}^{n} \left(\frac{1}{2k-1} - \frac{1}{2k} \right) = \frac{1}{2} s_{2n}.$$

Thus we can write

$$S_{3n} = s_{2n}/2. \tag{3.62}$$

Also

$$S_{3n-1} = s_{2n}/2 + 1/(4n) \text{ and } S_{3n-2} = s_{2n}/2 + 1/(4n-2). \tag{3.63}$$

From (3.62) and (3.63) we conclude that

$$\lim_{n \to \infty} S_{3n} = \lim_{n \to \infty} S_{3n-1} = \lim_{n \to \infty} S_{3n-2} = s_{2n}/2.$$

Thus we have proved that series (3.61) converges and its sum is equal to $s/2 = (\ln 2)/2$.

Hence, by rearrangement of a conditionally convergent series we got a convergent series whose sum does not agree with the sum of the initial series. $\triangle$

The previous example illustrates that the commutativity is not longer valid for arbitrary series. We introduce the "positive case"; that is, a result of Cauchy on rearrangement of absolutely convergent series.

Theorem 3.24 (Cauchy) *The rearrangement of the terms of an absolutely convergent series supplies another absolutely convergent series having the same sum as the original one.*

Proof. Consider $\sum a_n$ an absolutely convergent series. Furthermore, consider the positive, respectively, the negative, parts of its terms; more precisely,

$$a_n^+ = \begin{cases} a_n, & a_n \geq 0 \\ 0, & a_n < 0 \end{cases} \quad a_n^- = \begin{cases} -a_n, & a_n \leq 0 \\ 0, & a_n > 0. \end{cases} \tag{3.64}$$

We have

$$a_n = a_n^+ - a_n^-.$$

Consider the following two series with nonnegative terms

$$\sum a_n^+ \text{ and } \sum a_n^-. \tag{3.65}$$

Then both series in (3.65) converge, because $a_n^+ \leq |a_n|$ and $a_n^- \leq |a_n|$.

Let $\sum b_n$ be the rearranged series of $\sum a_n$. For it construct the series $\sum b_n^+$ of positive parts, respectively, the series $\sum b_n^-$ of negative parts, as in (3.64). Then

$$\sum a_n = \sum(a_n^+ - a_n^-) = \sum a_n^+ - \sum a_n^- = \sum b_n^+ - \sum b_n^-$$
$$= \sum(b_n^+ - b_n^-) = \sum b_n.$$

The first equality is obvious, the second equality follows from Theorem 3.3, and the third one follows from Theorem 3.6. For the other equalities we argue in the same way.   □

Corollary 3.15. *Suppose the series $\sum a_n$ is absolutely convergent. Then the series $\sum a_n^+$ and $\sum a_n^-$ are absolutely convergent, too.*

The converse statement is true because the difference of two convergent series both of them having nonnegative terms is a convergent series, Theorem 3.3.

Corollary 3.16. *For series $\sum a_n$ to be absolutely convergent it is necessary and sufficient that series (3.65) generated by it be convergent.*

Lemma 3.4. *If $\sum a_n$ is convergent but not absolutely, both series (3.65) generated by it are divergent but $a_n^+ \to 0$ and $a_n^- \to 0$.*

Proof. From Corollary 3.16 it follows that at least one of the series (3.65) generated by it is divergent, that is, $\sum a_n^+ = \infty$, because $a_n^+ \geq 0$.

$$\sum_{}^{n} a_k^- = \sum_{}^{n} a_k^+ - \sum_{}^{n} a_k. \tag{3.66}$$

We study the behavior of the right-hand side of (3.66) as $n \to \infty$. The second sum tends to a finite number, because $\sum a_n$ is convergent. The first sum increases to ∞. Therefore the sum in the left-hand side of (3.66) increases to ∞ as $n \to \infty$. As a conclusion, if one of the series (3.65) is divergent, under our assumptions, the other is divergent, too.

The sequences $(a_n^\pm)$ tend to zero because the series $\sum a_n$ is convergent. $\square$

We now introduce a famous result of Riemann on rearrangement of conditionally convergent series.

Theorem 3.25. (Riemann[23]) *Let $\sum_0^\infty a_n$ and $\sum_0^\infty b_n$ be two divergent series with positive terms whose general terms tend to zero; that is, $a_n, b_n \to 0$ as $n \to \infty$.*

Then for any $s \in [-\infty, \infty]$ one can construct a series

$$a_0 + a_1 + \cdots + a_{m_1} - b_0 - b_1 - \cdots - b_{n_1} \tag{3.67}$$
$$+ a_{m_1+1} + a_{m_1+2} + \cdots + a_{m_2} - b_{n_1+1} - b_{n_1+2} - \cdots - b_{n_2} + \ldots$$

whose sum is equal to s. Series (3.67) contains all the terms of $\sum_0^\infty a_n$ and $\sum_0^\infty b_n$ only once.

Proof. First we suppose that $s \in \mathbb{R}$. In this case the indices $n_1, n_2, \ldots$ and $m_1, m_2, \ldots$ can be chosen as the smallest natural numbers for which the corresponding inequalities are fulfilled.

(i) $\alpha_1 - \sum_0^{n_1} a_k > s$.
(ii) $\alpha_2 = \alpha_1 - \sum_0^{m_1} b_k < s$.
(iii) $\alpha_3 = \alpha_2 + \sum_{n_1+1}^{n_2} b_k > s$.

---

[23] Georg Frederich Bernhard Riemann, 1826–1866.

(iv) $\alpha_4 = \alpha_3 - \sum_{m_1+1}^{m_2} b_k < s, \ldots$.

At the pth step of this construction we indeed can choose the natural numbers n_p and m_p satisfying the pth inequality because the series $\sum_0^\infty a_n$ and $\sum_0^\infty b_n$ have positive terms terms and diverge. The fact that the series thus constructed converges to s follows from the above inequalities and from the assumption that $a_n, b_n \to 0$ as $n \to \infty$.

Suppose that $s = +\infty$. We can replace s in the right-hand side of the inequalities (i), (ii), (iii), ... by a divergent sequence of the form 2,1,4,3, 5, ... $\square$

Corollary 3.17. *Let $\sum a_n$ be a convergent series but not absolutely. Choose an $s \in [-\infty, \infty]$. Then there is a rearrangement of the series such that the resulting series converges to s.*

Proof. We just split the series $\sum a_n$ into two series $\sum a_n^+$ and $\sum a_n^-$ as indicated by (3.64); then apply Lemma 3.4 and Theorem 3.25. $\square$

3.3.5 The $W-Z$ method

A general method for proving and discovering combinatorial identities is presented in [134] by Wilf[24] and Zeilberger.[25]

Suppose two functions $F(n,k), G(n,k)$ are given for $n \in \mathbb{N}$, and $k \in \mathbb{Z}$ and the following condition is satisfied.

$$F(n+1,k) - F(n,k) = G(n,k+1) - G(n,k), \quad n \in \mathbb{N}, \ k \in \mathbb{Z}. \qquad (3.68)$$

Such a pair (F, G) is said to be a $W-Z$ pair. Under certain additional boundary conditions one obtains a simple evaluation of the sum

$$\sum_k F(n,k), \quad n \in \mathbb{N}.$$

One also obtains a simple evaluation of the associated sum

$$\sum_{n=0}^\infty G(n,k), \quad k \in \mathbb{Z}.$$

The boundary conditions are as follows.

(F_1) For each k, the limit
$$f_k = \lim_{n\to\infty} F(n,k) \qquad (3.69)$$

exists and is finite.

---

[24] Herbert Saul Wilf, 1931–.
[25] Doron Zeilberger, 1950–.

(G_1) For each $n \in \mathbb{N}$,

$$\lim_{k \to \pm\infty} G(n, k) = 0.$$

(G_2) We have

$$\lim_{L \to \infty} \sum_{n=0}^{\infty} G(n, -L) = 0.$$

Theorem 3.26. *Let (F, G) satisfy (3.68). If (G_1) holds, then we have the identity*

$$\sum_{k} F(n, k) = \text{const.} \quad n \in \mathbb{N}, \tag{3.70}$$

where "const." is found by putting $n = 0$. Furthermore, if (F_1), (G_2) hold, then we have the identity

$$\sum_{n=0}^{\infty} G(n, k) = \sum_{j \le k-1} (f_j - F(0, j)), \tag{3.71}$$

where f is defined by (3.69).

Proof. Sum both sides of (3.68) from $k = -L$ to $k = K$. This gives

$$\triangle_n \left(\sum_{k=-L}^{K} F(n, k) \right) = G(n, K+1) - G(n, -L),$$

where $\triangle_n$ is the forward difference operator that acts on n. Let $K, L \to \infty$ and use (G_1) to find that $\sum_k F(n, k)$ is independent of $n \in \mathbb{N}$, and (3.70) is proved.

Similarly, if we sum both sides of (3.68) from $n = 0$ to N, we obtain

$$F(N+1, k) - F(0, k) = \triangle_k \left(\sum_{n=0}^{N} G(n, k) \right),$$

where $\triangle_k$ is the forward difference operator that acts on k. Taking the limit as $N \to \infty$ and using (F_1) yields

$$f_k - F(0, k) = \triangle_k \left(\sum_{n=0}^{\infty} G(n, k) \right).$$

We sum from $-L$ to $k - 1$, let $L \to \infty$, and use (G_2), getting (3.71). $\square$

Remark. All W-Z-pairs can be constructed as follows. First choose any function $\Phi(n, k)$. Then set $F = \triangle_k \Phi$ and $G = \triangle_n \Phi$. $\triangle$

Examples. (a) We prove the (well-known) identity

$$\sum_{k=0}^{n} \binom{n}{k} = 2^n, \quad n \in \mathbb{N},$$

by the $W-Z$-method.

Consider

$$F(n,k) = \binom{n}{k}/2^n \text{ and } G(n,k) = -\binom{n}{k-1}/2^{n+1} = \frac{-k}{2(n-k+1)}F(n,k).$$

Then (3.68), (F$_1$), (G$_1$), and (G$_2$), are satisfied, and $f_k = 0$ for all k. Then (3.70) and (3.71) become the two identities

$$\sum_{k} \binom{n}{k}/2^n = 1, \quad n \in \mathbb{N},$$

and

$$\sum_{n \geq 0} \binom{n}{k-1}/2^{n+1} = 1, \quad k \geq 1.$$

(b) Following [45], we introduce a $W-Z$ proof to the next identity

$$\sum_{k=0}^{m}(x+m+1)(-1)^k \binom{x+y+k}{m-k}\binom{y+2k}{k}$$

$$-\sum_{k=0}^{m}\binom{x+k}{m-k}(-4)^k = (x-m)\binom{x}{m}. \qquad (3.72)$$

Consider

$$g(k) = \binom{x}{k}\frac{(x^2-2kx+k+k^2)}{(-2)^{k+1}(1+k)(x+k+2)(x+k+1)}$$

with the motives that

$$\sum_{k=0}^{m}\binom{x+k}{m-k}(-4)^k = (-2)^m(x+m+1)\left(\frac{1}{x+1}+\sum_{k=0}^{m-1}g(k)\right). \qquad (3.73)$$

Denote

$$T(m) = \sum_{k=0}^{m}(-1)^k \binom{x+y+k}{m-k}\binom{y+2k}{k}.$$

Then

$$a_0(m)T(m) + a_1(m)T(m+1) + a_2(m)T(m+2) + a_3(m)T(m+3) = 0,$$

where

$$a_0(m) = 2(x-m-1)(x-m-2),$$
$$a_1(m) = -(x-m-2)(2y-x+5m+11),$$
$$a_2(m) = (-xy+3ym-2xm+4m^2+8y-5x+21m+28),$$
$$a_3(m) = (m+3)(y+m+3).$$

It is easy to check that the sums of the right-hand sides of (3.72) and (3.73) both satisfy the recurrence relation. Moreover, both sides agree for $m = 0$, $1, 2.$ $\triangle$

We introduce a $W-Z$-style proof of the Abel identity.

Theorem 3.27. *For $n \geq 0$, $s \in \mathbb{R}$, and $r \in \mathbb{R} \setminus \{0\}$*

$$\sum_{k=0}^{n} \binom{n}{k}(r+k)^{k-1}(s-k)^{n-k} = \frac{(r+s)^n}{r}. \tag{3.74}$$

Proof. Denote

$$F_{n,k}(r,s) = \binom{n}{k}(r+k)^{k-1}(s-k)^{n-k},$$

$$a_n(r,s) = \sum_{k=0}^{n} \binom{n}{k}(r+k)^{k-1}(s-k)^{n-k},$$

$$G_{n,k}(r,s) = (s-n)\binom{n-1}{k-1}(k+r)^{k-1}(s-k)^{n-k-1}.$$

Inasmuch as

$$F_{n,k}(r,s) - sF_{n-1,k}(r,s) - (n+r)F_{n-1,k}(r+1, s-1)$$
$$+ (n-1)(r+s)F_{n-2,k}(r+1, s-1) = G_{n,k} - G_{n,k+1},$$

we have by summing from $k = 0$ to $k = n$,

$$a_n(r,s) - sa_n(r,s) - (n+r)a_{n-1}(r+1, s-1) + (n-1)(r+s)a_{n-2}(r+1, s-1) = 0. \tag{3.75}$$

Because $(r+s)^n/r$ also satisfies (3.75) with the same initial conditions $a_0(r,s) = 1/r$ and $a_1(r,s) = (r+s)/r$, identity (3.74) follows. $\square$

By the Abel identity theorem 3.27, one can easily prove the Cayley[26] theorem on the number of labeled trees.

Theorem 3.28. *The number of labeled trees on n vertices is n^{n-2}.*

Proof. Let $n \rightarrow n-2$, $r \rightarrow 1$, and $s \rightarrow n-1$ in (3.74). Setting $b_n = n^{n-2}$, the recurrence results,

$$b_n = \sum_{k=0}^{n-2} \binom{n-2}{k} b_{n+1}((n-k-1)b_{n-k-1}). \tag{3.76}$$

Let t_n be the number of labeled trees on n vertices; then

$$t_n = \sum_{k=0}^{n-2} \binom{n-2}{k} t_{n+1}((n-k-1)t_{n-k-1}). \tag{3.77}$$

---

[26] Arthur Cayley, 1821–1895.

Indeed every labeled tree on $\{1, 2, \dots\}$ rooted at vertex numbered by 1 generates a unique triple (U, V, W), where V is the rooted tree to which the vertex 2 belongs, in the forest resulting from deleting 1, U is the tree obtained from the original tree by deleting V, and W is the set of labels (1 included) participating in U. Now we sum over all possible $k = |W|$. Thus (3.77) follows. Because $t_1 = b_1$, (3.76) follows.   □

3.4 Series of functions

Let $(f_n)_n$ be a sequence of functions, where $f_n : A \to \mathbb{R}$, $\emptyset \neq A \subset \mathbb{R}$, and

$$s_n(x) = f_0(x) + f_1(x) + \cdots + f_n(x).$$

The *series of functions* associated with the sequence of functions $(f_n)_n$ is the pair of sequences of functions $((f_n)_n, (s_n)_n)$. An older but still largely used notation is the symbolic expression

$$f_0(x) + f_1(x) + \cdots + f_n(x) + \dots,$$

or, more concisely,

$$\sum_{n=0}^{\infty} f_n(x) \text{ or } \sum f_n(x) \text{ or even } \sum f_n.$$

The series $((f_n)_n, (s_n)_n)$ *converges* on a point $x \in A$ provided the numerical sequence $(s_n(x))$ converges. Let $s(x)$ be its limit. In this case we write

$$\sum_{n=0}^{\infty} f_n(x) = s(x).$$

$s(x)$ is said to be the *sum* of this series on x.

The concepts introduced earlier are valid if we consider a sequence of functions indexed by $\mathbb{N}^*$ instead of $\mathbb{N}$.

The set of points for which a given series of functions converges is said to be its *domain of convergence*.

The series of functions $\sum f_n$ is said to be an *absolutely convergent series* on a point $x \in A$ provided the numerical series $\sum f_n(x)$ is absolutely convergent, that is, provided $\sum |f_n(x)|$ converges.

The series of functions $\sum f_n$ is said to be a *uniform convergent series* on a subset of A provided the sequence of functions (s_n) converges uniformly on that set.

Example 4.1. We take a look on the uniform convergence of the series

$$1 + x + x^2 + \dots, \quad 0 < x < 1.$$

Denote $A =]0,1[$. The series converges pointwise on A, as a geometric series with $|x| < 1$. Its sum is $s(x) = 1/(1-x)$ and $s_n(x) = (1-x^n)/(1-x)$.

One way to check that the given series does not converge uniformly on A consists in considering a sequence, say (x_n), $x_n = 1 - 1/n$, $n \geq 1$ and then noting that the sequence $s_n(x_n)$ is unbounded. Indeed,

$$|s(x_n) - s_n(x_n)| = n\,(1 - 1/n)^n \xrightarrow{\;n\to\infty\;} \infty.$$

We note that the series converges uniformly on each interval $[-\alpha, \alpha]$, $0 < \alpha < 1$. $\triangle$

A similar result to Theorem 3.1 is the following general criterion of Cauchy.

Theorem 4.1. (Cauchy) *A series of functions $\sum f_n$ converges uniformly on a subset B of A if and only if for every positive ε there exists $n_\varepsilon \in \mathbb{N}$ such that for every natural $n > n_\varepsilon$, $p \in \mathbb{N}$, and $x \in B$ one has*

$$|f_{n+1}(x) + \cdots + f_{n+p}(x)| < \varepsilon.$$

Proof. It results from Theorem 2.1. $\square$

A useful criterion of uniform and absolute convergence is the next one.

Theorem 4.2. (Weierstrass) *Suppose there are given a series of functions $\sum f_n$, $f_n : A \to \mathbb{R}$ and a convergent numerical series $\sum a_n$ with $a_n \geq 0$ such that for all x in a subset B of A it holds $|f_n(x)| \leq a_n$. Then $\sum f_n$ converges absolutely and uniformly on B.*

Proof. Choose a positive ε. Because $\sum a_n$ is convergent, there exists $n_\varepsilon \in \mathbb{N}$ such that for every natural $n > n_\varepsilon$ and $p \in \mathbb{N}$ one has that

$$a_{n+1} + \cdots + a_{n+p} < \varepsilon.$$

Then for every natural $n > n_\varepsilon$, $p \in \mathbb{N}$, and $x \in B$,

$$|f_{n+1}(x) + \cdots + f_{n+p}(x)| \leq |f_{n+1}(x)| + \cdots + |f_{n+p}(x)|$$
$$\leq a_{n+1} + \cdots + a_{n+p} < \varepsilon.$$

By Theorem 4.1, we conclude that $\sum f_n$ converges absolutely and uniformly on B. $\square$

Examples. (a) Consider

$$\sum_{n \geq 1} \frac{\cos nx}{n^2}$$

Because

$$\left| \frac{\cos nx}{n^2} \right| \leq \frac{1}{n^2}, \quad \forall x \in \mathbb{R} \text{ and } \sum_{n \geq 1} \frac{1}{n^2} < \infty,$$

we conclude that the series of functions converges uniformly on $\mathbb{R}$.

(b) Consider

$$\sum_{n \geq 1} \frac{(-1)^{n+1}}{\sqrt{n} + \sqrt{x}} \quad \text{for } x \geq 0.$$

The series converges pointwise for each real x by the Leibniz theorem. Denote $f(x)$ its limit at x. Then

$$|f(x) - s_n(x)| \leq |f_{n+1}(x)| \leq 1/\sqrt{n+1} < \varepsilon, \quad \forall x \in \mathbb{R}.$$

So the series converges uniformly on $\mathbb{R}$. $\triangle$

A similar result to Theorem 3.21 is the following one.

Theorem 4.3. *Suppose we are given* $a_n : A \to [0, \infty[$ *and* $f_n : A \to \mathbb{R}$, $n \in \mathbb{N}$ *such that*

(a) $a_n(x) \geq a_{n+1}(x)$, *for all* $n \in \mathbb{N}$, *and* (a_n) *converges uniformly to the zero function an a subset B of A.*

(b) *There exists a positive constant c such that for all $n \in \mathbb{N}$ and $x \in B$,* $|f_0(x) + \cdots + f_n(x)| \leq c$.

Then $\sum a_n f_n$ *converges uniformly on B.*

Proof. We use the general criterion of the Cauchy theorem 4.1. Choose a positive ε. Then

$$a_{n+1}(x) f_{n+1}(x) + \cdots + a_{n+p}(x) f_{n+p}(x)$$
$$= a_{n+1}(x)(s_{n+1}(x) - s_n(x)) + \cdots + a_{n+p}(x)(s_{n+p}(x) - s_{n+p-1}(x))$$
$$= -a_{n+1}(x)s_n(x) + (a_{n+1}(x) - a_{n+2}(x))s_{n+1}(x) + \ldots$$
$$+ (a_{n+p-1}(x) - a_{n+p}(x))s_{n+p-1}(x) + a_{n+p}(x)s_{n+p}(x).$$

For every natural i and $x \in B$ we have $|s_i(x)| \leq c$ and $a_i(x) \geq a_{i+1}(x) \geq 0$. Therefore

$$|a_{n+1}(x) f_{n+1}(x) + \cdots + a_{n+p}(x) f_{n+p}(x)|$$
$$\leq c\, a_{n+1}(x) + c[a_{n+1}(x) - a_{n+2}(x)] + \cdots + c[a_{n+p-1}(x) - a_{n+p}(x)]$$
$$+ c\, a_{n+p}(x) = 2c\, a_{n+1}(x).$$

Because (a_n) converges uniformly on B to zero there exists a natural n_ε such that for every natural $n \geq n_\varepsilon$ and every $x \in B$, $a_n(x) < \varepsilon/(2c)$. Hence for every natural $n \geq n_\varepsilon$ and every $x \in B$,

$$|a_{n+1}(x) f_{n+1}(x) + \cdots + a_{n+p}(x) f_{n+p}(x)| < \varepsilon. \quad \square$$

3.4.1 Power series

Given a sequence (c_n) of complex numbers, the series

$$\sum_{n=0}^{\infty} c_n z^n = c_0 + \sum_{n=1}^{\infty} c_n z^n,$$

is called a *power series*. The numbers c_n are called the *coefficients* of the series; z is a complex number.

Theorem 4.4. (Cauchy–Hadamard[27]) *Given the power series $\sum c_n z^n$, set*

$$\alpha = \limsup_{n \to \infty} \sqrt[n]{|c_n|}, \quad R = \frac{1}{\alpha}.$$

(If $\alpha = 0$, $R = +\infty$; if $\alpha = \infty$, $R = 0$.) Then $\sum c_n z^n$ converges provided $|z| < R$ and diverges whenever $|z| > R$.

Remark. R is called the *radius of convergence* of $\sum c_n z^n$. Each power series converges at least for $z = 0$.   △

Proof. Put $x_n = c_n z^n$ and apply the root test

$$\limsup_{n \to \infty} \sqrt[n]{|x_n|} = |z| \limsup_{n \to \infty} \sqrt[n]{|c_n|} = \frac{|z|}{R}. \quad \square$$

Examples. (a) $\sum n^n z^n$, $R = 0$.

(b) $\sum z^n/n!$, $R = +\infty$.

(c) $\sum z^n$, $R = 1$. If $|z| = 1$, the series diverges, because z^n does not tend to 0 as $n \to \infty$.

(d) $\sum z^n/n$, $R = 1$. On the circle $|z| = 1$ the series diverges at $z = 1$; it converges at all other points of $|z| = 1$. The last assertion follows from Theorem 4.5.

(e) $\sum z^n/n^2$, $R = 1$. The series converges at all points of the circle $|z| = 1$, by the comparison test, because $|z^n/n^2| = 1/n^2$.   △

Theorem 4.5. *Suppose the radius of convergence of $\sum c_n z^n$ is 1, and suppose $c_0 \geq c_1 \geq c_2 \ldots$, $\lim_{n \to \infty} c_n = 0$. Then $c_n z^n$ converges at every point of the circle $|z| = 1$, except possibly at $z = 1$.*

Proof. Put $a_n = z^n$, $b_n = c_n$. The hypotheses of Theorem 3.21 are satisfied, because

$$|A_n| = \left| \sum_{k=0}^{n} z^k \right| = \left| \frac{1 - z^{n+1}}{1 - z} \right| \leq \frac{2}{|1 - z|},$$

if $|z| = 1$, $z \neq 1$. $\square$

---

[27] Jacques Solomon Hadamard, 1865–1963.

Theorem 4.6. (Abel) *Suppose the power series $\sum c_n x^n$ converges to a function f on $]a - R, a + R[$. Then $f(a - R)$ $(f(a + R))$ can be obtained as*

$$f(a - R) = \lim_{x \downarrow a - R} f(x) \quad (f(a + R) = \lim_{x \uparrow a + R} f(x))$$

if the power series converges at $a - R$ $(a + R)$.

3.4.2 Hypergeometric series

The power series

$$F(\alpha, \beta, \gamma, x) = 1 + \frac{(\alpha)_1 (\beta)_1}{1! \, (\gamma)_1} x + \frac{(\alpha)_2 (\beta)_2}{2! \, (\gamma)_2} x^2 + \cdots + \frac{(\alpha)_n (\beta)_n}{n! \, (\gamma)_n} x^n + \cdots$$

$$= 1 + \sum_{k=1}^{\infty} \frac{(\alpha)_k (\beta)_k}{k! \, (\gamma)_k} x^k, \tag{3.78}$$

for some numbers α, β, and γ such that all the fractions are well defined, is said to be the *hypergeometric* series. Here $(\alpha)_n = \alpha(\alpha + 1) \ldots (\alpha + n - 1)$ is the rising factorial.

Forty identities involving hypergeometric series are proved in [43] by the $W - Z$ method. We introduce only two of them here.

Theorem 4.7. *There hold*

$$F(-n, -4n, -1/2, -3n, -1) = \left(\frac{2^6}{3^3}\right)^n \frac{(3/8)_n (5/8)_n}{(1/3)_n (2/3)_n}, \tag{3.79}$$

$$F(-n, -3n, -1/2, -4n, 4) = \left(\frac{-3^3}{2^4}\right)^n \frac{(1/3)_n (5/6)_n}{(1/4)_n (3/4)_n}. \tag{3.80}$$

Proof. The proofs of identities (3.79) and (3.80) are based on the $W - Z$ method. One has to divide the summand by the right-hand side, and call it $F(n, k)$. Calling the supplied certificate $R(n, k)$, one defines $G(n, k) = R(n, k) F(n, k)$ and verifies the identity

$$F(n + 1, k) - F(n, k) = G(n, k + 1) - G(n, k).$$

The certificate to (3.79) is P/Q, where

$$P = -2(296n^3 - 216n^2 k + 771n^2 + 651n - 372nk + 52nk^2$$
$$- 155k + 177 - 4k^3 + 45k^2)(3n - k + 1)$$
$$Q = (8n + 9 - 2k)(8n + 7 - 2k)(8n + 5 - 2k)(n - k + 1)$$

and the certificate to (3.80) is

$$\frac{(344n^3 - 216n^2 k + 768n^2 + 522n - 282nk + 34nk^2 - 75k + 104 + 13k^2)}{4(6n + 7 - 2k)(6n + 5 - 2k)(6n + 3 - 2k)}$$
$$\cdot \frac{(4n - k + 1)k}{(n - k + 1)(3n + 1)}. \qquad \square$$

3.5 The Riemann Zeta function $\zeta(p)$

The *Zeta function* was first introduced by Euler and it is defined by

$$\zeta(p) = \sum_{n=1}^{\infty} \frac{1}{n^p}. \tag{3.81}$$

Usually it is called the *Zeta function* of Riemann.

The series is convergent when p is a complex number with $\operatorname{Re}(p) > 1$. This is so because for $\operatorname{Re}(p) > 1$ the series in the right-hand side of (3.81) is absolutely convergent by Theorem 3.10. Some special values of $\zeta(p)$ are well known. For example, the values $\zeta(2) = \pi^2/6$ and $\zeta(4) = \pi^4/90$ have been obtained by Euler. They are proved by (5.25) and (5.26) at page 230.

The convergence of the Riemann Zeta function for real p might be studied by other tools, too.

Theorem 5.1. *For real p the Riemann Zeta function converges if and only $p > 1$.*

Proof. The case $p = 1$ is clear by Exercise 3.17 or by Theorem 3.10. We study the other cases.

By Theorem 3.15 of Bertrand one has

$$\alpha = \lim_{n\to\infty} \ln n \cdot \left(n \left(\frac{a_n}{a_{n+1}} - 1 \right) - 1 \right) = \lim_{n\to\infty} \ln n \cdot \left(n \left(\left(1 + \frac{1}{n} \right)^p - 1 \right) - 1 \right)$$

$$= \lim_{n\to\infty} \ln n \cdot \left(\frac{(1+1/n)^p - 1}{1/n} - 1 \right) = \begin{cases} \infty, & p > 1, \\ 0, & p = 1, \\ -\infty, & p < 1. \end{cases} \quad \square$$

3.6 Exercises

3.1. Show that there exists a function $f : \mathbb{N}^* \to \mathbb{N}^*$ such that

$$f(f(n)) = n^2, \quad \forall n \in \mathbb{N}^*.$$

3.2. Consider the sequence $(a_n)_{n\geq 1}$ defined by

$$a_n = 1 + 1/2 + 1/3 + \cdots + 1/n, \quad n \in \mathbb{N}^*.$$

Show that $a_n \to \infty$.

3.3. Consider a numerical sequence (a_n) satisfying $|a_n - a_m| > 1/n$ for any $n < m$. Then the sequence (a_n) is unbounded.

3.4. Consider the sequence $(x_n)_n$ defined by

$$x_{n+1} = (1/2)(x_n + a/x_n), \quad n \in \mathbb{N}, \; x_0 = 1, \; a > 0. \qquad (3.82)$$

Show that it converges.

3.5. Suppose $|x| < 1$. Show that $\lim_n x^n = 0$.

3.6. Suppose $\alpha \in \mathbb{R}$ and define $x_n = n^\alpha/(1+p)^n$. Show that $\lim_n x_n = 0$.

3.7. Let (x_n) be a sequence defined by

$$x_1 = \sqrt{a}, \; x_2 = \sqrt{a + \sqrt{a}}, \; \ldots, \; x_{n+1} = \sqrt{a + x_n}, \; \ldots, \; a > 0.$$

Show that (x_n) converges.

3.8. Consider $a_1 = \sqrt{2}$ and $a_{n+1} = \sqrt{2}^{\,a_n}$, $n \geq 1$. Find $\lim a_n$.

3.9. Show that the sequence (x_n) defined by

$$x_1 = \sqrt{a_1}, \; x_2 = \sqrt{a_1 + \sqrt{a_2}}, \ldots, x_n = \sqrt{a_1 + \sqrt{a_2 + \cdots + \sqrt{a_n}}}, \ldots, \; a_i > 1,$$

converges if

$$\lim_{n\to\infty} (1/n)\ln(\ln a_n) < \ln 2.$$

3.10. Consider the sequences (a_n) and (b_n) defined by

$$a_1 = a \geq 0, \; b_1 = b \geq 0, \quad a_{n+1} = (a_n + b_n)/2, \quad b_{n+1} = \sqrt{a_n b_n} \quad n \geq 1.$$

Show that the sequences (a_n) and (b_n) converge and $\lim a_n = \lim b_n$. The common limit point is said to be the *arithmetic–geometric mean iteration of Gauss* to the numbers a and b. It is denoted by $M(a, b)$. Moreover, the sequence (a_n) is decreasing, whereas (b_n) is increasing.

3.11. (The Borchardt[28]–Pfaff[29] algorithm) Consider the following recursion

$$a_0 = 2\sqrt{3}, \; b_0 = 3, \quad a_{n+1} = \frac{2a_n b_n}{a_n + b_n}, \quad b_{n+1} = \sqrt{a_{n+1} b_n}.$$

Show that the sequences (a_n) and (b_n) converge to the same limit equal to 2π.

3.12. Find the limits

(a) $\displaystyle \lim_{n\to\infty} \left(\frac{1}{n^2 + 1} + \frac{1}{n^2 + 2} + \cdots + \frac{1}{n^2 + n} \right)$.

(b) $\displaystyle \lim_{n\to\infty} \frac{1 + 2^2 + \cdots + n^n}{n^n}$.

---

[28] Carl Wilhelm Borchardt, 1817–1880.
[29] Johan Friedrich Pfaff, 1764–1825.

3.13. Find the limits of the sequences defined by

(a) $x_{n+1} = \dfrac{x_n + 1}{x_n + 2}$, $x_1 = 0$, $n \in \mathbb{N}^*$.

(b) $x_{n+1} = \dfrac{1}{3}\left(2x_n + \dfrac{125}{x_n^2}\right)$, $x_1 > 0$, $n \in \mathbb{N}^*$.

(c) $x_{n+1} = 1 - x_n^2$, $x_1 = a \in {]}0, 1{[}$, $n \in \mathbb{N}^*$.

(d) $x_{n+1} = \dfrac{1}{1 + x_n}$, $x_1 = a \in {]}-1, 1{[}$, $n \in \mathbb{N}^*$.

(e) Set $x_1 = a \in {]}0, 1{[}$, and $x_{n+1} = 1 + px_n$, $n \in \mathbb{N}^*$. Find the values of $p \in [0, 1]$ for which exists $\lim x_n$.

3.14. Let (a_n) and (b_n) be two sequences of real numbers such that

$$\limsup_{n\to\infty} a_n = \limsup_{n\to\infty} b_n = +\infty.$$

Show that there exist m, n such that $|a_m - a_n| > 1$ and $|b_m - b_n| > 1$.

3.15. Let (a_n) and (b_n) be two sequences of real numbers. Show that

$$\liminf_{n\to\infty} a_n + \liminf_{n\to\infty} b_n \le \liminf_{n\to\infty}(a_n + b_n) \le \liminf_{n\to\infty} a_n + \limsup_{n\to\infty} b_n, \quad (3.83)$$

$$\liminf_{n\to\infty} a_n + \limsup_{n\to\infty} b_n \le \limsup_{n\to\infty}(a_n + b_n) \le \limsup_{n\to\infty} a_n + \limsup_{n\to\infty} b_n. \quad (3.84)$$

3.16. (a) Let $p > 1$ be given. Show that

$$\lim_{n\to\infty} \frac{p^n}{n} = +\infty.$$

(b) For $p \in \mathbb{N}$, show that

$$\lim_{n\to\infty} \frac{1^p + 2^p + \cdots + n^p}{n^{p+1}} = \frac{1}{p+1}.$$

(c) For $k \in \mathbb{N}^*$, show that

$$\lim_{n\to\infty} \frac{\ln n}{n^k} = 0.$$

(d) Show

$$\lim_{n\to\infty} \frac{1 + 2^2\sqrt{2} + 3^2\sqrt[3]{3} + \cdots + n^2\sqrt[n]{n}}{n(n+1)(2n+1)} = \frac{1}{6}.$$

(e) For $k \in \mathbb{N}$, show that

$$\lim_{n\to\infty} \frac{1^k + 3^k + \cdots + (2n-1)^k}{n^{k+1}} = \frac{2^k}{k+1}.$$

(f) For $p \in \mathbb{N}^*$, show that

$$\lim_{n\to\infty} \left(\frac{1^p + 2^p + \cdots + n^p}{n^p} - \frac{n}{p+1}\right) = \frac{1}{2}.$$

(g) Suppose there are given two sequences (a_n) and (b_n) with $\lim a_n = a$, $\lim b_n = b$. If

$$c_n = \frac{a_1 b_n + a_2 b_{n-1} + \cdots + a_{n-1} b_2 + a_n b_1}{n}, \quad n \in \mathbb{N}^*,$$

show that $\lim c_n = ab$.

3.17. Consider the limit

$$\lim_{n \to \infty} \left(1 + \frac{1}{2} + \cdots + \frac{1}{n} - \alpha \ln n \right)$$

and determine the set of constants α for which the above limit exists and is finite.

3.18. Suppose we are given positive numbers $a_1, a_2, \ldots, a_n$ so that $a_k \leq \sqrt{a_{k-1} a_{k+1}}$, for $k = 2, \ldots, n-1$. Find $\max\{a_1, a_2, \ldots, a_n\}$.

3.19. Define the sequence $(x_n)_{n \geq 1}$ by $x_1 = a > 1$ and $x_{n+1} = x_n^2 - x_n + 1$, for all $n \geq 1$. Find

$$\sum_{n=1}^{\infty} \frac{1}{x_n}.$$

3.20. Given a decreasing sequence (x_n) of positive terms. Show that the series

$$\sum_{n=1}^{\infty} \left(1 - \frac{x_n}{x_{n+1}} \right)$$

is convergent.

3.21. Study the convergence of the following series and find their sums if any.

(i) $\displaystyle\sum_{n=1}^{\infty} \frac{\sqrt{n}}{n}$.

(ii) $\displaystyle\sum_{n=1}^{\infty} \frac{n}{(n+1)2^n}$.

(iii) $\displaystyle\sum_{n=1}^{\infty} \frac{1}{2n-1}$.

(iv) $\displaystyle\sum_{n=1}^{\infty} \frac{\sqrt[3]{n^2}}{n}$.

(v) $\displaystyle\sum_{n=1}^{\infty} \frac{n+1}{n^2 3^n}$.

(vi) $\displaystyle\sum_{n=1}^{\infty} \frac{\sin(nx)}{5^n}$.

(vii) $\displaystyle\sum_{n=1}^{\infty} \frac{2}{3^n}$.

(viii) $\displaystyle\sum_{n=1}^{\infty} \frac{\cos(nx) - \cos(n+1)x}{n}$.

(ix) $\displaystyle\sum_{n=1}^{\infty} \frac{1}{n(n+1)}$.

(x) $\displaystyle\sum_{n=1}^{\infty} \frac{(-1)^{n-1}}{n+1} \left(2 + \frac{1}{n} \right)$.

(ix) $\displaystyle\sum_{n=1}^{\infty} \frac{1}{n(n+1)}.$

(x) $\displaystyle\sum_{n=1}^{\infty} \frac{(-1)^{n-1}}{n+1}\left(2+\frac{1}{n}\right).$

(xi) $\displaystyle\sum_{n=1}^{\infty} \arctan\frac{1}{2n^2}.$

(xii) $\displaystyle\sum_{n=1}^{\infty}(\sqrt[3]{n+2}-2\sqrt[3]{n+1}+\sqrt[3]{n}).$

(xiii) $\displaystyle\sum_{n=1}^{\infty} \ln\frac{n^2+5n+6}{n^2+5n+4}.$

(xiv) $\displaystyle\sum_{n=1}^{\infty} \frac{e^{-\sqrt{n}}}{\sqrt{n}}.$

(xv) $\displaystyle\sum_{n=1}^{\infty} \frac{1}{n^p}\sin\frac{\pi}{n}, \quad p\in\mathbb{R}.$

(xvi) $\displaystyle\sum_{n=1}^{\infty} \frac{(n+1)^{n^2}}{n^{n^2}3^n}.$

(xvii) $\displaystyle\sum_{n=1}^{\infty} \frac{\ln n}{n(\ln^4 n+1)}.$

3.22. Find the domains of convergence to the next series.

(a) $\displaystyle\sum_{n\geq1} \frac{1}{n^x}.$

(b) $\displaystyle\sum_{n\geq1} \frac{(-1)^n}{n^x}.$

(c) $\displaystyle\sum_{n\geq1} \tan^n x.$

(d) $\displaystyle\sum_{n\geq1} \frac{x^n}{n}.$

(e) $\displaystyle\sum_{n\geq1} \frac{(x-3)^n}{n^n}.$

(f) $\displaystyle\sum_{n\geq1} \frac{x^n}{1+x^{2n}}.$

(g) $\displaystyle\sum_{n\geq1} \frac{n!}{(x^2+1)(x^2+2)\cdots(x^2+n)}.$

3.23. Find the domains of convergence to the following power series.

(i) $\displaystyle\sum_{n=0}^{\infty} \frac{x^n}{n+1}.$

(ii) $\displaystyle\sum_{n=1}^{\infty} \frac{(-1)^n x^n}{\sqrt{n}}.$

(iii) $\displaystyle\sum_{n=0}^{\infty} \frac{x^n}{5^n}.$

(iv) $\displaystyle\sum_{n=1}^{\infty} n!\,x^n.$

(v) $\displaystyle\sum_{n=0}^{\infty} \frac{x^{n3}}{3^{n3}}.$

(vi) $\displaystyle\sum_{n-0}^{\infty} \frac{5^n x^n}{(2n+1)^2\sqrt{3^n}}.$

(vii) $\displaystyle\sum_{n=0}^{\infty} \frac{(x-1)^n}{2^n(n+3)}.$

(viii) $\displaystyle\sum_{n=0}^{\infty} 5^n(n^2+1)(x+2)^{2n}.$

(ix) $\displaystyle\sum_{n=1}^{\infty} \sqrt{\frac{(2n-1)!!}{(2n)!!}}\left(\frac{x-2}{3}\right)^n.$

(x) $\displaystyle\sum_{n=1}^{\infty} \frac{x^n}{a^n+b^n}, \quad a>0,\ b>0.$

3.24. Find the domains of convergence to the next series of functions.

$$\sum_{n=1}^{\infty} \frac{\sin^n x}{n^\alpha}, \quad \alpha \in \mathbb{R},$$

$$\sum_{n=2}^{\infty} \frac{(-1)^n}{\ln n} \left(\frac{1-x^2}{1+x^2} \right), \qquad \sum_{n=1}^{\infty} (-1)^{n+1} \left(\sqrt{x^2 + 1/n^2} - |x| \right).$$

3.25. Show that the series $\sum_{n=1}^{\infty} (-1)^{n+1}/(x^2+n)$ converges uniformly on $\mathbb{R}$, but it is nowhere absolutely convergent.

3.7 References and comments

Theorem 1.9 coincides with [47, Theorem 4.3.8.].

Theorem 1.10 appears in [67].

The notations big Oh, small oh, big omega, and big theta are discussed in detail in [79].

Theorem 1.24 appears in [83, p. 14].

There are many papers treating the statements of Proposition 1.26. We mention [28] and the references therein.

Theorem 1.28 is given in [2].

Theorems 1.29 and 1.30 appeared in [130].

Proposition 1.13 is taken from [40].

Theorem 1.31 appeared in [82].

Remark 1.3 may be found in [7, p. 67].

The exercise of Traian Lalescu, inserted here as Lemma 3.2, was published in 1901 in [136, Exercise 579]. Different approaches to Lalescu's exercise may be found in [128, Chapter 14], and [15, Exercise 3.20, page 437].

The notations $\left\{ {n \atop k} \right\}$ and $\left[{n \atop k} \right]$ for the Stirling numbers of the second kind, respectively, for the Stirling numbers of the first kind were introduced in [74] as mentioned in [60].

Proposition 1.21 is from [69].

For numerical series we recommend [115] and [99]. A proof of Theorem 3.12 may be found in [5, pp. 135–136]. The proof of the Raabe–Duhamel test is based on the comparison with a properly chosen generalized harmonic series. The geometric series is used for comparison by the ratio and root tests. The Raabe–Duhamel test is finer than the ratio and root tests. At the same time the last two are (often) easier to use. It would be nice to have a universal series of comparison, however, one does not exist [70, vol. 1, p. 442].

For sequences and series of functions we recommend [115] and [86].

Theorem 3.26 appeared in [134], Theorem 3.27 appeared in [44], and Theorem 3.28 appeared in [133]. Theorem 4.7 may be found in [43].

More facts on the Riemann Zeta function $\zeta(s)$ may be found in [126, §4.4.3].

For exercises on series we mention [41], [120], and [125]. Exercise 3.1 is taken from [96]. Exercises 3.3, 3.7, and 3.8 are from [117]. Exercises 3.19 and Exercise 3.20 are taken from [125].

4

Limits and Continuity

The aim of this chapter is to introduce notions and results on limits and continuity of functions.

4.1 Limits

4.1.1 The limit of a function

Let X and Y be metric spaces. Suppose $\emptyset \neq E \subset X$, f maps E into Y, and p is a limit point of E. We write

$$f(x) \to q \text{ as } x \to p$$

or equivalently,

$$\lim_{x \to p} f(x) = q$$

if there is a point $q \in Y$ with the following property: for every $\varepsilon > 0$ there exists a $\delta (= \delta(\varepsilon, p)) > 0$ such that

$$\rho(f(x), q) < \varepsilon$$

for all points $x \in E$ for which $0 < \rho(x, p) < \delta$.

Remark. It should be noted that $p \in X$, but that p does not need to be a point of E. Moreover, even if $p \in E$, we may have $f(p) \neq \lim_{x \to p} f(x)$. $\triangle$

We can recast this definition in terms of sequences.

Theorem 1.1. (Heine) *Let X, Y, E, f, and p be as in the definition given above. Then*

$$\lim_{x \to p} f(x) = q \qquad (4.1)$$

if and only if

$$\lim_{n \to \infty} f(p_n) = q \tag{4.2}$$

for every sequence $(p_n)_n$ *in* E *such that*

$$p_n \neq p, \text{ and } \lim_{n \to \infty} p_n = p. \tag{4.3}$$

Proof. Suppose (4.1) holds. Choose (p_n) in E satisfying (4.3). Let $\varepsilon > 0$ be given. Then there exists $\delta > 0$ such that $\rho(f(x), q) < \varepsilon$, if $x \in E$ and $0 < \rho(x, p) < \delta$. Also, there exists n_ε such that $n \geq n_\varepsilon$ implies $0 < \rho(p_n, p) < \delta$. Thus for $n \geq n_\varepsilon$ we have $\rho(f(p_n), q) < \varepsilon$, which shows that (4.2) holds.

Conversely, suppose (4.1) is false. Then there exists some $\varepsilon > 0$ such that for every $\delta > 0$ there is a point $x \in E$ (depending on δ), for which $\rho(f(x), q) > \varepsilon$ but $0 < \rho(x, p) < \delta$. Taking $\delta_n = 1/n$, $n \in \mathbb{N}^*$, we thus find a sequence in E satisfying (4.3) for which (4.2) is false. $\square$

Corollary 1.1. *If* f *has a limit at* p, *this limit is unique.*

Proof. It follows by (b) of Theorem 1.1 at page 73 and by Theorem 1.1. $\square$

Theorem 1.2. (Cauchy) *Let* X, Y, E, f, *and* p *be as in the definition given above. Moreover, suppose that* Y *is complete. Then* (4.1) *holds if and only if for every* $\varepsilon > 0$ *there is a neighborhood* V *of* p *such that for every* $u, v \in V \cap E \setminus \{p\}$ *we have*

$$\rho(f(u), f(v)) < \varepsilon. \tag{4.4}$$

Proof. Suppose (4.1) holds. Then there exists a neighborhood V of p so that for every $t \in V \cap E \setminus \{p\}$ we have $\rho(f(t), q) < \varepsilon/2$. Then for every $u, v \in V \cap E \setminus \{p\}$ we have

$$\rho(f(u), f(v)) \leq \rho(f(u), q) + \rho(f(v), q) \leq \varepsilon,$$

thus (4.4) holds.

Suppose that for every $\varepsilon > 0$ there is a neighborhood V of p such that for every $u, v \in V \cap E \setminus \{p\}$ we have (4.4). Suppose (x_n) is a sequence in E so that $x_n \neq p$ for every n and $x_n \to p$ as $n \to \infty$. There exists a rank n_ε so that $x_n \in V \cap E \setminus \{p\}$, for all natural $n \geq n_\varepsilon$. By (4.4) we have that $\rho(f(x_n), f(x_m)) < \varepsilon$, for $n, m \geq n_\varepsilon$. Thus $(f(x_n))$ is a fundamental sequence in a complete metric space, so it is convergent. Now we apply Theorem 1.1 of Heine. $\square$

We define the upper and the lower limit to functions similarly to the upper and the lower limit for sequences. Consider $X, Y = \mathbb{R}, E, f$, and p as in Theorem 1.1. Then the *upper limit of* f *at* p *is* q; that is,

$$\limsup_{x \to p} f(x) = q$$

if and only if there exists a sequence (t_n) in E with $t_n \neq p$, for all n and $t_n \to p$ as $n \to \infty$ so that $\lim_{n \to \infty} f(t_n) = q$ and for each sequence $(p_n)_n$ in E such that $p_n \neq p$, and $\lim_{n \to \infty} p_n = p$, one has $\limsup_{n \to \infty} f(p_n) \leq q$.

The *lower limit of f at p is* q; that is,

$$\liminf_{x \to p} f(x) = q$$

if and only if there exists a sequence (t_n) in E with $t_n \neq p$, for all n and $t_n \to p$ as $n \to \infty$ so that $\lim_{n \to \infty} f(t_n) = q$ and for each sequence $(p_n)_n$ in E such that $p_n \neq p$, and $\lim_{n \to \infty} p_n = p$ one has $\liminf_{n \to \infty} f(p_n) \geq q$. Equivalently, we may rephrase as follows. Let Z be the set of convergent sequences (in E) converging to p such that no term of any such sequence coincides with p. Consider the set U of limit points of sequences in Z. Then $\limsup_{x \to p} f(x) = \sup U$ and $\liminf_{x \to p} f(x) = \inf U$.

Now we state a convergence test for series of positive terms.

Theorem 1.3. (Ermakov[1] test) *Let f be a positive monotonic nonincreasing function defined on the nonnegative real axis and consider the series $\sum f(n)$. If*

$$\limsup_{x \to \infty} \frac{e^x \cdot f(e^x)}{f(x)} < 1,$$

the series converges. If

$$\liminf_{x \to \infty} \frac{e^x \cdot f(e^x)}{f(x)} > 1,$$

the series diverges.

Examples. (1) One has

$$\lim_{x \to \infty} (1 + 1/x)^x = e.$$

(2) It holds

$$\lim_{x \to -\infty} (1 + 1/x)^x = e. \tag{4.5}$$

(3) We have

$$\lim_{x \to 0} (1 + x)^{1/x} = e.$$

(4) Suppose $u(x) \to \infty$ for $x \to a$, $a \in \overline{\mathbb{R}}$. Then

$$\lim_{x \to a} (1 + 1/u(x))^{u(x)} = e.$$

(5) Suppose $u(x) \to -\infty$ for $x \to a$, $a \in \overline{\mathbb{R}}$. Then

$$\lim_{x \to a} (1 + 1/u(x))^{u(x)} = e.$$

(6) Suppose $u(x) \to 0$ for $x \to a$, $a \in \overline{\mathbb{R}}$. Then

$$\lim_{x \to a} (1 + u(x))^{1/u(x)} = e.$$

---

[1] Vasiliĭ Petrovich Ermakov (Василий Петрович Ермаков), 1845–1922

(7) Set $a > 0$. Then

$$\lim_{x \to 0} \frac{a^x - 1}{x} = \ln a.$$

(8) Let a, b, c, and d be positive numbers. Then

$$\lim_{x \to 0} \frac{a^x - b^x}{c^x - d^x} = \frac{\ln(a/b)}{\ln(c/d)}.$$

(9) Let $a_1, a_2, \ldots, a_n$ be positive numbers. Find

$$\lim_{x \to 0} \left(\frac{a_1^x + a_2^x + \cdots + a_n^x}{n} \right)^{1/x}.$$

(10) Let $a_1, a_2, \ldots, a_n$ be positive numbers. Find

$$\lim_{x \to 0} \left(\frac{a_1^{x+1} + a_2^{x+1} + \cdots + a_n^{x+1}}{a_1 + a_2 + \cdots + a_n} \right)^{1/x}.$$

(11) Show that

$$\lim_{x \to 0} \frac{\sin x}{x} = 1. \tag{4.6}$$

Solutions. (1) By (vi) of Proposition 1.16 at page 101 we have that the limit holds for every strictly increasing and divergent sequence of reals. Then by Theorem 1.1 the proof ends.

(2) Set $z = -1 - x$. If $x \to -\infty$, then $z \to \infty$. We have $(1 + 1/x)^x = (1 + 1/z)^{z+1}$. Thus (2) follows.

(3) Set $z = 1/x$. The result follows from (1) and (2).

(4) - (6) The limits follow from (vi)–(viii) in Proposition 1.16 at page 101 and Theorem 1.1.

(7) Set $a^x - 1 = y$. Then $x \to 0$ if and only if $y \to 0$. Therefore

$$\lim_{x \to 0} \frac{a^x - 1}{x} = \lim_{y \to 0} \frac{y}{(\ln(y+1))/\ln a} = \ln a \cdot \lim_{y \to 0} \frac{y}{\ln(y+1)}$$

$$= \ln a \cdot \lim_{y \to 0} \frac{1}{\ln(1 + y)^{1/y}} = \ln a.$$

(8) Successively we have

$$\lim_{x \to 0} \frac{a^x - b^x}{c^x - d^x} = \lim_{x \to 0} \frac{(a^x - 1)/x - (b^x - 1)/x}{(c^x - 1)/x - (d^x - 1)/x} = \frac{\ln a - \ln b}{\ln c - \ln d}.$$

(9) By (6) and (7) we have

$$\lim_{x \to 0} \left(\frac{a_1^x + a_2^x + \cdots + a_n^x}{n} \right)^{1/x} = \lim_{x \to 0} \left(1 + \frac{a_1^x - 1 + \cdots + a_n^x - 1}{n} \right)^{1/x}$$

$$= \lim_{x \to 0} \left(\left(1 + \frac{\sum_{k=1}^{n}(a_k^x - 1)}{n} \right)^{\frac{n}{\sum_{k=1}^{n}(a_k^x - 1)}} \right)^{\left(\sum_{k=1}^{n} \frac{a_k^x - 1}{x} \right) \frac{1}{n}}$$

$$= e^{(\ln a_1 + \cdots + \ln a_n)/n} = \sqrt[n]{a_1 \ldots a_n}.$$

(10) Similarly to (9) we have

$$\lim_{x \to 0} \left(\frac{a_1^{x+1} + a_2^{x+1} + \cdots + a_n^{x+1}}{a_1 + a_2 + \cdots + a_n} \right)^{1/x}$$

$$= \lim_{x \to 0} \left(1 + \frac{a_1(a_1^x - 1) + \cdots + a_n(a_n^x - 1)}{a_1 + a_2 + \cdots + a_n} \right)^{1/x}$$

$$= \lim_{x \to 0} \left(\left(1 + \frac{\sum a_k(a_k^x - 1)}{\sum a_k} \right)^{\frac{\sum a_k}{\sum a_k(a_k^x - 1)}} \right)^{\left(\sum \frac{a_k^x - 1}{x} \right) \frac{1}{\sum a_k}}$$

$$= e^{\frac{a_1 \ln a_1 + \cdots + a_n \ln a_n}{a_1 + \cdots + a_n}} = a_1^{\frac{a_1}{a_1 + \cdots + a_n}} \ldots a_n^{\frac{a_n}{a_1 + \cdots + a_n}}.$$

(11) By Exercise 1.51 at page 39, we have that

$$0 < |\sin x| < |x| < |\tan x|, \quad 0 < |x| < \frac{\pi}{2}.$$

From here it follows that

$$\cos x < \frac{\sin x}{x} < 1.$$

Passing to the limit, the result is obvious.   △

We introduce some arithmetic properties of the limits of functions. Suppose $Y = \mathbb{R}^k$ and $f, g : E \to \mathbb{R}^k$. Define $f + g$ and $\langle f, g \rangle$ by

$$(f + g)(x) = f(x) + g(x), \quad \langle f, g \rangle(x) = \langle f(x), g(x) \rangle, \quad x \in E,$$

and if λ is a real number, $(\lambda f)(x) = \lambda f(x)$.

Theorem 1.4. *Suppose $E \subset X$ is a metric space, $p \in E'$, f, g are functions defined on E with values in $\mathbb{R}^k$, and*

$$\lim_{x \to p} f(x) = A, \quad \lim_{x \to p} g(x) = B.$$

Then

(a) $\lim_{x \to p}(f + g)(x) = A + B$.
(b) $\lim_{x \to p} \langle f, g \rangle(x) = \langle A, B \rangle$.

(c) *If $Y = \mathbb{R}$, and $B \neq 0$, then $\lim_{x \to p} f(x)/g(x) = A/B$.*

Proof. It follows from Theorem 1.1 and Theorem 1.3 at page 75.   □

The equality $f(x) = O(g(x))$ means that $|f(x)| < Ag(x)$ for all x close to a given limit point. We write $(x^3 + 1)^2 = O(1)$ for $x \to 1$.

The equality $f(x) = o(g(x))$ means that $f(x)/g(x) \to 0$ for all x close to a given limit point. We write $(x + 1)^2 = o(x^3)$ for $x \to +\infty$.

4.1.2 Right-hand side and left-hand side limits

Let f be defined on $]a, b[$, $a < b$. Consider any point x such that $a \leq x < b$. We write

$$f(x+) = q$$

if $f(t_n) \to q$ as $n \to \infty$ for all sequences (t_n) in $]x, b[$ such that $t_n \to x$ and say "the right-hand side limit of f at x is q." Sometimes one denotes

$$\lim_{\substack{t \to x, t > x}} f(x) = \lim_{\substack{t \to x \\ t > x}} f(x) = \lim_{t \downarrow x} f(x) = f(x+).$$

Consider any point x such that $a < x \leq b$. We write

$$f(x-) = q$$

if $f(t_n) \to q$ as $n \to \infty$ for all sequences (t_n) in $]a, x[$ such that $t_n \to x$ and say "the left-hand side limit of f at x is q." Sometimes one denotes

$$\lim_{\substack{t \to x, t < x}} f(x) = \lim_{\substack{t \to x \\ t < x}} f(x) = \lim_{t \uparrow x} f(x) = f(x-).$$

Theorem 1.5. *Let I be a nonempty interval and $p \in I$. If $f : I \to \mathbb{R}$, then*

$$\lim_{x \to p} f(x) = q \iff f(p-) = f(p+) = q.$$

Proof. The necessity part is immediate.

Suppose that $f(x-) = f(x+) = q$. We use the characterization in Theorem 1.1 and consider a sequence (x_n) in I, $x_n \to p$ and having infinitely many terms greater than p (denoted by (y_n)), and infinitely many terms less than p (denoted by (z_n)). By hypothesis, $f(y_n) \to q$ and $f(z_n) \to q$. Then $f(x_n) \to q$.   □

4.2 Continuity

Suppose X and Y are metric spaces, $E \subset X$, $p \in E$, and f maps E into Y. Then f is said to be *continuous* at p if for every $\varepsilon > 0$, there exists a $\delta > 0$ such that

$$\rho(f(x), f(p)) < \varepsilon,$$

for all points $x \in E$ for which $\rho(x, p) < \delta$.

If f is continuous at every point of E, then f is said to be *continuous on E*.

Remark. It should be noted that f has to be defined at the point p in order to be continuous at p. If p is an isolated point (page 63) of E, then our definition implies that function f is continuous at p.   △

Theorem 2.1. *In the above setting, assume that p is a point in E. Then f is continuous at p if and only if $\lim_{x \to p} f(x) = f(p)$.*

Proof. If $\lim_{x \to p} f(x) = f(p)$, for every $\varepsilon > 0$ there exists a $\delta(= \delta(\varepsilon, p)) > 0$ such that $\rho(f(x), f(p)) < \varepsilon$ for all points $x \in E$ for which $0 < \rho(x, p) < \delta$. If $x = p$, $\rho(f(x), f(p)) < \varepsilon$ holds obviously. The other side runs easily.   □

Corollary 2.2. *Suppose all the assumptions of Theorem 2.1 are satisfied. Then f is continuous at p if and only if for every sequence (p_n) so that $p_n \to p$ one has $\lim_{n \to \infty} f(p_n) = f(p)$.*

Proof. It follows from Theorem 2.1 and Theorem 1.1.   □

Now we prove that a continuous function of a continuous function is continuous, more precisely that the following theorem holds.

Theorem 2.2. *Suppose X, Y, and Z are metric spaces, $E \subset X$, f maps E in Y, g maps the range of f, $f(E)$, into Z, and h maps E into Z defined by*

$$h(x) = g(f(x)), \quad x \in E.$$

If f is continuous at a point $p \in E$ and if g is continuous at $f(p)$, then h is continuous at p.

Proof. Let $\varepsilon > 0$ be given. Because g is continuous at $f(p)$, there exists $\eta > 0$ such that if $\rho(y, f(p)) < \eta$ and $y \in f(E)$, we have

$$\rho(g(y), g(f(p))) < \varepsilon .$$

Because f is continuous at p, there exists $\delta > 0$ such that $\rho(f(x), f(p)) < \eta$, if $\rho(x, p) < \delta$ and $x \in E$. It follows that

$$\rho(h(x), h(p)) = \rho(y(f(x)), g(f(p))) < \varepsilon,$$

if $\rho(x, p) < \delta$ and $x \in E$. Thus h is continuous at p.   □

The next result is a largely used characterization of continuity by open sets.

Theorem 2.3. *A mapping f of a metric space X into a metric space Y is continuous on X if and only if $f^{-1}(V)$ is open in X for every open set V in Y.*

Proof. Suppose f is continuous on X and V is an open set in Y. We have to show that every point of $f^{-1}(V)$ is an interior point of $f^{-1}(V)$. Suppose $p \in X$, $f(p) \in V$. Because V is open, there exists $\varepsilon > 0$ such that $y \in V$ if $\rho(f(p), y) < \varepsilon$, and because f is continuous at p there exists $\delta > 0$ such that $\rho(f(x), f(p)) < \varepsilon$ if $\rho(x, p) < \delta$. Thus $x \in f^{-1}(V)$ as soon as $\rho(x, p) < \delta$.

Conversely, suppose $f^{-1}(V)$ is open in X for every open V in Y. Let $p \in X$ and $\varepsilon > 0$. Let

$$V = \{y \in Y \mid \rho(y, f(p)) < \varepsilon\}.$$

Then V is open and hence $f^{-1}(V)$ is open. Therefore, there exists $\delta > 0$ such that $x \in f^{-1}(V)$ as soon as $\rho(p, x) < \delta$. However, if $x \in f^{-1}(V)$, then $f(x) \in V$, so that $\rho(f(x), f(p)) < \varepsilon$. $\square$

Corollary 2.3. *A mapping f of a metric space X into a metric space Y is continuous on X if and only if $f^{-1}(C)$ is closed in X for every closed set C in Y.*

Proof. It follows from the previous theorem and (e) of Theorem 1.5 at page 9. $\square$

Theorem 2.4. *Let f and g be complex-valued continuous functions on a metric space X. Then $f + g$, fg, and f/g are continuous on X $(g(x) \neq 0$ for all $x \in X)$.*

Proof. At isolated points of X there is nothing to prove. At limit points, the statement follows by Theorems 1.4 and 2.1. $\square$

Theorem 2.5. (a) *Let $f_1, \ldots, f_k$ be real functions on a metric space X, and let f be the mapping of X into $\mathbb{R}^k$ defined by*

$$f(x) = (f_1(x), \ldots, f_k(x)), \quad x \in X.$$

Then f is continuous if and only if each of the functions $f_1(x), \ldots, f_k(x)$ is continuous.
(b) *If f and g are continuous mappings of X into $\mathbb{R}^k$, then $f + g$ and $\langle f, g \rangle$ are continuous on X.*

Functions $f_1, \ldots, f_k$ are called *components* of f.

Proof. Part (a) follows from the inequalities

$$|f_j(x) - f_j(y)| \leq \|f(x) - f(y)\|_2 = \left(\sum_{i=1}^{k} |f_i(x) - f_i(y)|^2 \right)^{1/2},$$

for $j = 1, \ldots, k$. Part (b) follows from (a) and Theorem 1.4. $\square$

An important result in analysis is the following uniform approximation theorem of Weierstrass.

Theorem 2.6. (Weierstrass) *Suppose f is a continuous function on $[a, b]$. Then there exists a sequence of polynomials (P_n) so that*

$$\lim_{n \to \infty} P_n(x) = f(x) \tag{4.7}$$

uniformly on $[a, b]$; in other words, for every $\varepsilon > 0$ there is a polynomial P such that

$$\|P - f\|_\infty < \varepsilon. \tag{4.8}$$

The statement of the above theorem can be rephrased as: the set of polynomials on $[a, b]$ is dense in the space of continuous functions on $[a, b]$ endowed with the uniform norm $(C[a, b], \| \cdot \|_\infty)$.

Due to its importance in approximation theory and not only, the uniform approximation theorem of Weierstrass was and still is the object of special attention from many great names in mathematics. They produced many interesting proofs. Below we introduce one of them given by S. N. Bernstein.[2] Because we need the concept of uniform continuity, the proof of this theorem is introduced at page 157.

It holds an even stronger result; that is, a polynomial that approximates and interpolates simultaneously.

Theorem 2.7. *Let $x_1, x_2, \ldots, x_n$ be distinct points of $[a, b]$ and f a continuous function on $[a, b]$. Then there exists a polynomial P that passes through the points $(x_1, f(x_1)), \ldots, (x_n, f(x_n))$ satisfying* (4.8).

Proof. From Theorem 2.6 we have that there is a polynomial Q so that $\|Q - f\|_\infty < \varepsilon$. With $\lambda_k = f(x_k) - Q(x_k)$, $1 \le k \le n$, let

$$R(x) = \sum_{k=1}^{n} \lambda_k \prod_{i=1, \, i \neq k}^{n} \frac{x - x_i}{x_k - x_i}, \text{ and } P(x) = Q(x) + R(x).$$

Then

$$|R(x)| \le \varepsilon \sup_{a \le x \le b} \sum_{k=1}^{n} \prod_{\substack{i=1 \\ i \neq k}}^{n} \left| \frac{x - x_i}{x_k - x_i} \right| = A\varepsilon,$$

$$|P(x) - f(x)| = |Q(x) + R(x) - f(x)| \le |Q(x) - f(x)| + |R(x)| \le (1 + A)\varepsilon,$$

which implies that $P(x_k) = f(x_k)$. $\square$

4.2.1 Continuity and compactness

A mapping f of a set E into $\mathbb{R}^k$ is said to be *bounded* if there is a real number m such that $\|f(x)\| < m$, for all $x \in E$.

---

[2] Sergey Natanovich Bernstein (Сергей Натанович Бернстейн), 1880–1968.

Theorem 2.8. *Suppose f is a continuous mapping from a compact metric space X into a metric space Y. Then $f(X)$ is compact.*

Proof. Let $\{V_\alpha\}$ be an open covering of $f(X)$. Because f is continuous, each of the sets $f^{-1}(V_\alpha)$ is open, by Theorem 2.3. Because X is compact, there are finitely many indices, say $\alpha_1, \ldots, \alpha_n$, such that

$$X \subset f^{-1}(V_{\alpha_1}) \cup f^{-1}(V_{\alpha_2}) \cup \cdots \cup f^{-1}(V_{\alpha_n}). \qquad (4.9)$$

Because $f(f^{-1}(E)) = E$, for every $E \subset Y$, (4.9) implies that

$$f(X) \subset V_{\alpha_1} \cup V_{\alpha_2} \cup \cdots \cup V_{\alpha_n}.$$

This completes the proof. $\square$

Theorem 2.9. *Suppose f is a continuous mapping from a compact metric space X into $\mathbb{R}^k$. Then $f(X)$ is closed and bounded. Thus f is bounded.*

Proof. By Theorem 2.8 we have that $f(X)$ is compact. Theorem 3.13 at page 69 guarantees that $f(X)$ is closed and bounded. $\square$

Theorem 2.10. *Suppose f is a continuous real function on a compact metric space X, and*
$$M = \sup_{p \in X} f(p), \quad m = \inf_{p \in X} f(p).$$
Then there exist points $p, q \in X$ such that $f(p) = M$, $f(q) = m$.

Proof. By Theorem 2.9, $f(X)$ is a closed and bounded set of real numbers. Hence $f(X)$ contains its sup and inf, by Theorem 2.27 and Corollary 2.19, both at page 29. $\square$

Theorem 2.11. *Suppose f is a continuous bijective mapping of a compact metric space X onto a metric space Y. Then the inverse function f^{-1} defined on Y is a continuous mapping of Y onto X.*

Proof. Applying Theorem 2.3 to f^{-1} in place of f, we see that it suffices to prove that $f(V)$ is an open set in Y for every open set V in X. Fix such a set V. The complement $X \setminus V$ is closed in X, hence compact (Theorem 3.6, page 67). It follows that $f(X \setminus V)$ is a compact subset of Y which implies that it is closed in Y. Because f is bijective, $f(X \setminus V) = Y \setminus f(V)$. Hence $f(V)$ is open. $\square$

4.2.2 Uniform continuous mappings

Let f be a mapping on a metric space X with values in a metric space Y. We say that f is *uniformly continuous* on X if for every $\varepsilon > 0$ there exists $\delta(= \delta(\varepsilon)) > 0$ such that
$$\rho(f(p), f(q)) < \varepsilon$$

for all $p, q \in X$ for which $\rho(p, q) < \delta$.

Remark. Uniform continuity is the property of a function on a set, whereas continuity can be defined at a single point. Asking whether a given function is uniformly continuous at a certain point is meaningless. Generally, if f is continuous on X, it is possible to find, for each $\varepsilon > 0$ and each point $p \in X$, a number $\delta > 0$ having the property specified in definition of continuity; thus $\delta = \delta(\varepsilon, p)$.

However, if f is uniformly continuous on X, then it is possible, for each $\varepsilon > 0$, to find one number $\delta > 0$ that is acceptable for all points $p \in X$.

It is trivial to see that every uniformly continuous function is continuous. That the two concepts are equivalent on compact sets is a nontrivial fact and follows from the next theorem. $\triangle$

Theorem 2.12. (Cantor) *Let f be a continuous mapping of a compact metric space X into a metric space Y. Then f is uniformly continuous on X.*

Proof. Let $\varepsilon > 0$ be given. Because f is continuous, we can associate with each point $p \in X$, a positive number $\phi(p)$ such that

$$x \in X, \ \rho(p, x) < \phi(p) \implies \rho(f(p), f(x)) < \varepsilon/2. \tag{4.10}$$

Let $I(p)$ be the ball defined as $I(p) = B(p, \phi(p)/2)$. Because $p \in I(p)$, the family of all sets $I(p)$ is an open covering of X and because X is compact, there is a finite set of points $p_1, \ldots, p_n$ in X, such that

$$X = I(p_1) \cup \cdots \cup I(p_n). \tag{4.11}$$

Set

$$\delta = (1/2) \min\{\phi(p_1), \ldots, \phi(p_n)\}(> 0).$$

Now, let x and p be points of X such that $\rho(x, p) < \delta$. By (4.11), there is an integer m, $1 \le m \le n$, such that $p \in I(p_m)$. It follows that $\rho(p, p_m) < \phi(p_m)/2$, and also that

$$\rho(x, p_m) \le \rho(x, p) + \rho(p, p_m) < \delta + (1/2)\phi(p_m) \le \phi(p_m).$$

This, together with (4.10) implies that

$$\rho(f(p), f(x)) \le \rho(f(p), f(p_m)) + \rho(f(p_m), f(x)) < \varepsilon. \quad \square$$

Proof of Theorem 2.6. The interval $[a, b]$ can be transformed by a first-degree polynomial into the interval $[0, 1]$ in a bijective way (and thus the function and its inverse are continuous), therefore we consider our problem as been defined on $[0, 1]$.

We prove that there exists a sequence of polynomials fulfilling (4.7) uniformly. This sequence is defined by

$$B_m(f)(x) = \sum_{k=0}^{m} p_{m,k}(x) f(k/m),\tag{4.12}$$

where

$$p_{m,k}(x) = \binom{m}{k} x^k (1-x)^{m-k}, \quad x \in [0,1].$$

For every $x \in [0,1]$, we have

$$p_{m,k}(x) \geq 0, \quad \sum_{k=0}^{m} p_{m,k}(x) = 1.\tag{4.13}$$

We show that

$$\lim_{m \to \infty} B_m(f) = f, \quad \text{uniformly on} \quad [0,1].\tag{4.14}$$

Function f is continuous on the compact $[0,1]$; then it is uniformly continuous on $[0,1]$. So, given $\varepsilon > 0$ there exists $\delta = \delta(\varepsilon)$ such that for every $x, x_0 \in [0,1]$ with $|x - x_0| < \delta$, we have

$$|f(x) - f(x_0)| < \varepsilon/2.\tag{4.15}$$

Set $M = \|f\|_\infty$. By the identity in (4.13), we have

$$f(x) - B_m(f)(x) = \sum_{k=0}^{m} p_{m,k}(x) \left(f(x) - f(k/m) \right).$$

Using the inequality in (4.13), one has

$$|f(x) - B_m(f)(x)| \leq \sum_{k=0}^{m} p_{m,k}(x) |f(x) - f(k/m)|$$

$$\leq \sum_{k \in I_m} p_{m,k}(x) |f(x) - f(k/m)| + \sum_{k \in J_m} p_{m,k}(x) |f(x) - f(k/m)|,$$

where

$$I_m = \{k \mid |k/m - x| < \delta\} \text{ and } J_m = \{k \mid |k/m - x| \geq \delta\}.$$

Then by (4.15) it follows

$$|f(x) - B_m(f)(x)| < \frac{\varepsilon}{2} \sum_{k \in I_m} p_{m,k}(x) + 2M \sum_{k \in J_m} p_{m,k}(x).\tag{4.16}$$

We may increase the first sum to 1. We deal now with the second sum. From $|k/m - x| \geq \delta$ we write $1 \leq \delta^{-2}(k/m - x)^2$. Thus

$$\sum_{k \in J_m} p_{m,k}(x) \leq \frac{1}{\delta^2} \sum_{k \in J_m} \left(\frac{k}{m} - x \right)^2 p_{m,k}(x) \leq \frac{1}{\delta^2} \sum_{k=0}^{m} \left(\frac{k}{m} - x \right)^2 p_{m,k}(x),$$

$$\tag{4.17}$$

and furthermore,

$$\sum_{k=0}^{m} (k/m - x)^2 \, p_{m,k}(x) \tag{4.18}$$

$$= \sum_{k=0}^{m} (k/m)^2 p_{m,k}(x) - 2x \sum_{k=0}^{m} (k/m) p_{m,k}(x) + x^2 \sum_{k=0}^{m} p_{m,k}(x)$$

$$= B_m(e_2)(x) - 2x B_m(e_1)(x) + x^2,$$

where

$$e_1(x) = x, \quad e_2(x) = x^2, \quad x \in [0,1].$$

On the other hand,

$$B_m(e_1)(x) = \sum_{k=1}^{m} \frac{k}{m} \binom{m}{k} x^k (1-x)^{m-k} = \sum_{k=1}^{m} \binom{m-1}{k-1} x^k (1-x)^{m-k}.$$

We change the index of summation by $k - 1 = j$ and thus we have

$$B_m(e_1)(x) = \sum_{j=0}^{m-1} \binom{m-1}{j} x^{j+1}(1-x)^{m-j-1} = x(x+1-x)^{m-1} = x.$$

By the identity $k^2 = k(k-1) + k$ we have

$$B_m(e_2)(x) = \sum_{k=1}^{m} \frac{k(k-1)}{m^2} \binom{m}{k} x^k (1-x)^{m-k} + \frac{1}{m} \sum_{k=1}^{m} \frac{k}{m} p_{m,k}(x)$$

$$= \frac{m-1}{m} \sum_{k=2}^{m} \frac{(m-2)!}{(k-2)!(m-k)!} x^k (1-x)^{m-k} + \frac{x}{m}.$$

Taking $j = k - 2$, we have

$$B_m(e_2)(x) = \frac{m-1}{m} \sum_{j=0}^{m-2} \frac{(m-2)!}{(j)!(m-j-2)!} x^{j+2}(1-x)^{m-2-j} + \frac{x}{m}$$

$$= \frac{m-1}{m} x^2 (x+1-x)^{m-2} + \frac{x}{m} = \frac{m-1}{m} x^2 + \frac{x}{m} = x^2 + \frac{x(x-1)}{m}.$$

Taking into account these results and (4.18), we can write the identity

$$\sum_{k=0}^{m} (k/m - x)^2 \, p_{m,k}(x) = x(x-1)/m.$$

We substitute this result in (4.17) and because $x(1-x) \le 1/4$, we further get

$$\sum_{k=0}^{m} p_{m,k}(x) \le (1/\delta^2) x(1-x)/m \le 1/(4m\delta^2). \tag{4.19}$$

We substitute (4.19) into (4.16) and thus we can write

$$|f(x) - B_m(f)(x)| \leq \varepsilon/2 + M/(2m\delta^2).$$

Now, if $m > M/(\varepsilon\delta^2)$, we obtain the desired inequality

$$|f(x) - B_m(f)(x)| < \varepsilon, \quad \forall x \in [0,1]. \quad \square$$

4.2.3 Continuity and connectedness

A set E in a metric space X is said to be *connected* if there do not exist two disjoint open sets A and B, $A \cap E \neq \emptyset$, $B \cap E \neq \emptyset$ such that $E \subset A \cup B$.

Theorem 2.13. *On the real axis, a set is connected if and only if it is an interval.*

Similar to Theorem 2.8, we have the following.

Theorem 2.14. *Suppose f is a continuous mapping of a connected metric space X into a metric space. Then $f(X)$ is connected.*

Proof. If $f(X)$ is not connected, there are disjoint open sets V and W in Y, both of which intersect $f(X)$, such that $f(X) \subset V \cup W$. Because f is continuous, the sets $f^{-1}(V)$ and $f^{-1}(W)$ are open in X; they are clearly disjoint and nonempty, and their union is X. This implies that X is not connected, in contradiction to the hypothesis. $\square$

Let $I \subset \mathbb{R}$ be a nonempty interval and $f : I \to \mathbb{R}$ be a mapping. We say that f is a *Darboux*[3] *function* if for any $a, b \in I$, $a < b$ and any λ between $f(a)$ and $f(b)$, there is $c \in]a, b[$ such that $f(c) = \lambda$. We denote the set of Darboux functions on an interval I by $\mathcal{D}_I$.

Now we introduce the Darboux property of continuous functions also called the *intermediate value property*.

Theorem 2.15. (Darboux) *Let f be a continuous real function on the interval $[a, b]$. If $f(a) < f(b)$ and if λ is a number such that $f(a) < \lambda < f(b)$, there exists a point $c \in]a, b[$ such that $f(c) = \lambda$.*

A similar result holds if $f(a) > f(b)$.

Proof. By Theorem 2.13, $[a, b]$ is connected; hence, Theorem 2.14 shows that $f([a, b])$ is a connected subset of $\mathbb{R}$, and the assertion follows if we appeal again to Theorem 2.13. $\square$

---

[3] Jean Gaston Darboux, 1842–1917.

4.2.4 Discontinuities

Suppose x is a point in the domain of definition of a function f at which f is not continuous. Then we say that f is *discontinuous* at x, or that f has a discontinuity at x.

Let f be defined on $]a, b[$. If f is discontinuous at a point $x \in]a, b[$, and if $f(x+)$ and $f(x-)$ exist, then f is said to have a *discontinuity of the first kind* or a *simple discontinuity*. Otherwise the discontinuity is said to be of *the second kind*. There are two ways in which a function can have a simple discontinuity: either $f(x+) \neq f(x-)$ (in which case the value $f(x)$ is immaterial), or $f(x+) = f(x-) \neq f(x)$.

Examples. (a) Define

$$f(x) = \begin{cases} 1, & x \in \mathbb{Q}, \\ 0, & x \in \mathbb{R} \setminus \mathbb{Q}. \end{cases}$$

Then f has a discontinuity of the second kind at the point x, because neither $f(x+)$ nor $f(x-)$ exist.
(b) Define

$$f(x) = \begin{cases} x, & x \in \mathbb{Q}, \\ 0, & x \in \mathbb{R} \setminus \mathbb{Q}. \end{cases}$$

Then f is continuous at $x = 0$ and has a discontinuity of the second kind at every other point.
(c) Define

$$f(x) = \begin{cases} \sin(1/x), & x \neq 0, \\ 0, & x = 0. \end{cases}$$

Neither $f(0+)$ nor $f(0-)$ exist, therefore f has a discontinuity of the second kind at $x = 0$. However, f is continuous at every point $x \neq 0$. $\triangle$

4.2.5 Monotonic functions

Consider $f :]a, b[\to \mathbb{R}$, $a < b$. Then f is said to be *monotonically increasing* on $]a, b[$ if $a < x < y < b$ implies $f(x) \leq f(y)$. If the last inequality is reversed, we obtain the definition of a *monotonically decreasing* function. The set of *monotonic* functions consists of both the increasing and the decreasing functions.

Consider again $f :]a, b[\to \mathbb{R}$, $a < b$. Then f is said to be *strictly monotonically increasing* or *strictly increasing* on $]a, b[$ if $a < x < y < b$ implies $f(x) < f(y)$. If the last inequality is $>$, then we obtain the definition of a *strictly monotonically decreasing* or *strictly decreasing* function. The set of *strictly monotonic* functions consists of both the strictly increasing and the strictly decreasing functions.

Some arithmetic properties of monotonic functions are introduced below.

Theorem 2.16. *Consider* $f, g : I \to \mathbb{R}$, *where* I *is a nonempty interval.*

(a) $f + g$ *is monotonically increasing (decreasing) on* I *whenever* f *and* g *are monotonically increasing (decreasing) on* I.

(b) *Suppose* $\alpha > 0$. *Then* αf *is monotonically increasing (decreasing) on* I *whenever* f *is monotonically increasing (decreasing) on* I.

(c) *Suppose* $\alpha < 0$. *Then* αf *is monotonically increasing (decreasing) on* I *whenever* f *is monotonically decreasing (increasing) on* I.

(d) *Suppose* f *and* g *are (strictly) monotonic functions both either positive or negative. Then* fg *is (strictly) monotonic.*

(e) *Suppose* f *is strictly increasing (decreasing) on* I. *Then* $f^{-1} : f(I) \to \mathbb{R}$ *is strictly increasing (decreasing) on* $f(I)$.

Proof. The sentences (a)–(d) are obvious.

(e) Suppose f is strictly increasing. Then f is one-to-one. Take $f : I \to f(I)$. Then this mapping is onto. Thus we can consider $f^{-1} : f(I) \to \mathbb{R}$. Suppose f^{-1} is not strictly increasing. Then there exist $u, v \in f(I)$ so that $u < v$ implies $f^{-1}(u) \geq f^{-1}(v)$. There exist two unique points $x, y \in I$ such that $f^{-1}(u) = x$ and $f^{-1}(v) = y$. Then

$$u < v \implies x = f^{-1}(u) \geq f^{-1}(v) = y \implies u = f(x) \geq f(y) = v.$$

We got a contradiction, therefore the inverse function is strictly increasing. $\square$

Remarks. (a) The sum of two monotonic functions need not be a monotonic function. Let $f(x) = x$, $g(x) = -x^2$, $x \geq 0$. Then $f + g$ is neither increasing nor decreasing on $[0, \infty[$. So, the set of monotonic functions does not form a vector space.

(b) The product of two monotonic functions need not be a monotonic function. Indeed, consider $f, g : \mathbb{R} \to \mathbb{R}$ defined by

$$f(t) = t, \quad g(t) = \operatorname{sign} t.$$

Then $(fg)(t) = |t|$, which is not monotonic on $\mathbb{R}$. $\triangle$

Theorem 2.17. *Suppose* $I, J,$ *and* K *are nonempty intervals,* f *maps* I *in* J, g *maps the range of* f, $f(I)$, *into* K, *and* f *and* g *are both monotonically increasing (decreasing) on* I, *respectively on* $f(I)$. *Then function* h *defined by*

$$h(x) = g(f(x)), \quad x \in I$$

is monotonically increasing.

Theorem 2.18. *Let* f *be monotonically increasing on* $]a, b[$. *Then* $f(x+)$ *and* $f(x-)$ *exist at every point* $x \in]a, b[$. *More precisely,*

$$\sup_{a < t < x} f(t) = f(x-) \leq f(x) \leq f(x+) = \inf_{x < t < b} f(t). \tag{4.20}$$

Furthermore, if $a < x < y < b$, then

$$f(x+) \leq f(y-). \tag{4.21}$$

Proof. The set of numbers $f(t)$ with $a < t < x$ is bounded above by $f(x)$. Therefore it has a least upper bound that we denote by A. It is obvious that $A \leq f(x)$. We have to show that $A = f(x-)$.

Let $\varepsilon > 0$ be given. It follows from the definition of A as supremum that there exists $\delta > 0$ such that $a < x - \delta < x$ and

$$A - \varepsilon < f(x - \delta) \leq A. \tag{4.22}$$

Because f is monotonic, we have

$$f(x - \delta) \leq f(t) \leq A, \quad x - \delta < t < x. \tag{4.23}$$

Combining (4.22) and (4.23), we obtain $|f(t) - A| < \varepsilon$, $x - \delta < t < x$. Hence $f(x-) = A$.

The second half of (4.20) is proven in precisely the same way. Next, if $a < x < y < b$ we deduce from (4.20) that

$$f(x+) = \inf_{x<t<b} f(t) = \inf_{x<t<y} f(t). \tag{4.24}$$

Similarly

$$f(y-) = \sup_{a<t<y} f(t) = \sup_{x<t<y} f(t). \tag{4.25}$$

Comparing (4.24) and (4.25), we get (4.21). □

Corollary 2.4. *A monotonic function has no discontinuity of the second kind.*

Theorem 2.19. *Let f be monotonic on $]a, b[$. Then the set of points of $]a, b[$ on which f is discontinuous is at most countable.*

Proof. Suppose, for the sake of definiteness, that f is increasing, and let E be the set of points at which f is discontinuous.

With every point $x \in E$, we associate a rational number $r(x)$ such that

$$f(x-) < r(x) < f(x+).$$

Because $x_1 < x_2$, it follows that $f(x_1+) \leq f(x_2-)$. We note that $r(x_1) \neq r(x_2)$, if $x_1 \neq x_2$.

We have thus established a one-to-one correspondence between the set E and a subset of the rational numbers. The latter is countable. □

Theorem 2.20. *Suppose $f : I \to \mathbb{R}$ is continuous and strictly increasing (decreasing). Then $f^{-1} : f(I) \to \mathbb{R}$ is continuous and strictly increasing (decreasing) on $f(I)$.*

Proof. Suppose f is strictly increasing on I. By (e) in Theorem 2.16 we proved that f^{-1} is strictly increasing on $f(I)$. We choose an arbitrary element $y_0 = f(x_0) \in f(I)$ and show that f^{-1} is continuous on y_0. We distinguish three cases.

Suppose $x_0 \in \text{int}(I)$. Then for every positive ε one can find x_1 and x_2 so that

$$x_0 - \varepsilon < x_1 < x_0 < x_2 < x_0 + \varepsilon.$$

Denote $y_1 = f(x_1)$ and $y_2 = f(x_2)$. Then $y_1 < y_0 < y_2$, so $U = \,]y_1, y_2[$ is an open neighborhood of y_0. For $y \in U \cap f(I)$ there is unique $x \in \,]x_1, x_2[$ with $y = f(x)$. Then

$$|f^{-1}(y) - f^{-1}(y_0)| = |x - x_0| < 2\varepsilon,$$

and thus f^{-1} is continuous on every interior point in I.

Suppose x_0 is the leftmost point in I. Then for every positive ε there exists $x_2 \in I$ such that $x_0 < x_2 < x_0 + \varepsilon$. Denote $y_2 = f(x_2)$. Then $]x_0 - \varepsilon, x_2[$ is an open neighborhood of x_0. For $y \in U \cap f(I)$ there is unique $x \in \,]x_1, x_2[$ with $y = f(x)$. Then

$$|f^{-1}(y) - f^{-1}(y_0)| = |x - x_0| < \varepsilon,$$

and thus f^{-1} is continuous at y_0.

The case when x_0 is the rightmost point in I runs similarly. $\square$

Theorem 2.21. *Suppose $f : I \to \mathbb{R}$ is monotonic. Then f is continuous on I if and only if the range of f is an interval.*

Proof. If f is continuous, $f(I)$ is an interval by Theorem 2.14.

Suppose f is increasing and the range of f is an interval. Then the conclusion follows by (4.20). $\square$

Corollary 2.5. *Consider two real intervals I and J. A monotonic and onto function $f : I \to J$ is continuous.*

A function is *nowhere monotone* if there exists no interval where the function is monotone.

4.3 Periodic functions

Consider a function $f : \mathbb{R} \to \mathbb{R}$. Function f is said to be *periodic* provided there exists a real nonzero constant a satisfying $f(x) = f(x + a)$, for every $x \in \mathbb{R}$. Such a nonzero constant a is said to be a *period* of f. Note that if a function has a period a, then any positive multiplier of a is a period of that function as well.

4.4 Darboux functions

We recall the definition of a Darboux function from page 160. Let $I \subset \mathbb{R}$ be a nonempty interval and $f : I \to \mathbb{R}$ be a mapping. We say that f is a Darboux function if for any $a, b \in I$, $a < b$ and any λ between $f(a)$ and $f(b)$, there is $c \in]a, b[$ such that $f(c) = \lambda$.

Remark. Geometrical the Darboux property says that for every $a, b \in I$, $a < b$ and every λ between $f(a)$ and $f(b)$, the parallel to the Ox axis through $y = \lambda$ meets the restriction of f to the open interval $]a, b[$ at least in one point. $\triangle$

Proposition 4.1. *For a mapping* $f : I \to \mathbb{R}$ *the next statements are equivalent.*

(a) f *is a Darboux function.*
(b) *If* $J \subset I$ *is an interval, then* $f(J)$ *is an interval (the image of an interval is an interval).*
(c) *If* $a, b \in I$, $a < b$, $f([a, b])$ *is an interval.*

Proof. (a) $\implies$ (b). For every $y_1 = f(t_1)$, $y_2 = f(t_2) \in f(J)$ (we may suppose that $y_1 < y_2$) and every $\lambda \in]y_1, y_2[$ by (a), there is $c \in]t_1, t_2[$ so that $\lambda = f(c)$. Because J is an interval and $t_1, t_2 \in J$, it follows that $c \in J$ and therefore $\lambda = f(c) \in f(J)$. Thus $f(J)$ is an interval.
(b) $\implies$ (c). It is trivial.
(c) $\implies$ (a). Let $a, b \in I$, $a < b$ and any λ between $f(a)$ and $f(b)$. Then $\lambda \in f([a, b])$ and by (c), there is $c \in [a, b]$ such that $\lambda = f(c)$. We note that $c \notin \{a, b\}$ and $\lambda = f(c)$. Hence f is a Darboux mapping on I. $\square$

Corollary 4.6. *Let* $f : I \to \mathbb{R}$ *and* $g : f(I) \to \mathbb{R}$ *be two Darboux mappings. Then their composition* $g \circ f$ *is a Darboux mapping.*

Proof. If f, g are Darboux mappings and J is a subinterval of I, $(g \circ f)(J) = g(f(J))$ is an interval. Thus, by Proposition 4.1, the mapping $g \circ f$ is a Darboux mapping. $\square$

Corollary 4.7. *Let* $f : I \to \mathbb{R}$ *be a Darboux mapping whose range is an at most countable set. Then* f *is constant.*

Proof. $f(I)$ is an interval and is at most countable. Thus $|f(I)| - 1$; that is, f is constant. $\square$

Corollary 4.8. *Let* $f : I \to \mathbb{Q}$ *be a continuous function. Then* f *is constant.*

Proof. Function f is a Darboux function. Then we apply Corollary 4.7 $\square$

Corollary 4.9. *Let* $f : I \to \mathbb{R}$ *be a Darboux mapping. Then* f *is one-to-one if and only if is strictly monotonic.*

Proof. Sufficiency. If f is strictly monotonic, it is one-to-one.

Necessity. Suppose $f : I \to \mathbb{R}$ is a Darboux and one-to-one function. If f is not strictly monotonic, there exist $t_1 < t_2 < t_3$ such that either $f(t_1) < f(t_2) > f(t_3)$ or $f(t_1) > f(t_2) < f(t_3)$. Let us suppose that the first case holds. The other case runs similarly. The first case has two subcases.

- $f(t_1) < f(t_3) < f(t_2)$. Then there is $c \in]t_1, t_2[$ with $f(c) = f(t_3)$. Thus function f is not injective which is a contradiction.
- $f(t_3) < f(t_1) < f(t_2)$. Reasoning as before, we find $c \in]t_2, t_3[$ with $f(c) = f(t_1)$. Thus the injectivity of f is contradicted. $\quad\square$

Corollary 4.10. *Let $f : I \to \mathbb{R}$ be a Darboux mapping. Then f is also a Darboux function on any subinterval $J \subset I$.*

Proof. It follows from (ii) of Proposition 4.1. $\quad\square$

Proposition 4.2. *Suppose $f : I \to \mathbb{R}$ is a Darboux mapping. Then*

(a) $|f|$, f^2, and $\sqrt{|f|}$ *are Darboux mappings on I.*
(b) *If $f(t) \neq 0$ for every $t \in I$, then $1/f$ is a Darboux mapping on I.*

Proof. Apply Corollary 4.6 to mapping f and to mappings $g_1(t) = |t|$, $g_2(t) = t^2$, $g_3(t) = \sqrt{|t|}$, respectively, $g(t) = 1/t$. $\quad\square$

Remark. We introduce two Darboux functions whose sum is no longer a Darboux function.

$$f(t) = \begin{cases} \sin(1/t), & t \neq 0, \\ 0, & t = 0, \end{cases} \qquad g(t) = \begin{cases} -\sin(1/t), & t \neq 0, \\ 1, & t = 0, \end{cases}$$

are two Darboux functions whereas their sum

$$(f + g)(t) = \begin{cases} 0, & t \neq 0, \\ 1, & t = 0, \end{cases}$$

is not. $\quad\triangle$

Corollary 4.11. *Let $f : I \to \mathbb{R}$ be a Darboux mapping. Then f has no discontinuity point of the first kind.*

Theorem 4.1. (Sierpinski[4]) *Every mapping $f : \mathbb{R} \to \mathbb{R}$ is a sum of two discontinuous Darboux mappings; that is, there exist f_1 and f_2 Darboux and discontinuous functions such that*

$$f(t) = f_1(t) + f_2(t), \quad \forall t \in \mathbb{R}.$$

---

[4] Wacław Sierpinski, 1882–1969.

4.5 Lipschitz functions

Let I be a real interval and $f : I \to \mathbb{R}$ be a function. Then f is said to be *Lipschitz*[5] on I if there exists a nonnegative L such that

$$|f(x) - f(y)| \le L|x - y|, \quad \forall x, y \in I.$$

Remark. Every Lipschitz function on an interval is uniformly continuous on it.   △

Let A be a nonempty set and $f : A \to A$ be a mapping. A point $x \in A$ is said to be a *fixed point* of f if $f(x) = x$.

Let (X, ρ) be a metric space and $f : X \to X$ be a mapping. f is said to be a *contraction* if there exists a constant $\alpha \in]0, 1[$ so that

$$\rho(f(x), f(y)) \le \alpha \rho(x, y), \quad \text{for every } x, y \in X.$$

In this very case the mapping f is said to be an α-*contraction*, too.

Remark. Every α-contraction is a Lipschitz function.   △

Theorem 5.1. (Banach fixed point theorem) *Let* (X, ρ) *be a nonempty complete metric space and let* $f : X \to X$ *be an* α-*contraction. Then* f *has a unique fixed point.*

Proof. Set an arbitrary, but fixed, point $x_0 \in X$, and define the sequence $(x_n)_{n \ge 1}$ by $x_{n+1} = f(x_n)$, $n = 0, 1, \ldots$. For every $n \in \mathbb{N}^*$ and every integer p, $p \ge 1$, we have the estimates

$$\rho(x_{n+1}, x_n) = \rho(f(x_n), f(x_{n-1})) \le \alpha \rho(x_n, x_{n-1}) \le \cdots \le \alpha^n \rho(x_1, x_0),$$

and by the triangle inequality

$$
\begin{aligned}
\rho(x_{n+p}, x_n) &\le \rho(x_{n+p}, x_{n+p-1}) + \rho(x_{n+p-1}, x_{n+p-2}) + \cdots + \rho(x_{n+1}, x_n) \\
&\le (\alpha^{n+p-1} + \alpha^{n+p-2} + \cdots + \alpha^n)\rho(x_1, x_0) \\
&= \alpha^n(\alpha^{p-1} + \alpha^{p-2} + \cdots + 1)\rho(x_1, x_0) \\
&\le \frac{\alpha^n}{1 - \alpha}\rho(x_1, x_0).
\end{aligned}
\tag{4.26}
$$

Thus the sequence (x_n) is a Cauchy sequence. Because X is complete, the sequence (x_n) is convergent; that is, it has a limit in X. Set $x \in X$ as the limit of (x_n). We now prove that x is a fixed point of f. Passing p to $+\infty$ in (4.26), we get

$$\rho(x, x_n) \le \frac{\alpha^n}{1 - \alpha}\rho(x_1, x_0). \tag{4.27}$$

---

[5] Rudolf Otto Sigismund Lipschitz, 1832–1903.

Now, substituting x_n by $f(x_{n-1})$ and passing n to $+\infty$, we get $\rho(x, f(x)) = 0$. Thus $x = f(x)$; that is, x is a fixed point of f.

Suppose that there are points x and y so that $x = f(x)$ and $y = f(y)$. Then

$$\rho(x, y) = \rho(f(x), f(y)) \leq \alpha\rho(x, y)$$

implies that $(1 - \alpha)\rho(x, y) \leq 0$. Hence $x = y$; that is, the fixed point of the function f is unique. $\square$

Remark. We notice that for any starting point $x_0 \in X$ we get precisely the same limit point and this limit point is the unique fixed point of f. $\triangle$

Corollary 5.12. *Let (X, ρ) be a complete metric space, A be a nonempty closed subset of it, and $f : A \to A$ be an α-contraction. Then f has a unique fixed point in A.*

The sequence (x_n) thus obtained is said to be the *sequence of successive approximations* of x. Based on (4.27), one can establish the *speed of convergence* of the sequence (x_n) to its limit x.

Example. Suppose we have to find the real roots of the equation

$$x^3 + 2x - 1 = 0. \tag{4.28}$$

First we note that if we denote the left-hand side of (4.28) by $g(x)$, we get a polynomial function; this function is strictly increasing. Hence it has at most one real root. At the same time from

$$\lim_{x \to -\infty} g(x) = -\infty \quad \text{and} \quad \lim_{x \to +\infty} g(x) = +\infty,$$

and, moreover, taking into account Theorem 2.15, we conclude that (4.28) has precisely one real root.

In order to find this real root we rewrite the given equation as

$$x = 1/(x^2 + 2). \tag{4.29}$$

Denote the right-hand side of (4.29) by $f(x)$ and thus we get a rational function defined on $\mathbb{R}$.

We are checking the assumption of Theorem 5.1. For $x, y \in \mathbb{R}$ we have

$$|f(x) - f(y)| = \frac{|x + y|}{(2 + x^2)(2 + y^2)} |x - y|.$$

However

$$|t| = \sqrt{2t^2}/\sqrt{2} \leq (2 + t^2)/(2\sqrt{2}).$$

Thus

$$|x + y| \leq |x| + |y| \leq (2 + x^2 + 2 + y^2)/(2\sqrt{2}) \leq (2 + x^2)(2 + y^2)/(2\sqrt{2}).$$

We conclude that function f is a contraction. From (4.29) it easily follows that the solution has to be strictly positive. Because function f is strictly decreasing on the positive semi-axis and $f([0,1]) = [1/3, 1/2] \subset [0,1]$, it follows that it is a contraction on the compact interval $[0,1]$. Thus we may apply the Banach fixed point theorem and if we start from $x_0 = 0$, then

$$x_1 = 0.5, \quad x_2 = 0.444444\ldots, \quad x_3 = 0.455056\ldots, \quad x_4 = 0.453088\ldots,$$
$$x_5 = 0.453455\ldots, \quad x_6 = 0.453386\ldots, \quad x_7 = 0.453399\ldots. \quad \triangle$$

4.6 Convex functions

4.6.1 Convex functions

Let I be a real interval and $f : I \to \mathbb{R}$ be a function. Then f is said to be *convex* (on I) provided

$$x, y \in I, \quad \alpha \in [0,1] \implies f((1-\alpha)x + \alpha y) \leq (1-\alpha)f(x) + \alpha f(y). \quad (4.30)$$

f is said to be *strictly convex* provided

$$x, y \in I, \quad \alpha \in [0,1] \implies f((1-\alpha)x + \alpha y) < (1-\alpha)f(x) + \alpha f(y).$$

Figure 4.1 represents a convex function.

If

$$x, y \in I, \quad \alpha \in [0,1] \implies f((1-\alpha)x + \alpha y) \geq (1-\alpha)f(x) + \alpha f(y),$$

then f is said to be *concave*. f is said to be *strictly concave* provided

$$x, y \in I, \quad \alpha \in [0,1] \implies f((1-\alpha)x + \alpha y) > (1-\alpha)f(x) + \alpha f(y).$$

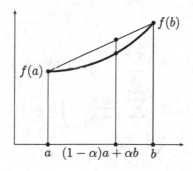

Fig. 4.1. A convex function

A function that is simultaneously convex and concave is said to be an *affine* function.

Remarks. Two geometrical characterizations of convexity are supplied below.
(a) Consider a convex function $f : [a, b] \to \mathbb{R}$ and the points $A = (a, f(a))$ and $B = (b, f(b))$. Then the equation of the AB straight line is given by

$$y(x) = f(a) + \frac{f(b) - f(a)}{b - a}(x - a).$$

Consider $\alpha \in [0, 1]$ and $\lambda = (1 - \alpha)a + \alpha b$. Then

$$f(\lambda) - y(\lambda) = f((1 - \alpha)a + \alpha b) - [(1 - \alpha)f(a) + \alpha f(b)].$$

Hence, the left-hand side is nonpositive if and only if f is convex; that is, the AB straight line lies above the graph of f if and only if f is convex.
(b) Define the *epigraph* of $f : I \to \mathbb{R}$ as

$$\operatorname{epi} f = \{(x, \alpha) \mid \alpha \geq f(x),\ x \in I\}.$$

Then f is convex if and only if $\operatorname{epi} f$ is a convex set. $\triangle$

Lemma 4.1. (Jensen[6]) *Suppose f is a convex function. Then*

$$f(\alpha_1 x_1 + \cdots + \alpha_n x_n) \leq \alpha_1 f(x_1) + \cdots + \alpha_n f(x_n) \qquad (4.31)$$

for every $n \in \mathbb{N}^*$, $x_1, \ldots, x_n \in I$, and $\alpha_1, \ldots, \alpha_n \geq 0$ with $\alpha_1 + \cdots + \alpha_n = 1$.

Proof. If $n = 2$, the claim is true. Suppose $n \geq 2$ and the claim is true for $n - 1$. Consider $x_1, \ldots, x_n \in I$, $\alpha_1, \ldots, \alpha_n \geq 0$, with $\alpha_1 + \cdots + \alpha_n = 1$. Then $\sum_{i=1}^{n-1} \alpha_i/(1 - \alpha_n) = 1$, so by hypothesis

$$f\left(\sum_{i=1}^{n-1} \frac{\alpha_i}{1 - \alpha_n} x_i\right) \leq \sum_{i=1}^{n-1} \frac{\alpha_i}{1 - \alpha_n} f(x_i).$$

Hence

$$f\left(\sum_1^n \alpha_i x_i\right) = f\left((1 - \alpha_n) \sum_1^{n-1} \frac{\alpha_i}{1 - \alpha_n} x_i + \alpha_n x_n\right)$$

$$\leq (1 - \alpha_n) f\left(\sum_1^{n-1} \frac{\alpha_i}{1 - \alpha_n} x_i\right) + \alpha_n f(x_n)$$

$$\leq (1 - \alpha_n) \sum_1^{n-1} \frac{\alpha_i}{1 - \alpha_n} f(x_i) + \alpha_n f(x_n) = \sum_1^n \alpha_i f(x_i). \quad \square$$

---

[6] Johan Ludwig William Valdemar Jensen, 1859–1925.

Corollary 6.13. *Suppose* $f : I \to \mathbb{R}$ *is a convex mapping. Then*

$$f\left(\frac{\alpha_1 x_1 + \cdots + \alpha_n x_n}{\alpha_1 + \cdots + \alpha_n}\right) \leq \frac{\alpha_1 f(x_1) + \cdots + \alpha_n f(x_n)}{\alpha_1 + \cdots + \alpha_n},$$

for each $x_1, \ldots, x_n \in I$, $\alpha_1, \ldots, \alpha_n \geq 0$, *with* $\alpha_1 + \cdots + \alpha_n > 0$.

Let f be a real function defined on an interval I. Then

$$s_{f,x_0}(x) = s_{x_0}(x) = \frac{f(x) - f(x_0)}{x - x_0}, \quad x \in I \setminus \{x_0\}$$

is said to be the *slope* of f.

Theorem 6.1. *Suppose* $f : I \to \mathbb{R}$ *is a convex mapping. Then*

(a) s_{f,x_0} *is increasing on* $I \setminus \{x_0\}$.
(b) f *is strictly convex on* $I \implies s_{f,x_0}$ *is strictly increasing on* $I \setminus \{x_0\}$.
(c) *Suppose there exist two points* $x_1, x_2 \in I$, $x_1 < x_2$ *and* $\lambda_0 \in]0,1[$ *such that*

$$f((1 - \lambda_0)x_1 + \lambda_0 x_2) = (1 - \lambda_0)f(x_1) + \lambda_0 f(x_2).$$

Then

$$f((1 - \lambda)x_1 + \lambda x_2) = (1 - \lambda)f(x_1) + \lambda f(x_2), \quad \forall \lambda \in [0,1], \quad (4.32)$$

$$f(x) = \frac{x_2 - x}{x_2 - x_1}f(x_1) + \frac{x - x_1}{x_2 - x_1}f(x_2), \quad \forall x \in [x_1, x_2]. \quad (4.33)$$

(d) f *is Lipschitz on any compact interval* J *contained in* int (I); f *is continuous at any point in* int (I).
(e) *Either* f *is monotonic on* int (I) *or there exists* $x_0 \in$ int (I) *so that* f *is decreasing on* $] - \infty, x_0[\cap I$ *and increasing on* $]x_0, +\infty] \cap I$.

Proof. (a) Consider $x_1, x_2, x_3 \in I$ with $x_1 < x_2 < x_3$. Then there exists $\lambda \in]0,1[$ such that $x_2 = (1 - \lambda)x_1 + \lambda x_3$, namely $\lambda = (x_2 - x_1)/(x_3 - x_1)$. By the convexity of f we write

$$f(x_2) \leq \frac{x_3 - x_2}{x_3 - x_1}f(x_1) + \frac{x_2 - x_1}{x_3 - x_1}f(x_3). \quad (4.34)$$

Subtract successively $f(x_1), f(x_2)$, and $f(x_3)$ from (4.34), then multiply each inequality by $1/(x_2 - x_1)$, $(x_3 - x_1)/[(x_2 - x_1)(x_3 - x_2)]$, respectively $1/(x_3 - x_1)$. Thus we get

$$\frac{f(x_2) - f(x_1)}{x_2 - x_1} \leq \frac{f(x_3) - f(x_1)}{x_3 - x_1}, \quad (4.35)$$

$$\frac{f(x_1) - f(x_2)}{x_1 - x_2} \leq \frac{f(x_3) - f(x_2)}{x_3 - x_2}, \quad (4.36)$$

$$\frac{f(x_1) - f(x_3)}{x_1 - x_3} \leq \frac{f(x_2) - f(x_3)}{x_2 - x_3}. \quad (4.37)$$

Considering inequalities (4.35)–(4.37) in terms of slope, one gets

$$s_{x_1}(x_2) \le s_{x_1}(x_3), \quad s_{x_2}(x_1) \le s_{x_2}(x_3), \quad s_{x_3}(x_1) \le s_{x_3}(x_2). \tag{4.38}$$

Inequalities (4.38) say precisely that the slope function is increasing on I.
(b) The conclusion follows because we now have strict inequality in (4.34), which is preserved along the proof.
(c) Suppose there exists a scalar $\lambda \in [0,1]$ so that

$$f((1-\lambda)x_1 + \lambda x_2) < (1-\lambda)f(x_1) + \lambda f(x_2).$$

Suppose $\lambda < \lambda_0$. The case $\lambda > \lambda_0$ runs similarly, therefore we neglect it. Take $\mu = (1-\lambda_0)/(1-\lambda) \in \,]0,1[$. We have $\lambda_0 = \mu\lambda + (1-\lambda)\cdot 1$ and $(1-\lambda_0)x_1 + \lambda_0 x_2 = \mu[(1-\lambda)x_1 + \lambda x_2] + (1-\mu)x_2$. Then it follows that

$$f((1-\lambda_0)x_1 + \lambda_0 x_2) \le \mu f((1-\lambda)x_1 + \lambda x_2) + (1-\mu)f(x_2)$$
$$< \mu[(1-\lambda)f(x_1) + \lambda f(x_2)] + (1-\mu)f(x_2) = (1-\lambda_0)f(x_1) + \lambda_0 f(x_2).$$

Thus we contradicted our hypothesis, therefore (4.32) holds.
    Take $x \in [x_1, x_2]$ and $\lambda = (x - x_1)/(x_2 - x_1)$ in (4.32) and (4.33) follows.
(d) Choose $u, x, y, v \in J$ with $u < x < y < v$. Then by Theorem 2.18 and by (a) we have

$$-\infty < s_u(u+) \le s_u(x) = \frac{f(x) - f(u)}{x - u} \tag{4.39}$$
$$\le \frac{f(y) - f(x)}{y - x} \le \frac{f(y) - f(v)}{y - v} \le s_v(v-) < +\infty.$$

Thus $|f(y) - f(x)| \le M|x - y|$, where $M = \max\{|s_u(u+)|, |s_v(v+)|\}$. Therefore f is Lipschitz on J.
    Consider a point $x_0 \in \mathrm{int}\,(I)$. Then there exists a compact interval J satisfying $x_0 \in \mathrm{int}\,(J) \subset I$. Because f is Lipschitz on J, as we just proved, f is continuous at x_0.
(e) If $s_x(x+) \ge 0$ for some $x \in \mathrm{int}\,(I)$, then from (4.39) it follows that f is increasing on $[x, \infty[\,\cap I$. Similarly, if $s_x(x-) \le 0$ for some $x \in \mathrm{int}\,(I)$, then f is decreasing on $]-\infty, x] \cap I$.
    Now, if $s_x(x+) \ge 0$ for every $x \in \mathrm{int}\,(I)$, we conclude that f is increasing on $\mathrm{int}\,(I)$. If $s_x(x+) \le 0$ for every $x \in \mathrm{int}\,(I)$, then $s_x(x-) \le 0$ for every $x \in \mathrm{int}\,(I)$. Thus f is decreasing on $\mathrm{int}\,(I)$.
    Suppose there exist $u, v \in \mathrm{int}\,(I)$ such that $u < v$ and $s_u(u+) < 0 < s_v(v+)$. Denote $z = \inf\{x \in I \mid s_x(x+) \ge 0\} \in [u, v]$. Furthermore $s_x(x+) < 0$ for every $x \in I$, $x < z$. Thus f is decreasing on $]-\infty, z] \cap I$.   □
    Suppose $f : I \to \mathbb{R}$ is a convex and nonmonotonic function. Then x_0 in (e) of Theorem 6.1 is a global minimum of f on I. More generally, consider a metric space X, M a nonempty subset of it, and a mapping $f : M \to \mathbb{R}$. A point $p \in M$ is said to be a *global minimum* of f if $f(p) \le f(x)$, for all $x \in M$.

Corollary 6.14. *Suppose* $f : I \to \mathbb{R}$ *is a convex and nonmonotonic function. Then* f *has a point of global minimum.*

Remark. We proved by (d) that a convex function is continuous on the interior of its domain of definition. By an example one shows that there exists a convex function discontinuous at the ends of the interval of definition. The example follows

$$f(x) = \begin{cases} 1, & x \in \{0, 1\} \\ 0, & x \in]0, 1[. \end{cases} \quad \triangle$$

Theorem 6.2. *Every increasing convex function of a convex function is convex.*

Proof. Consider I an interval, $f : I \to \mathbb{R}$ a convex function, $g : J \to \mathbb{R}$ an increasing convex function such that $f(I) \subset J$, and $h = g \circ f$. Then for every $\lambda \in [0, 1]$ and $x, y \in I$ one has

$$h((1 - \lambda)x + \lambda y) = g(f((1 - \lambda)x + \lambda y)) \leq g((1 - \lambda)f(x) + \lambda f(y))$$
$$\leq (1 - \lambda)g(f(x)) + \lambda g(f(y)) = (1 - \lambda)h(x) + \lambda h(y). \quad \square$$

4.6.2 Jensen convex functions

Let I be a real interval and $f : I \to \mathbb{R}$ be a function. Then f is said to be *Jensen convex* or *J-convex* provided

$$x, y \in I \implies f((x + y)/2) \leq (f(x) + f(y))/2. \tag{4.40}$$

Obviously, a convex function is J-convex. A converse is supplied by the next result.

Proposition 6.3. *A continuous and J-convex function* $f : I \to \mathbb{R}$ *is convex.*

Proof. From (4.40), by induction, we get

$$f((x_1 + \cdots + x_{2^k})/2^k) \leq (f(x_1) + \cdots + f(x_{2^k}))/2^k, \tag{4.41}$$

for every $k \in \mathbb{N}^*$, and $x_1, \ldots, x_{2^k} \in I$.

Consider $x, y \in I$ and $\alpha \in]0, 1[$. We show that (4.30) is satisfied. Write α as

$$\alpha = \alpha_1/2 + \alpha_2/2^2 + \cdots + \alpha_k/2^k + \ldots,$$

where $\alpha_i \in \{0, 1\}$, for all $i \in \mathbb{N}^*$. Denoting

$$\alpha^{(k)} = \alpha_1/2 + \alpha_2/2^2 + \cdots + \alpha_k/2^k = (\alpha_1 2^{k-1} + \cdots + \alpha_k)/2^k = \beta_k/2^k,$$

we have

$$\lim_{k \to \infty} \alpha^{(k)} = \alpha, \quad 1 - \alpha^{(k)} = (2^k - \beta_k)/2^k.$$

Consider in (4.41) that

$$x_1 = x_2 = \cdots = x_{\beta_k} = x, \quad x_{\beta_k+1} = x_{\beta_k+2} = \cdots = x_{2^k} = y.$$

Then we get

$$f(\alpha^{(k)}x + (1 - \alpha^{(k)})y) = f\left((\beta_k x + (2^k - \beta_k)y)/2^k\right)$$
$$\leq (\beta_k f(x) + (2^k - \beta_k)f(y))/2^k = \alpha^{(k)}f(x) + (1 - \alpha^{(k)})f(y).$$

Now passing $k \to \infty$ and taking into account the continuity of f, the conclusion follows at once.   □

Corollary 6.15. *Consider a continuous function* $f : I \to \mathbb{R}$. *Then* f *is convex if and only if it is J-convex.*

Consider two n-tuples of real numbers $(x_1, x_2, \ldots, x_n)$ and $(y_1, y_2, \ldots, y_n)$ satisfying $x_1 \geq x_2 \geq \cdots \geq x_n$ and $y_1 \geq y_2 \geq \cdots \geq y_n$. If

$$x_1 \geq y_1,$$
$$x_1 + x_2 + \cdots + x_m \geq y_1 + y_2 + \cdots + y_m, \quad 2 \leq m \leq n - 1,$$
$$x_1 + x_2 + \cdots + x_n = y_1 + y_2 + \cdots + y_n,$$

then we say that $(x_1, x_2, \ldots, x_n)$ *majorizes* $(y_1, y_2, \ldots, y_n)$ and we denote $(x_1, x_2, \ldots, x_n) \succ (y_1, y_2, \ldots, y_n)$.

Now, consider two n-tuples of real numbers $(x_1, x_2, \ldots, x_n)$ and $(y_1, y_2, \ldots, y_n)$ satisfying $x_1 \geq x_2 \geq \cdots \geq x_n$ and $y_1 \geq y_2 \geq \cdots \geq y_n$. If

$$x_1 \leq y_1,$$
$$x_1 + x_2 + \cdots + x_m \leq y_1 + y_2 + \cdots + y_m, \quad 2 \leq m \leq n - 1,$$
$$x_1 + x_2 + \cdots + x_n = y_1 + y_2 + \cdots + y_n,$$

then we say that $(x_1, x_2, \ldots, x_n)$ *minorizes* $(y_1, y_2, \ldots, y_n)$ and we denote $(x_1, x_2, \ldots, x_n) \prec (y_1, y_2, \ldots, y_n)$.

Theorem 6.3. (Karamata[7] inequality *or the* majorization inequality) *If* $(x_1, x_2, \ldots, x_n) \succ (y_1, y_2, \ldots, y_n)$ *and* f *is a real-valued convex function defined on an interval containing the two n-tuples, then*

$$f(x_1) + f(x_2) + \cdots + f(x_n) \geq f(y_1) + f(y_2) + \cdots + f(y_n). \qquad (4.42)$$

If f *is concave, then*

$$f(x_1) + f(x_2) + \cdots + f(x_n) \leq f(y_1) + f(y_2) + \cdots + f(y_n). \qquad (4.43)$$

Corollary 6.16. *Given that a function* f *satisfies* (4.42) *for any two n-tuples* $(x_1, x_2, \ldots, x_n)$ *and* $(y_1, y_2, \ldots, y_n)$ *so that* $(x_1, x_2, \ldots, x_n) \succ (y_1, y_2, \ldots, y_n)$, *then* f *is J-convex.*

---

[7] Jovan Karamata, 1902–1967.

Proof. Consider an n-tuple $(x_1, x_2, \ldots, x_n)$ and its arithmetic mean $m = (x_1 + x_2 + \cdots + x_n)/n$. Note that $(x_1, x_2, \ldots, x_n) \succ (m, m, \ldots, m)$. Applying (4.42) we find that

$$f(x_1) + f(x_2) + \cdots + f(x_n) \geq nf(m)$$

$$\implies \frac{f(x_1) + f(x_2) + \cdots + f(x_n)}{n} \geq f\left(\frac{x_1 + x_2 + \cdots + x_n}{n}\right). \quad \square$$

Corollary 6.17. *Given that a function f satisfies (4.43) for any two n-tuples $(x_1, x_2, \ldots, x_n)$ and $(y_1, y_2, \ldots, y_n)$ so that $(x_1, x_2, \ldots, x_n) \succ (y_1, y_2, \ldots, y_n)$, show that f is J-concave.*

Proof. Similar to the proof of Corollary 6.16. $\square$

Corollary 6.18. *Consider $a_1 \geq a_2 \geq \cdots \geq a_n > 0$, $S = a_1 + a_2 + \cdots + a_n$, and f a convex function defined on an interval containing S and $S - (n-1)a_1$. Then*

$$\sum_{i=1}^{n} f(S - (n-1)a_i) \geq \sum_{i=1}^{n} f(a_i), \qquad (4.44)$$

$$(S - (n-1)a_1)^m + (S - (n-1)a_2)^m + \cdots + (S - (n-1)a_n)^m$$

$$(4.45)$$

$$\geq a_1^m + a_2^m + \cdots + a_n^m, \quad m \in \mathbb{N}^*, \ m > 1.$$

Proof. Note that

$$(S - (n-1)a_n, S - (n-1)a_{n-1}, \ldots, S - (n-1)a_1) \succ (a_1, a_2, \ldots, a_n).$$

We can apply the Karamata inequality and (4.44) follows immediately.
    Consider $f(x) = x^m$, $m \in \mathbb{N}^*$, $m > 1$ in (4.44) and (4.45) follows. $\square$

Corollary 6.19. *Let a, b, c be the lengths of the sides to a triangle. Then*

$$(b+c-a)^{b+c-a} + (c+a-b)^{c+a-b} + (a+b-c)^{a+b-c} \geq a^a + b^b + c^c, \qquad (4.46)$$

$$(b+c-a)^{b+c-a}(c+a-b)^{c+a-b}(a+b-c)^{a+b-c} \geq a^a b^b c^c. \qquad (4.47)$$

Proof. Both inequalities are symmetrical. So we may suppose that $a \geq b \geq c$. It is clear that

$$(a+b-c, a+c-b, b+c-a) \succ (a, b, c).$$

We take the convex function $f(x) = x^x$ in (4.42) and (4.46) follows.
    Consider the convex function $f(x) = x \ln x$ in (4.42) and then we get (4.47). $\square$

Theorem 6.4. (Popoviciu[8] generalized inequality) *If f is a convex function on I and $a_1, a_2, \ldots, a_n \in I$, then*

$$f(a_1) + f(a_2) + \cdots + f(a_n) + n(n-2)f(a) \geq (n-1)(f(b_1) + \cdots + f(b_n)),$$

where $a = (a_1 + \cdots + a_n)/n$ and $b_i = (na - a_i)/(n-1)$ for all i.

Theorem 6.5. (Popoviciu–Cârtoaje[9]–Zhao inequality) *Let n be a natural number, and let m be an integer. Let $I \subset \mathbb{R}$ be an interval, and $a_1, a_2, \ldots, a_n \in I$. Consider a convex function $f : I \to \mathbb{R}$. Then*

$$m \sum_{1 \leq i_1 < i_2 < \ldots < i_m \leq n} f\left(\frac{a_{i_1} + a_{i_2} + \ldots + a_{i_m}}{m}\right)$$

$$\leq \binom{n-2}{m-1} \sum_{i=1}^{n} f(a_i) + n\binom{n-2}{m-2} f\left(\frac{a_1 + a_2 + \ldots + a_n}{n}\right).$$

Here the sum $\sum_{1 \leq i_1 < i_2 < \ldots < i_m \leq n} f((a_{i_1} + a_{i_2} + \ldots + a_{i_m})/m)$ is considered as being null $m \leq 0$ and for $m > n$.

Let $a_1, a_2, \ldots, a_n$ be positive numbers and $p = (p_1, \ldots, p_n) \in \mathbb{R}^n$. The p-mean of $a = (a_1, a_2, \ldots, a_n)$ is defined by

$$m(a, p) = \frac{1}{n!} \sum_{\sigma \in S_n} a_{\sigma(1)}^{p_1} a_{\sigma(2)}^{p_2} \cdots a_{\sigma(n)}^{p_n}, \tag{4.48}$$

where S_n is the set of all permutations of $\{1, 2, \ldots, n\}$. So, the summation sign means to sum $n!$ terms. The arithmetic mean is written as

$$m(a, (1, 0, \ldots, 0)) = \frac{1}{n}(a_1 + \cdots + a_n),$$

and the geometric mean as

$$m(a, (1/n, 1/n, \ldots, 1/n)) = a_1^{1/n} a_2^{1/n} \ldots a_n^{1/n} = \sqrt[n]{a_1 \cdots a_n}.$$

Theorem 6.6. (Muirhead[10] inequality) *Let $a_1, a_2, \ldots, a_n$ be positive numbers and $p, q \in \mathbb{R}^n$. If $p \succ q$, then $m(a, p) \geq m(a, q)$. Moreover, if $p \neq q$, equality holds if and only if $a_1 = a_2 = \cdots = a_n$.*

Corollary 6.20. *Given are $a_1, a_2, \ldots, a_n$ positive numbers. Because $(1, 0, \ldots, 0) \succ (1/n, 1/n, \ldots, 1/n)$, it holds*

$$\sqrt[n]{a_1 \cdots a_n} \leq \frac{a_1 + \cdots + a_n}{n}.$$

Given are x, y, z positive numbers. Because $(3, 0, 0) \succ (1, 1, 1)$, it holds $x^3 + y^3 + z^3 \geq 3xyz$, $x, y, z \geq 0$.

---

[8] Tiberiu Popoviciu, 1906–1975.

[9] Vasile Cârtoaje.

[10] Muirhead.

4.7 Functions of bounded variations

Consider an interval $I = [a, b]$ and a *partition* P of I, that is, a finite set of points $P = \{x_0, x_1, \ldots, x_n\}$ satisfying $a = x_0 \leq x_1 \leq \cdots \leq x_n = b$. Denote the *norm of the partition* P by $\|P\| = \max\{x_{i+1} - x_i \mid i = 0, 1, \ldots, n-1\}$.

The *variation of* $f : I \to \mathbb{R}$ in respect to P is the nonnegative number

$$V(f; P) = \sum_{i=0}^{n-1} |f(x_{i+1}) - f(x_i)|.$$

Figure 4.2 represents the variation of a function.

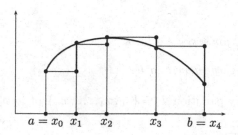

$$a = x_0 \quad x_1 \qquad x_2 \qquad\qquad x_3 \qquad b = x_4$$

Fig. 4.2. Variation of a function

Proposition 7.4. *Let P_1 and P_2 be two partitions of the interval $[a, b]$ with $P_1 \subset P_2$. Then $V(f; P_1) \leq V(f; P_2)$.*

Proof. It is enough to prove it if $P_2 \setminus P_1 = \{x\}$. Suppose $x_i < x < x_{i+1}$. Then

$$V(f; P_2) - V(f; P_1) = |f(x_{i+1}) - f(x)| + |f(x) - f(x_i)| - |f(x_{i+1}) - f(x_i)| \geq 0. \quad \square$$

The element in $[0, \infty]$ defined as

$$\overset{b}{\underset{a}{\vee}} f = \sup\{V(f; P) \mid P \text{ is a partition of } [a, b]\}$$

is said to be the *total variation of* f *on* $[a, b]$.

Remark. Let $f : [a, b] \to \mathbb{R}$ be a monotonic function. Because for every partition P of $[a, b]$, $V(f; P) = |f(b) - f(a)|$, it follows that $\overset{b}{\underset{a}{\vee}} f = |f(b) - f(a)|$. $\triangle$

Proposition 7.5. *Suppose it is given $f : [a, b] \to \mathbb{R}$. Then $\overset{b}{\underset{a}{\vee}} f = 0$ if and only if f is constant on $[a, b]$.*

Proof. It is obvious that if f is constant on $[a, b]$, for each partition P of $[a, b]$ we have $V(f; P) = 0$. So $\bigvee\limits_a^b f = 0$.

Suppose $\bigvee\limits_a^b f = 0$. Consider $a < x \leq b$ and $P = \{a, x, b\}$. Then

$$|f(x) - f(a)| \leq V(f; P) \leq \bigvee\limits_a^b f = 0. \quad \square$$

Theorem 7.1. *Suppose it is given* $f, g : [a, b] \to \mathbb{R}$ *and* $\alpha \in \mathbb{R}$. *Then*

(a) $\bigvee\limits_a^b (f + g) \leq \bigvee\limits_a^b f + \bigvee\limits_a^b g$.

(b) $\bigvee\limits_a^b (\alpha f) = |\alpha| \bigvee\limits_a^b f$.

(c) $\bigvee\limits_a^b |f| \leq \bigvee\limits_a^b f$.

(d) $\bigvee\limits_a^b f = \bigvee\limits_a^c f + \bigvee\limits_c^b f$, *for every* $c \in [a, b]$.

(e) $\bigvee\limits_u^v f \leq \bigvee\limits_a^b f$, *for every* $[u, v] \subset [a, b]$.

Proof. (a) For every partition $P = \{x_0, x_1, \ldots, x_n\}$ of $[a, b]$,

$$V(f + g, P) = \sum_{i=0}^{n-1} |f(x_{i+1}) + g(x_{i+1}) - f(x_i) - g(x_i)|$$

$$\leq \sum_{i=0}^{n-1} |f(x_{i+1}) - f(x_i)| + \sum_{i=0}^{n-1} |g(x_{i+1}) - g(x_i)| \leq \bigvee\limits_a^b f + \bigvee\limits_a^b g.$$

(b) For every partition $P = \{x_0, x_1, \ldots, x_n\}$ of $[a, b]$, one has

$$V(\alpha f, P) = |\alpha| V(f, P) \leq |\alpha| \bigvee\limits_a^b f.$$

(c) For every partition $P = \{x_0, x_1, \ldots, x_n\}$ of $[a, b]$,

$$V(|f|, P) = \sum_{i=0}^{n-1} ||f(x_{i+1})| - |f(x_i)|| \leq \sum_{i=0}^{n-1} |f(x_{i+1}) - f(x_i)| \leq \bigvee\limits_a^b f.$$

(d) For $c \in \{a, b\}$, it is obvious. Suppose $a < c < b$. Let P_1 be a partition of $[a, c]$, P_2 a partition of $[c, b]$, and $P = P_1 \cup P_2$. Then

$$V(f, P_1) + V(f, P_2) = V(f, P) \leq \bigvee\limits_a^b f.$$

Passing to supremum in respect to P_1 and P_2, it follows that $\bigvee\limits_a^c f + \bigvee\limits_c^b f \leq \bigvee\limits_a^b f$.

On the other side consider $c \in [a, b]$, P a partition of $[a, b]$, $P_1 = (P \cup \{c\}) \cap [a, c]$, and $P_2 = (P \cup \{c\}) \cap [c, b]$. Then

$$V(f,P) \le V(f, P \cup \{c\}) = V(f, P_1) + V(f, P_2) \le \overset{c}{\underset{a}{\vee}} f + \overset{b}{\underset{c}{\vee}} f.$$

Passing to supremum in respect to P, the conclusion follows.

(e) Suppose we are given $[u, v] \subset [a, b]$. Then

$$\overset{v}{\underset{u}{\vee}} f \le \overset{u}{\underset{a}{\vee}} f + \overset{v}{\underset{u}{\vee}} f + \overset{b}{\underset{v}{\vee}} f = \overset{b}{\underset{a}{\vee}} f. \quad \square$$

A function $f : [a, b] \to \mathbb{R}$ is said to be of *finite variation on* $[a, b]$ whenever $\overset{b}{\underset{a}{\vee}} f < \infty$.

Remarks 7.1. (a) Suppose $f : [a, b] \to \mathbb{R}$ is a monotonic function. We already saw that for every partition of $[a, b]$, $\overset{b}{\underset{a}{\vee}} f = |f(b) - f(a)|$. That is $\overset{b}{\underset{a}{\vee}} f < \infty$, that is, f is of finite variation on $[a, b]$. Thus we conclude that a monotonic and discontinuous function on $[a, b]$ is of finite variation on $[a, b]$.

(b) Suppose $f : [a, b] \to \mathbb{R}$ is a Lipschitz function and L is its Lipschitz constant. Let P be a partition of $[a, b]$. Then

$$V(f, P) = \sum_{i=0}^{n-1} |f(x_{i+1}) - f(x_i)| \le L \sum_{i=0}^{n-1} (x_{i+1} - x_i) = L(b - a).$$

Thus $\overset{b}{\underset{a}{\vee}} f \le L(b - a)$, that is, f is of finite variation on $[a, b]$.

(c) A function of finite variation is bounded. Indeed, consider $f : [a, b] \to \mathbb{R}$ a function of finite variation. Then for every $x \in [a, b]$, one has

$$|f(x)| \le |f(a)| + |f(x) - f(a)| \le |f(a)| + \overset{b}{\underset{a}{\vee}} f,$$

thus f is bounded on $[a, b]$. $\triangle$

We saw earlier that a monotonic and discontinuous function on $[a, b]$ is of finite variation on this interval. Below by an example we show that there is a continuous function without being of finite variation.

Example 7.1. Let $f : [0, 2] \to [0, 2]$ be defined by

$$f(x) = \begin{cases} x \sin \dfrac{\pi}{x}, & 0 < x \le 2 \\ 0, & x = 0. \end{cases}$$

Consider $P = \{0, 2/(2n-1), 2/(2n-3), \dots, 2/5, 2/3, 2\}$. Then

$$V(f, P) = \frac{2}{2n-1} + \left(\frac{2}{2n-3} + \frac{2}{2n-1} \right) + \cdots + \left(\frac{2}{5} + \frac{2}{3} \right) + \left(2 + \frac{2}{3} \right)$$

$$> 1/2 + 1/3 + \cdots + 1/n. \tag{4.49}$$

By Exercise 3.2 in Section 3.6, the right-hand side in (4.49) tends to ∞ as $n \to \infty$. So f is a continuous function without being of finite variation. $\triangle$

Remark. Earlier we saw that given two real functions f and g of finite variation on $[a, b]$, then $f + g$ and αf (for every real α) are of finite variation on $[a, b]$. Then the set of finite variation functions on $[a, b]$ turns into a vector space. This vector space is a normed space, because one can define a norm on it by

$$\|f\|_v = |f(a)| + \overset{b}{\underset{a}{\bigvee}} f,$$

where f is of finite variation on $[a, b]$. $\triangle$

Proposition 7.6. *Suppose it is given* $f : [a, b] \to \mathbb{R}$ *of bounded variation on* $[a, b]$. *Then there exists a sequence of partitions* (P_n) *with* $\|P_n\| \to 0$ *so that* $V(f, P_n) \to \overset{b}{\underset{a}{\bigvee}} f.$

Proof. From Theorem 2.7 at page 18 we have that for every $\varepsilon (= 1/n)$ we can find a partition P_n of $[a, b]$ so that

$$\overset{b}{\underset{a}{\bigvee}} f - 1/n < V(f, P_n).$$

Suppose Q_n is a partition of $[a, b]$ satisfying $P_n \subset Q_n$ and $\|Q_n\| < 1/n$. Then

$$\overset{b}{\underset{a}{\bigvee}} f - 1/n < V(f, P_n) \leq V(f, Q_n) \leq \overset{b}{\underset{a}{\bigvee}} f < \overset{b}{\underset{a}{\bigvee}} f + 1/n.$$

Thus $V(f, P_n) \to \overset{b}{\underset{a}{\bigvee}} f$, as $n \to \infty$. $\square$

Theorem 7.2. *Suppose it is given* $f : [a, b] \to \mathbb{R}$ *continuous and of finite variation on* $[a, b]$. *Then for every sequence of partitions* (P_n) *of* $[a, b]$ *with* $\|P_n\| \to 0$ *we have*

$$\lim_{n \to \infty} V(f, P_n) = \overset{b}{\underset{a}{\bigvee}} f.$$

Theorem 7.3. (Jordan[11]) *Suppose it is given* $f : [a, b] \to \mathbb{R}$. *Then* f *is of finite variation on* $[a, b]$ *if and only if it is a difference of two increasing function.*

Proof. An increasing function is of finite variation by (a) in Remarks 7.1 and by (a) and (b) in Theorem 7.1 the difference of two monotone functions is of finite variation. Thus f is a function of finite variation.

Suppose f is of finite variation. Define $[a, b] \ni t \to \varphi(t) = \overset{t}{\underset{a}{\bigvee}} f$. By (d) in Theorem 7.1, function φ is increasing. We show that function $\psi = \varphi - f$ is also increasing. For it consider $t \in [a, b]$ and $h > 0$ with $t + h \in [a, b]$. Then

---

[11] Camille Jordan, 1838–1922.

$$\psi(t+h) - \psi(t) = (\varphi(t+h) - f(t+h)) - (\varphi(t) - f(t))$$

$$= (\overset{t+h}{\underset{a}{V}} f - \overset{t}{\underset{a}{V}} f) - (f(t+h) - f(t)) = \overset{t+h}{\underset{t}{V}} f - (f(t+h) - f(t))$$

$$\geq |f(t+h) - f(t)| - (f(t+h) - f(t)) \geq 0. \quad \square$$

Theorem 7.3 is still valid if we suppose that the functions φ and ψ are increasing and positive because one can add a large enough positive constant to both functions. Also, the theorem is valid if functions φ and ψ are decreasing.

Corollary 7.21. *Consider two functions of finite variations on $[a, b]$. Then their product is of finite variation on $[a, b]$.*

Proof. Let f and g be the two functions of finite variations on $[a, b]$. Then there exist increasing functions $\varphi_1, \varphi_2, \psi_1, \psi_2$ such that $f = \varphi_1 - \psi_1$ and $g = \varphi_2 - \psi_2$. We have

$$fg = (\varphi_1\varphi_2 + \psi_1\psi_2) - (\varphi_1\psi_2 + \varphi_2\psi_1)$$

and we note that

$$\varphi_1\varphi_2, \quad \psi_1\psi_2, \quad \varphi_1\psi_2, \quad \varphi_2\psi_1, \quad \varphi_1\varphi_2 + \psi_1\psi_2, \quad \varphi_1\psi_2 + \varphi_2\psi_1$$

are all increasing functions. The conclusion follows by Theorem 7.3. $\quad \square$

Corollary 7.22. *Let f be a function of finite variation on $[a, b]$. Then there exist $f(t+)$ for all $t \in [a, b[$ and $f(t-)$ for every $t \in]a, b]$.*

Proof. By Theorem 7.3 we have that f can be represented as a difference of two increasing functions and by Theorem 2.18 we have the existence of the side limits. $\quad \square$

Lemma 4.2. *Suppose f is a function of finite variation on $[a, b]$ and there exists a constant c with $\inf_{t \in [a,b]} |f(t)| \geq c > 0$. Then $1/f$ is of finite variation on $[a, b]$.*

Proof. Consider a partition $P = \{a = x_0 < x_1 < \cdots < x_{n-1} < x_n = b\}$ of $[a, b]$. Then

$$V(1/f, P) = \sum_{k=1}^{n} |1/f(x_k) - 1/f(x_{k-1})| \leq (1/c^2) \sum_{k=1}^{n} |f(x_k) - f(x_{k-1})|$$

$$= (1/c^2) V(f, P) \leq (1/c^2) \overset{b}{\underset{a}{V}} f.$$

Thus

$$\overset{b}{\underset{a}{V}}(1/f) \leq (1/c^2) \overset{b}{\underset{a}{V}} f. \quad \square$$

Remark. A useful method to find the total variation of a differentiable function with an integrable derivative is supplied by Theorem 1.9 at page 260. $\quad \triangle$

Most of the earlier results are valid for vector-valued functions.

Consider $f : [a, b] \to \mathbb{R}^k$. The *variation of* $f : I \to \mathbb{R}^k$ *in respect to the* partition P is the nonnegative number

$$V(f; P) = \sum_{i=0}^{n-1} \| f(x_{i+1}) - f(x_i) \|.$$

Proposition 7.7. *Let P_1 and P_2 be two partitions of the interval $[a, b]$ with $P_1 \subset P_2$. Then $V(f; P_1) \leq V(f; P_2)$.*

The element in $[0, \infty]$ defined as

$$\bigvee_a^b f = \sup\{V(f; P) \mid P \text{ is a partition of } [a, b]\}$$

is said to be the *total variation on* f *on* $[a, b]$.

Proposition 7.8. *Suppose we are given $f : [a, b] \to \mathbb{R}^k$. Then $\bigvee_a^b f = 0$ if and only if f is constant on $[a, b]$.*

Theorem 7.4. *Suppose we are given $f, g : [a, b] \to \mathbb{R}^k$ and $\alpha \in \mathbb{R}$. Then*

(a) $\bigvee_a^b (f + g) \leq \bigvee_a^b f + \bigvee_a^b g.$

(b) $\bigvee_a^b (\alpha f) = |\alpha| \bigvee_a^b f.$

(c) $\bigvee_a^b \|f\| \leq \bigvee_a^b f.$

(d) $\bigvee_a^b f = \bigvee_a^c f + \bigvee_c^b f,$ *for every $c \in [a, b]$.*

(e) $\bigvee_u^v f \leq \bigvee_a^b f,$ *for every $[u, v] \subset [a, b]$.*

A function $f : [a, b] \to \mathbb{R}^k$ is said to be of *finite variation on* $[a, b]$ whenever $\bigvee_a^b f < \infty$.

Many properties of vector-valued functions of bounded variations can be reduced to the real case, as follows from the next result.

Theorem 7.5. *Suppose we are given $f : [a, b] \to \mathbb{R}^k$, where $f = (f_1, \ldots, f_k)$. Then f is of bounded variation on $[a, b]$ if and only if each of the functions f_i is of bounded variation on $[a, b]$. Moreover,*

$$V(f_i, P) \leq V(f, P) \leq \sum_{i=1}^k V(f_i, P), \quad i \in \{1, \ldots, n\}.$$

4.8 Continuity of sequences of functions

We saw by (b) in Examples 2.1 at page 109 that the pointwise limit of a sequence of continuous functions is not necessarily continuous even if the terms of the sequence are defined and converges on a compact interval.

Theorem 8.1. *Suppose $f_n : [a,b] \to \mathbb{R}$ are continuous, $n \in \mathbb{N}$, and the sequence (f_n) converges uniformly to a function f on $[a,b]$. Then f is continuous on $[a,b]$.*

Proof. Set $\varepsilon > 0$. Because the sequence $(f_n)_n$ converges uniformly to f on $A = [a,b]$, there exists $n_\varepsilon \in \mathbb{N}$ such that for all $n \geq n_\varepsilon$ one has $|f_n(t) - f(t)| < \varepsilon/3$ for all $t \in A$. The function f_{n_ε} is continuous, so there exists $\delta = \delta(\varepsilon) > 0$ such that $|t - t_0| < \delta$, $t \in A$, implies $|f_{n_\varepsilon}(t) - f_{n_\varepsilon}(t_0)| < \varepsilon/3$. So for $|t - t_0| < \delta$, $t \in A$, one has

$$|f(t) - f(t_0)| \leq |f(t) - f_{n_\varepsilon}(t)| + |f_{n_\varepsilon}(t) - f_{n_\varepsilon}(t_0)| + |f_{n_\varepsilon}(t_0) - f(t_0)| \leq \varepsilon.$$

Hence f is continuous on an arbitrary $t_0 \in A$. $\square$

Remark. The proof does not use the fact that the domains of definition are a compact interval. Much weaker conditions are satisfactory as we show in Theorem 8.3. $\triangle$

Consider $I \subset \mathbb{R}$. A sequence of functions $f_n : I \to \mathbb{R}$ is said to *converge uniformly Cauchy* on I if for every positive ε one finds a rank n_ε such that $|f_m(x) - f_n(x)| < \varepsilon$, for every $m, n \geq n_\varepsilon$ and $x \in I$.

Theorem 8.2. *Consider $I \subset \mathbb{R}$ and a sequence of functions $f_n : I \to \mathbb{R}$. Then (f_n) converges uniformly if and only if it converges uniformly Cauchy.*

Proof. It appears at page 110 by Theorem 2.1. $\square$

Theorem 8.3. *A function $f : X \to Y$ from a metric space X into a metric space Y is continuous if there is a sequence (f_n) of continuous functions that converges uniformly to f.*

A sufficient condition for uniform convergence is supplied by the next result.

Theorem 8.4. (Dini[12]) *Suppose $f_n : [a,b] \to \mathbb{R}$ are continuous, $n \in \mathbb{N}$, $f_0(t) \leq f_1(t) \leq f_2(t) \leq \ldots$, for every $t \in [a,b]$, and f, the pointwise limit of the sequence $(f_n)_n$, is continuous on $[a,b]$. Then $(f_n)_n$ converges uniformly to f on $[a,b]$.*

Proof. Suppose the sequence $(f_n)_n$ does not converge uniformly to f on $[a,b]$. Then there exists a positive ε such that for every natural n^ε one can find a natural $n \geq n^\varepsilon$ and an $x \in [a,b]$ with $|f_n(x) - f(x)| \geq \varepsilon$. Denote

---

[12] Ulisse Dini, 1845–1918.

by $x_{n^\varepsilon} \in [a, b]$ such a point. Thus we get a sequence $(x_{n_k^\varepsilon})_k$ in $[a, b]$ of such points. Because $f(x) \geq f_n(x)$ for all $x \in [a, b]$ and $n \in \mathbb{N}$, it holds

$$f(x_{n_k^\varepsilon}) - f_{n_k^\varepsilon}(x_{n_k^\varepsilon}) \geq \varepsilon. \tag{4.50}$$

We can select a convergent subsequence of the sequence $(x_{n_k^\varepsilon})_k$, keeping the notation. Denote by $x_0 \in [a, b]$ the limit of the convergent subsequence; that is, $\lim_{k \to \infty} x_{n_k^\varepsilon} = x_0$. Pick a natural i. Because f and f_i are continuous, for k sufficiently large, we write

$$|(f(x_{n_k^\varepsilon}) - f(x_0)) - (f_i(x_{n_k^\varepsilon}) - f_i(x_0))|$$
$$= |(f(x_{n_k^\varepsilon}) - f_i(x_{n_k^\varepsilon})) - (f(x_0) - f_i(x_0))| < \varepsilon/2.$$

Thus

$$f(x_{n_k^\varepsilon}) - f_i(x_{n_k^\varepsilon}) - [f(x_0) - f_i(x_0)] < \varepsilon/2.$$

By inequality (4.50), we can write that

$$f(x_0) - f_i(x_0) > \varepsilon/2.$$

But this inequality contradicts the pointwise convergence of $(f_n)_n$ to f on $[a, b]$.   □

4.9 Continuity of series of functions

Theorem 9.1. *Consider continuous and nonnegative functions* $f_n : [a, b] \to \mathbb{R}$, $n \in \mathbb{N}$, *such that* $s : [a, b] \to \mathbb{R}$ *given by* $s(x) = \lim_{n \to \infty} s_n(x)$ *is continuous, where* $s_n(x) = \sum_{k=0}^{n} f_k(x)$. *Then the series* $\sum f_n$ *converges uniformly on* $[a, b]$.

Proof. We remark that the sequence of partial sums (s_n) is continuous and increasing on $[a, b]$. Then by Theorem 8.4 we conclude that (s_n) converges uniformly to s. Hence the series of functions $\sum f_n$ converges uniformly on $[a, b]$.   □

Theorem 9.2. *Suppose there is given* $\sum f_n$ *a uniform convergent series on a subset* $A \subset \mathbb{R}$. *Moreover, we suppose that every function* f_n *is continuous on a point* x_0 *in* A. *Then the series* $\sum f_n$ *is continuous at* x_0.

Proof. It follows at once by Theorem 8.1.   □

Examples. (a) Find the sum of the series

$$1 + \sum_{n \geq 1} x^n/n!.$$

Solution. The series converges absolutely for every $x \in \mathbb{R}$ by the ratio test and uniformly on every compact interval and by Theorem 4.2 at page 135 of Weierstrass. Denote

$$f(x) = 1 + \sum_{n \geq 1} x^n/n!.$$

Then f is continuous on $\mathbb{R}$. By Theorem 3.3 at page 111 of Mertens we write

$$f(x)f(y) = 1 + \left(\sum_{n \geq 1} \frac{x^n}{n!} \right) \left(1 + \sum_{n \geq 1} \frac{y^n}{n!} \right) = \sum_{n \geq 1} \frac{(x+y)^n}{n!} = f(x+y).$$

Note that $f(1) = e$. By Exercise 4.7 we conclude that $f(x) = e^x$.

(b) Show that the series

$$1 + \sum_{n \geq 1} (-1)^n \frac{x^{2n}}{(2n)!}, \quad \sum_{n=0}^{\infty} (-1)^n \frac{x^{2n+1}}{(2n+1)!}, \tag{4.51}$$

are absolutely and uniformly for all $x \in \mathbb{R}$. Denote C, respectively S, their sums. Show

$$C(x+y) = C(x)C(y) - S(y)S(y), \tag{4.52}$$
$$S(x+y) = S(x)C(y) + C(x)S(y). \tag{4.53}$$

Solution. We treat series (4.52). The second series is discussed similarly.

By the ratio criterion the first series converges absolutely for each real x. By Theorem 4.2 at page 135 of Weierstrass it converges uniformly on each compact interval. Thus C is continuous on $\mathbb{R}$. The general term of $C(x)C(y)$ is

$$a_n = (-1)^n \left[\frac{y^{2n}}{(2n)!} + \frac{x^2}{2!} \frac{y^{2n-2}}{(2n-2)!} + \cdots + \frac{x^{2i}}{(2i)!} \frac{y^{2n-2i}}{(2n-2i)!} + \cdots + \frac{x^{2n}}{(2n)!} \right]$$
$$= \frac{(-1)^n}{(2n)!} \sum_{i=0}^{n} \binom{2n}{2i} x^{2i} y^{2n-2i}.$$

The general term of order $n - 1$ of $S(x)S(y)$ is

$$b_n = (-1)^{n-1} \left[\frac{x}{1!} \frac{y^{2n-1}}{(2n-1)!} + \frac{x^3}{3!} \frac{y^{2n-3}}{(2n-3)!} + \cdots + \frac{x^{2n-1}}{(2n-1)!} \frac{y}{1!} \right]$$
$$= \frac{(-1)^n}{(2n)!} \sum_{i=0}^{n-1} \binom{2n}{2i+1} x^{2i+1} y^{2n-2i-1}.$$

By Theorem 3.3 at page 111, the series $\sum a_n$ and $\sum b_n$ are absolutely convergent and their sums are $C(x)C(y)$, respectively, $S(x)S(y)$. Moreover

$$c_n = a_n - b_n = (-1)^n \frac{(x+y)^n}{n!}.$$

The series $\sum c_n$ is absolutely convergent and its sum is $C(x)C(y) - S(x)S(y)$. On the other side it is equal to $C(x+y)$. Thus (4.52) is established.   △

4.10 Exercises

4.1. Consider the series

$$\sum_{n=1}^{\infty} \frac{1}{n^p}.$$

Applying the Ermakov test, show that for $p > 1$, the series converges whereas for $p \leq 1$, it diverges.

4.2. Consider $f : \mathbb{R}^2 \to \mathbb{R}$ defined by

$$f(x,y) = \begin{cases} \dfrac{x^3 + y^3}{x^2 + y^2}, & (x,y) \neq (0,0) \\ 0, & (x,y) = (0,0). \end{cases}$$

Study its continuity on $\mathbb{R}^2$.

4.3. (Cauchy) Let $\varphi : \mathbb{R} \to \mathbb{R}$ be a continuous function satisfying

$$\varphi(x + y) = \varphi(x) + \varphi(y), \quad \forall\, x, y \in \mathbb{R}. \tag{4.54}$$

Find the set φ of continuous functions fulfilling (4.54).

4.4. Let $\varphi : \mathbb{R} \to \mathbb{R}$ be a continuous function satisfying, for a given real a,

$$\varphi(x + y) = \varphi(x) + \varphi(y) + a, \quad \forall\, x, y \in \mathbb{R}. \tag{4.55}$$

Find the set φ of continuous functions fulfilling (4.55).

4.5. Let $\varphi :]0, \infty[\to \mathbb{R}$ be a continuous function satisfying

$$\varphi(xy) = \varphi(x) + \varphi(y), \quad \forall\, x, y \in]0, \infty[. \tag{4.56}$$

Find the set φ of continuous functions fulfilling (4.56).

4.6. Let $\varphi : \mathbb{R} \to \mathbb{R}$ be a nonzero continuous function satisfying

$$\varphi(x + y) = \varphi(x)\varphi(y), \quad \forall\, x, y \in \mathbb{R}. \tag{4.57}$$

Find the set φ of continuous functions fulfilling (4.57).

4.7. Let $\varphi : [0, \infty] \to \mathbb{R}$ be a nonzero continuous function satisfying

$$\varphi(xy) = \varphi(x)\varphi(y), \quad \forall\, x, y \in]0, \infty[. \tag{4.58}$$

Determine the set φ of continuous functions fulfilling (4.58).

4.8. Let $\varphi : \mathbb{R} \setminus \{-1, 1\} \to \mathbb{R}$ be a continuous function satisfying

$$\varphi(x) + \varphi(y) = \varphi\left(\frac{x+y}{1+xy}\right), \quad \forall\, x, y \in \mathbb{R} \setminus \{-1, 1\} \ \text{ with } \ 1 + xy \neq 0. \tag{4.59}$$

Determine the set φ of continuous functions fulfilling (4.59).

4.9. Show that a function $f : \mathbb{R} \to \mathbb{R}$ satisfying one of the following relations satisfies the other

$$f(x + y) = f(x) + f(y), \quad x, y \in \mathbb{R},$$
$$f(xy + x + y) = f(xy) + f(x) + f(y), \quad x, y \in \mathbb{R}.$$

4.10. Find all continuous functions $f : \mathbb{R} \to [0, \infty[$ such that

$$f^2(x + y) - f^2(x - y) = 4 \cdot f(x)f(y), \tag{4.60}$$

for all real numbers x, y.

4.11. Set $t \in {]0, 1[}$. Find all functions $f : \mathbb{R} \to \mathbb{R}$ that are continuous at 0 and for all real x it holds $f(x) - 2f(tx) + f(t^2x) = x^2$.

4.12. Show that every continuous and bounded function $f : \mathbb{R} \to \mathbb{R}$ has a fixed point, that is, a point x such that $f(x) = x$.

4.13. Suppose $f : [0, 1] \to [0, 1]$ is a continuous function. Then the sequence of iterates $x_{n+1} = f(x_n)$ converges if and only if

$$\lim_{n \to \infty} (x_{n+1} - x_n) = 0. \tag{4.61}$$

4.14. Find all functions $f : \mathbb{R} \to \mathbb{R}$ such that $f(x^3 + x) \le x \le (f(x))^3 + f(x)$, $\forall x \in \mathbb{R}$.

4.15. Let f be a continuous and increasing function defined on an interval $[a, b]$ and such that $f(a) \ge a$ and $f(b) \le b$. Choose an arbitrary $x_1 \in [a, b]$ and consider the sequence (x_n) obtained as $x_{n+1} = f(x_n)$, $n \ge 1$. Show that the limit $\lim x_n = x^*$ exists and that $f(x^*) = x^*$.

4.16. Is it true that if $f : [0, 1] \to [0, 1]$ is

(a) Monotonically increasing
(b) Monotonically decreasing

then there is an $x \in [0, 1]$ for which $f(x) = x$?

4.17. Let f be continuous and nowhere monotone on $[0, 1]$. Show that the set of points on which f attains local minima is dense in $[0, 1]$.

4.18. Find all continuous functions $f : \mathbb{R} \to \mathbb{R}$ satisfying

$$f(f(x)) + x = 0, \quad x \in \mathbb{R}. \tag{4.62}$$

4.19. Find all strictly increasing functions $f : \mathbb{R} \to \mathbb{R}$ satisfying

$$f(f(x)) = x, \quad x \in \mathbb{R}. \tag{4.63}$$

4.20. Find the set of all strictly increasing functions $f : \mathbb{N} \to \mathbb{N}$ with $f(2) = 2$ and

$$f(n \cdot m) = f(n) \cdot f(m), \quad n, m \in \mathbb{N}. \tag{4.64}$$

4.21. Suppose $f, g : \mathbb{R} \to \mathbb{R}$ are two functions such that f is continuous, g is monotonic, and $f(x) = g(x)$, for all $x \in \mathbb{Q}$. Show that $f(x) = g(x)$, for all $x \in \mathbb{R}$.

4.22. (Cauchy) Find the set of increasing functions $\phi : \mathbb{R} \to \mathbb{R}$ satisfying $\phi(x + y) = \phi(x) + \phi(y)$, for all $x, y \in \mathbb{R}$.

4.23. Consider a function $f : [a, b] \to [a, b]$ such that $|f(x) - f(y)| < |x - y|$ for all $x, y \in [a, b]$. Set $a_1 \in [a, b]$ and define $a_{n+1} = (a_n + f(a_n))/2$, for $n = 1, 2, \ldots$. Show that $\lim_{n \to \infty} a_n = a_0$ and $f(a_0) = a_0$.

4.24. (Schur inequality for increasing functions) Consider $I \subset [0, \infty[$ an interval, $a, b, c \in I$, and f a positive and increasing function defined on I. Show

$$f(a)(a - b)(a - c) + f(b)(b - a)(b - c) + f(c)(c - a)(c - b) \geq 0. \tag{4.65}$$

4.25. (Hermite[13]) Let $n \in \mathbb{N}^*$. Show that for every real x

$$\lfloor x \rfloor + \lfloor x + 1/n \rfloor + \cdots + \lfloor x + (n - 1)/n \rfloor = \lfloor nx \rfloor, \tag{4.66}$$

where $\lfloor t \rfloor$ represents the floor of t.

4.26. Let $f : \mathbb{R} \to \mathbb{R}$ be a continuous and periodic function. Show that f has a fixed point.

4.27. Let $f : \mathbb{R} \to \mathbb{R}$ be a continuous function satisfying $f(x) = f(x + 1/n)$, for all $x \in \mathbb{Q}$ and $n \in \mathbb{N}^*$. Show that function f is constant.

4.28. Pick a a real number. Consider the function $f : \mathbb{R} \to \mathbb{R}$ so that

$$f(x + a) = 1/2 + \sqrt{f(x) - (f(x))^2}, \quad \forall x \in \mathbb{R}. \tag{4.67}$$

Show that

(a) f is periodic.
(b) Give an example of such a function for $a = 1$.

4.29. Consider a nonempty interval I and a Darboux and one-to-one function $f : I \to \mathbb{R}$. Show that its inverse function is a Darboux function.

---

[13] Charles Hermite, 1822–1901.

4.30. Let f be a convex function on $[a, b]$. Suppose

$$a \leq x_1 < x_2 < \cdots < x_{n-1} < x_n \leq b.$$

Show that

$$(x_2 - x_1)(f(x_1) + f(x_2)) + \cdots + (x_n - x_{n-1})(f(x_{n-1}) + f(x_n))$$
$$\leq (x_n - x_1)(f(x_1) + f(x_n)).$$

4.31. Consider the function

$$f(x) = \begin{cases} 0, & t = 0 \\ x^k \sin(1/x), & t \in \,]0, 2/\pi], \ k \in \mathbb{N}. \end{cases}$$

Is this function of bounded variation? Prove it.

4.32. Consider the series

$$\sum_{n=1}^{\infty} \left(\frac{nx}{1 + n^2 x^2} - \frac{(n-1)x}{1 + (n-1)^2 x^2} \right), \quad x \in [0, 1].$$

Prove that the series converges nonuniformly to a continuous function.

4.33. Consider the series

$$f(x) - \sum_{n=1}^{\infty} \frac{f(x)}{n(n+1)}, \quad \text{where } x \in \mathbb{R} \text{ and } f(x) = \begin{cases} 1, & x \in \mathbb{Q} \\ 0, & x \in \mathbb{R} \setminus \mathbb{Q}. \end{cases}$$

Prove that the series converges uniformly to a continuous function.

4.11 References and comments

Many parts of the present chapter follow [115] and [87]. We indicate some sources more precisely.

Theorem 2.6 of uniform approximation of a continuous function on a compact interval appears in many books. We mention just one [115, p. 146].

Theorem 2.7 has been published in [129].

Theorems 2.20 and 2.21 and Corollary 2.5 may be found in [87].

A proof of Corollary 4.11 may be found in [87, p. 52].

A proof of Theorem 4.1 may be found in [87, p. 46-48].

A modern survey on results about Banach fixed point theorem is the recent paper [116].

Here we indicate some references on the Karamata inequality [73], [103], [107], [108], and [109].

Theorem 6.4 has been published in [137].

Theorem 6.5 appeared at http://www.mathlinks.ro/Forum/viewtopic.php ?t=21786&highlight=popoviciu.

Corollaries 6.18 and 6.19 are from [75] and [96].

Example 7.1 may be found in [115].

Theorem 7.2 may be found in [87, p. 24]; theorem 7.3 may be found in [87, p. 27]. Theorem 7.5 may be found in [115, p. 117]

Theorem 8.3 may be found in [67, p. 28], and theorem 8.4 has been published in [70, p. 22]. Theorem 8.1 follows from Theorem 8.3.

Exercises 4.9–4.11, 4.14, 4.19, 4.20, 4.23, 4.24, 4.28, are from [96].

Exercise 4.15 appears as [117, Exercise 11, page 9].

Exercise 4.16 is Exercise 1, first day, and Exercise 4.17 is Exercise 2, second day, IMC, 2000.

Exercise 4.25 is from the paper [65].

Exercise 4.30 may be found in [136, vol. XVIII, nr. 3, 1967, p. 108].

Exercises 4.32 and 4.33 are from [52].

5

Differential Calculus on $\mathbb{R}$

This chapter is devoted to introducing some basic results on differential calculus on the real axis.

5.1 The derivative of a real function

Let f be defined on $[a, b]$ and real-valued. For any $x \in [a, b]$ form the quotient

$$\phi(t) = \frac{f(t) - f(x)}{t - x}, \quad (a < t < b, \ t \neq x) \tag{5.1}$$

and define

$$f'(x) = \lim_{t \to x} \phi(t) = \lim_{h \to 0} \frac{f(h + x) - f(x)}{h} \tag{5.2}$$

provided this limit exists.

We thus associate with the function f a function f' whose domain of definition is the set of points x at which limit (5.2) exists; f' is called the *derivative* or the *first-order derivative* of f.

If f' is defined at a point x, we say that f is *differentiable* at x. If f' is defined at every point of a set $E \subset [a, b]$, we say that f is *differentiable on* E.

Theorem 1.1. *Let f be defined on $[a, b]$. If f is differentiable at a point $x \in [a, b]$, then f is continuous at x.*

Proof. We have

$$f(t) - f(x) = \frac{f(t) - f(x)}{t - x}(t - x) \to f'(x) \cdot 0 = 0 \text{ as } t \to x. \quad \square$$

Remark. The converse is not true. Consider $f(t) = |t|$, $t \in \mathbb{R}$. Take $t = 0$. $\triangle$

The behavior of the quotient in (5.1) is studied in Exercise 5.1.

A stronger result than Theorem 1.1 holds, namely the following.

Theorem 1.2. (Weierstrass) *There exists a continuous function on $\mathbb{R}$ having no point of differentiability.*

Sometimes a continuous function having no point of differentiability on an interval is said to be a *Weierstrass function* on that interval.

Proof. There are available many nice proofs to this theorem exhibiting an abundance of such functions. Basically, the methods of proving this theorem consist in finding a convenient zigzag function, passing it to the limit, getting "very small teeth," and proving that the limiting function is continuous without being differentiable. One pretty short proof may be found in [115, p. 141]. But it appeals to series of functions. We follow a more geometrical approach.

We construct on the interval $[0, 1]$ a satisfactory function that is extended afterwards to the real line. That we get the desired function.

Consider a polygonal line represented by some straight lines so that the union of their domains of definition is the interval $[0, 1]$. Let (x_1, y_1) and (x_2, y_2) be consecutive vertices of the polygonal line where $x_1 < x_2$ and $y_1 \neq y_2$. Let $h = (x_2 - x_1)/3$ and $k = \lambda(y_2 - y_1)$, for a fixed positive λ. To define the next polygonal line we replace the segment joining (x_1, y_1) and (x_2, y_2) by a zigzag joining the consecutive vertices

$$(x_1, y_1), \quad (x_1 + h, y_2 - k), \quad (x_2 - h, y_1 + k), \quad (x_2, y_2).$$

The differences in ordinate of the successive vertices are

$$(1 - \lambda)(y_2 - y_1), \quad (2\lambda - 1)(y_2 - y_1), \quad (1 - \lambda)(y_2 - y_1).$$

We choose λ so that the greatest absolute difference in the successive is $\mu|y_2 - y_1|$, where

$$\mu = 1 - \lambda < 1. \tag{5.3}$$

This decrease on the absolute difference yields continuity in the limit.

We denote the slope of the segment joining (x_1, y_1) to (x_2, y_2) by m and obtain for the slopes of the three consecutive segments of the zigzag, in order,

$$3(1 - \lambda)m, \quad 3(2\lambda - 1)m, \quad 3(1 - \lambda)m.$$

We choose λ so that the minimum steepness of absolute slope of the three segments is $\nu|m|$, where

$$\nu = 3(1 - 2\lambda) > 1. \tag{5.4}$$

This increase in steepness yields nondifferentiability in the limit.

To fulfill (5.3) and (5.4) any $\lambda \in {]}0, 1/3{[}$ is satisfactory. We note that under conditions (5.3) and (5.4) the entire zigzag on the open interval ${]}x_1, x_2{[}$ remains between the horizontal line $y = y_1$ and $y = y_2$.

For the construction of a Weierstrass function we fix $\lambda \in {]}0, 1{[}$. We start with the graph of function $y = x$, $x \in [0, 1]$. The first zigzag polygon is constructed by the above method taking $(x_1, y_1) = (0, 0)$ and $(x_2, y_2) = (1, 1)$.

The second zigzag polygon is obtained from the first by using the zigzag construction on each third

$$[0, 1/3], \quad [1/3, 2/3], \quad [2/3, 1]$$

of the interval $[0, 1]$. We iterate the scheme by applying the zigzag construction to each segment of the previous zigzag polygon to obtain the next zigzag polygon. The abscissae of the vertices of the nth zigzag polygon are the ternary points $x_{n,i} = i/3^n$, $i = 0, 1, 2, \ldots, 3^n$. These vertices are vertices of all succeeding zigzag polygons, thus they will be points on the graph of the expected Weierstrass function.

We denote $f_0(x) = x$, f_1 is the first zigzag function whose graph is the first zigzag polygon, f_2 is the second zigzag function whose graph is the second zigzag polygon, and so on. Thus we get a sequence of continuous functions.

We observe that for the difference in function f_n values at successive ternary points are

$$|f_n(x_{n,i}) - f_n(x_{n,i-1})| \leq \mu^n, \quad i = 1, 2, \ldots, 3^n, \tag{5.5}$$

where μ is given by (5.3). For the slope of the segment joining the corresponding vertices we have

$$\frac{|f_n(x_{n,i}) - f_n(x_{n,i-1})|}{x_{n,i} - x_{n,i-1}} \geq \nu^n, \quad i = 1, 2, \ldots, 3^n. \tag{5.6}$$

We intend to show that the sequence of functions (f_n) converges uniformly on $[0, 1]$. A necessary and sufficient condition for uniform convergence is given by the general criterion of the Cauchy theorem 2.1 at page 110; that is, for every $\varepsilon > 0$ there exists $n_\varepsilon \in \mathbb{N}$ so that for any $n \geq n_\varepsilon$, $p \in \mathbb{N}^*$, and any $t \in [0, 1]$, $|f_{n+p}(t) - f_n(t)| < \varepsilon$. Fix a t in $[0, 1]$ and $n, p \in \mathbb{N}$. If $t \in \{x_{n,i} \mid i = 0, 1, \ldots, 3^n\}$, then $t \in \{x_{m,i} \mid i = 0, 1, \ldots, 3^m\}$, for all $m > n$. Thus, the criterion is satisfied obviously. Suppose $t \notin \{x_{n,i} \mid i = 0, 1, \ldots, 3^n\}$. Then we can find the smallest intervals $[x_{n,i-1}, x_{n,i}]$ and $[x_{n+p,j-1}, x_{n+p,j}]$ so that

$$t \in [x_{n+p,j-1}, x_{n+p,j}] \subset [x_{n,i-1}, x_{n,i}].$$

We have

$$|f_n(t) - f_n(x_{n,i})| < |f_n(x_{n,i-1}) - f_n(x_{n,i})| \leq \mu^n,$$
$$|f_n(x_{n,i}) - f_{n+p}(x_{n+p,j})| \leq |f_n(x_{n,i-1}) - f_n(x_{n,i})| \leq \mu^n,$$
$$|f_{n+p}(x_{n+p,j}) - f_{n+p}(t)| \leq |f_{n+p}(x_{n+p,j}) - f_{n+p}(x_{n+p,j-1})| \leq \mu^{n+p}.$$

Furthermore we have

$$|f_n(t) - f_{n+p}(t)| \leq |f_n(t) - f_n(x_{n,i})| + |f_n(x_{n,i}) - f_{n+p}(x_{n+p,j})|$$
$$+ |f_{n+p}(x_{n+p,j}) - f_{n+p}(t)| \leq 2\mu^n + \mu^{n+p} < 3\mu^n.$$

Because $3\mu^n \to 0$ as $n \to \infty$, we have that indeed there exists n_ε with the required features. Hence the sequence of functions (f_n) converges uniformly. On the other hand each function in $\{f_n\}$ is continuous. So, by Theorem 8.1 at page 183, the sequence of functions (f_n) tends uniformly to a continuous function, say f.

To end the proof we have to show that f is not differentiable on $[0,1]$. First we prove that f is not differentiable on any ternary point; then we prove the nondifferentiability of f on any other point.

Let t be a ternary point of f_n. Then t is of the form $i/3^n$ and because $i/3^n = i3^m/3^{n+m}$, t is a ternary point for all f_k, $k \geq n$. Therefore it holds that $f(t) = f_n(i/3^n)$. On any neighborhood of t it follows from (5.6) that no bound can be placed on the steepness of the segment of the graph of f with one endpoint at $(t, f(t))$. Thus f is not differentiable at any ternary point.

If t is not a ternary point (such a point does exist; $\sqrt{2}/2 \in \,]0,1]$ is irrational and does not belong to any set of the form $\{i/3^n \mid i = 0, 1, \ldots, 3^n\}$ containing only rational numbers), let a_n and b_n be the endpoints of the smallest interval of ternary points of f_n containing t. Thus we get the sequences (a_n) and (b_n) such that $a_n < t < b_n$, $n \in \mathbb{N}$. The steepness of one of the two segments joining $(t, f(t))$ to the points $(a_n, f(a_n))$ and $(b_n, f(b_n))$ is at least as great as steepness of the segments of the segment C joining $(a_n, f(a_n))$ to $(b_n, f(b_n))$. If $f(a_n) < f(b_n)$, then, if $(t, f(t))$ lies above C, the segment to $(a_n, f(a_n))$ is steeper, if $(t, f(t))$ lies on C, the segments are equally steep, and if $(t, f(t))$ lies below C, the segment to $(b_n, f(b_n))$ is steeper. A similar argument holds if $f(a_n) > f(b_n)$. Analytically, in all cases, we have

$$\frac{|f(b_n) - f(t)|}{b_n - t} + \frac{|f(t) - f(a_n)|}{t - a_n} > \frac{|f(b_n) - f(t)| + |f(t) - f(a_n)|}{b_n - a_n}$$

$$\geq \frac{|f(b_n) - f(a_n)|}{b_n - a_n} \geq \nu^n.$$

It follows that at least one of the slopes

$$\frac{f(b_n) - f(t)}{b_n - t} \quad \text{or} \quad \frac{f(t) - f(a_n)}{t - a_n}$$

is greater than $\nu^n/2$. Again we see that no bound can be placed on the steepness of the segments with one endpoint at $(t, f(t))$ in any neighborhood of t. Thus we conclude that f is nowhere differentiable on $[0,1]$.

The last step of the proof consists in extending function f from $[0,1]$ to $\mathbb{R}$ keeping its continuity and nondifferentiability. First we extend our function to $[0,2]$ by $f(x) = f(2-x)$, $x \in [0,1]$. Then we extend it to $\mathbb{R}$. This can be realized easily taking

$$f(x-2) = f(x). \quad \square$$

We introduce some arithmetic properties of differentiable functions.

Theorem 1.3. *Suppose f and g are defined on $[a, b]$ and are differentiable at a point $x \in [a, b]$. Then $f + g$, $f \cdot g$, and f/g are differentiable at x, and*

(a) $(f + g)'(x) = f'(x) + g'(x)$.

(b) $(f \cdot g)'(x) = f'(x)g(x) + f(x)g'(x)$.

(c) $\left(\dfrac{f}{g} \right)'(x) = \dfrac{f'(x)g(x) - f(x)g'(x)}{g^2(x)}$ *(we assume that $g(x) \neq 0$).*

Proof. (a) is trivial.

(b) Let $h = f \cdot g$. Then

$$\frac{h(t) - h(x)}{t - x} = \frac{f(t)g(t) - f(x)g(x)}{t - x}$$

$$= \frac{f(t)g(t) - f(x)g(t) + f(x)g(t) - f(x)g(x)}{t - x}$$

$$= g(t)\frac{f(t) - f(x)}{t - x} + f(x)\frac{g(t) - g(x)}{t - x}.$$

Passing $t \to x$, the result follows.

(c) We have

$$\frac{\left(\dfrac{f}{g} \right)(t) - \left(\dfrac{f}{g} \right)(x)}{t - x} = \frac{f(t)g(x) - g(t)f(x)}{(t - x)g(x)g(t)}$$

$$= \frac{1}{g(t)g(x)} \left(\frac{f(t) - f(x)}{t - x}g(x) - f(x)\frac{g(t) - g(x)}{t - x} \right).$$

Letting $t \to x$, the result follows. $\square$

Theorem 1.4. (The chain rule) *Suppose f is continuous on $[a, b]$, $f'(x)$ exists at some point $x \in [a, b]$, g is defined on a closed interval I that contains the range of f, and g is differentiable at $f(x)$. If*

$$h(t) = g(f(t)), \quad a \leq t \leq b,$$

h is differentiable at x, and

$$h'(x) = g'(f(x)) \cdot f'(x). \tag{5.7}$$

Proof. Let $y = f(x)$. Based on the definition of the derivative, we have

$$f(t) - f(x) = (t - x)(f'(x) + u(t)) \tag{5.8}$$

$$g(s) - g(y) = (s - y)(g'(y) + v(s)), \tag{5.9}$$

where $t \in [a, b]$, $s \in I$, $u(t) \to 0$ as $t \to x$, $v(s) \to 0$ as $s \to y$. First we use (5.9) and then (5.8). We obtain

$$h(t) - h(x) = g(f(t)) - g(f(x)) = (f(t) - f(x))(g'(y) + v(s))$$
$$= (t - x)(f'(x) + u(t))(g'(y) + v(s))$$

or, if $t \neq x$,

$$\frac{h(t) - h(x)}{t - x} = (g'(y) + v(s))(f'(x) + u(t)). \tag{5.10}$$

Let $t \to x$ and see that $s \to y$, by the continuity of f. Thus the right-hand side of (5.10) tends to $g'(y)f'(x)$, which is actually (5.7). $\square$

Examples 1.1. (a) Consider

$$f(x) = \begin{cases} x \sin(1/x), & x \neq 0 \\ 0, & x = 0. \end{cases}$$

Then $f'(x) = \sin(1/x) - (1/x)\cos(1/x)$, $x \neq 0$, but $f'(0)$ does not exist.
(b) Consider

$$f(x) = \begin{cases} x^2 \sin(1/x), & x \neq 0 \\ 0, & x = 0. \end{cases}$$

Then $f'(x) = 2x \sin(1/x) - \cos(1/x)$, $x \neq 0$. Now $f'(0) = 0$ but there does not exist $\lim_{x \to 0} f'(x)$. So f' is discontinuous on 0.
(c) The above cases are particular cases of one of the next. Let $a \geq 0$, $c > 0$ be constants and $f : [-1, 1] \to \mathbb{R}$ a function defined by

$$f(x) = \begin{cases} |x|^a \sin(|x|^{-c}), & x \neq 0 \\ 0, & x = 0. \end{cases}$$

Then

(c_1) f is continuous $\iff a > 0$.
(c_2) $\exists\, f'(0) \iff a > 1$.
(c_3) f' is bounded $\iff a \geq 1 + c$.
(c_4) f' is continuous $\iff a > 1 + c$.
(c_5) $\exists\, f''(0) \iff a > 2 + c$.
(c_6) f'' is bounded $\iff a \geq 2 + 2c$.
(c_7) f'' is continuous $\iff a > 2 + 2c$. $\triangle$

Regarding the derivative of the inverse function we introduce the next result.

Theorem 1.5. *Let $I, J \subset \mathbb{R}$ be compact intervals and $f : I \to J$ be a continuous and bijective mapping. Suppose f is differentiable on a point $x_0 \in I$ and $f'(x_0) \neq 0$. Then the inverse function $f^{-1} : J \to I$ is differentiable at $y_0 = f(x_0)$ and it holds*

$$(f^{-1}(y_0))' = \frac{1}{f'(x_0)}.$$

Proof. Because function f is continuous and bijective and interval I is compact, it follows that f^{-1} is continuous (Theorem 2.11, page 156). Therefore

$$x_n \in I \setminus \{x_0\}, \ x_n \to x_0 \iff y_n = f(x_n) \in J \setminus \{y_0\}, \ y_n \to y_0. \quad (5.11)$$

Take $y = f(x)$. Then

$$\frac{f^{-1}(y) - f^{-1}(y_0)}{y - y_0} = \frac{1}{\dfrac{f(x) - f(x_0)}{x - x_0}}, \quad \forall x \in I \setminus \{x_0\}.$$

From (5.11) it follows that $y \to y_0 \iff x \to x_0$, hence

$$\lim_{y \to y_0} \frac{f^{-1}(y) - f^{-1}(y_0)}{y - y_0} = \lim_{x \to x_0} \frac{1}{\dfrac{f(x) - f(x_0)}{x - x_0}} = \frac{1}{f'(x_0)}. \quad \square$$

Proposition 1.1. *Let A be a matrix with n columns and m rows whose entries a_{ij} are differentiable functions on a point a in an interval I; that is, $a_{ij} : I \to \mathbb{R}$. Thus*

$$A(x) = \begin{bmatrix} a_{11}(x) & \dots & a_{1n}(x) \\ \dots & \dots & \dots \\ a_{m1}(x) & \dots & a_{mn}(x) \end{bmatrix}, \quad x \in I.$$

Then

$$A'(a) = \begin{bmatrix} a'_{11}(a) & \dots & a'_{1n}(a) \\ \dots & \dots & \dots \\ a'_{m1}(a) & \dots & a'_{mn}(a) \end{bmatrix}.$$

5.2 Mean value theorems

Let f be a real function defined on a metric space X. We say that f has a *local maximum* at a point $p \in X$ if there exists $\delta > 0$ such that $f(q) \leq f(p)$ for all $q \in B(p, \delta)$. We say that f has a *strict local maximum* at a point $p \in X$ if p is a local maximum to f and there exists $\delta > 0$ such that $f(q) < f(p)$ for all $q \in B(p, \delta) \setminus \{p\}$. Analogously, we say that f has a *local minimum* at a point $p \in X$ if there exists $\delta > 0$ such that $f(q) \geq f(p)$ for all $q \in B(p, \delta)$. f has a *strict local minimum* at a point $p \in X$ if p is a local minimum to f and there exists $\delta > 0$ such that $f(q) > f(p)$ for all $q \in B(p, \delta) \setminus \{p\}$.

Theorem 2.1. (Fermat[1] stationary principle) *Let f be defined on $[a, b]$. If f has a local maximum point $x \in \,]a, b[$ and if there exists $f'(x)$, then $f'(x) = 0$. The analogous statement for local minimum is also true.*

---

[1] Pierre de Fermat, 1601–1665.

Proof. The idea is suggested in Figure 5.1. Choose δ in accordance with the above definition and so that $a < x - \delta < x < x + \delta < b$. If $x - \delta < t < x$,

$$\frac{f(t) - f(x}{t - x} \geq 0.$$

Letting $t \to x$, we get that $f'(x) \geq 0$.

If $x < t < x + \delta$,

$$\frac{f(t) - f(x}{t - x} \leq 0.$$

Letting $t \to x$, we get that $f'(x) \leq 0$. Hence $f'(x) = 0$.   □

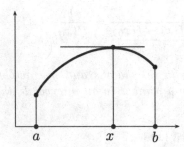

Fig. 5.1. Fermat theorem

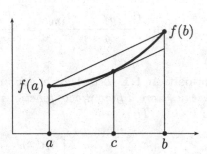

Fig. 5.2. Lagrange theorem

We introduce the mean value theorem of Cauchy.

Theorem 2.2. (Cauchy) *If f and g are continuous real functions on $[a, b]$ that are differentiable in $]a, b[$, then there exists a point $x \in]a, b[$ at which*

$$(f(b) - f(a))g'(x) = (g(b) - g(a))f'(x).$$

Proof. Set

$$h(t) = (f(b) - f(a))g(t) - (g(b) - g(a))f(t), \quad a \leq t \leq b.$$

Then h is continuous on $[a, b]$ and differentiable on $]a, b[$, and

$$h(a) = f(b)g(a) - f(a)g(b) = h(b). \tag{5.12}$$

We have to show that $h'(x) = 0$ for some $x \in]a, b[$.

If h is constant, $h'(x) = 0$ for every $x \in]a, b[$.

If $h(t) > h(a)$ for some $t \in]a, b[$, let x be a point in $[a, b]$ at which h attains its maximum value, Theorem 2.10 at page 156. By (5.12) it follows that $x \in]a, b[$, and by Theorem 2.1 we get the conclusion.

If $h(t) < h(a)$ for some $t \in]a, b[$, the same reasoning applies if we choose as x a point in $[a, b]$ at which h attains its minimum value.   □

We introduce the mean value theorem of Lagrange.

Theorem 2.3. (Lagrange) *If f is a continuous real function on $[a, b]$ that is differentiable in $]a, b[$, then there exists a point $c \in]a, b[$ at which*

$$f(b) - f(a) = (b - a)f'(c).$$

Proof. Take $g(t) = t$ in Theorem 2.2. The idea is suggested by Figure 5.2. □

Corollary 2.1. (Napier[2]) *For $0 < a < b$,*

$$\frac{1}{b} < \frac{\ln b - \ln a}{b - a} < \frac{1}{a}.$$

Theorem 2.4. *Suppose f is differentiable on $]a, b[$.*

(a) *If $f'(t) \geq 0$ for every $t \in]a, b[$, f is monotonically increasing on $]a, b[$.*
(b) *If $f'(t) = 0$ for every $t \in]a, b[$, f is constant on $]a, b[$.*
(c) *If $f'(t) \leq 0$ for every $t \in]a, b[$, f is monotonically decreasing on $]a, b[$.*

Proof. All conclusions can be read off from the relation

$$f(x_2) - f(x_1) = (x_2 - x_1)f'(x),$$

which is valid for each pair of numbers $x_1, x_2 \in]a, b[$ and for some x between x_1 and x_2. □

The next result immediately follows from Theorem 2.3.

Theorem 2.5. (Rolle[3]) *Suppose the assumptions of Theorem 2.3 are satisfied. Moreover, we consider $f(a) = f(b)$. Then there exists $x \in]a, b[$ so that $f'(x) = 0$.*

Below we introduce a generalization of the Bernoulli inequality, Exercise 1.26 at page 35.

Proposition 2.2. (Bernoulli) *Let I be the interval $]-1, \infty[$ and $\alpha \in \mathbb{R} \setminus \{0, 1\}$. Then*

$$(1 + x)^\alpha < 1 + \alpha x, \quad \forall \alpha \in]0, 1[, \quad x \in I \setminus \{0\}, \tag{5.13}$$
$$(1 + x)^\alpha > 1 + \alpha x, \quad \forall \alpha \in]-\infty, 0[\cup]1, \infty[, \quad x \in I \setminus \{0\}. \tag{5.14}$$

Proof. Consider $f : I \to \mathbb{R}$ defined as

$$f(x) = (1 + x)^\alpha - 1 - \alpha x.$$

Then f is differentiable and its derivative is $f'(x) = \alpha(1 + x)^{\alpha - 1} - \alpha$, for all $x \in I$. Then $f'(x) = 0 \iff x = 0$. There are two cases.
• If $0 < \alpha < 1$, then $f'(x) < 0$ for all $x > 0$ and $f'(x) > 0$ for all $x \in]-1, 0[$, which means that 0 is the global maximum point for f. Therefore $f(x) < f(0)$ for all $x \in I \setminus \{0\}$; that is, (5.13).
• If $\alpha \in]-\infty, 0[\cup]1, \infty[$, then $f'(x) > 0$ for all $x > 0$ and $f'(x) < 0$ for all $x \in]-1, 0[$, meaning that 0 is the global minimum point for f. Therefore $f(x) > f(0)$ for all $x \in I \setminus \{0\}$; that is, (5.14). □

---

[2] John Napier, 1550–1617.
[3] Michel Rolle, 1652–1719.

Proposition 2.3. *For $m = 1, 2, \ldots$, the inequality*

$$\left(1 + \frac{1}{m}\right)^m \le e\left(1 - \frac{1 - 2/e}{m}\right) \tag{5.15}$$

holds, where the constant $1 - 2/e$ is the best possible; that is, the largest value of β in

$$\left(1 + \frac{1}{m}\right)^m \le e\left(1 - \frac{\beta}{m}\right) \tag{5.16}$$

for which (5.16) holds.

Proof. Inequality (5.16) is equivalent to $\beta \le m - (m/e)(1 + 1/m)^m$.
Let

$$f(x) = \frac{1}{x} - \frac{1}{ex}(1 + x)^{1/x}, \quad x \in]0, 1].$$

It is obvious that f is decreasing on the interval $]0, 1]$. Consequently, $\beta = f(1) = 1 - 2/e$ is the optimal value satisfying (5.16), so (5.15) holds.   □

Proposition 2.4. *For $m = 1, 2, \ldots$, the inequality*

$$\left(1 + \frac{1}{m}\right)^m \le e \bigg/ \left(1 + \frac{1}{m}\right)^{1/\ln 2 - 1} \tag{5.17}$$

holds, where the constant $1/\ln 2 - 1$ is the best possible.

Proof. The inequality

$$(1 + 1/m)^m \le e/(1 + 1/m)^\alpha \tag{5.18}$$

is equivalent to

$$\alpha \le 1/\ln(1 + 1/m) - m.$$

Let

$$f(x) = 1/\ln(1 + x) - 1/x, \quad x \in]0, 1].$$

Because the function f is decreasing on $]0, 1]$, $\alpha = f(1) = 1/\ln 2 - 1$ is the best value satisfying (5.18), and thus (5.17) holds.   □
Combining the two previous propositions we get the following lemma.

Lemma 5.1. *For $m = 1, 2, \ldots$, the inequality*

$$\left(1 + \frac{1}{m}\right)^m \le e\left(1 - \frac{\beta}{m}\right) \bigg/ \left(1 + \frac{1}{m}\right)^\alpha \tag{5.19}$$

holds, where $0 \le \alpha \le 1/\ln 2 - 1$, $0 \le \beta \le 1 - 2/e$, and $e\beta + 2^{1+\alpha} = e$.

Proof. Inequality (5.19) is equivalent to

$$\beta \le m - (m/e)(1 + 1/m)^{m+\alpha}. \tag{5.20}$$

If

$$f(x) = 1/x - 1/(ex)(1 + x)^{1/x+\alpha}, \quad x \in]0,1], \quad 0 \le \alpha \le 1/\ln 2 - 1,$$

then f is decreasing on $]0,1]$. Consequently, $\beta = f(1) = 1 - (1/e)2^{1+\alpha}$ is the optimal value satisfying (5.20). Moreover, $0 \le \beta \le 1 - (1/e)$, and $e\beta + 2^{1+\alpha} = e$. So, inequality (5.19) holds.   □

Remark. If $\alpha = 0$, then $\beta = 1 - (2/e)$, and we obtain Proposition 2.3. If $\beta = 0$, then $\alpha = (1/\ln 2) - 1$, and we obtain Proposition 2.4.   △

Remark. Although the coefficient e in Proposition 3.23 at page 118 is optimal, Carleman's inequality can be refined. Here we mention two results.   △

Theorem 2.6. *For every sequence* $(a_n)_{n\ge 1}$ *with nonnegative terms so that* $\sum_{n=1}^{\infty} a_n < \infty$, *we have*

$$\sum_{n=1}^{\infty}(a_1 a_2 \dots a_n)^{1/n} \le e \sum_{n=1}^{\infty}\left(1 - \frac{1 - 2/e}{n}\right) a_n, \tag{5.21}$$

and

$$\sum_{n=1}^{\infty}(a_1 a_2 \dots a_n)^{1/n} \le e \sum_{n=1}^{\infty} \frac{a_n}{(1 + 1/m)^{1/\ln 2 - 1}}. \tag{5.22}$$

Theorem 2.7. *For every sequence* $(a_n)_{n\ge 1}$ *with nonnegative terms so that* $\sum_{n=1}^{\infty} a_n < \infty$, *we have*

$$\sum_{n=1}^{\infty}(a_1 a_2 \dots a_n)^{1/n} \le e \sum_{n=1}^{\infty}\left(\left(1 - \frac{\beta}{n}\right) \Big/ \left(1 + \frac{1}{n}\right)^{\alpha}\right) \cdot a_n, \tag{5.23}$$

where $0 \le \alpha \le 1/\ln 2 - 1$, $0 \le \beta \le 1 - 2/e$, *and* $e\beta + 2^{1+\alpha} = e$.

5.2.1 Consequences of the mean value theorems

Lemma 5.2. *Consider* $I = [a, b[$, $a < b \le +\infty$. *Suppose* $f, g : I \to \mathbb{R}$ *are continuous functions. Moreover,*

(i) *f and g are differentiable on $]a, b[$ and $f'(t) \le g'(t)$ for every $t \in]a, b[$.*
(ii) *$f(a) \le g(a)$.*

Then $f(t) \le g(t)$ *for each* $t \in I$.

Proof. Define $h = g - f$. Function h is differentiable on $]a, b[$ and $h'(t) \geq 0$ for any $t \in]a, b[$. From Theorem 2.4 it follows that h is increasing and hence

$$h(t) \geq h(a) \geq 0, \text{ for all } t \in [a, b[. \quad \square$$

Lemma 5.3. *Consider $I = [a, b]$, $a < b$. Suppose $f, g : I \to \mathbb{R}$ are continuous functions. Moreover,*

(i) *f and g are differentiable on $]a, b[$.*
(ii) *$|f'(t)| \leq g'(t)$ for every $t \in]a, b[$.*

Then
$$|f(b) - f(a)| \leq g(b) - g(a).$$

Proof. The conclusion is equivalent to

$$g(a) - g(b) \leq f(b) - f(a) \leq g(b) - g(a); \tag{5.24}$$

that is,

$$g(b) - f(b) - (g(a) - f(a)) \geq 0 \text{ and } f(b) + g(b) - (f(a) + g(a)) \geq 0.$$

These two inequalities suggest that we consider the next auxiliary functions

$$h_1, h_2 : I \to \mathbb{R}, \quad \begin{cases} h_1(t) = g(t) - f(t) - (g(a) - f(a)), \\ h_2(t) = f(t) + g(t) - (f(a) + g(a)). \end{cases}$$

We note that $h_1(a) = h_2(a) = 0$ and

$$h_1'(t) = g'(t) - f'(t) \geq 0, \quad h_2'(t) = f'(t) + g'(t) \geq 0, \quad \forall t \in [a, b].$$

Then $h_1(t) \geq h_1(a)$ and $h_2(t) \geq h_2(a)$, on $[a, b]$, and hence (5.24) follows. $\square$

Lemma 5.4. *Let $I \subset \mathbb{R}$ be a nonempty interval, $t_0 \in I$. Consider $f : I \to \mathbb{R}$ a continuous function on I and differentiable on $I \setminus \{t_0\}$.*

(a) *If f' has a left-hand side limit at t_0, then ϕ in (5.1) has a left-hand side limit at t_0 and*
$$\lim_{\substack{t \to t_0 \\ t < t_0}} f'(t) = \lim_{\substack{t \to t_0 \\ t < t_0}} \phi(t).$$

(b) *If f' has a right-hand side limit at t_0, then ϕ in (5.1) has a right-hand side limit at t_0 and*
$$\lim_{\substack{t \to t_0 \\ t > t_0}} f'(t) = \lim_{\substack{t \to t_0 \\ t > t_0}} \phi(t).$$

(c) *If f' has a finite limit at t_0, then f is differentiable at t_0 and f' is continuous at t_0.*

Proof. (a) Let $t < t_0$, $t \in I$. By the mean value theorem 2.3 there exists $c_t \in {]}t, t_0{[}$ such that

$$\phi(t) = \frac{f(t) - f(t_0)}{t - t_0} = f'(c_t).$$

Letting $t \to t_0$, the conclusion follows.
(b) Similar to (a).
(c) It follows from (a) and (b). $\quad\square$

We denote

$$f'_l(t_0) = \lim_{\substack{t \to t_0 \\ t < t_0}} \phi(t), \quad f'_r(t_0) = \lim_{\substack{t \to t_0 \\ t > t_0}} \phi(t). \tag{5.25}$$

Then $f'_l(t_0)$ is said to be the *left-hand side derivative* of f at t_0, and $f'_r(t_0)$ is said to be the *right-hand side derivative* of f at t_0, if any.

Remark. The continuity assumption in Lemma 5.4 is basic. Consider the following function

$$f(x) = \begin{cases} \arctan \dfrac{1 - x}{1 + x}, & x \in \mathbb{R} \setminus \{-1\} \\ 0, & x = -1. \quad \triangle \end{cases}$$

Lemma 5.5. *Let $I \subset \mathbb{R}$ be a nonempty interval and $f : I \to \mathbb{R}$ be a differentiable function on I. If f' is bounded on I, then f is Lipschitz on I.*

Proof. Because f' is bounded on I, there is a positive L such that $|f'(t)| \leq L$, for any $t \in I$. Then for any $t_1, t_2 \in I$, $t_1 < t_2$ there is $c \in {]}t_1, t_2{[}$ such that

$$|f(t_2) - f(t_1)| = |f'(c)(t_2 - t_1)| \leq L(t_2 - t_1),$$

hence function f is Lipschitz on I. $\quad\square$

Theorem 2.8. (Denjoy[4]–Bourbaki[5]) *Let $I \subset \mathbb{R}$ be an interval, $a, b \in I$, $a < b$, and $f : I \to \mathbb{R}$ a function on I. Suppose*

(i) *Function f is continuous on $[a, b]$.*
(ii) *There is an at most countable set $A \subset {]}a, b{[}$ such that f is right-hand side differentiable on ${]}a, b{[} \setminus A$.*

Then

$$\inf_{t \in {]}a,b{[} \setminus A} \phi(t+) \leq \frac{f(b) - f(a)}{b - a} \leq \sup_{t \in {]}a,b{[} \setminus A} \phi(t+),$$

where ϕ is the quotient in (5.1).

---

[4] Arnaud Denjoy, 1884–1974.
[5] Nikolas Bourbaki, collective pseudonym of several French mathematicians, 1939↑

Proof. We prove that

$$\frac{f(b) - f(a)}{b - a} \le \sup_{t \in \,]a,b[\,\setminus A} \phi(t+).$$

Denote $M = \sup_{t \in \,]a,b[\,\setminus A} \phi(t+)$. The other side follows in a similar way.

If $M = +\infty$, the inequality is obvious. Suppose $M < \infty$ and consider the function $g : I \to \mathbb{R}$, $g(t) = Mt - f(t)$. We remark that function g fulfills the assumptions of Lemma 5.7. Thus g is increasing on I. Then $g(b) \ge g(a)$; that is, $M(b - a) \ge f(b) - f(a)$. $\square$

Corollary 2.2. *Suppose* $f, g : I \to \mathbb{R}$ *are continuous on a nonempty interval* I. *Then*

(a) f *is constant on* I *if and only if there is an at most countable set* $A \subset I$ *such that* f *is right-hand side differentiable on* $I \setminus A$ *and* $\phi(t+) = 0$, *for every* $t \in I \setminus A$.
(b) $f - g$ *is constant on* I *if and only if there is an at most countable set* $A \subset I$ *such that* f *and* g *are right-hand side differentiable on* $I \setminus A$ *and* $\phi_f(t+) = \phi_g(t+) = 0$, *for every* $t \in I \setminus A$, *where* ϕ_f *and* ϕ_g *are the* ϕ *functions corresponding to* f, *respectively to* g.

Remark. The right-hand side differentiability from the Denjoy–Bourbaki theorem may be substituted by the left-hand side differentiability. $\triangle$

Lemma 5.6. *Let* $I \subset \mathbb{R}$ *be a nonempty interval and* $f : I \to \mathbb{R}$ *a function. Suppose*

(i) *Function* f *is continuous on* I.
(ii) *There is an at most countable set* $A \subset I$ *such that for every* $s \in I \setminus A$ *and* $\delta > 0$ *there is* $t \in \,]s, s + \delta[$ *satisfying* $f(t) \ge f(s)$.

Then function f *is increasing on* I.

Proof. Consider $a, b \in I$, $a < b$. We show that $f(a) \le f(b)$.

Let $\lambda \notin f(A)$, $\lambda < f(a)$. Consider the set $S_\lambda = \{s \in [a, b] \mid \lambda \le f(s)\}$. Note that

- $S_\lambda \ne \emptyset$, because $a \in S_\lambda$.
- S_λ is bounded above, because $S_\lambda \subset [a, b]$.

Hence there exists $\sup S_\lambda$ $(\in [a, b])$. Denote it by M; that is, $M = \sup S_\lambda$. Because M is a limit point of S_λ, there is $s_n \in S_\lambda$ such that $s_n \to M$. From

$$\left.\begin{array}{r} s_n \in S_\lambda \\ f \text{ continuous at } M \end{array}\right\} \implies [\lambda \le f(s_n) \implies \lambda \le f(M)]. \tag{5.26}$$

Thus $M \in S_\lambda$.

Now we show that $M = b$. Suppose $M < b$. Then M is a limit point for the complement of S_λ, so there is $t_n \notin S_\lambda$, $t_n \to M$. It follows that $f(t_n) < \lambda$, and, due to the continuity of f,

$$f(M) \leq \lambda. \tag{5.27}$$

From (5.26) and (5.27) it follows that $f(M) = \lambda$ and thus $M \notin A$.

On the other side, because $M < b$ and (ii), there is $t \in]M, b[$ such that $\lambda = f(M) \leq f(t)$ and so $t \in S_\lambda$ and $t > M$, that is, a contradiction. Thus $M = b$.

Hence for any $\lambda \notin f(A)$ such that $\lambda < f(a)$ it follows that $\lambda \leq f(b)$. So we have that $f(a) \leq f(b)$. $\quad\square$

Lemma 5.7. *Let* $I \subset \mathbb{R}$ *be a nonempty interval and* $f : I \to \mathbb{R}$ *a function. Suppose*

(i) *Function* f *is continuous on* I.
(ii) *There is an at most countable set* $A \subset I$ *such that there exists* $\phi(t+)$ *(* ϕ *defined by (5.1)) for any* $t \in I \setminus A$.
(iii) $\phi(t+) \geq 0$ *for any* $t \in I \setminus A$.

Then f *is increasing on* I.

Proof. From (iii) it follows that for every $\varepsilon > 0$ and $t \in I \setminus A$ there is a positive $\delta = \delta(\varepsilon, t)$ such that for every $h \in]0, \delta[$ it holds

$$|(f(t + h) - f(t))/h - \phi(t+)| < \varepsilon,$$

or

$$-\varepsilon < \phi(t+) - \varepsilon < (f(t + h) - f(t))/h < \phi(t+) + \varepsilon.$$

Therefore $f(t + h) - f(t) + h\varepsilon > 0$, for every $h \in]0, \delta[$, $t \in I \setminus A$; that is,

$$f(t + h) + (t + h)\varepsilon - (f(t) + t\varepsilon) > 0, \quad \text{for every } h \in]0, \delta[, \ t \in I \setminus A.$$

Denote $g_\varepsilon(t) = f(t) + t\varepsilon$. Then the last inequality is equivalent with

$$g_\varepsilon(t + h) - g_\varepsilon(t) > 0, \quad \text{for every } h \in]0, \delta[, \ t \in I \setminus A.$$

From Lemma (5.6) it follows that g_ε is increasing on I; that is, for any $t_1, t_2 \in I$, $t_1 < t_2$ we have

$$g_\varepsilon(t_1) \leq g_\varepsilon(t_2) \iff f(t_1) \leq f(t_2) + \varepsilon(t_2 - t_1).$$

Letting $\varepsilon \to 0$ we get that $f(t_1) \leq f(t_2)$, meaning that f is increasing. $\quad\square$

Lemma 5.8. *Let* f *be a differentiable function with bounded derivative on* $[a, b]$. *Then* f *is of bounded variation on* $[a, b]$.

Proof. We have that there exists a constant $M > 0$ so that $|f'(t)| \leq M$ for all $t \in [a, b]$. Then by the Lagrange mean value theorem it follows

$$|f(x) - f(y)| = |f'(c)(x - y)| \leq M|x - y|.$$

Thus f is a Lipschitz function. Now by (b) of Remarks 7.1 at page 179 we conclude that f is of bounded variation on $[a, b]$. $\quad\square$

5.3 The continuity and the surjectivity of derivatives

We have already seen by (b) of Examples 1.1 that a function f may have a derivative f' which exists at every point, but is discontinuous at some point. However, not every function is a derivative. In particular, derivatives that exist at every point of an interval have one important property in common with functions that are continuous on an interval: intermediate values are assumed.

Theorem 3.1. (Darboux) *Consider f a real differentiable function defined on $[a, b]$ and suppose $f'(a) < \lambda < f'(b)$. Then there is a point $x \in]a, b[$ such that $f'(x) = \lambda$.*

A similar result holds of course if $f'(a) > f'(b)$.

Proof. First approach. Put $c = (a + b)/2$. If $a \leq t \leq c$, define $\alpha(t) = a$, $\beta(t) = 2t - a$. If $c \leq t \leq b$, define $\alpha(t) = 2t - b$, $\beta(t) = b$. Then $a \leq \alpha(t) < \beta(t) \leq b$ in $]a, b[$. Define

$$g(t) = \frac{f(\beta(t)) - f(\alpha(t))}{\beta(t) - \alpha(t)}, \quad a < t < b.$$

Then g is continuous on $]a, b[$, $g(t) \to f'(a)$, as $t \to a$, $g(t) \to f'(b)$, as $t \to b$, and so Theorem 2.14 from page 160 implies that $g(t_0) = \lambda$ for some $t_0 \in]a, b[$. Set t_0. By Theorem 2.3 page 199 there is a point x such that $\alpha(t_0) < x < \beta(t_0)$ and such that $f'(x) = g(t_0)$. Hence $f'(x) = \lambda$.

Second approach. Define $g : [a, b] \to \mathbb{R}$ by $g(t) = f(t) - \lambda t$. Then g is differentiable on $[a, b]$, $g'(a) = f'(a) - \lambda < 0$, and $g'(b) = f'(b) - \lambda > 0$. Thus neither a nor b can be a point where g attains even a local minimum. Because g is continuous on $[a, b]$ by Theorem 2.10 at page 156, there exists a point $x \in]a, b[$ where g attains a local minimum. Then $0 = g'(x) = f'(x) - \lambda$. $\square$

Corollary 3.3. *If function f is differentiable on $[a, b]$, then f' have no discontinuity of the first kind on $[a, b]$.*

Theorem 3.2. *Suppose that $f : \mathbb{R} \to \mathbb{R}$ is differentiable everywhere and suppose that*

$$\lim_{|x| \to \infty} \frac{f(x)}{|x|} = \infty.$$

Then $\{f'(x) \mid x \in \mathbb{R}\} = \mathbb{R}$.

Proof. Let t be an arbitrary real number. Define $g(x) = f(x) - tx$, $x \in \mathbb{R}$. Then g is differentiable on $\mathbb{R}$ and $g'(x) = f'(x) - t$. We have

$$\lim_{|x| \to \infty} g(x)/|x| = \lim_{|x| \to \infty} (f(x)/|x| - t \cdot \text{sign}\, x) = \infty.$$

Thus there exists a positive $r > t$ so that $|g(x)| > |t|$, for all $|x| > r$. Then g attains a global minimum on the interval $[-r, r]$ at, say, $\xi \in [-r, r]$. So, $0 = g'(\xi) = f'(\xi) - t$. Hence $f'(\xi) = t$. $\square$

5.4 L'Hospital theorem

A very useful result is the following.

Theorem 4.1. (L'Hospital[6]) *Assume f and g are real and differentiable on* $]a, b[$ *and* $g'(x) \neq 0$ *for all* $x \in]a, b[$, *where* $-\infty \leq a < b \leq +\infty$. *Suppose*

$$\frac{f'(x)}{g'(x)} \to A \ (\in [-\infty, \infty]) \ \text{as} \ x \to a. \tag{5.28}$$

If

$$f(x) \to 0 \ \text{and} \ g(x) \to 0 \ \text{as} \ x \to a, \tag{5.29}$$

or if

$$g(x) \to +\infty \ \text{as} \ x \to a, \tag{5.30}$$

then

$$\frac{f(x)}{g(x)} \to A \ \text{as} \ x \to a. \tag{5.31}$$

Proof. We first consider the case when $-\infty \leq A < +\infty$. Choose a real number q such that $A < q$, and choose r such that $A < r < q$. By (5.28) there is a point $c \in]a, b[$ such that $a < x < c$ implies

$$f'(x)/g'(x) < r.$$

If $a < x < y < c$, then the Cauchy theorem 2.2 shows that there is a point $t \in]x, y[$ such that

$$\frac{f(x) - f(y)}{g(x) - g(y)} = \frac{f'(t)}{g'(t)} < r. \tag{5.32}$$

Suppose (5.29) holds. Letting $x \to a$ in (5.32) we see that

$$f(y)/g(y) \leq r < q \quad (a < y < c). \tag{5.33}$$

Next, suppose (5.30) holds. Keeping y fixed in (5.32) we can choose a point $c_1 \in]a, y[$ such that $g(x) > g(y)$ and $g(x) > 0$ if $a < x < c_1$. Multiplying (5.32) by $(g(x) - g(y))/g(x)$, we get

$$\frac{f(x)}{g(x)} < r - r\frac{g(y)}{g(x)} + \frac{f(y)}{g(x)}, \quad (a < x < c_1). \tag{5.34}$$

If we let $x \to a$ in (5.34), (5.30) shows that there is a point $c_2 \in]a, c_1[$ such that

$$f(x)/g(x) < q \quad (a < x < c_2). \tag{5.35}$$

Summing up, (5.33) and (5.35) show that for any q, subject only to the condition $A < q$, there is a point c_2 such that $f(x)/g(x) < r$ if $a < x < c_2$.

In the same manner, if $-\infty < A \leq +\infty$, and p is chosen so that $p < A$, we can find a point c_2 such that

$$p < f(x)/g(x) \quad a < x < c_2,$$

and (5.31) follows from these two statements.   □

---

[6] Guillaume François Antoine de L'Hospital, marquis de Sainte-Mesme, 1661–1704.

5.5 Higher-order derivatives and the Taylor formula

Suppose $f : A \to \mathbb{R}$ and f is differentiable on an interval $I \subset A$. If there exists the derivative of f' at a point $x \in I$, it is said to be the *second-order derivative* of f at x. We denote it by $f''(x)$ or $f^{(2)}(x)$. In this case we also say that f is *twice differentiable*. Analogously we obtain $f'''(x)$ or $f^{(3)}(x)$, $f^{(4)}(x)$, ..., $f^{(n)}(x)$ each of which is the derivative of the preceding one, if any. $f^{(n)}(x)$ is said to be the nth *order derivative* of f. By definition, $f^{(0)} = f$.

Let n be a natural number and I be an interval. By C^n or $C^n(I)$ we denote the set of functions having continuous derivatives up to the nth order on I. In this case we say that f *is of class* C^n or f *belongs to class* C^n. By C^∞ or $C^\infty(I)$ we denote the set of functions having derivatives of all orders on I. Now we say that f *is of class* C^∞ or f *belongs to class* C^∞.

We introduce a generalization of Theorem 4.1.

Corollary 5.4. *Assume* $n \in \mathbb{N}^*$, f *and* g *are real and* n *times differentiable on* $]a, b[$, *where* $-\infty \le a < b \le +\infty$. *Suppose* $f^{(k)}$, $g^{(k)}$ *are continuous on* $]a, b[$ *for* $k = 1, \ldots, n - 1$ *and*

$$\frac{f^{(n)}(x)}{g^{(n)}(x)} \to A \ (\in [-\infty, \infty]) \quad as \ \ x \to a.$$

If

$$f^{(k)}(x) \to 0 \quad and \quad g^{(k)}(x) \to 0 \ \ as \ \ x \to a, \quad \forall k = 1, \ldots, n - 1$$

or if

$$g^{(k)}(x) \to +\infty \ \ as \ \ x \to a, \quad \forall k = 1, \ldots, n - 1$$

then

$$f(x)/g(x) \to A \ \ as \ \ x \to a.$$

Proof. Apply Theorem 4.1 to functions $f^{(n-1)}$ and $g^{(n-1)}$. We get

$$f^{(n-1)}(x)/g^{(n-1)}(x) \to A \ \ as \ \ x \to a.$$

Successively apply Theorem 4.1 to functions $f^{(n-2)}$ and $g^{(n-2)}$, ..., f and g. Then we get

$$\lim_{x \downarrow a} f^{(n-2)}(x)/g^{(n-2)}(x) = A, \ldots, \ \lim_{x \downarrow a} f(x)/g(x) = A. \quad \square$$

Example 5.1. Suppose one desires computing

$$\lim_{x \to 0} \frac{1 - \cos x}{x^2}. \tag{5.36}$$

Solution. First approach. Denote $f(x) = 1 - \cos x$ and $g(x) = x^2$, for $x \in \mathbb{R}$. Then $f(0) = f'(0) = 0, f''(0) = 1$ and $g(0) = g'(0) = 0, g''(0) = 2$. Therefore the limit in (5.36) is equal to $1/2$.

Second approach. Recall (4.6) at page 150. Then we have

$$\lim_{x \to 0} \frac{1 - \cos x}{x^2} = \lim_{x \to 0} \frac{2 \sin^2(x/2)}{x^2} = \lim_{x \to 0} \frac{1}{2} \left(\frac{\sin(x/2)}{x/2} \right)^2$$
$$= (1/2) \lim_{t \to 0} (\sin t/t)^2 = 1/2. \quad \triangle$$

Theorem 5.1. (Leibniz) *Let u and v be two functions having derivatives up to the nth order on an interval. Then on that interval*

$$(uv)^{(n)} = u^{(n)}v + \binom{n}{1} u^{(n-1)}v' + \cdots + \binom{n}{n-1} u'v^{(n-1)} + uv^{(n)}, \quad \forall n \in \mathbb{N}.$$

Proof. By induction. □

Theorem 5.2. (Rolle) *Let k be a positive integer and $f \in C^k[a,b]$. If $f^{(k)}$ has at least $k+1$ roots in $[a,b]$, not necessarily distinct, then f has a root in $[a,b]$.*

As an application of Theorem 5.2, we prove Theorem 1.27 at page 96.

Proof. The idea is to study the quadratic polynomial

$$\frac{d^{n-k-1}}{d\,x^{n-k-1}} \left(x^{n-k+1} \cdot \left(\frac{d^{k-1}}{d\,x^{k-1}} A(x) \right)_{x \to x^{-1}} \right)$$

having real roots (by the Rolle theorem). Then we just have to look at its discriminant. Successively we have

$$A(x) = a_n x^n + a_{n-1} x^{n-1} + \cdots + a_{n-m} x^{n-m} + \cdots + a_1 x + a_0,$$

$$\frac{d^{k-1}}{d\,x^{k-1}} A(x) = \frac{n!}{(n-k+1)!} a_n x^{n-k+1} + \ldots$$

$$+ \frac{(n-m)!}{(n-m-k+1)!} a_{n-m-k+1} x^{n-m-k+1} + \cdots + (k-1)! a_{k-1},$$

$$\left(\frac{d^{k-1}}{d\,x^{k-1}} A(x) \right)_{x \to x^{-1}} = \frac{n!}{(n-k+1)!} a_n x^{k-n-1}$$

$$+ \cdots + \frac{(n-m)!}{(n-m-k+1)!} a_{n-m-k+1} x^{k-n+m-1} + \cdots + (k-1)! a_{k-1},$$

$$x^{n-k+1} \left(\frac{d^{k-1}}{d\,x^{k-1}} A(x) \right)_{x \to x^{-1}} = \frac{n!}{(n-k+1)!} a_n$$

$$+ \cdots + \frac{(n-m)!}{(n-m-k+1)!} a_{n-m-k+1} x^m + \cdots + (k-1)! a_{k-1} x^{n-k+1},$$

$$\frac{d^{n-k-1}}{d\,x^{n-k-1}} \left(x^{n-k+1} \cdot \left(\frac{d^{k-1}}{d\,x^{k-1}} A(x) \right)_{x \to x^{-1}} \right)$$

$$= \frac{(k+1)!(n-k-1)!}{2!} a_{k+1} + k!\,(n-k)!\,a_k x + \frac{(k-1)!(n-k+1)!}{2!} a_{k-1} x^2.$$

The discriminant of the above written second-degree polynomial is nonnegative because it has real roots. Hence

$$a_k^2 \geq \left(\frac{k+1}{k}\right)\left(\frac{n-k+1}{n-k}\right) a_{k-1} a_{k+1},$$

and thus the Theorem 1.27 at page 96 is proved.   □

Theorem 5.3. *Let k be a positive integer and $f \in C^{(k)}[a,b]$. If $f^{(k)}$ has at least $k+1$ roots in $[a,b]$, not necessarily distinct, then*

$$\max_{x \in [a,b]} |f(x)| \leq \frac{(b-a)^n}{n!} \max_{x \in [a,b]} |f^{(k)}(x)|.$$

Proof. Let $x_1, x_2, \ldots, x_n$ be the zeros of function f in $[a,b]$. Some of these zeros are multiple. Consider $x_0 \in [a,b] \setminus \{x_1, x_2, \ldots, x_n\}$. Define the polynomial

$$p(x) = f(x_0) \frac{\prod_{k=1}^{n}(x - x_k)}{\prod_{k=1}^{n}(x_0 - x_k)}$$

and the function $g(x) = f(x) - p(x)$. Function g vanishes at $x_0, x_1, \ldots, x_n$. We apply Theorem 5.2 and thus there exists a point $\xi \in [a,b]$ so that $g^{(n)}(\xi) = 0$. Then

$$0 = g^{(n)}(\xi) = f^{(n)}(\xi) - p^{(n)}(\xi) = f^{(n)}(\xi) - n! f(x_0) / \prod_{k=1}^{n}(x_0 - x_k)$$

and

$$|f(x_0)| = \frac{|f^{(n)}(\xi)| \prod_{k=1}^{n}(x_0 - x_k)}{n!} \leq \frac{|f^{(n)}(\xi)|(b-a)^n}{n!}$$

$$\leq \frac{(b-a)^n}{n!} \max_{x \in [a,b]} |f^{(n)}(x)|.$$

Because $x_0 \in [a,b] \setminus \{x_1, x_2, \ldots, x_n\}$ is arbitrary, the conclusion follows.   □

Proposition 5.5. *The following identities hold on $\mathbb{R}$*

$$(e^x)^{(n)} = e^x, \quad \sin^{(n)}(x) = \sin\left(x + n\frac{\pi}{2}\right),$$

$$\cos^{(n)}(x) = \cos\left(x + n\frac{\pi}{2}\right), \quad \forall n \in \mathbb{N}.$$

Suppose f is a real function on an interval I, $a \in I$, $n \in \mathbb{N}$, $f^{(n-1)}$ is continuous on I, and $f^{(n)}$ exists at a. For $x \in I$ we define the *Taylor*[7] *polynomial* of degree n at a of f by

$$T_n(x) = f(a) + \frac{x-a}{1!} f'(a) + \frac{(x-a)^2}{2!} f''(a) + \cdots + \frac{(x-a)^n}{n!} f^{(n)}(a).$$

---

[7] Brook Taylor, 1685–1731.

Denote for every $x \in I$,

$$R_n(x) = f(x) - T_n(x),\qquad(5.37)$$

the *remainder* of the Taylor polynomial $T_n(x)$. Then we have

$$f(x) = f(a) + \frac{x-a}{1!}f'(a) + \cdots + \frac{(x-a)^n}{n!}f^{(n)}(a) + R_n(x), \quad x \in I. \quad(5.38)$$

Identity (5.38) is called the *Taylor formula about a*.

Sometimes, the Taylor formula (5.38) is written as

$$f(a+h) = f(a) + \frac{h}{1!}f'(a) + \cdots + \frac{(h)^n}{n!}f^{(n)}(a) + R_n(h), \quad a, a+h \in I.$$

We note the remainder (5.37) of the Taylor polynomial depends on four variables, namely f, n, a, x. Often we neglect some of them, depending on what we intend to emphasize.

Proposition 5.6. *Under the above-mentioned assumptions the following relations hold.*

$$T_n(a) = f(a), \quad T_n'(a) = f'(a), \dots, T_n^{(n)}(a) = f^{(n)}(a), \quad T_n^{(n+1)}(a) = 0,$$
$$R_n(a) = 0, \quad R_n'(a) = 0, \dots, R_n^{(n)}(a) = 0.$$

Lemma 5.9. *Under the above-mentioned assumptions it holds*

$$\lim_{x \to a} \frac{R_n(x)}{(x-a)^n} = 0.$$

Proof. From Theorem 4.1 we have

$$\lim_{x \to a} \frac{R_n(x)}{(x-a)^n} = \lim_{x \to a} \frac{R_n'(x)}{n(x-a)^{n-1}} = \cdots = \lim_{x \to a} \frac{R_n^{(n)}(x)}{n!} = 0. \quad \square$$

Remark. For $a = 0$ in (5.38), results

$$f(x) = f(0) + \frac{x}{1!}f'(0) + \cdots + \frac{x^n}{n!}f^{(n)}(0) + R_n(x), \quad x \in I, \quad(5.39)$$

called the *Maclaurin* formula. $\triangle$

We list the Maclaurin formula for some functions.

$$e^x = 1 + \frac{x}{1!} + \frac{x^2}{2!} + \cdots + \frac{x^n}{n!} + R_n(x), \quad x \in \mathbb{R},$$
$$\sin x = \frac{x}{1!} - \frac{x^3}{3!} + \frac{x^5}{5!} + \cdots + (-1)^{n+1}\frac{x^{2n-1}}{(2n-1)!} + R_{2n-1}(x), \quad x \in \mathbb{R},$$

$$\cos x = 1 - \frac{x^2}{2!} + \frac{x^4}{4!} + \cdots + (-1)^n \frac{x^{2n}}{(2n)!} + R_{2n}(x), \quad x \in \mathbb{R},$$

$$\ln(1+x) = \frac{x}{1} - \frac{x^2}{2} + \frac{x^3}{3} - \frac{x^4}{4} \cdots + (-1)^{n+1} \frac{x^n}{n} + R_n(x), \quad x > -1$$

$$(1+x)^\alpha = 1 + \frac{\alpha}{1!} x + \frac{\alpha(\alpha-1)}{2!} x^2 + \cdots \qquad\qquad (5.40)$$

$$+ \frac{\alpha(\alpha-1)\ldots(\alpha-n+1)}{n!} x^n + R_n(x), \quad \alpha \in \mathbb{R}, \ x > -1.$$

Remark. From Lemma 5.9 it follows that for given x and a there exist $p \in \mathbb{N}^*$ and $k \in \mathbb{R}$ such that

$$f(x) = f(a) + \frac{x-a}{1!} f'(a) + \cdots + \frac{(x-a)^n}{n!} f^{(n)}(a) + (x-a)^p k. \quad \triangle \quad (5.41)$$

Lemma 5.10. *Suppose* $n \in \mathbb{N}^*$; *there exists* $f^{(n+1)}$ *on an interval* I, $p \in \mathbb{N}$, *and* $a, x \in I$, *arbitrary, but fixed, with* $a \neq x$. *Then the remainder of the Taylor polynomial can be written under the Schlömilch*[8] *form; that is, there exists a* ξ *between* a *and* x *satisfying*

$$R_n(x) = \frac{(x-a)^p (x-\xi)^{n-p+1}}{n! p} f^{(n+1)}(\xi). \qquad\qquad (5.42)$$

Proof. Define

$$\varphi(t) = f(t) + \frac{x-t}{1!} f'(t) + \cdots + \frac{(x-t)^n}{n!} f^{(n)}(t) + (x-t)^p k, \quad t \in I.$$

Function φ is differentiable on I; moreover $\varphi(x) = f(x)$ and $\varphi(a) = f(x)$. Then based on Theorem 2.5 there exists a ξ between a and x with $\varphi'(\xi) = 0$. Thus we get k. Substituting it in (5.41), we get (5.42). $\quad\square$

Remark. If we set $\xi = a + \theta(x-a)$, $0 < \theta < 1$, the remainder (5.42) can be written as

$$R_n(x) = \frac{(x-a)^{n+1}(1-\theta)^{n-p+1}}{n! p} f^{(n+1)}(a + \theta(x-a)). \quad \triangle$$

For the Maclaurin formula we have

$$R_n(x) = \frac{x^{n+1}(1-\theta)^{n-p+1}}{n! p} f^{(n+1)}(\theta x).$$

We mention two particular cases of the remainder to the Taylor formula.

(a) For $p = 1$, we get the remainder under the Cauchy form

$$R_n(x) = \frac{(x-a)(x-\xi)^n}{n!} f^{(n+1)}(\xi).$$

---

[8] Oscar Xavier Schlömilch, 1823–1901.

(b) For $p = n + 1$, we get the remainder under the Lagrange form

$$R_n(x) = \frac{(x-a)^{n+1}}{(n+1)!} f^{(n+1)}(\xi). \tag{5.43}$$

Remarks. (a) Point ξ in (5.42) depends on a, x, n, and p.
(b) Maclaurin formula (5.39) with the remainder under Lagrange form for $x \in I$ is

$$f(x) = f(0) + \frac{x}{1!} f'(0) + \cdots + \frac{x^n}{n!} f^n(0) + \frac{x^{n+1}}{(n+1)!} f^{(n+1)}(\xi),$$

where $\xi = \theta x,\ 0 < \theta < 1$. $\triangle$

Example. Consider the remainder of the function in (5.40); that is,

$$R_n(x) = \frac{x^{n+1}(1-\theta)^{n-p+1}}{n!p} \alpha(\alpha-1)\cdots(\alpha-n)(1+\theta x)^{\alpha-n-1}.$$

Under the Cauchy form, that is, $p = 1$, we get

$$R_n(x) = \frac{\alpha(\alpha-1)\cdots(\alpha-n)}{n!} x^{n+1} \left(\frac{1-\theta}{1+\theta x}\right)^n (1+\theta x)^{\alpha-1}.$$

If, moreover, $|x| < 1$, we have

$$\left(\frac{1-\theta}{1+\theta x}\right)^n < 1, \quad |1+\theta x| < 2, \quad \text{and} \quad \lim_{n \to \infty} \frac{\alpha(\alpha-1)\cdots(\alpha-n)}{n!} x^{n+1} = 0.$$

Then

$$|R_n(x)| \le 2^{\alpha+1} \left|\frac{\alpha(\alpha-1)\dots(\alpha-n)}{n!} x^{n+1}\right| \to 0, \quad \text{as} \quad n \to \infty.$$

The conclusion is that under the above-mentioned assumptions the Taylor polynomial locally approximates the function.

To calculate $\sqrt[3]{30}$, we can take

$$\sqrt[3]{30} = \sqrt[3]{27+3} = 3\sqrt[3]{1+1/9},$$

that is, $\alpha = 1/3$ and $x = 1/9$, and then use (5.40). $\triangle$

Example. Consider the series $\sum_{n=1}^{\infty} 1/n^p$ and suppose we have to study its convergence.

We already discussed this question in Theorem 5.1 at page 139 and Exercise 4.1 in Section 4.10. Here we use the Gauss test theorem 3.15 at page 121. Denote $a_n = n^{-p}$. Then by (5.40) we have

$$\frac{a_n}{a_{n+1}} = \left(1 + \frac{1}{n}\right)^p = 1 + \frac{p}{n} + \frac{p(p-1)}{2n^2}\left(1 + \frac{\theta_n}{n}\right)^{p-2} = 1 + \frac{p}{n} + \frac{\beta_n}{n^{1+1/2}},$$

where $\beta_n = p(p-1)/(2\sqrt{n})\,(1+\theta_n/n)^{p-2}$.

We conclude that the series converges if and only if $p > 1$. $\triangle$

5.6 Convex functions and differentiability

Theorem 6.1. *Suppose I is a nonempty interval, $f : I \to \mathbb{R}$ is convex, $a = \inf I$, and $b = \sup I$. Then*

(a) *f has side derivatives on $]a, b[$ and for every $t_1, t_2 \in]a, b[$, $t_1 < t_2$, we have*

$$f'_l(t_1) \leq f'_r(t_1) \leq f'_l(t_2) \leq f'_r(t_2).$$

(b) *If $a \in I$ ($b \in I$), then f is right-hand differentiable in a (respectively, it is left-hand differentiable in b) and*

$$f'_r(a) \leq f'_l(t) \qquad (\text{respectively } f'_r(t) \leq f'_l(b)), \quad t \in \text{int}\,(I).$$

(c) *There is an at most countable set $A \subset I$ such that f is differentiable on $I \setminus A$.*

Proof. (a) Suppose $t \in]a, b[$. Because $s_{f,t}$ is increasing on $I \setminus \{t\}$ (by (a) in Theorem 6.1 at page 171), it follows that $s_{f,t}$ has finite side limits at t, and hence f has side derivatives at t.

Suppose $t_1, t_2 \in]a, b[$, $t_1 < t_2$, and choose u, v, w satisfying $a < u < t_1 < v < t_2 < w < b$. Then from

$$s_{f,t_1}(u) \leq s_{f,t_1}(v) = s_{f,v}(t_1) \leq s_{f,v}(t_2) = s_{f,t_2}(v) \leq s_{f,t_2}(w)$$

it follows

$$f'_l(t_1) = s_{f,t_1}(t_1-) \leq s_{f,t_1}(t_1+) = f'_r(t_1)$$
$$\leq s_{f,t_2}(t_2-) = f'_l(t_2) \leq s_{f,t_2}(t_2+) = f'_r(t_2).$$

(b) If $a \in I$, we repeat the previous proof for $a < t_1 < u$.
(c) From (a) it follows that $f'_l(t)$ is increasing on $]a, b[$. Hence there is an at most countable set A such that $f'_l(t)$ is continuous on $]a, b[\setminus A$.

Choose $t_0 \in]a, b[\setminus A$. Then $f'_l(\cdot)$ is continuous on t_0, and for $t > t_0$

$$f'_l(t_0) \leq f'_r(t_0) \leq f'_l(t).$$

Letting $t \to t_0$, we get $f'_l(t_0) = f'_r(t_0)$; that is, f is differentiable at t_0. Hence f is differentiable on $I \setminus A$.   □

Corollary 6.5. *Suppose $f : I \to \mathbb{R}$ is convex, $a = \inf I$, $b = \sup I$, and $t_0 \in]a, b[$. Then*

$$f(t) \geq f(t_0) + m(t - t_0)$$

for every $t \in I$ and $m \in [f'_l(t_0), f'_r(t_0)]$.

Proof. If $t > t_0$,

$$s_{f,t_0}(t) = (f(t) - f(t_0))/(t - t_0) \geq \inf_{s \in I,\ s > t_0} s_{f,s}(s) = f'_r(t_0) \geq m.$$

Similarly, if $t < t_0$,

$$s_{f,t_0}(t) = (f(t) - f(t_0))/(t - t_0) \leq \sup_{s \in I, \, s < t_0} s_{f,s}(s) = f'_l(t_0) \leq m.$$

The conclusion follows.   □

Corollary 6.6. *Suppose $f : I \to \mathbb{R}$ is differentiable on I. Then f is convex if and only if*

$$f(t) \geq f(t_0) + f'(t_0)(t - t_0), \quad \text{for any } t, t_0 \in I. \tag{5.44}$$

Proof. The necessity part follows from the previous corollary.
   Sufficiency. From (5.44) for any $a, b \in I$ and $\alpha \in [0,1]$ follow

$$f(a) \geq f((1 - \alpha)a + \alpha b) + f'((1 - \alpha)a + \alpha b)\alpha(a - b),$$
$$f(b) \geq f((1 - \alpha)a + \alpha b) - f'((1 - \alpha)a + \alpha b)(1 - \alpha)(a - b).$$

Multiply the first inequality by $(1 - \alpha)$, the second by α, and then add them.
Then $(1 - \alpha)f(a) + \alpha f(b) \geq f((1 - \alpha)a + \alpha b)$, hence f is convex.   □

Corollary 6.7. *Let I be an open and nonempty interval, and $f : I \to \mathbb{R}$ be a differentiable function. Then f is convex if and only if*

$$[f'(x) - f'(y)](x - y) \geq 0, \quad x, y \in I. \tag{5.45}$$

Proof. Suppose f is convex. Then by Corollary 6.6 we have

$$f'(x)(y - x) \leq f(y) - f(x), \quad f'(y)(x - y) \leq f(x) - f(y) \quad \forall x, y \in I.$$

Adding them, we get inequality (5.45).
   Suppose that inequality (5.45) holds. It means that f' is increasing. Consider $x, y \in I$ with $x < y$, $\lambda \in \,]0,1[$, and $x_\lambda = (1 - \lambda)x + \lambda y$. We have

$$
\begin{aligned}
(1 - \lambda)f(x) + \lambda f(y) - f(x_\lambda) &= (1 - \lambda)[f(x) - f(x_\lambda)] + \lambda[f(y) - f(x_\lambda)] \\
&= (1 - \lambda)f'(x)(x - x_\lambda) + \lambda f'(y)(y - x_\lambda) \\
&= \lambda(1 - \lambda)f'(x)(x - y) + \lambda(1 - \lambda)f'(y)(y - x) \\
&= \lambda(1 - \lambda)(x - y)[f'(x) - f'(y)] \geq 0. \quad □
\end{aligned}
$$

We introduce the Fermat theorem for convex functions.

Corollary 6.8. *Suppose $f : I \to \mathbb{R}$ is convex and differentiable on I. Then for any point $t_0 \in \text{int}\,(I)$ the following statements are equivalent*

(a) $(t_0, f(t_0))$ is the global minimum of f.
(b) $(t_0, f(t_0))$ is a local minimum of f.
(c) $f'(t_0) = 0$.

Proof. (a) $\Longrightarrow$ (b) is obvious.

(b) $\Longrightarrow$ (c) follows without any convexity assumption, by Theorem 2.1.

(c) $\Longrightarrow$ (a). This implication follows from the previous corollary. $\square$

Theorem 6.2. *A differentiable function* $f : I \to \mathbb{R}$ *is convex (strictly convex) if and only if its first derivative is increasing (respectively, strictly increasing).*

Proof. The necessity part follows from Theorem 6.1.

Sufficiency. Suppose that there exists a differentiable function whose first-order derivative is increasing and it is not convex. Then there exist $a, b, c \in I$, $a < b < c$, and $s_{f,b}(a) > s_{f,b}(c)$; that is,

$$(f(a) - f(b))/(a - b) > (f(c) - f(b))/(c - b).$$

By the Lagrange mean value theorem it follows that there are $t_1 \in {]a, b[}$ and $t_2 \in {]b, c[}$ with $f'(t_1) > f'(t_2)$. But this contradicts that the first derivative is increasing. $\square$

Corollary 6.9. *A differentiable function* $f : I \to \mathbb{R}$ *is concave (strictly concave) if and only if its derivative is decreasing (strictly decreasing).*

Proof. Apply the previous theorem to $-f$. $\square$

Corollary 6.10. *(Jensen) Let* $f : I \to \mathbb{R}$ *be a function having a second-order derivative on* I. *Then* f *is convex (concave) if and only if* $f'' \geq 0$ *($f'' \leq 0$).*

Proof. The claim follows from the remark that f' is increasing (decreasing) if and only if f'' is positive (negative). $\square$

Corollary 6.11. *Let* $f : I \to \mathbb{R}$ *be a function with a second-order derivative on* I. *Then* f *is strictly convex (strictly concave) if and only if* $f'' > 0$ *($f'' < 0$).*

5.6.1 Inequalities

We start with the Young generalized inequality.

Proposition 6.7. *Take* $n \in \mathbb{N}^*$, $y_k > 0$, $p_k > 0$, *for* $k = 1, \ldots, n$, *and* $\sum_{k=1}^{n} 1/p_k = 1$. *Then*

$$\prod_{k=1}^{n} y_k \leq \sum_{k=1}^{n} \frac{1}{p_k} y_k^{p_k}. \tag{5.46}$$

For $n = 2$ the above inequality reduces to (2.10) at page 52.

Proof. Consider the function $f(x) = \exp(x)$, $x \in \mathbb{R}$. Because $f''(x) > 0$, for every $x \in \mathbb{R}$, we conclude that function f is convex. Based on the Jensen inequality, (4.31) at the page 170, taking $\alpha_k = 1/p_k$ and $x_k = \ln y_k^{p_k}$, we may write

$$\exp\left(\sum_{k=1}^{n}\frac{1}{p_k}\ln y_k^{p_k}\right) \le \sum_{k=1}^{n}\frac{1}{p_k}\exp\left(\ln y_k^{p_k}\right).$$

Furthermore

$$\exp\left(\sum_{k=1}^{n}\frac{1}{p_k}\ln y_k^{p_k}\right) = \exp\left(\sum_{k=1}^{n}\ln y_k\right) \quad\text{and}\quad \sum_{k=1}^{n}\frac{1}{p_k}\exp\left(\ln y_k^{p_k}\right) = \sum_{k=1}^{n}\frac{1}{p_k}y_k^{p_k},$$

and the generalized Young inequality follows.   □

We introduce the generalized mean inequality.

Proposition 6.8. *Take $n \in \mathbb{N}^*$, $x_i > 0$, and $\alpha_i \ge 0$, $i = 1,\dots,n$ satisfying $\alpha_1 + \cdots + \alpha_n = 1$. Then*

$$x_1^{\alpha_1}x_2^{\alpha_2}\cdots x_n^{\alpha_n} \le \alpha_1 x_1 + \cdots + \alpha_n x_n. \tag{5.47}$$

Proof. Consider the function $f(x) = \ln x$, $x > 0$. Because $f''(x) < 0$, for every $x > 0$, we infer that function f is (strictly) concave. So $\ln\left(\sum_{i=1}^{n}\alpha_i x_i\right) \ge \sum_{i=1}^{n}\alpha_i \ln x_i$, and (5.47) follows.   □

Corollary 6.12. *Taking $\alpha_1 = \cdots = \alpha_n = 1/n$ in (5.47) it follows*

$$\sqrt[n]{x_1 x_2 \cdots x_n} \le (x_1 + x_2 + \cdots + x_n)/n.$$

Corollary 6.13. *Substituting x_i by $1/x_i$ in (5.47) we get*

$$(\alpha_1/x_1 + \cdots + \alpha_n/x_n)^{-1} \le x_1^{\alpha_1}x_2^{\alpha_2}\cdots x_n^{\alpha_n}. \tag{5.48}$$

Corollary 6.14. *Taking $\alpha_1 = \cdots = \alpha_n = 1/n$ in (5.48) it follows*

$$\frac{n}{1/x_1 + \cdots + 1/x_n} \le \sqrt[n]{x_1 \cdots x_n}.$$

5.7 Differentiability of sequences and series of functions

In this section we are concerned with the following question. Does the limit (if any) of the derivatives of the terms to a convergent sequence of functions coincide with the derivative of the limit of the sequence? The answer is no. But under some extra assumptions the answer is positive.

We need the following general result.

Theorem 7.1. *Let $f_n \to f$ uniformly on a set E in a metric space. Let x be a limit point of E and suppose that*

$$\lim_{t \to x} f_n(t) = a_n, \quad n \in \mathbb{N}.$$

Then (a_n) converges and

$$\lim_{t \to x} f(t) = \lim_{n \to \infty} a_n;$$

That is,

$$\lim_{t \to x} \lim_{n \to \infty} f_n(t) = \lim_{n \to \infty} \lim_{t \to x} f_n(t).$$

Theorem 7.2. *Suppose there is given (f_n) a sequence of functions differentiable on $[a, b]$ such that $(f_n(t))$ converges for some t in $[a, b]$. If, moreover, (f_n') converges uniformly on $[a, b]$, then (f_n) converges uniformly on $[a, b]$ to a function f, and*

$$f'(x) = \lim_{n \to \infty} f_n'(x), \quad \forall\, x \in [a, b]. \tag{5.49}$$

Proof. Choose $\varepsilon > 0$. There exists $n_\varepsilon \in \mathbb{N}$ so that for all naturals $n, m \geq n_\varepsilon$, one has

$$|f_n(t) - f_m(t)| < \varepsilon/2, \tag{5.50}$$

$$|f_n'(z) - f_m'(z)| < \varepsilon/(2(b-a)), \quad \forall\, z \in [a, b]. \tag{5.51}$$

Now, by the mean value theorem of Lagrange, theorem 2.3, applied to $f_n - f_m$ and by (5.51), we have

$$|f_n(x) - f_m(x) - f_n(u) + f_m(u)| \leq |x - u|\varepsilon/(2(b-a)) \leq \varepsilon/2 \tag{5.52}$$

for all $n, m \geq n_\varepsilon$ and $x, u \in [a, b]$. Obviously

$$|f_n(x) - f_m(x)| \leq |f_n(x) - f_m(x) - f_n(t) + f_m(t)| + |f_n(t) - f_m(t)|,$$

which, by (5.50) and (5.52), implies that for all $n, m \geq n_\varepsilon$ and $x, t \in [a, b]$,

$$|f_n(x) - f_m(x)| < \varepsilon.$$

The last inequality means that (f_n) converges uniformly on $[a, b]$. Then we denote $f : [a, b] \to \mathbb{R}$ its limit.

We have to prove relation (5.49). First we show that f is differentiable. Fix an $x \in [a, b]$. Consider

$$\phi_n(u) = \frac{f_n(u) - f_n(x)}{u - x}, \quad \phi(u) = \frac{f(u) - f(x)}{u - x}, \quad u \in [a, b], \quad u \neq x. \tag{5.53}$$

Then for all natural n

$$\lim_{u \to x} \phi_n(u) = f_n'(x). \tag{5.54}$$

From (5.52) it follows that (ϕ_n) converges uniformly on $[a, b] \setminus \{x\}$. But (f_n) converges uniformly to f, so by (5.53) we get

$$\lim_{n \to \infty} \phi_n(u) = \phi(u) \tag{5.55}$$

uniformly on $[a, b] \setminus \{x\}$. We apply Theorem 7.1 to (ϕ_n), (5.54), and (5.55) to get $\lim_{u \to x} \phi(u) = \lim_{n \to \infty} f_n'(x)$, which is (5.49). $\square$

Corollary 7.15. *Given is* (f_n) *a sequence of functions continuously differentiable on* $[a, b]$, *such that* $(f_n(t))$ *converges for some* t *in* $[a, b]$. *If, moreover,* (f'_n) *converges uniformly on* $[a, b]$, *then* (f_n) *converges uniformly on* $[a, b]$ *to a function* f, *and*

$$f'(x) = \lim_{n \to \infty} f'_n(x) \quad \forall\, x \in [a, b].$$

Regarding the differentiability of a series of functions, the following theorem is useful.

Theorem 7.3. *Consider* $f_n : [a, b] \to \mathbb{R}$, f_n *differentiable and* f'_n *continuous on* $[a, b]$ *for all* $n \in \mathbb{N}$, $\sum f_n(t)$ *converges for some* $t \in [a, b]$, *and* $\sum f'_n$ *converges uniformly to* g. *Then* $\sum f_n$ *converges uniformly to* f, f *is differentiable, and*

$$f' = g.$$

Proof. It follows from Corollary 7.2. □

Corollary 7.16. *There hold*

$$x + 2x^2 + 3x^3 + \cdots + nx^n + \cdots = \frac{x}{(1-x)^2}, \quad |x| < 1, \tag{5.56}$$

$$x + 2^2 x^2 + 3^2 x^3 + \cdots + n^2 x^n + \cdots = \frac{x + x^2}{(1-x)^3}, \quad |x| < 1. \tag{5.57}$$

5.8 Power series and Taylor series

We have introduced the concept of power series at page 137. We discussed Taylor polynomials on page 210. The highest possible degree of a Taylor polynomial is given by the highest possible order of derivatives of function associated with it. No doubt, each polynomial is a truncated power series; its coefficients vanish starting with a given index. A polynomial is defined on the real line, whereas a power series, generally, is defined as we already saw on a nonempty interval of the real line. If a function has derivatives of any order, what kind of Taylor "polynomial" can we assign to it? Our intuition suggests that it could be a "polynomial" of unbounded high degree, that is, a power series. In this section we carry forth some results related to this question.

A similar result to Theorem 4.4 at page 137 holds. Consider (a_n), $n \in \mathbb{N}$, a sequence, $f_n(x) = a_n(x - x_0)^n$, $f_0(x) = a_0$, a sequence of polynomials, where x_0 is a fixed real number. We are interested in finding properties of the series of functions

$$\sum_{n=0}^{\infty} f_n(x) = a_0 + a_1(x - x_0) + a_2(x - x_0)^2 + \dots, \tag{5.1}$$

usually called a *power series*. We notice series (5.1) converges at least on one point, namely, x_0. The *domain of convergence* is the set of points on which it converges. How large this set is follows from the next theorem.

Theorem 8.1. *Given series (5.1), set*

$$\alpha = \limsup_{n\to\infty} \sqrt[n]{|a_n|}, \quad R = \frac{1}{\alpha}.$$

(If $\alpha = 0$, $R = +\infty$; if $\alpha = \infty$, $R = 0$.)
    Then (5.1) converges absolutely for every x with $|x - x_0| < R$ and diverges for every x with $|x - x_0| > R$. For every x in $[x_0 - r, x_0 + r]$ with $0 < r < R$, series (5.1) converges uniformly.

Remark. R is said to be the *radius of convergence* of (5.1). $\triangle$

Proof of Theorem 8.1. It is analogous to the one supplied by Theorem 4.4 at page 137. $\square$
    The previous theorem guarantees that if the radius of a power series is positive, it is uniformly convergent on any interval $[x_0 - r, x_0 + r]$, provided $0 < r < R$.

Theorem 8.2. *Suppose the series $\sum a_n R^n$ converges and its sum is s. Then the power series $\sum a_n(x - x_0)^n$ converges uniformly for $x \in [x_0, x_0 + R]$ and if f is its sum,*

$$\lim_{x\to x_0+R} f(x) = s.$$

Proof. It follows at once by Theorem 8.1. $\square$

Theorem 8.3. *Consider $\sum a_n(x - x_0)^n$ a power series and $R > 0$ its radius of convergence. Then the radius of convergence to the series*

$$a_1 + 2a_2(x - x_0) + \cdots + na_n(x - x_0)^{n-1} + \ldots \tag{5.2}$$

is R.

Proof. Because $\lim \sqrt[n]{n} = 1$, we have

$$\limsup_{n\to\infty} \sqrt[n]{n|a_n|} = \limsup_{n\to\infty} \sqrt[n]{|a_n|},$$

hence the series (5.1) and (5.2) have the same radii of convergence. $\square$

Corollary 8.1. *Consider $k \in \mathbb{N}^*$, $\sum a_n(x - x_0)^n$ a power series and $R > 0$ its radius of convergence. Then the series*

$$k!a_k + (k+1)!a_{k+1}(x - x_0) + \cdots + n(n-1)\ldots(n-k+1)a_n(x - x_0)^{n-k} + \ldots$$

has the same radius of convergence as (5.1), $k \in \mathbb{N}^*$.

    Consider series (5.1) and functions s_n defined by

$$s_n(x) = a_0 + a_1(x - x_0) + \cdots + a_n(x - x_0)^n, \quad n \in \mathbb{N}.$$

s_n is said to be the *partial sum* of series (5.1).

Theorem 8.4. *Given is a series* (5.1) *with* $R > 0$ *its radius of convergence. Let* $f :]x_0 - R, x_0 + R[\to \mathbb{R}$ *be the sum of series* (5.1). *Then* f *has derivatives of all orders in* $]x_0 - R, x_0 + R[$. *Its derivatives are given by*

$$f^{(k)}(x) = \sum_{n=k}^{\infty} n(n-1)\dots(n-k+1)a_n(x-x_0)^{n-k}.$$

In particular,

$$f^{(k)}(x_0) = k!a_k. \tag{5.3}$$

Proof. It follows from Theorem 9.2 at page 184 and Theorems 7.3 and 8.3. □

Suppose we are given a function f having derivatives of all orders in $]x_0 - R,\ x_0 + R[$, $R > 0$. The power series (5.1) associated with it, where the coefficients a_n are determined by (5.3), is said to be the *Taylor series* of the function f about x_0.

Remark. A function necessarily has to have derivatives of all orders about a point in order to admit a Taylor series. Is this Taylor series equal to the function? The answer is no; that is, there exists a function whose Taylor series does not coincide with it.

Consider $f : \mathbb{R} \to \mathbb{R}$ a function defined as

$$f(x) = \begin{cases} e^{-1/x^2}, & x \neq 0 \\ 0, & x = 0. \end{cases} \tag{5.4}$$

Set $x_0 = 0$. It can easily be proved that f has derivatives of all orders and $f^{(n)}(0) = 0$, $n \geq 0$. So the coefficients to the Taylor series associated with (5.4) are all zero. Therefore the radius of convergence is $R = \infty$. The Taylor series is equal to zero on the whole real line. But (5.4) is zero only at origin. Hence the functions coincide on a single point. △

Consider a real function f on an interval I. Suppose

(i) f has derivatives of all orders.
(ii) The Taylor series associated with f converges about a point $x_0 \in I$ and its radius of convergence R is positive. Admit $J =]x_0 - R, x_0 + R[\subset I$.
(iii) The sum of the Taylor series defined on J coincides with f on J.

Then f is said to be an *analytic function* at x_0. If f is analytic at each $x_0 \in I$, f is said to be *analytic on* I.

We need a criterion to establish the analyticity of a function.

Theorem 8.5. *Let* $f :]a, b[\to \mathbb{R}$ *be a function of class* C^∞, *two constants* $R > 0$ *and* $M > 0$ *such that for every* $x \in]x_0 - R, x_0 + R[\subset]a, b[$ *and* $n \in \mathbb{N}$, $|f^{(n)}(x)| \leq M^n$. *Then* f *is an analytic function at* x_0.

Proof. The general term of the Taylor series associated with f about x_0 can be estimated as

$$\left| \frac{f^{(n)}(x_0)}{n!} (x - x_0)^n \right| \leq \frac{M^n R^n}{n!}, \quad |x - x_0| < R.$$

The numerical series $\sum M^n R^n / n!$ converges. Thus the Taylor series associated with f about x_0 has a positive radius of convergence. The Taylor polynomial is precisely the partial sum of the Taylor series. Thus, using the Taylor polynomial of degree n at x_0 of f with the remainder under the Lagrange form, namely (5.43), we have

$$|f(x) - s_n(x)| = |f(x) - T_n(x)| = |R_n(x)| \leq M^n R^n / n!.$$

Therefore

$$\lim_{n \to \infty} [f(x) - s_n(x)] = 0;$$

that is, to say f coincides with the Taylor series. Hence f is analytic. $\square$

Theorem 8.6. *Let I be an open interval and $x_0 \in I$. If $f(x) = \sum a_n(x - x_0)^n$ is an analytic function about x_0, then $f(x) = \sum a_n(x - x_0)^n$ for all $x \in I \cap J$, where J is the domain of convergence of the Taylor series.*

Corollary 8.2. *Suppose f is analytic on an open interval I, $x_0 \in I$, and $f^{(n)}(x_0) = 0$, for all $n \in \mathbb{N}$. Then $f(x) = 0$ on I.*

Corollary 8.3. *Suppose f and g are analytic on an open interval I, $x_0 \in I$, and $f^{(n)}(x_0) = g^{(n)}(x_0)$, for all $n \in \mathbb{N}$. Then $f(x) = g(x)$ on I.*

5.8.1 Operations with power series

Consider two power series

$$\sum a_n(x - x_0)^n \text{ and } \sum b_n(x - x_0)^n \tag{5.5}$$

about the same point x_0. Suppose $\min\{R_1, R_2\} > 0$, where R_1, R_2 are the radii of convergence of the two power series.

Let α, β be two real numbers. By Theorem 3.3 at page 111 we have that

$$\sum (\alpha a_n + \beta b_n)(x - x_0)^n,$$

converges for $|x - x_0| < R$, where R is at most $\min\{R_1, R_2\}$.

Theorem 8.7. *Consider series (5.5) about x_0 having R_1, respectively, R_2, radii of convergence. If $c_n = \sum_{k=0}^{n} a_k b_{n-k}$, the product power series about x_0, $\sum c_n(x - x_0)^n$, converges for $|x - x_0| < R$, for some R, $R \leq \min\{R_1, R_2\}$.*

Proof. By Theorem 8.1 series (5.5) converges absolutely on the interior of the domains of convergence. Furthermore, by Theorem 3.3 at page 111, $\sum c_n(x - x_0)^n$, converges absolutely on $|x - x_0| < R$, for some $R \leq \min\{R_1, R_2\}$. $\square$

Power series $\sum c_n(x - x_0)^n$ is said to be the *product power series* of $\sum a_n(x - x_0)^n$ and $\sum b_n(x - x_0)^n$.

Theorem 8.8. *Consider the power series $\sum a_n(x-x_0)^n$ about x_0. Let $R_1 > 0$ be its radius of convergence and f its sum on $|x - x_0| < R_1$. Suppose $a_0 \neq 0$. Then there exists a power series $\sum b_n(x - x_0)^n$ about x_0 with $R_2 > 0$ radius of convergence and such that if g is its sum, $f(x)g(x) = 1$, for all $|x - x_0| < \min\{R_1, R_2\}$.*

Proof. Suppose for a while that $R_2 > 0$. By Theorem 8.7 we have that the product power series of $\sum a_n(x - x_0)^n$ and $\sum b_n(x - x_0)^n$ converges on $|x - x_0| < \min\{R_1, R_2\}$. On the interior of the domains of convergence the power series turn into analytic functions. Then, by Corollary 8.3, we get a system of linear equations with unknowns b_n,

$$a_0 b_0 = 1, \quad a_0 b_n + a_1 b_{n-1} + \cdots + a_{n-1} b_1 + a_n b_0 = 0, \quad n \geq 1.$$

Successively, we have

$$b_0 = 1/a_0, \quad b_1 = -a_1 b_0/a_0 = -a_1/a_0^2, \tag{5.6}$$
$$b_n = -(a_1 b_{n-1} + \cdots + a_{n-1} b_1 + a_n b_0)/a_0, \quad n \geq 2. \tag{5.7}$$

Now, we show that the radius of convergence of $\sum b_n$ is positive. We note that in multiplying each term of a power series with a nonzero constant its radius of convergence remains unchanged. So, we multiply the series $\sum a_n(x - x_0)^n$ by $1/a_0$ and we get a power series with the constant term equal to 1. We keep the notation; that is, we denote this new series also by $\sum a_n(x - x_0)^n$, with $a_0 = 1$. By (5.6) and (5.7) we have that

$$b_0 = 1, \quad b_1 = -a_1, \quad b_n = -(a_1 b_{n-1} + \cdots + a_{n-1} b_1 + a_n), \quad n \geq 2. \tag{5.8}$$

Let $0 < r < R_1$. Then

$$\limsup_{n \to \infty} \sqrt[n]{|a_n|} < 1/r$$

and so there is a natural ν such that for all $n \geq \nu$,

$$\sqrt[n]{|a_n|} < 1/r \implies |a_n| r^n < 1.$$

Consider $M > \max\{1, |a_1|r, \ldots, |a_\nu|r^\nu\}$. Thus

$$M > \begin{cases} |a_n|r^n, & n \leq \nu; \\ |a_n|r^n (< 1), & n > \nu \end{cases}$$

imply that $|a_n| r^n < M$, for all $n \in \mathbb{N}$.

Successively, by induction and using (5.8) and the earlier estimate, we find

$$|b_0| = 1, \quad |b_1| = |a_1| < M/r, \quad |b_2| \leq |a_1||b_1| + |a_2| < M(M+1)/r^2,$$
$$|b_3| \leq |a_1||b_2| + |a_2||b_1| + |a_3| < M(M+1)^2/r^3, \ldots$$
$$|b_n| \leq |a_1||b_{n-1}| + |a_2||b_{n-2}| + \cdots + |a_n|$$
$$< M(M+1)^{n-1}/r^n = (M/(r(M+1)))\,((M+1)/r)^n, \ldots.$$

Therefore

$$R_2 = \frac{1}{\limsup_{n\to\infty} \sqrt[n]{|b_n|}} \geq \frac{1}{\limsup_{n\to\infty} \sqrt[n]{\dfrac{M}{r(M+1)}} \left(\dfrac{M+1}{r}\right)} = \frac{r}{M+1}.$$

Thus the radius of convergence to $\sum b_n$ is positive. $\square$

Theorem 8.9. *Consider $f(x) = \sum a_n(x - x_0)^n$ an analytic function given as a power series about x_0 with $R > 0$ its radius of convergence. Suppose we are given a function $\varphi : [\alpha, \beta] \to \mathbb{R}$ such that for every $t \in [\alpha, \beta]$, $|\varphi(t) - x_0| < R$. Then the series*

$$\sum a_n(\varphi(t) - x_0)^n$$

converges for every $t \in [\alpha, \beta]$ and its sum is equal to $f(\varphi(t))$.

Example. We show by (5.13) the Taylor expansion of e^{-t^2}, $t \in \mathbb{R}$. Substituting x by $-t^2$ into the Taylor expansion of e^x we have

$$e^{-t^2} = \sum_{n=0}^{\infty} (-1)^n \frac{t^{2n}}{n!}. \quad \triangle$$

Recall the nth Bell number B_n is defined as the number of partitions to a set of n elements, Exercise 1.16 at page 34.

Proposition 8.1. (a) *The exponential generating function of the sequence (B_n)*

$$f(x) = \sum_{n=0}^{\infty} \frac{B_n}{n!} x^n \tag{5.9}$$

is

$$f(x) = e^{e^x - 1}. \tag{5.10}$$

(b) *There holds the equality*

$$B_n = \frac{1}{e} \sum_{n=1}^{\infty} \frac{k^n}{n!}. \tag{5.11}$$

(c) *The sequence (B_n) is log-convex.*

Proof. (a) Because the number of partitions of a set of n elements is less than the number of subsets of a set of n elements, $B_n < 2^n$. Then (5.9) converges for every real x. Moreover,

$$f'(x) = e^x f(x), \quad f(0) = 1. \tag{5.12}$$

Thus (5.12) is a differential equation with an initial condition. By [64], (5.12) has a unique solution, namely (5.10).

(b) (5.10) can be written as

$$f(x) = (1/e)(1 + e^x/1! + e^{2x}/2! + e^{3x}/3! + \dots).$$

Therefore

$$f^{(n)}(x) = \frac{1}{e}\left(\sum_{k=1}^{\infty} \frac{k^n}{k!} e^{kx}\right), \quad n \geq 1.$$

Because $B_n = f^{(n)}(0)$, (5.11) follows.   □

5.8.2 The Taylor expansion of some elementary functions

Consider $f(x) = e^x$. We know that the exponential function is defined and continuous on the real line and $f'(x) = e^x$, $x \in \mathbb{R}$. Then $f^{(n)}(x) = e^x$, for all $x \in \mathbb{R}$ and $n \in \mathbb{N}$. If $|x| < R$, $|f^{(n)}(x)| < e^R$, and by Theorem 8.5, the exponential function is analytic on the real line. Because $f^{(n)}(0) = 1$, we can write

$$e^x = \exp(x) = 1 + \frac{x}{1!} + \frac{x^2}{2!} + \dots + \frac{x^n}{n!} + \dots = \sum_{n=0}^{\infty} \frac{x^n}{n!}, \quad x \in \mathbb{R}. \quad (5.13)$$

Analogously, if $f(x) = \sin x$, we see that this function is defined and continuous on the real line, and together with its derivatives it is bounded in absolute value by 1. Then by Theorem 8.5, the sinus function is analytic on the real line. It is easy to see that $f^{(n)}(x) = \sin(x + n\pi/2)$, and consequently, $f^{(n)}(0) = \sin(n\pi/2)$, $n \in \mathbb{N}$. Hence for each $x \in \mathbb{R}$,

$$\sin x = \frac{x}{1!} - \frac{x^3}{3!} + \frac{x^5}{5!} - \dots + (-1)^n \frac{x^{2n+1}}{(2n+1)!} + \dots = \sum_{n=0}^{\infty} \frac{(-1)^n x^{2n+1}}{(2n+1)!}. \quad (5.14)$$

Reasoning as before we find

$$\cos x = 1 - \frac{x^2}{2!} + \frac{x^4}{4!} - \dots + (-1)^n \frac{x^{2n}}{(2n)!} + \dots = \sum_{n=0}^{\infty} \frac{(-1)^n x^{2n}}{(2n)!}, \quad x \in \mathbb{R}. \quad (5.15)$$

Consider $f(x) = \ln(1 + x)$, $x > -1$. We find that $f'(x) = 1/(1 + x)$, $x > -1$. We can use Theorem 3.8 at page 115 or directly find the Taylor series to the function $f'(x) = 1/(1 + x)$. In any case for $|x| < 1$, we get

$$\ln(1+x) = x - \frac{x^2}{2} + \frac{x^3}{3} - \frac{x^4}{4} + \dots + \frac{(-1)^n x^{n+1}}{n+1} + \dots = \sum_{n=0}^{\infty} \frac{(-1)^n x^{n+1}}{n+1}. \quad (5.16)$$

We notice that for $x = 1$ the middle term of (5.16) converges to $\ln 2$, by Corollary 1.13 page 105.

Consider $f(x) = (1+x)^\alpha = \exp(\alpha \ln(1+x))$, $|x| < 1$. Our desire is to find its Taylor series. Because the exponential function and the logarithmic function have derivatives of all orders on their domains of definition and because by composition the differentiability is preserved, f has derivatives of all orders on the open interval $]-1,1[$. Successively, we have

$$f'(x) = e^{\alpha \ln(1+x)} \frac{\alpha}{1+x} = \alpha(1+x)^{\alpha-1}, \quad f''(x) = \alpha(\alpha-1)(1+x)^{\alpha-2}, \ldots,$$
$$f^{(n)}(x) = \alpha(\alpha-1)\ldots(\alpha-n+1)(1+x)^{\alpha-n}, \quad n \in \mathbb{N}^*.$$

The Taylor series about zero is

$$1 + \frac{\alpha}{1!}x + \frac{\alpha(\alpha-1)}{2!}x^2 + \cdots + \frac{\alpha(\alpha-1)\ldots(\alpha-n+1)}{n!}x^n + \ldots. \quad (5.17)$$

Because

$$\lim_{n\to\infty}\left|\frac{\alpha(\alpha-1)\ldots(\alpha-n)}{(n+1)!}\frac{n!}{\alpha(\alpha-1)\ldots(\alpha-n+1)}\right| = 1,$$

the radius of convergence to (5.17) is equal to 1. Hence $f(x) = (1+x)^\alpha$ is analytic for $|x| < 1$ and on this set

$$(1+x)^\alpha = 1 + \sum_{n=1}^{\infty} \frac{\alpha(\alpha-1)\ldots(\alpha-n+1)}{n!}x^n. \quad (5.18)$$

Series (5.18) for $|x| < 1$ is said to be the *binomial series*. For natural α one gets the Newton binomial formula.

Consider $f(x) = \arcsin x$, for $|x| \leq 1$. We know that this function has derivatives of all orders provided $|x| < 1$ and $f'(x) = (1-x^2)^{-1/2}$. From now on we can use the binomial series and finally we get

$$\arcsin x = x + \frac{1}{2}\frac{x^3}{3} + \frac{1\cdot 3}{2\cdot 4}\frac{x^5}{5} + \cdots + \frac{(2n-1)!!}{(2n)!!}\frac{x^{2n+1}}{2n+1} + \ldots, \quad |x| < 1.$$

Consider $f(x) = \arctan x$, for $x \in \mathbb{R}$. This function has derivatives of all orders. The power series associated with it converges for $|x| < 1$. Hence f is analytic for $|x| < 1$ and

$$\arctan x = x - \frac{x^3}{3} + \frac{x^5}{5} - \cdots + (-1)^n\frac{x^{2n+1}}{2n+1} + \ldots, \quad |x| < 1. \quad (5.19)$$

Relation (5.19) can be used to compute some special numbers; that is, [8]. According to theorem 3.22 at page 126, the series

$$1 - \frac{1}{3} + \frac{1}{5} - \frac{1}{7} + \cdots = \sum_{k=0}^{\infty} \frac{(-1)^k}{2k+1}$$

converges. Then by Theorem 8.2, series (5.19) converges at $x = 1$. Thus we can write

$$\frac{\pi}{4} = 1 - \frac{1}{3} + \frac{1}{5} - \frac{1}{7} + \dots .$$

The above identity is said to be the *Leibniz–Gregory*[9] *formula.*

If we substitute in (5.19) $x = 1/\sqrt{3}$, we get

$$\frac{\pi}{6} = \frac{1}{\sqrt{3}} \left(1 - \frac{1}{3 \cdot 3} + \frac{1}{5 \cdot 9} - \dots + (-1)^n \frac{1}{(2n+1)3^n} + \dots \right).$$

Another way to compute π even faster is the next one. One has

$$\tan \left(2 \arctan \frac{1}{5} \right) = \frac{5}{12}, \quad \tan \left(4 \arctan \frac{1}{5} \right) = \frac{120}{119},$$

$$\tan \left(4 \arctan \frac{1}{5} - \arctan \frac{1}{239} \right) = 1.$$

So

$$\frac{\pi}{4} = 4 \arctan \frac{1}{5} - \arctan \frac{1}{239}, \tag{5.20}$$

known as the Machbin[10] formula. Expanding, we can write

$$\frac{\pi}{4} = 4 \left(\frac{1}{5} - \frac{1}{3 \cdot 5^3} + \frac{1}{5 \cdot 5^5} - \dots \right) - \left(\frac{1}{239} - \frac{1}{3 \cdot 239^3} + \dots \right).$$

Remark. Nowadays (5.20) can easily be deduced from elementary considerations on complex arithmetic. Consider

$$\frac{(5+i)^4}{239+i} = 2(1+i)$$

and take the arguments in both sides. △

Remark. We just mention that (5.13) is still valid for complex argument, that is if $x \in \mathbb{C}$. Using it, one can define *hyperbolic functions.* We introduce the next four hyperbolic functions

$$\sinh x = \frac{e^x - e^{-x}}{2}, \quad \cosh x = \frac{e^x + e^{-x}}{2}, \quad \tanh x = \frac{\sinh x}{\cosh x}, \quad \forall x \in \mathbb{R}$$

and

$$\coth x = \frac{\cosh x}{\sinh x}, \quad \forall x \in \mathbb{R} \setminus \{0\}.$$

All these functions can also be defined on $\mathbb{C}$. In this case for tanh and coth we have to avoid certain points in $\mathbb{C}$ where they are not defined [126]. There hold the following relations

---

[9] James Gregory, 1638–1675.
[10] John Machbin, 1680–1751.

$$\sin x = -i \sinh ix, \quad \cos x = \cosh ix,$$
$$\tan x = -i \tanh ix, \quad \cot x = i \coth ix. \quad \triangle$$

Warning. Most of the previous results of this section hold in a more general setting, namely on the field of complex numbers $\mathbb{C}$. We state a result on $\mathbb{C}$ that is very useful in getting Ramanujan[11] formulas by the $W-Z$ method.

A function defined on $\mathbb{C}$ is said to be *entire* if is differentiable $\mathbb{C}$.
A function is of *exponential type* if

$$f(z) = O(1)e^{\tau |z|}$$

for some $\tau < \infty$.

Theorem 8.10. (Carlson[12]) *If $f(z)$ is an entire function of exponential type and for some $c < \pi$,*

$$f(iy) = O(1)e^{c|y|},$$

and f vanishes identically on the nonnegative integers, then f is identically zero.

5.8.3 Bernoulli numbers and polynomials

The *Bernoulli*[13] *numbers* $\mathfrak{b}_n$ are defined by the implicit recurrence

$$\sum_{k=0}^{n} \binom{n+1}{k} \mathfrak{b}_k = \begin{cases} 1, & n = 0, \\ 0, & n \in \mathbb{N}^*. \end{cases}$$

or

$$\mathfrak{b}_0 = 1 \quad \text{and} \quad \sum_{k=0}^{n} \binom{n}{k} \mathfrak{b}_k = \mathfrak{b}_n + \begin{cases} 1, & n = 1, \\ 0, & n \in \mathbb{N}^* \setminus \{1\}. \end{cases}$$

The first Bernoulli numbers are supplied by Table 5.1.

Table 5.1. Some Bernoulli numbers

n	0	1	2	3	4	5	6	7	8
$\mathfrak{b}_n$	1	$-1/2$	$1/6$	0	$-1/30$	0	$1/42$	0	$-1/30$

The Bernoulli numbers have appeared as trying to give closed form to the sum

$$s_m(n) = 0^m + 1^m + 2^m + \cdots + n^m,$$

---

[11] Srinivasa Aiyangar Ramanujan, 1887–1920.
[12] Fritz David Carlson, 1888–1952.
[13] Jakob Bernoulli, 1654–1705.

where $m, n \in \mathbb{N}$ and 0^0 is considered as 1. Obviously

$$s_0(n) = n, \quad s_1(n) = \frac{1}{2}n^2 - \frac{1}{2}n, \quad s_2(n) = \frac{1}{3}n^3 - \frac{1}{2}n^2 + \frac{1}{6}n, \ldots .$$

By induction on m, it can be shown [60, §6.5] that

$$s_m(n) = \frac{1}{m+1}\left(\mathfrak{b}_0 n^{m+1} + \binom{m+1}{1}\mathfrak{b}_1 n^m + \cdots + \binom{m+1}{m}\mathfrak{b}_m n\right)$$

$$= \frac{1}{m+1}\sum_{k=0}^{m}\binom{m+1}{k}\mathfrak{b}_k n^{m+1-k}.$$

Consider the exponential generating function of the Bernoulli numbers

$$f(x) = \sum_{n=0}^{\infty}\mathfrak{b}_n \frac{x^n}{n!}.$$

Then

$$(e^x - 1)f(x) = \left(\sum_{k=1}^{\infty}\frac{x^k}{k!}\right)\left(\sum_{n=0}^{\infty}\mathfrak{b}_n\frac{x^n}{n!}\right) = \sum_{n=0}^{\infty}\left(\sum_{k=0}^{n}\binom{n+1}{k}\mathfrak{b}_k\right)\frac{x^n}{n!} = x.$$

Thus the generating function of the Bernoulli numbers is

$$\sum_{n=0}^{\infty}\mathfrak{b}_n\frac{x^n}{n!} = \frac{x}{e^x - 1}.$$

We have

$$1 + \sum_{n\geq 2}\mathfrak{b}_n\frac{x^n}{n!} = \frac{x}{e^x - 1} + \frac{x}{2} = \frac{x}{2}\coth\frac{x}{2}.$$

Because $x \to x\coth x$ is an even function, it follows

$$\mathfrak{b}_{2n+1} = 0, \ n \in \mathbb{N}^* \ \text{ and } \ x\coth x = \sum_{n\geq 0}4^n\mathfrak{b}_{2n}\frac{x^{2n}}{(2n)!}.$$

Recall $\cot x = i\coth ix$. Therefore

$$x\cot x = ix\coth ix = \sum_{n\geq 0}(-4)^n\mathfrak{b}_{2n}\frac{x^{2n}}{(2n)!}. \tag{5.21}$$

Because

$$x\cot x - 1 - 2\sum_{k\geq 1}\frac{x^2}{k^2\pi^2 - x^2} - 1 - 2\sum_{n\geq 1}\frac{x^{2n}}{\pi^{2n}}\zeta(2n). \tag{5.22}$$

From (5.21) and (5.22), we get

$$\zeta(2n) = \frac{(-1)^{n-1}2^{2n-1}\pi^{2n}\mathfrak{b}_{2n}}{(2n)!}, \quad n \in \mathbb{N}^*. \tag{5.23}$$

Therefore

$$\frac{|\mathfrak{b}_{2n}|}{(2n)!} = \frac{2}{(2\pi)^{2n}}\,\zeta(2n) < \frac{4}{(2\pi)^{2n}} = O((2\pi)^{-2n}), \quad n \in \mathbb{N}^*. \tag{5.24}$$

From (5.23) there follow

$$\zeta(2) = 1 + \frac{1}{2^2} + \frac{1}{3^2} + \frac{1}{4^2} + \cdots = \frac{\pi^2}{6}, \tag{5.25}$$

$$\zeta(4) = 1 + \frac{1}{2^4} + \frac{1}{3^4} + \frac{1}{4^4} + \cdots = \frac{\pi^4}{90}. \tag{5.26}$$

Because $\tan x = \cot x - 2\cot 2x$, we have

$$x\tan x = \sum_{n\geq 0}(-4)^n\mathfrak{b}_{2n}\frac{x^{2n}}{(2n)!} - \sum_{n\geq 0}(-4)^n\mathfrak{b}_{2n}\frac{(2x)^{2n}}{(2n)!}$$

$$= \sum_{n\geq 0}(-1)^{n+1}4^n(4^n-1)\mathfrak{b}_{2n}\frac{x^{2n}}{(2n)!}.$$

The *Bernoulli polynomials* B_n are defined by

$$B_n(x) = \sum_{k=0}^{n}\binom{n}{k}\mathfrak{b}_k x^{n-k}.$$

Proposition 8.2. *We have*

$$B'_m(x) = mB'_{m-1}(x), \quad m \in \mathbb{N}^*,$$
$$B_m(1) = B_m(0) = (-1)^m\mathfrak{b}_m, \quad m > 1. \tag{5.27}$$

Proposition 8.3. *The function* $B_{2m} : [0,1] \to \mathbb{R}$ *attains its extreme values at* $x = 0$ *and* $x = 1/2$.

Proposition 8.4. *One has*

$$B_{2m}\left(\frac{1}{2}\right) = \left(2^{1-2m} - 1\right)\mathfrak{b}_{2m}.$$

Corollary 8.4. *One has*

$$|B_{2m}(\{x\})| \leq |\mathfrak{b}_{2m}|.$$

5.9 Some elementary functions introduced by recurrences

There are many ways to introduce elementary functions. Here we present one way based on recurrences.

5.9.1 The square root function

Theorem 9.1. *Consider $x \geq 0$ and let $(x_n)_{n\geq 0}$ be the sequence defined by*

$$x_0 = 0, \quad x_{n+1} = \frac{1}{2}\left(x_n + \frac{x}{x_n}\right), \quad n \geq 0. \tag{5.28}$$

Then $x_n > 0$ for all $n \geq 0$, the sequence (x_n) converges to a number $p \geq 0$, satisfying $p^2 = x$, and p is the unique nonnegative number fulfilling $p^2 = x$.

Proof. By induction it is clear that $x_n > 0$ for all $n \geq 0$. We already saw by Exercise 3.4 in Section 3.6 that the sequence (x_n) converges to a number $p \geq 0$, satisfying $p^2 = x$. Now, suppose that p_1 is a nonnegative number fulfilling $p_1^2 = x$. Then $0 = p^2 - p_1^2 = (p - p_1)(p + p_1)$ and it follows that $p_1 = p$. $\square$

The *square root* function $\sqrt{} : [0, \infty[\to [0, \infty[$ assigns to each nonnegative number x the number p in Theorem 9.1, namely

$$\sqrt{x} = p = \lim_{n\to\infty} x_n.$$

Corollary 9.5. *One has*

$$\sqrt{0} = 0, \quad \sqrt{1} = 1, \quad \text{and} \quad \lim_{x\to\infty} \sqrt{x} = \infty.$$

Proof. By theorem 9.1 we have that there exists the number $p = \sqrt{0}$ and $p^2 = 0$. Because $0^2 = 0$, by the same theorem, we have $p = 0$. The case $\sqrt{1} = 1$ runs similarly.

Consider an arbitrary $\varepsilon > 0$. There exists $x_0 > \varepsilon^2$. Let $x \geq x_0$. Then $\sqrt{x} > c$, because otherwise

$$\sqrt{x} \leq \varepsilon \implies x \leq \varepsilon\sqrt{x} \implies x \leq \varepsilon\sqrt{x} \leq \varepsilon^2 < x_0 \implies x_0 \leq x < x_0,$$

a contradiction. $\square$

By induction on the nonnegative integers n we define the 2^nth root function $\sqrt[2^n]{} : [0, \infty[\to [0, \infty[$ as

$$\sqrt[2^0]{x} = x \quad \text{and} \quad \sqrt[2^{n+1}]{x} = \sqrt{\sqrt[2^n]{x}}. \tag{5.29}$$

By induction on n one can prove that

$$\left(\sqrt[2^n]{x}\right)^{2^n} = x \quad \text{and} \quad \sqrt[2^n]{xy} = \sqrt[2^n]{x} \cdot \sqrt[2^n]{y}, \quad x, y \geq 0.$$

5.9.2 The logarithm function

The model to define the logarithm function is the limit

$$\lim_{h\to 0} \frac{x^h - 1}{h} = \ln x, \quad x > 0.$$

Lemma 5.11. *Consider $x > 0$ and let $(z_n)_{n\geq 0}$ be the sequence defined by*

$$z_0 = x \quad \text{and} \quad z_{n+1} = \sqrt{z_n}, \quad n \geq 0. \tag{5.30}$$

Then $z_n = \sqrt[2^n]{x}$ and the sequence (z_n) satisfies

$$0 < z_n < z_{n+1} < 1 \quad \text{whenever} \quad x < 1,$$
$$1 < z_{n+1} < z_n \quad \text{whenever} \quad x > 1,$$
$$z_n = 1 \quad \text{whenever} \quad x = 1,$$

and $\lim z_n = 1$.

Proof. We prove by induction on n. By (5.30) and (5.29) we have that $z_n = \sqrt[2^n]{x}$.

Suppose $x < 1$. By induction we have that $0 < z_n < 1$, and thus $z_n < \sqrt{z_n}$. Thus $z_n < z_{n+1}$.

Suppose $x > 1$. By induction we have that $z_n > 1$, and thus $z_n > \sqrt{z_n}$. Thus $z_n > z_{n+1}$.

Suppose $x = 1$. By induction we get that $z_n = 1$, for all $n \geq 0$.

In any case the sequence (z_n) is monotonic and bounded, thus convergent. Denote $p = \lim z_n$. Because $z_{n+1}^2 = z_n$, we get that $p^2 = 1$. There exists only one positive solution of it, namely $p = 1$. Hence $\lim z_n = 1$. $\square$

Theorem 9.2. *Consider $x > 0$ and let $(x_n)_{n\geq 0}$ and $(y_n)_{n\geq 0}$ be the sequences defined by*

$$x_n = 2^n (z_n - 1) \quad \text{and} \quad y_n = 2^n (1 - 1/z_n), \quad n \geq 0,$$

where (z_n) is the sequence defined by (5.30). Then

$$y_{n-1} < y_n < x < x_n < x_{n-1} \quad \text{whenever } x \neq 1, \quad n \geq 1.$$

Sequences (x_n) and (y_n) are convergent to the same limit p and

$$y_n < p < x_n \quad \text{whenever } x \neq 1, \ n \geq 0. \tag{5.31}$$

If $x = 1$, then $p = 0$.

Proof. If $x = 1$, $z_n = 1$ and thus $x_n = y_n = 0$. Therefore $p = 0$.

Suppose $x \neq 1$ and $n \geq 1$. Consider $x < 1$. Then

$$y_{n-1} < y_n \iff 1 - 1/z_n^2 < 2(1 - 1/z_n) \iff (z_n - 1)^2 > 0.$$

Because $0 < z_n < 1$, $y_n < z_n$. Also

$$x_n > x_{n-1} \iff 2(z_n - 1) < z_n^2 - 1 \iff (z_n - 1)^2 > 0.$$

The case $x > 1$ is discussed analogously. Thus the sequences (x_n) and (y_n) are convergent. Because $x_n = y_n \cdot z_n$, by Lemma (5.11) we obtain that $\lim x_n = \lim y_n$. $\square$

The *logarithm* function $\ln :]0, \infty[\to \mathbb{R}$ assigns to each positive number x the number p in Theorem 9.2, namely

$$\ln x = p = \lim_{n \to \infty} 2^n (z_n - 1).$$

Corollary 9.6. *One has* $\ln 1 = 0$. *If* $x > 0$ *and* $x \neq 1$, *then*

$$1 - 1/x < \ln x < x - 1 \quad \text{and} \quad 2 \left(1 - 1/\sqrt{x} \right) < \ln x < 2(\sqrt{x} - 1).$$

Proof. By (5.31) we have

$$2^n \left(1 - 1/z_n \right) < \ln x < 2^n (z_n - 1). \quad \square$$

Theorem 9.3. *The logarithmic function satisfies*

$$\ln(xy) = \ln x + \ln y, \quad x, y > 0, \tag{5.32}$$
$$\ln(1/x) = -\ln x, \quad x > 0, \tag{5.33}$$
$$\ln x < \ln y, \quad 0 < x < y. \tag{5.34}$$

Proof. Consider $x, y > 0$. Denote

$$q = xy, \quad z_n = \sqrt[2n]{x}, \quad z_n' = \sqrt[2n]{y}, \quad z_n'' = \sqrt[2n]{q}.$$

Then $z_n'' = z_n z_n'$. By Lemma 5.11 we have

$$\ln q = \lim 2^n (z_n'' - 1) = \lim \left(2^n (z_n - 1) z_n' + 2^n (z_n' - 1) \right)$$
$$= \lim 2^n (z_n - 1) \cdot \lim z_n' + \lim 2^n (z_n' - 1) = \ln x + \ln y.$$

Consider $x > 0$. By Corollary 9.6 and by (5.32) we have

$$0 = \ln 1 = \ln \left(x \cdot \frac{1}{x} \right) = \ln x + \ln \frac{1}{x}.$$

Thus (5.33) follows.

Consider $0 < x < y$. Then by Corollary 9.6, (5.32), and (5.33) we have

$$0 < 1 - \frac{x}{y} < \ln \frac{y}{x} = \ln y + \ln \frac{1}{x} = \ln y - \ln x.$$

Thus (5.34) follows. $\quad \square$

Corollary 9.7. *There hold*

$$\lim_{x \to \infty} \ln x = \infty, \quad \lim_{x \to \infty} \frac{\ln x}{x} = 0, \tag{5.35}$$

$$\lim_{x \downarrow 0} \ln x = -\infty, \quad \lim_{x \to 1} \ln x = 0, \quad \text{and} \quad \lim_{x \to 1} \frac{\ln x}{x - 1} = 1. \tag{5.36}$$

Proof. From (5.32) and Corollary 9.6 we have $0 = \ln 1 < 2$. Consider an arbitrary $\varepsilon > 0$. Then there exists a positive integer n so that $n \geq \varepsilon/\ln 2$. Consider $x > 2^n$. So,

$$\ln x > \ln 2^n = n \ln 2 \geq \varepsilon$$

and the first equality in (5.35) follows. From Corollary 9.6 we have that for $x > 1$

$$2\left(1/x - 1/(x\sqrt{x})\right) < \ln x/x < 2\left(1/\sqrt{x} - 1/x\right).$$

From here the second equality in (5.35) is clear.

Consider an arbitrary $\varepsilon \in \mathbb{R}$. Because $\lim_{y \to \infty} \ln y = \infty$, there exists a positive y_0 such that for each $y > y_0$ one has $\ln y > -\varepsilon$. Let $\delta = 1/y_0$ and $x > 0$, with $x < \delta$. Then

$$1/x > 1/\delta = y_0 \implies -\ln x = \ln(1/x) > -\varepsilon \iff \ln x < \varepsilon,$$

and the first equality in (5.36) follows. By Corollary 9.6, the second equality in (5.36) follows. Also by Corollary 9.6, we have

$$\left.\begin{array}{l} 1/x < \ln x/(x-1) < 1, \quad \text{for } x > 1 \\ 1/x > \ln x/(x-1) > 1, \quad \text{for } x < 1 \end{array}\right\} \implies \lim_{x \to 1} \frac{\ln x}{x-1} = 1. \quad \square$$

Theorem 9.4. *The logarithmic function is continuous, bijective, and differentiable with* $(\ln x)' = 1/x$, $x > 0$.

Proof. Consider $x, h > 0$ and $t = h/x$. By Corollary 9.7 we have

$$\lim_{h \to x} \ln h = \lim_{h \to x} \ln\left(x\frac{h}{x}\right) = \ln x + \lim_{h \to x} \ln \frac{h}{x} = \ln x + \lim_{t \to 1} \ln t = \ln x.$$

Because the logarithmic function is strictly increasing, it is one-to-one. Consider an arbitrary $y \in \mathbb{R}$. By Corollary 9.7 and continuity it follows that there exist $0 < u < v$ so that $\ln u < y < \ln v$. The logarithmic function being continuous, it is a Darboux function. Thus there exists an $x \in [u, v]$ so that $\ln x = y$. Thus the logarithmic function is onto.

Consider $x > 0$ and $u = 1 + h/x > 0$. By Theorem 9.3 and Corollary 9.7, we have

$$\lim_{h \to 0} \frac{\ln(x+h) - \ln x}{h} = \frac{1}{x} \lim_{h \to 0} \frac{\ln(1+h/x)}{h/x} = \frac{1}{x} \lim_{u \to 0} \frac{\ln u}{u-1} = \frac{1}{x}. \quad \square$$

Consider $a > 0$ and $a \neq 1$. The *logarithm in base a* function $\log_a :\ ]0, \infty[\to \mathbb{R}$ is defined by

$$\log_a x = \frac{1}{\ln a} \ln x.$$

We immediately note that the function $\log_a(\cdot)$ is continuous, bijective, and differentiable. Regarding its monotonicity we also note that it is strictly increasing whenever $a > 1$ and strictly decreasing whenever $0 < a < 1$.

5.9.3 The exponential function

By Theorem 9.4 we saw that the logarithmic function is continuous, bijective, and differentiable on $]0, \infty[$. Then it has an inverse called the *exponential* function and denoted $\exp : \mathbb{R} \to]0, \infty[$. It satisfies $\ln(\exp x) = x$, for all $x \in \mathbb{R}$ and $\exp(\ln t) = t$, for all $t > 0$. A common notation of exponential function is $e^x = \exp(x)$, for all $x \in \mathbb{R}$.

Theorem 9.5. *The exponential function is strictly increasing, differentiable with $\exp'(x) = \exp x$, $x \in \mathbb{R}$, and satisfies the relation*

$$\exp(x + y) = \exp x \cdot \exp y, \quad \forall x, y \in \mathbb{R}. \tag{5.37}$$

Moreover, $\exp 0 = 1$, $\lim_{x \to \infty} \exp x = \infty$, and $\lim_{x \to -\infty} \exp x = 0$.

Proof. The logarithmic function is strictly increasing and continuous, thus by Theorem 2.20 at page 163, we have the exponential function is strictly increasing and continuous. By Theorem 1.5 we have that if $x \in \mathbb{R}$ and $\ln t = x$, then

$$(\exp x)' = 1/(\ln t)' = t = \exp x.$$

Consider $x, y \in \mathbb{R}$. Denote $u = \exp x$ and $v = \exp y$. Then

$$\exp(x + y) = \exp(\ln u + \ln v) = \exp(\ln(uv)) = uv = \exp x \cdot \exp y,$$

and thus (5.37) is proved.

One has $\exp 0 = 1$ because $\ln 1 = 0$.

Consider $\varepsilon > 0$. Denote $x_0 = \ln \varepsilon$. Choose $x < x_0$. Then $\exp x < \exp x_0 = \exp(\ln \varepsilon) = \varepsilon$. Thus for every positive ε one can find a real x so that $0 < \exp x < \varepsilon$. Hence $\lim_{x \to -\infty} \exp x = 0$. Analogously it can be proved that $\lim_{x \to \infty} \exp x = \infty$. $\square$

Consider $\alpha \in \mathbb{R}$. The function $f_\alpha :]0, \infty[\to]0, \infty[$ defined by

$$f_\alpha(x) = x^\alpha, \quad \text{where } x^\alpha = \exp(\alpha \ln x)$$

is said to be the *power* function (of exponent α).

Consider $a > 0$. The function $g_a : \mathbb{R} \to]0, \infty[$ defined by

$$g_a(x) = a^x, \quad \text{where } a^x = \exp(x \ln a)$$

is said to be the *exponential* function (of base a).

The logarithmic function and the exponential function of base e are strictly increasing and continuous, therefore it follows that the functions $]0, \infty[\ni x \mapsto x^\alpha$ and $\mathbb{R} \ni x \mapsto a^x$ are continuous. Moreover, the functions are strictly increasing for $\alpha > 0$ and $a > 1$ and strictly decreasing for $\alpha < 0$ and $0 < a < 1$.

By (5.32) and (5.37) one has

$$(xy)^\alpha = x^\alpha \cdot y^\alpha, \quad \forall x, y > 0 \text{ and } a^{x+y} = a^x \cdot a^y, \quad \forall x, y \in \mathbb{R}.$$

For $\alpha > 0$, by Corollary 9.7, the definition of the exponential function, and Theorem 9.5 we have that

$$\lim_{x \to 0} f_\alpha(x) = \lim_{y \to -\infty} \exp(y) = 0.$$

Therefore we can extend function f by continuity to $\tilde{f}_\alpha : [0, \infty[\to [0, \infty[$ given as

$$\tilde{f}_\alpha(x) = \begin{cases} 0, & x = 0 \\ f_\alpha(x), & x > 0. \end{cases}$$

Remarks. (i) For $\alpha = n \in \{0, 1, 2, \dots\}$ from the definition of $\tilde{f}_\alpha$ we recover the definition of the power function because $\exp(n \ln x) = \exp(\ln x^n) = x^n$, $x > 0$.
(ii) For $\alpha \in \mathbb{R}$ we have $\ln x^\alpha = \ln(\exp(\alpha \ln x)) = \alpha \ln x$, with $x > 0$. For $\alpha = 1/2$, we have $\ln x^{1/2} = (1/2) \ln x$. Because $\sqrt{x} \cdot \sqrt{x} = x$, it follows that $2 \ln \sqrt{x} = \ln x$. From here $\ln \sqrt{x} = (1/2) \ln x$ implies $\ln \sqrt{x} = \ln x^{1/2}$. Thus

$$\sqrt{x} = x^{1/2}, \quad x \geq 0. \tag{5.38}$$

From here it follows that the square root function is continuous on $[0, \infty[$.
(iii) The relation (5.38) suggests the following definition

$$\sqrt[n]{x} = x^{1/n} = \exp\left(\frac{1}{n} \ln x\right), \quad x > 0, \ n = 1, 2, \dots \quad \triangle$$

Theorem 9.6. *One has*

$$e^x = \lim_{h \to 0} (1 + hx)^{1/h}, \quad x \in \mathbb{R}, \text{ and } e = \lim_{n \to \infty} \left(1 + \frac{1}{n}\right)^n, \quad n = 1, 2, \dots.$$

Proof. Suppose $x \in \mathbb{R} \setminus \{0\}$. Choose $h \neq 0$ so that $1 + hx > 0$. Then

$$\lim_{h \to 0} (1 + hx)^{1/h} = \lim_{h \to 0} \exp\left(\frac{1}{h} \ln(1 + hx)\right) = \lim_{h \to 0} \exp\left(x \frac{\ln(1 + hx)}{hx}\right)$$

$$= \exp\left(x \lim_{y \to 1} \frac{\ln y)}{y - 1}\right) = \exp(x \cdot 1) = e^x.$$

Suppose $x = 0$. Then

$$e^0 = \exp(0) = 1 = \lim_{h \to 0} (1 + h \cdot 0)^{1/h}.$$

Suppose $x = 1$ and $h = 1/n$, $n = 1, 2, \dots$. Then we get the second equality. $\square$

Theorem 9.7. *The exponential function* $\mathbb{R} \ni x \mapsto e^x$ *is differentiable and* $(e^x)' = e^x$, *for all real* x.

Proof. We have already introduced a proof for this statement by Theorem 9.5. Here we present a different proof. Note that

$$\lim_{h \to 0} \frac{e^h - 1}{h} = \lim_{t \to 1} \frac{t - 1}{\ln t} = 1$$

by the substitution $h = \ln t$, $t \geq 0$. By (5.37) we have

$$\lim_{h \to 0} \frac{e^{x+h} - e^x}{h} = \lim_{h \to 0} e^x \cdot \frac{e^h - 1}{h} = e^x. \quad \square$$

Corollary 9.8. (i) *Suppose* $a > 0$ *and* $a \neq 1$. *Then the logarithmic function* $]0, \infty[\ni x \mapsto \log_a x$ *is differentiable and*

$$(\log_a x)' = \frac{1}{x \ln a}, \quad x > 0.$$

(ii) *Suppose* $\alpha \in \mathbb{R}$. *Then the power function* $]0, \infty[\ni x \mapsto x^\alpha$ *is differentiable and*

$$(x^\alpha)' = \alpha x^{\alpha - 1}, \quad x > 0.$$

(iii) *Suppose* $a > 0$. *Then the exponential function* $\mathbb{R} \ni x \mapsto a^x$ *is differentiable and*

$$(a^x)' = a^x \ln a, \quad x \in \mathbb{R}.$$

Remarks. (i) By Theorem 4.1 of L'Hospital and Theorem 9.5 for $x > 0$ we have

$$\lim_{h \to 0} (x^h - 1)/h = \lim_{h \to 0} (x^h \ln x)/1 = \ln x \lim_{h \to 0} \exp(h \ln x) = \ln x.$$

(ii) By (i) we have

$$\lim_{n \to \infty} n(\sqrt[n]{x} - 1) = \ln x, \quad x > 0. \quad \triangle$$

5.9.4 The arctangent function

Choose a real x. The restriction of the tangent function at the $]-\pi/2, \pi/2[$ is one-to-one and onto. Therefore there exists a unique $y \in]-\pi/2, \pi/2[$ so that $\tan y = x$. By

$$\tan y = \frac{2\tan(y/2)}{1 + \tan^2(y/2)}$$

it immediately follows that

$$\tan(y/2) = \frac{\tan y}{1 + \sqrt{1 + \tan^2 y}}$$

and
$$\tan(y/2^{n+1}) = \frac{\tan(y/2^n)}{1 + \sqrt{1 + \tan^2(y/2^n)}}, \qquad n = 0, 1, 2, \dots.$$

Denote $z_n = \tan y/2^n$. We have $z_0 = x$, and $z_{n+1} = z_n/(1 + \sqrt{1 + z_n^2})$. By $x_n = 2^n \cdot z_n$ and whenever $x \neq 0$, that is, $y \neq 0$, we have

$$x_n = y \cdot \frac{\tan y/2^n}{y/2^n} \xrightarrow{n \to \infty} y = \arctan x.$$

Lemma 5.12. *Consider a real x and $(z_n)_{n \geq 0}$ a sequence defined by*

$$z_0 = x \text{ and } z_{n+1} = z_n/1 + \sqrt{1 + z_n^2}, \quad n \geq 0. \qquad (5.39)$$

Then all the terms of the sequence $(z_n)_{n \geq 0}$ are positive for $x > 0$, negative for $x < 0$, respectively, zero for $x = 0$. Moreover, $\lim z_n = 0$.

Proof. The first claim follows by induction. By recurrence it follows that (z_n) is monotonic. More precisely, it is decreasing for $x > 0$ and increasing for $x < 0$. Thus (z_n) is convergent. Denote $z = \lim z_n$. Passing to the limit in recurrence we get $z = z/(1 + \sqrt{1 + z^2})$. This equality is satisfied by zero only. $\square$

Theorem 9.8. *Consider a real x and two sequences $(x_n)_{n \geq 0}$ and $(y_n)_{n \geq 0}$ defined by*
$$x_n = 2^n \cdot z_n \text{ and } y_n = 2^n \cdot z_n/\sqrt{1 + z_n^2}, \qquad (5.40)$$
where (z_n) is defined by (5.39). Then

(a) *If $x > 0$, we have*

$$0 < y_{n-1} < y_n < x_n < x_{n-1}, \quad n \geq 1,$$

the sequences (x_n) and (y_n) are convergent to the same limit, let it be l, and
$$y_n < l < x_n, \quad n \geq 0. \qquad (5.41)$$

(b) *If $x < 0$, we have*

$$0 < x_{n-1} < x_n < y_n < y_{n-1}, \quad n \geq 1,$$

the sequences (x_n) and (y_n) are convergent to the same limit, let it be l, and
$$x_n < l < y_n, \quad n \geq 0. \qquad (5.42)$$

(c) *If $x = 0$, then $x_n = y_n = 0$ for $n \geq 0$ and thus $\lim x_n = \lim y_n = 0$.*

Proof. Suppose $x > 0$ and n is a positive and integer number. By Lemma 5.12, z_{n-1} and thus y_{n-1}. The inequality $y_{n-1} < y_n$ is equivalent to

$$\sqrt{1 + z_{n-1}^2} < 1 + z_{n-1}^2.$$

The inequality $y_n < x_n$ follows from

$$y_n = x_n / \sqrt{1 + z_n^2}. \tag{5.43}$$

Then we have

$$x_n = 2^n \cdot z_n = \frac{2^n z_{n-1}}{1 + \sqrt{1 + z_{n-1}^2}} < 2^{n-1} \cdot z_{n-1} = x_{n-1}.$$

From Lemma 5.12 and (5.43) it follows that there exist $\lim x_n$ and $\lim y_n$ and they coincide. Let l be the common limit. Thus (5.41) is obvious.

The claims (b) and (c) are proved similarly. $\square$

The function $\arctan : \mathbb{R} \to \mathbb{R}$ assigning to each real number x the limit l in Theorem 9.8 is said to be the *arctangent* function.

Corollary 9.9. *One has* $\arctan 0 = 0$. *Suppose* $x \neq 0$. *Then*

$$|x| / \sqrt{1 + x^2} < |\arctan x| < |x|,$$

$$\frac{2|x|}{\sqrt{2(1 + x^2 + \sqrt{1 + x^2})}} < |\arctan x| < \frac{2|x|}{1 + \sqrt{1 + x^2}}.$$

Theorem 9.9. *The arctangent function satisfies the relation*

$$\arctan u + \arctan v = \arctan \frac{u + v}{1 - uv}, \quad u, v \in \mathbb{R}, \ uv \neq 1.$$

Theorem 9.10. *The arctangent function is odd, increasing, and bounded on* $\mathbb{R}$.

5.10 Functions with primitives

5.10.1 The concept of a primitive function

Let $I \subset \mathbb{R}$ be an interval and $f : I \to \mathbb{R}$ be given. A function $F : I \to \mathbb{R}$ is said to be a *primitive* (an *antiderivative*) of f on I if F is differentiable on I and

$$F'(x) = f(x), \quad \text{for every } x \subset I.$$

In this case we say that function f *has a primitive* (function). We denote by $\mathcal{P}_I$ the class of real functions defined on I with a primitive.

Remarks 10.1. (i). If $f \in \mathcal{P}_I$, then function f has an infinity of primitives. That is, if F is a primitive of f, then $F + c$ is also a primitive of f, for any real constant c.

An even stronger statement holds: any two primitives of f on I differ by a constant. Indeed, for any two primitives F_1, F_2 of f,

$$F_1'(x) = F_2'(x) = f(x), \quad \forall x \in I.$$

By the Lagrange mean value theorem there is a real constant c such that

$$F_1(x) = F_2(x) + c, \quad \forall x \in I.$$

Suppose $f : I \to \mathbb{R}$ has a primitive and F is a primitive of it. Then we write

$$\int f(x)dx = \{F + c \mid c \in \mathbb{R}\} = F + \mathbb{R}$$

for the set of primitives to f. This set of primitives of f is said to be the *indefinite integral* of f.

(ii). Suppose $f : I \to \mathbb{R}$ has a primitive, $u \in I$, and c is a real number. Then there exists a unique primitive F_c of f on I satisfying $F_c(u) = c$. Indeed, let F be an arbitrary primitive of f. Then $F_c : I \to \mathbb{R}$ defined by $F_c(x) = F(x) - (F(u) - c)$ satisfies all the requirements. $\triangle$

Theorem 10.1. *A function f has a primitive on an interval I if and only if f has a primitive on every compact interval $[a, b] \subset I$.*

Proof. It is clear that if f has a primitive on an interval I, it has a primitive on every compact interval $[a, b] \subset I$.

Suppose f has a primitive on every compact interval $[a, b] \subset I$. Consider a sequence $I_n = [a_n, b_n]$, $n \geq 0$, of compact intervals fulfilling

$$I_n \subset I_{n+1}, \quad \forall n \geq 0, \text{ and } \cup_{n \geq 0} I_n = I.$$

Let $F_0 : [a_0, b_0] \to \mathbb{R}$ be a primitive of f on $[a_0, b_0]$, $u \in [a_0, b_0]$ a fixed point, and $c_0 = F_0(u)$. By (ii) of Remarks 10.1 for every natural $n \geq 1$ there exists a unique primitive F_n on I_n of f so that $F_n(u) = c_0$. Furthermore, for $m > n \geq 1$ let F_m and F_n be the primitives of f on I_m, respectively, I_n. Then on I_n the two primitives F_m and F_n coincide. Thus we can define

$$F : I \to \mathbb{R}, \quad F(x) = F_n(x), \quad x \in I_n.$$

Then F is a primitive of f on I. $\square$

Theorem 10.2. *Suppose function $f : I \to \mathbb{R}$ has a primitive on the interval I. Then f satisfies the Darboux property; that is, $\mathcal{P}_I \subset \mathcal{D}_I$.*

Proof. Let F be a primitive of f, $a, b \in I$, $a < b$, and λ be a point between $f(a)$ and $f(b)$. Suppose $f(a) < \lambda < f(b)$. Then the mapping $G : [a, b] \to \mathbb{R}$, $G(x) = F(x) - \lambda x$ is differentiable and

$$G'(a) = f(a) - \lambda < 0 < f(b) - \lambda = G'(b).$$

G is continuous, thus there exists a point of minimum of G on $[a, b]$; denote it by $c \in [a, b]$.

We show that $c \notin \{a, b\}$. Indeed, from

$$G'(a) = \lim_{\substack{x \to a \\ x > a}} \frac{G(x) - G(a)}{t - a} < 0$$

it follows that there is $r > 0$ such that $G(t) < G(a)$, for every $t \in {]a, a + r[}$. Thus $c \neq a$. Similarly, from $G'(b) > 0$ we infer that $c \neq b$. Thus c is a point of minimum of G on $]a, b[$. Based on the Fermat theorem we may write $G'(c) = 0$; that is, $f(c) = \lambda$. Hence f is a Darboux mapping.   □

Corollary 10.10. *A primitive function of $f \in \mathcal{P}_I$ with $f : I \to \mathbb{R} \setminus \{0\}$ is strictly monotone.*

Proof. Let F be a primitive of f. Then $F'(t) = f(t) \neq 0$, for every $t \in I$. Because f is a Darboux mapping, either $f > 0$ or $f < 0$ on I. Hence F is strictly monotonic on I.   □

Corollary 10.11. *A function $f \in \mathcal{P}_I$ has no discontinuity point of the first kind.*

Proof. The claim follows from Theorem 10.2 and from Corollary 4.11 at page 166.   □

Corollary 10.12. *Let $f : I \to \mathbb{R}$ be discontinuous and monotonic. Then $f \notin \mathcal{P}_I$.*

Proof. Let t_0 be a discontinuity point of f. Because f is monotonic, t_0 is a discontinuity point of the first kind. Then by Corollary 10.11, $f \notin \mathcal{P}_I$.   □

Example. We show that there is a Darboux function with no primitives.
    Consider the following mapping,

$$f_m : \mathbb{R} \to \mathbb{R}, \quad f_m(t) = \begin{cases} \sin(1/t), & t \neq 0, \\ m, & t = 0. \end{cases}$$

It is a Darboux function if and only if $m \in [-1, 1]$ (because for any interval $I \ni 0$, $|I| > \aleph_0$, $f_m(I) = [-1, 1] \cup \{m\}$).
    On the other hand

$$f_m \in \mathcal{P}_I \iff m = 0.$$

Hence for any $m \in [-1, 0[\cup]0, 1]$ it follows that $f_m \in \mathcal{D}_I \setminus \mathcal{P}_I$.   △

5.10.2 The existence of primitives for continuous functions

By Theorem 10.1 we saw that a mapping $f : I \to \mathbb{R}$ has a primitive if and only if it has a primitive on any compact $[a,b] \subset I$. Hence it is important to know the criteria of having primitives on compact intervals.

By a *partition* P of $[a,b]$ we mean a finite set of points $x_0, x_1, \ldots, x_n$, satisfying

$$a = x_0 \leq x_1 \leq x_2 \leq \cdots \leq x_{n-1} \leq x_n = b.$$

Denote $I_k = [x_{k-1}, x_k]$, $\Delta x_k = x_k - x_{k-1}$, $k = 1, 2, \ldots, n$. Consider a bounded function $f : [a,b] \to \mathbb{R}$ and denote

$$m = \inf_{[a,b]} f, \quad M = \sup_{[a,b]} f, \quad m_k = \inf_{I_k} f, \quad M_k = \sup_{I_k} f, \quad k = 1, 2, \ldots, n.$$

The real number

$$L(f, P) = \sum_{k=1}^{n} m_k \Delta x_k$$

is said to be the *lower Darboux sum* of the bounded mapping $f : [a,b] \to \mathbb{R}$ with respect to the partition P.

Remark. Because $m \leq m_k \leq M_k \leq M$, it follows that

$$m(b-a) \leq L(f, P) \leq M(b-a),$$

for any partition P of the interval $[a,b]$. Thus the set of lower Darboux sums is bounded. $\triangle$

Then it makes sense to define

$$\underline{I} = \sup_P L(f, P)$$

as the *lower Darboux integral* of the bounded function $f : [a,b] \to \mathbb{R}$. We denoted it by

$$\underline{\int_a^b} f \quad \text{or} \quad \underline{\int_a^b} f(x)dx.$$

Proposition 10.5. *Suppose* $f : [a,b] \to \mathbb{R}$ *is a bounded function. Define*

$$F : I \to \mathbb{R}, \quad F(x) = \underline{\int_a^x} f.$$

Then

(a) $F(b) = F(c) + \underline{\int_c^b} f$, *for every* $c \in [a,b]$.

(b) *There is a positive* L *so that* $|F(x') - F(x'')| \leq L|x' - x''|$, *for all* $x', x'' \in I$.

Proof. We remark that $\int_c^c f = 0$.

(a) Choose $c \in]a, b[$, and let P_1 and P_2 be two arbitrary partitions of $I_1 = [a, c]$, respectively, $I_2 = [c, b]$. Set $P = P_1 \cup P_2$. Then

$$L(f, P_1) + L(f, P_2) = L(f, P) \leq F(b),$$

$$\sup_{P_1} L(f, P_1) + L(f, P_2) \leq F(b), \quad F(c) + \sup_{P_2} L(f, P_2) \leq F(b)$$

$$\implies F(c) + \underline{\int_c^b} f \leq F(b).$$

In order to prove the reverse inequality let P be an arbitrary partition of $[a, b]$.

- If $c \in P$, then taking $P_1 = P \cap [a, c]$ and $P_2 = P \cap [c, b]$, it results

$$L(f, P) = L(f, P_1) + L(f, P_2) \leq F(c) + \underline{\int_c^b} f. \tag{5.44}$$

- If $c \notin P$, then taking $P^* = P \cup \{c\}$, it follows

$$L(f, P) \leq L(f, P^*) \leq F(c) + \underline{\int_c^b} f. \tag{5.45}$$

Passing to the supremum in (5.44) and (5.45), the conclusion follows.
(b) Consider $x', x'' \in I$, $x' < x''$, and $L = \sup_{x \in I} |f(x)|$. Then (a) implies

$$|F(x') - F(x'')| = |\int_{x'}^{x''} f| \leq L(x'' - x') = L|x' - x''|. \quad \square$$

Theorem 10.3. *Every continuous function $f : I \to \mathbb{R}$ has a primitive; that is,*

$$C(I) \subset \mathcal{P}_I.$$

Proof. Taking into account claim (ii) of Remarks 10.1 saying that $f \in \mathcal{P}_I \iff f \in \mathcal{P}_{[a,b]}$, $\forall [a, b] \subset I$, it is enough to prove that every continuous function $f : [a, b] \to \mathbb{R}$ has a primitive.

Therefore consider a continuous function $f : [a, b] \to \mathbb{R}$ and introduce

$$F : [a, b] \to \mathbb{R}, \quad F(x) = \int_a^x f.$$

This definition makes sense because f is bounded on $[a, b]$, by Theorem 2.9 at page 156.

We show that F is a primitive of f. Take an arbitrary but fixed $x_0 \in [a, b]$. We show that F has a right-hand derivative on x_0 which is equal to $f(x_0)$;

similarly, we show that F has a left-hand side derivative on t_0 which is equal to $f(x_0)$.

So, choose $x_0 \in [a, b[$. From the right-hand side continuity at x_0 of f it follows that for every $\varepsilon > 0$ there exists $\delta > 0$ such that for any $x \in]x_0, x_0 + \delta[\subset]x_0, b[$ we have

$$f(x_0) - \varepsilon < f(x) < f(x_0) + \varepsilon,$$

$$(f(x_0) - \varepsilon)(x - x_0) < \int_{\underline{x_0}}^{x} f < (f(x_0) + \varepsilon)(x - x_0).$$

Then it follows

$$\left| \frac{F(x) - F(x_0)}{x - x_0} - f(x_0) \right| < \varepsilon, \quad \text{for every } x \in]x_0, x_0 + \delta[;$$

that is, F is right-hand side differentiable on x_0 and the right-hand side derivative of F on x_0 is equal to $f(x_0)$.

The left-hand side case runs similarly. $\square$

We can state the *fundamental formula of calculus*.

Corollary 10.13. *Suppose $f : [a, b] \to \mathbb{R}$ is continuous. Then f has a primitive $F : [a, b] \to \mathbb{R}$ and*

$$F'(x) = f(x), \quad \forall x \in [a, b].$$

A question arises: does the inverse of a function having a primitive have a primitive? The answer is supplied by the following.

Corollary 10.14. *Let I be a compact interval and suppose that the function $f : I \to \mathbb{R}$ is injective and it has a primitive. Then $f^{-1} : f(I) \to \mathbb{R}$ is continuous, hence it has a primitive.*

Proof. Function f has a primitive, thus it has the Darboux property, and thus it has no discontinuity points of the first kind.

Function f is injective and has the Darboux property, thus it is strictly monotonic; thus it has no discontinuity points of the second kind. Therefore f is continuous and strictly monotonic on the compact interval I. Then based on Theorem 2.11 page 156 we conclude that f^{-1} is continuous on $f(I)$. Hence it has a primitive. $\square$

5.10.3 Operations with functions with primitives

We show that $\mathcal{P}_I$ is a real linear space.

Proposition 10.6. *Suppose $f, g \in \mathcal{P}_I$ and $\alpha, \beta \in \mathbb{R}$. Then $\alpha f + \beta g \in \mathcal{P}_I$. Moreover, if $\alpha^2 + \beta^2 > 0$, then*

$$\int (\alpha f + \beta g) = \alpha \cdot \int f + \beta \cdot \int g.$$

Proof. Choose $F \in \int f$ and $G \in \int g$. Then $H = \alpha F + \beta G$ is differentiable and $H' = \alpha F' + \beta G' = \alpha f + \beta g = h$. It follows that

$$\alpha \cdot \int f + \beta \cdot \int g \subset \int (\alpha f + \beta g), \quad \text{for every } \alpha, \beta \in \mathbb{R}.$$

Suppose that $\alpha^2 + \beta^2 > 0$. Then we may consider $\beta \neq 0$. Consider $H \in \int h$, $F \in \int f$. Then the function

$$G = \frac{H - \alpha F}{\beta} \quad \text{is differentiable and} \quad G' = \frac{h - \alpha f}{\beta} = g.$$

Thus $G \in \int g$ and $H = \alpha F + \beta G \in \alpha \cdot \int f + \beta \cdot \int g$. Hence

$$\int (\alpha f + \beta g) \subset \alpha \cdot \int f + \beta \cdot \int g, \quad \text{if } \alpha^2 + \beta^2 > 0. \quad \square$$

Remark. The above proposition fails if $\alpha = \beta = 0$. Indeed, if $\alpha = \beta = 0$, then for any $f, g \in \mathcal{P}_I$,

$$\int (\alpha f + \beta g) = \int 0 = \mathbb{R} \neq \{0\} = \alpha \cdot \int f + \beta \cdot \int g. \quad \triangle$$

Remark. The product of two functions in $\mathcal{P}_I$ does not necessarily belong to $\mathcal{P}_I$. Indeed, consider

$$f(x) = \begin{cases} \sin^2 \dfrac{1}{x} - \sin \dfrac{1}{x}, & x \neq 0, \\ 1/2, & x = 0, \end{cases} \quad \text{and} \quad g(x) = \begin{cases} \sin^2 \dfrac{1}{x} + \sin \dfrac{1}{x}, & x \neq 0, \\ 1/2, & x = 0. \end{cases}$$

Then $f, g \in \mathcal{P}_I$ because $f = f_1 - f_2$ and $g = f_1 + f_2$, where

$$f_1(x) = \frac{1}{2} - \frac{1}{2} \begin{cases} \cos \dfrac{2}{x}, & x \neq 0, \\ 0, & x = 0, \end{cases} \quad \text{and} \quad f_2(x) = \begin{cases} \sin \dfrac{1}{x}, & x \neq 0, \\ 0, & x = 0. \end{cases}$$

On the other hand the mapping

$$(f \cdot g)(x) = \begin{cases} -(1/4) \sin^2(2/x), & x \neq 0, \\ 1/4, & \text{otherwise} \end{cases}$$

does not fulfill the Darboux property because

$$(f \cdot g)(\mathbb{R}) = [-1/4, 0] \cup \{1/4\}. \quad \triangle$$

A question arises: when does a product of two functions in $\mathcal{P}_I$ belong to $\mathcal{P}_I$? An answer is supplied by the next proposition.

Proposition 10.7. *Consider* $f : I \to \mathbb{R}$, $f \in \mathcal{P}_I$, *and* $g : I \to \mathbb{R}$, $g \in C^1(I)$. *Then* $f \cdot g \in \mathcal{P}_I$.

Proof. Consider $F \in \int f$ and $F_1 = F \cdot g$. Then F_1 is differentiable and $F_1' = fg + Fg'$. Because Fg' is continuous, then $Fg' \in \mathcal{P}_I$. Choose $G \in \int Fg'$. Then $H = F_1 - G$ is differentiable on I and

$$H' = F_1' - G' = fg + Fg' - Fg' = fg,$$

hence $fg \in \mathcal{P}_I$ and $H \in \int fg$.   □

Another question arises: does the conclusion of the previous proposition hold if we suppose that g is continuous on I? The answer is negative and it follows from the next example.

Example. The function

$$f : \mathbb{R} \to \mathbb{R}, \quad f(x) = \begin{cases} x \sin(1/x^2), & x \neq 0, \\ 0, & x = 0 \end{cases}$$

is continuous. The function

$$g : \mathbb{R} \to \mathbb{R}, \quad g(x) = \begin{cases} (1/x) \sin(1/x^2), & x \neq 0, \\ 0, & x = 0 \end{cases}$$

has a primitive. Indeed, we write $g = g_2 - g_1$, where $g_1, g_2 \in \mathcal{P}_I$,

$$g_1(x) = \begin{cases} x \cos(1/x^2), & x \neq 0, \\ 0, & x = 0, \end{cases}$$

$$g_2(x) = G'(x), \quad G(x) = \begin{cases} (x^2/2) \cos(1/x^2), & x \neq 0, \\ 0, & x = 0. \end{cases}$$

Suppose that $fg \in \mathcal{P}_I$. Then from

$$(fg)(x) = \begin{cases} 1/2, & x \neq 0, \\ 0, & x = 0 \end{cases} - \frac{1}{2} \begin{cases} \cos(2/x^2), & x \neq 0, \\ 0, & x = 0 \end{cases}$$

it follows that the function

$$h(x) = \begin{cases} 1/2, & x \neq 0, \\ 0, & x = 0 \end{cases}$$

has a primitive. This function is not a Darboux one, therefore it has no primitive. Hence the assumption that $fg \in \mathcal{P}_I$ does not hold.   △

Remark. Adding an extra assumption, the conclusion of the previous remark can be improved. That is, if $f : I \to \mathbb{R} \setminus \{0\}$, the product has a primitive.   △

Proposition 10.8. *Suppose* $I = [a, b]$, $f : I \to \mathbb{R} \setminus \{0\}$ *with* $f \in \mathcal{P}_I$, *and* $g \in C(I)$. *Then* $f \cdot g \in \mathcal{P}_I$.

Proof. Choose $F \in \int f$. Because $f = F'$ has a constant sign and it has the Darboux property, F is strictly monotonic. Then F^{-1} is differentiable on $F(I)$ and thus $g \circ F^{-1}$ is continuous on $F(I)$. It follows that $g \circ F^{-1}$ has a primitive on $F(I)$. Let G be a primitive of $g \circ F^{-1}$. Then

$$(G \circ F)' = (G' \circ F) \cdot F' = (g \circ F' \circ F) \cdot F' = gf'.$$

Hence the product fg has a primitive. $\square$

5.11 Exercises

5.1. Consider a function $f :]-1, 1[\to \mathbb{R}$ such that there exists $f'(0)$. Choose two sequences $(\alpha_n)_n$ and $(\beta_n)_n$ satisfying $-1 < \alpha_n < \beta_n < 1$ such that $\alpha_n \to 0$ and $\beta_n \to 0$ as $n \to \infty$. Define

$$d_n = \frac{f(\beta_n) - f(\alpha_n)}{\beta_n - \alpha_n}.$$

Show that

(a) If $\alpha_n < 0 < \beta_n$, $\lim_{n \to \infty} d_n = f'(0)$.
(b) If $0 < \alpha_n$ and the sequence $(\beta_n/(\beta_n - \alpha_n))_n$ is bounded, $\lim_{n \to \infty} d_n = f'(0)$.
(c) If f' is continuous on $]-1, 1[$, $\lim_{n \to \infty} d_n = f'(0)$.
(d) There exists a differentiable function f on $]-1, 1[$ (with discontinuous derivative) such that $\alpha_n \to 0$, $\beta_n \to 0$, and there exists $\lim_{n \to \infty} d_n$, but $\lim_{n \to \infty} d_n \neq f'(0)$.

5.2. Let f and g be nonconstant and differentiable functions defined on an interval $]a, b[$. Suppose that for every $x \in]a, b[$,

$$f(x) + g(x) \neq 0 \quad \text{and} \quad f(x)g'(x) - f'(x)g(x) = 0.$$

Show that g has no zero on $]a, b[$ and f/g is constant on $]a, b[$.

5.3. Show that there is a unique positive number a such that $a^x \geq x^a$ for all $x > 0$. Find the number a.

5.4. Let f be continuous on $[0, 1]$ and differentiable on $]0, 1[$. Suppose $f(0) - f(1) = 0$. Show that there exists $\xi \in]0, 1[$ so that $f(\xi) = f'(\xi)$.

5.5. For every real number x_1, construct the sequence $x_1, x_2, \ldots$ by setting

$$x_{n+1} = x_n (x_n + 1/n), \quad n = 1, 2, \ldots.$$

Prove that there exists exactly one value of x_1 which gives $0 < x_n < x_{n+1} < 1$, for all $n = 1, 2, \ldots$.

5.6. For all $0 < x < \pi/2$, show that $0 < \sin x < x < \tan x$.

5.7. Find the limits

(a) $\lim_{x \to \infty} x^{\ln x}/(\ln x)^x$.

(b) $\lim_{x \downarrow 0} x^x$.

(c) $\lim_{x \to \infty} \left(\sqrt[3]{x^3 + x^2 + x + 1} - \sqrt{x^2 + x + 1} \cdot (\ln(e^x + x))/x \right)$.

5.8. Consider $a \in C^1(\mathbb{R})$ with $a(x) \geq 1$ for all $x \in \mathbb{R}$, and $f \in C^2(\mathbb{R})$ satisfying

$$f(0) = f'(0) = 0 \text{ and } (a(x)f'(x))' + f(x) \geq 0, \quad \forall x \in \mathbb{R}.$$

Show that $f(3) \geq 0$.

5.9. Find all differentiable functions $f : \mathbb{R} \to \mathbb{R}$ satisfying the relation

$$f'\left(\frac{x+y}{2}\right) = \frac{f(y) - f(x)}{y - x}, \quad x, y \in \mathbb{R}, \, x \neq y.$$

5.10. Prove that any nonnull function $f : \mathbb{R} \to \mathbb{R}$ differentiable at $x = 0$ and satisfying the relation

$$f(x+y) = f(x) \cdot f(y), \quad x, y \in \mathbb{R},$$

is indefinite differentiable at each $x \in \mathbb{R}$.

5.11. Let $f : [0,1] \to \mathbb{R}$ be a differential function satisfying $f'(0) = 1$ and $f'(1) = 1$. Show that f' has a fixed point in $]0, 1[$.

5.12. Let f be a real-valued function with $n+1$ derivatives on $\mathbb{R}$. Show that for each pair of real numbers a, b, $a < b$, such that

$$\ln\left(\frac{f(b) + f'(b) + \cdots + f^{(n)}(b)}{f(a) + f'(a) + \cdots + f^{(n)}(a)}\right) = b - a$$

there is a number $c \in]a, b[$ for which $f^{(n+1)}(c) = f(c)$.

5.13. Set $m, n \in \mathbb{N}^*$. Let $a_1, \ldots, a_n$ be positive numbers. Show that

$$\frac{a_1^m + a_2^m + \cdots + a_n^m}{n} \geq \left(\frac{a_1 + a_2 + \cdots + a_n}{n}\right)^m.$$

5.14. Show that for every triangle ABC the following inequalities hold.

$$\sin A + \sin B + \sin C \leq 3\sqrt{3}/2, \qquad \sin A \sin B \sin C \leq 3\sqrt{3}/8,$$
$$\cos A + \cos B + \cos C \leq 3/2, \qquad \cos A \cos B \cos C \leq 1/8.$$

5.15. Show that the series $\sum_{n=1}^{\infty} e^{-n} \sin(nx)$ converges at every real x and its sum has a continuous derivative on $\mathbb{R}$.

5.16. Show that the function

$$f : \mathbb{R} \to \mathbb{R}, \quad f(x) = \begin{cases} \dfrac{\sin x}{x}, & x \neq 0, \\ 0 & x = 0 \end{cases}$$

has no primitive on $\mathbb{R}$.

5.12 References and comments

Many parts of the present chapter follow [115] and [87]. We indicate more precisely some different sources.

The proof of Theorem 1.2 is contained in [18].

Propositions 2.3 and 2.4, Lemma 5.1, as well as Theorems 2.6 and 2.7 appeared in [138].

Theorem 2.8 may be found in [86, vol.2, p. 77].

A new proof to Theorem 3.1 may be found in [102].

New results on the Taylor remainder can be found, for example in [56].

The mean inequality, Corollary 6.12, enjoyed a lot of attention and many proofs. One proof may be found in [106, Part 2, Chapter 2].

Theorem 7.1 appears in [115, p. 135].

In many textbooks the elementary functions (the exponential function, the sinus function, the cosines function, the logarithmic function, and the arc tangent function) are introduced as series of functions as we already did by (5.13)–(5.16), and (5.19). An interesting way to introduce the square root function, the logarithmic function, the exponential function, the arc tangent function, the sine function, and the cosines function may be found in [90]. These functions are introduced only by recurrent sequences.

In [14, §5.3] or [126, Ch. 5] are proofs of Theorem 8.10.

Exercises 5.5, 5.9, 5.10 may be found in [96].

Exercise 5.11 is from [117].

Exercise 5.12 appeared at the first IMC, 1994, second day as Problem 3.

6

Integral Calculus on $\mathbb{R}$

The aim of the present chapter is to introduce some basic results on integral calculus on the real axis.

6.1 The Darboux–Stieltjes integral

6.1.1 The Darboux integral

A *partition* P of $[a, b]$ (as we have defined it at page 177) is a set of points $x_0, x_1, \ldots, x_n$, satisfying $a = x_0 \le x_1 \le x_2 \le \cdots \le x_{n-1} \le x_n = b$. Denote

$$I_k = [x_{k-1}, x_k], \quad \Delta x_k = x_k - x_{k-1}, \quad k = 1, 2, \ldots, n.$$

For any partition P we denote $\|P\| = \max_{1 \le k \le n} \Delta x_k$ and call it the *norm* of P.

Let f be a real-valued bounded function defined on $[a, b]$. Corresponding to a given partition P of the interval $[a, b]$, we set

$$m_k = \inf_{x \in I_k} f(x), \quad M_k = \sup_{x \in I_k} f(x),$$

$$L(P, f) = \sum_{k=1}^{n} m_k \Delta x_k, \quad U(P, f) = \sum_{k=1}^{n} M_k \Delta x_k.$$

Moreover, we put

$$\underline{\int_a^b} f(x)dx = \sup_P L(P, f), \quad \overline{\int_a^b} f(x)dx = \inf_P U(P, f), \tag{6.1}$$

where, as indicated, the sup and the inf are considered over all partitions of the interval $[a, b]$. The $\underline{\int_a^b} f(x)dx$ is said to be the *lower Darboux integral* of the function f on the interval $[a, b]$, and $\int_a^{\overline{b}} f(x)dx$ is said to be the *upper Darboux integral* of the function f on the interval $[a, b]$. We remark for a bounded function f ($m \leq f(x) \leq M$, for all $x \in [a, b]$) the lower and the upper integral exist because

$$m \leq f(x) \leq M \implies m(b - a) \leq L(P, f) \leq U(P, f) \leq M(b - a).$$

If the lower and the upper Darboux integral of f are equal, we say that f is *Darboux integrable* on $[a, b]$, we write $f \in \mathcal{D}$, and denote the common value in (6.1) by

$$\int_a^b f(x)dx.$$

In this case (as we show in the next subsection) we may write

$$m(b - a) \leq L(P, f) \leq \int_a^b f(x)dx \leq U(P, f) \leq M(b - a).$$

6.1.2 The Darboux–Stieltjes integral

We now introduce another notion of integrability. Let α be a monotonically increasing function on the interval $[a, b]$. Consider a partition $P = \{x_0, x_1, \ldots, x_n\}$ of $[a, b]$ so that $a = x_0 \leq x_1 \leq \cdots \leq x_{n-1} \leq x_n = b$, and set

$$\Delta\alpha_k = \alpha(x_k) - \alpha(x_{k-1}), \quad k = 1, 2, \ldots, n.$$

Because α is monotonically increasing, $\Delta\alpha_k \geq 0$. For a bounded function f on $[a, b]$, we set

$$L(P, f, \alpha) = \sum_{k=1}^{n} m_k \Delta\alpha_k, \quad U(P, f, \alpha) = \sum_{k=1}^{n} M_k \Delta\alpha_k.$$

m_k and M_k have been defined earlier. Moreover, we consider

$$\int_{\underline{a}}^{b} f(x)d\alpha = \sup_P L(P, f, \alpha) \text{ and } \int_a^{\overline{b}} f(x)d\alpha = \inf_P U(P, f, \alpha), \tag{6.2}$$

where, as earlier, the sup and the inf are considered over all partitions of the interval $[a, b]$. The $\int_{\underline{a}}^{b} f(x)d\alpha$ is said to be the *lower Darboux–Stieltjes*[1] *integral* of the function f with respect to α, on the interval $[a, b]$, and $\int_a^{\overline{b}} f(x)d\alpha$ is said to be the *upper Darboux–Stieltjes integral* of the function f with respect to α on the interval $[a, b]$.

---

[1] Thomas Jan Stieltjes, 1856–1894.

A partition P^* of $[a, b]$ is said to be a *refinement* of P if $P \subset P^*$. Suppose two partitions P_1 and P_2 are given on $[a, b]$. Their common refinement P^* is defined by $P^* = P_1 \cup P_2$.

Lemma 6.1. *Let P^* be a refinement of P. Then*

$$L(P, f, \alpha) \le L(P^*, f, \alpha) \tag{6.3}$$

and

$$U(P, f, \alpha) \ge U(P^*, f, \alpha). \tag{6.4}$$

Proof. If $P = P^*$, in (6.3) and (6.4) we have equalities. Suppose for the beginning that P^* contains just one point more than P. Let t be this extra point and suppose that $x_{k-1} < t < x_k$. Set

$$s_1 = \sup_{x \in [x_{k-1}, t]} f(x), \quad s_2 = \sup_{x \in [t, x_k]} f(x).$$

Then $s_1 \le M_k$ and $s_2 \le M_k$. Therefore it results

$$U(P^*, f, \alpha) - U(P, f, \alpha)$$
$$= s_1(\alpha(t) - \alpha(x_{k-1})) + s_2(\alpha(x_k) - \alpha(t)) - M_k(\alpha(x_k) - \alpha(x_{k-1}))$$
$$= (s_1 - M_k)(\alpha(t) - \alpha(x_{k-1})) + (s_2 - M_k)(\alpha(x_k) - \alpha(t)) \le 0.$$

If P^* contains m points more than P, we repeat the above reasoning m times. Thus (6.4) is completely proved.

Inequality (6.3) is proved in a similar way.   □

Theorem 1.1. *It holds*

$$\underline{\int_a^b} f(x) d\alpha \le \overline{\int_a^b} f(x) d\alpha.$$

Proof. Choose P_1 and P_2 two arbitrary partitions and let P^* be their common refinement. By Lemma 6.1,

$$L(P_1, f, \alpha) \le L(P^*, f, \alpha) \le U(P^*, f, \alpha) \le U(P_2, f, \alpha).$$

Then

$$L(P_1, f, \alpha) \le U(P_2, f, \alpha).$$

Keeping P_2 fixed and taking the supremum in respect to P_1 in the above inequality, we have

$$\underline{\int_a^b} f(x) d\alpha \le U(P_2, f, \alpha).$$

Taking the infimum in respect to P_2 in the above inequality, we complete the proof.   □

If the lower and the upper Darboux–Stieltjes integrals of f are equal, we say that f is *Darboux–Stieltjes integrable* with respect to α, on $[a, b]$, we write $f \in \mathcal{D}(\alpha)$, and denote the common value in (6.2) by

$$\int_a^b f(x)d\alpha \ \text{ or } \ \int_a^b f(x)d\alpha(x).$$

If $\alpha(x) = x$ for all $x \in [a, b]$,

$$\int_a^b f(x)dx = \int_a^b f(x)d\alpha(x),$$

meaning that the Darboux integral is a special case of the Darboux–Stieltjes integral.

The Darboux–Stieltjes integrability can be characterized by the next result.

Theorem 1.2. $f \in \mathcal{D}(\alpha)$ *on* $[a, b]$ *is Darboux–Stieltjes integrable if and only if for every* $\varepsilon > 0$ *there exists a partition* P *such that*

$$U(P, f, \alpha) - L(P, f, \alpha) < \varepsilon. \tag{6.5}$$

Proof. Condition (6.5) is sufficient because from

$$L(P, f, \alpha) \le \underline{\int_a^b} f(x)d\alpha \le \overline{\int_a^b} f(x)d\alpha \le U(P, f, \alpha)$$

it follows that

$$0 \le \overline{\int_a^b} f(x)d\alpha - \underline{\int_a^b} f(x)d\alpha \le \varepsilon;$$

that is, $f \in \mathcal{D}(\alpha)$.

Suppose that

$$\overline{\int_a^b} f(x)d\alpha = \underline{\int_a^b} f(x)d\alpha = \int_a^b f(x)d\alpha$$

and let $\varepsilon > 0$ be given. Then we can find two partitions P_1 and P_2 such that

$$U(P_2, f, \alpha) - \int_a^b f(x)d\alpha \le \varepsilon/2 \ \text{ and } \ \int_a^b f(x)d\alpha - L(P_1, f, \alpha) \le \varepsilon/2.$$

Let P be the common refinement of P_1 and P_2. Then

$$U(P, f, \alpha) \le U(P_2, f, \alpha) \le \int_a^b f(x)d\alpha + \frac{\varepsilon}{2} \le L(P_1, f, \alpha) + \varepsilon \le L(P, f, \alpha) + \varepsilon;$$

that is, inequality (6.5) holds. $\square$

Theorem 1.3. *Suppose f is monotonic on $[a, b]$ and α is increasing and continuous on $[a, b]$. Then $f \in \mathcal{D}(\alpha)$.*

Proof. Suppose $f(a) = f(b)$. Then $f(x) = f(a)$, for all $x \in [a, b]$. We have

$$L(P, f, \alpha) = U(P, f, \alpha) = f(a)(\alpha(b) - \alpha(a)), \quad \text{for all partitions } P.$$

Therefore

$$\int_{\underline{a}}^{b} f(x)d\alpha = \overline{\int_a^b} f(x)d\alpha = f(a)(\alpha(b) - \alpha(a)),$$

and so $f \in \mathcal{D}(\alpha)$.

Suppose $f(a) \neq f(b)$. Suppose also f is monotonically increasing. The other case runs similarly. Let $\varepsilon > 0$ be given. Because α is continuous on $[a, b]$, it is uniformly continuous on $[a, b]$. Then there exists a partition P such that

$$\Delta\alpha_k = \frac{\alpha(b) - \alpha(a)}{n}, \quad k = 1, 2, \ldots, n \text{ and } \frac{\alpha(b) - \alpha(a)}{n}(f(b) - f(a)) < \varepsilon.$$

We have $M_k = f(x_k)$ and $m_k = f(x_{k-1})$ for $k = 1, 2, \ldots, n$. Hence

$$\begin{aligned}
U(P, f, \alpha) - L(P, f, \alpha) &= \frac{\alpha(b) - \alpha(a)}{n} \sum_{k=1}^{n} (f(x_k) - f(x_{k-1})) \\
&= \frac{\alpha(b) - \alpha(a)}{n}(f(b) - f(a)) < \varepsilon
\end{aligned}$$

for n large enough. Then by Theorem 1.2, $f \in \mathcal{D}(\alpha)$. $\square$

Theorem 1.4. *Suppose f is continuous on $[a, b]$. Then $f \in \mathcal{D}(\alpha)$. Moreover, to every $\varepsilon > 0$ there corresponds a $\delta > 0$ such that*

$$\left| \sum_{k=1}^{n} f(t_k)\Delta\alpha_k - \int_a^b f d\alpha \right| < \varepsilon \tag{6.6}$$

for every partition $P = \{x_0, \ldots, x_n\}$ of $[a, b]$ with $\|P\| < \delta$, and for every choice of points $t_k \in [x_{k-1}, x_k]$.

Proof. Let $\varepsilon > 0$ be given. Choose $\eta > 0$ such that $(\alpha(b) - \alpha(a))\eta < \varepsilon$. Function f is uniformly continuous on $[a, b]$, so there exits a positive δ such that

$$|f(x) - f(t)| < \eta \tag{6.7}$$

whenever $|x - t| < \delta$ and $x, t \in [a, b]$,

Choose a partition P such that $\|P\| < \delta$. By (6.7) we have $M_k - m_k \leq \eta$, for all $k = 1, 2, \ldots, n$. Thus

$$U(P, f, \alpha) - L(P, f, \alpha) = \sum_{k=1}^{n}(M_k - m_k)\Delta\alpha_k = \eta(\alpha(b) - \alpha(a)) < \varepsilon.$$

By the characterization Theorem 1.2, we conclude that $f \in \mathcal{D}(\alpha)$.

Because the two numbers $\sum_{k=1}^{n} f(t_k)\Delta\alpha_k$ and $\int_a^b f d\alpha$ lie between $U(P, f, \alpha)$ and $L(P, f, \alpha)$, inequality (6.6) is proved.   □

Theorem 1.4 is a useful tool to find the limits of certain sequences; see Exercise 6.1.

Theorem 1.5. (Leibniz–Newton formula) *Suppose f is continuous on $[a, b]$. Let F be a primitive of f on $[a, b]$. Then*

$$F(b) - F(a) = \int_a^b f(x)dx.$$

Proof. From Theorem 1.4 we have that f is integrable. Then

$$\overline{\int_a^b} f(x)dx = \underline{\int_a^b} f(x)dx.$$

The formula follows by (a) in Proposition 10.5 at page 242.   □

Theorem 1.6. *Let $f : [a, b] \to \mathbb{R}$ be a nondecreasing function with a primitive. Then a primitive is a convex function.*

Proof. By Corollary 2.4 at page 163 function f has no discontinuity of the second kind. Function f is with a primitive, thus by Corollary 10.11 at page 241 it has no discontinuity of the first kind. Hence f is continuous on $[a, b]$. Denote $g(x) = \int_a^x f(t)dt$ a primitive of f, $x \in [a, b]$. Consider a positive h and an x satisfying $a \leq x - h < x + h \leq b$. Then

$$\frac{g(x + h) - g(x)}{h} = \frac{1}{h}\int_x^{x+h} f(t)dt \geq f(x), \tag{6.8}$$

$$\frac{g(x) - g(x - h)}{h} = \frac{1}{h}\int_{x-h}^x f(t)dt \leq f(x). \tag{6.9}$$

Consider $a \leq x < y \leq b$, $\lambda \in [0, 1]$, and $z = \lambda x + (1 - \lambda)y$. From (6.8) and (6.9) we have

$$g(y) - g(z) \geq (y - z)f(z), \tag{6.10}$$

$$g(z) - g(x) \geq (z - x)f(z). \tag{6.11}$$

Multiply inequality (6.10) by $1 - \lambda$ and (6.11) by λ; then add them. It follows that $(1 - \lambda)g(y) - g(z) + \lambda g(x) \geq 0$; that is, g is convex on $[a, b]$.   □

Several arithmetic properties of the Darboux–Stieltjes integral follow.

Theorem 1.7. (a) *Suppose* $f_1, f_2 \in \mathcal{D}(\alpha)$ *on* $[a, b]$. *Then* $f_1 + f_2 \in \mathcal{D}(\alpha)$, $c \cdot f \in \mathcal{D}(\alpha)$ *for every constant* c, *and*

$$\int_a^b (f_1 + f_2) d\alpha = \int_a^b f_1 d\alpha + \int_a^b f_2 d\alpha, \qquad (6.12)$$

$$\int_a^b c \cdot f d\alpha = c \cdot \int_a^b f d\alpha.$$

(b) *Suppose* $f_1(x) \leq f_2(x)$, *for all* $x \in [a, b]$. *Then* $\int_a^b f_1 d\alpha \leq \int_a^b f_2 d\alpha$.

(c) *Suppose* $f \in \mathcal{D}(\alpha)$ *on* $[a, b]$ *and* $c \in]a, b[$. *Then* $f \in \mathcal{D}(\alpha)$ *on* $[a, c]$ *and* $[c, b]$, *and*

$$\int_a^b f d\alpha = \int_a^c f d\alpha + \int_c^b f d\alpha.$$

(d) *Suppose* $f \in \mathcal{D}(\alpha)$ *on* $[a, b]$ *and* $f(x) \geq 0$, *for every* $x \in [a, b]$. *Then* $\sqrt{f} \in \mathcal{D}(\alpha)$ *on* $[a, b]$.

(e) *Suppose* $f \in \mathcal{D}(\alpha)$ *on* $[a, b]$ *and the bounds of* f *are either positive or negative. Then* $1/f \in \mathcal{D}(\alpha)$ *on* $[a, b]$.

(f) *Suppose* $f, g \in \mathcal{D}(\alpha)$ *on* $[a, b]$. *Then* $f \cdot g \in \mathcal{D}(\alpha)$.

(g) *Suppose* $f \in \mathcal{D}(\alpha)$ *on* $[a, b]$. *Then*

$$\left| \int_a^b f d\alpha \right| \leq \int_a^b |f| \alpha.$$

If $|f(x)| < M$ *on* $[a, b]$, *then*

$$\left| \int_a^b f d\alpha \right| \leq M(\alpha(b) - \alpha(a)).$$

(h) *Suppose* $f, g \in \mathcal{D}(\alpha)$ *on* $[a, b]$ *and* f *is nonnegative on* $[a, b]$. *Denote* $m = \inf_{[a,b]} g(x)$ *and* $M = \sup_{[a,b]} g(x)$. *Then there exists* $\gamma \in [m, M]$ *such that*

$$\int_a^b f \cdot g d\alpha = \gamma \cdot \int_a^b f d\alpha.$$

(i) *(Cauchy–Buniakovski–Schwarz inequality for integrals) Suppose* $f, g \in \mathcal{D}(\alpha)$ *on* $[a, b]$. *Then*

$$\left(\int_a^b f \cdot g d\alpha \right)^2 \leq \left(\int_a^b f^2 d\alpha \right) \cdot \left(\int_a^b g^2 d\alpha \right).$$

(j) *Suppose* $f \in \mathcal{D}(\alpha_1)$ *and* $f \in \mathcal{D}(\alpha_2)$ *on* $[a, b]$. *Then* $f \in \mathcal{D}(\alpha_1 + \alpha_2)$ *and*

$$\int_a^b f d(\alpha_1 + \alpha_2) = \int_a^b f d\alpha_1 + \int_a^b f d\alpha_2.$$

(k) *Suppose $f \in \mathcal{D}(\alpha)$ on $[a, b]$ and $c > 0$ is a constant. Then*

$$\int_a^b f d(c\alpha) = c \cdot \int_a^b f d\alpha.$$

Proof. (a) We start showing that if $f = f_1 + f_2$, $f \in \mathcal{D}(\alpha)$. For, consider any partition P of $[a, b]$. Then

$$L(P, f_1, \alpha) + L(P, f_2, \alpha) \le L(P, f, \alpha) \le U(P, f, \alpha) \le U(P, f_1, \alpha) + U(P, f_2, \alpha).$$
$$(6.13)$$

There exist partitions P_1 and P_2 such that $U(P_k, f_k, \alpha) - L(P_k, f_k, \alpha) < \varepsilon$, $k = 1, 2$. Let P be their common refinement. The above inequalities hold by passing to partition P. Then by (6.13) it results $U(P, f, \alpha) - L(P, f, \alpha) < 2\varepsilon$; that is, $f \in \mathcal{D}(\alpha)$ on $[a, b]$.

We prove the equality in (6.12). Using the same P we have $U(P, f_k, \alpha) < \int_a^b f d\alpha + \varepsilon$, $k = 1, 2$. Thus (6.13) implies

$$\int_a^b f d\alpha \le U(P, f_k, \alpha) < \int_a^b f_1 d\alpha + \int_a^b f_2 d\alpha + 2\varepsilon.$$

Because $\varepsilon > 0$ has been chosen arbitrary, we get

$$\int_a^b f d\alpha \le \int_a^b f_1 d\alpha + \int_a^b f_2 d\alpha. \qquad (6.14)$$

Replace f_1 and f_2 by $-f_1$, respectively, $-f_2$; the inequality in (6.14) is reversed, hence (6.12) is proved.

(e) Suppose $M \ge m > 0$. Then if m_k' and M_k' are the infimum, respectively, the supremum of $1/f$ on I_k, we have

$$M_k' - m_k' \le 1/m_k' - 1/M_k' \le (1/m^2)(M_k - m_k).$$

The conclusion follows by Theorem 1.2.

(f) We may suppose that $f, g \ge 0$ on $[a, b]$. Then $0 \le m \le f(x) \le M$ and $0 \le m' \le g(x) \le M'$ for every $x \in [a, b]$. Choose an arbitrary ε. We admit that $M \cdot M' > 0$, because otherwise at least one function is the null function on $[a, b]$ and therefore the product is also null, and the conclusion follows. Suppose $M' \ge M$. For $\varepsilon/(2M)$, respectively, $\varepsilon/(2M')$, there exists a positive δ such that both

$$U(f, P, \alpha) - L(f, P, \alpha) \le \varepsilon/(2M) \text{ and } U(g, P, \alpha) - L(g, P, \alpha) \le \varepsilon/(2M')$$

for every partition P such that its norm is less than δ. We have

$$m_k' \le g(x) \le M_k' \text{ and } m_k \le f(x) \le M_k, \quad \forall x \in I_k.$$

Then $m_k m_k' \le f(x)g(x) \le M_k M_k'$ for every $x \in I_k$. Denote

$$M_k'' = \sup_{x \in I_k} f(x)g(x) \text{ and } m_k'' = \inf_{x \in I_k} f(x)g(x).$$

Then

$$m_k m_k' \le m_k'' \le f(x)g(x) \le M_k'' \le M_k M_k'.$$

We have

$$M_k'' - m_k'' \le M_k M_k' - m_k m_k' \le (M_k - m_k)M_k' + (M_k' - m_k')m_k$$
$$\le ((M_k - m_k) + (M_k' - m_k'))M.$$

From here we further have

$$U(f \cdot g, P, \alpha) - L(f \cdot g, P, \alpha) \le M \sum ((M_k - m_k) + (M_k' - m_k'))\Delta\alpha_k$$
$$\le M((U(f, P, \alpha) - L(f, P, \alpha)) + (U(g, P, \alpha) - L(g, P, \alpha))) \le M \cdot \varepsilon/M = \varepsilon.$$

(h) If $m \le g(x) \le M$, then $mf(x) \le g(x)f(x) \le Mf(x)$ and

$$m \int_a^b f d\alpha \le \int_a^b g \cdot f d\alpha \le M \int_a^b f d\alpha.$$

Thus $\gamma \in [m, M]$.

(i) For every real λ the square $(f + \lambda g)^2$ is integrable and nonnegative on $[a, b]$. Thus

$$0 \le \int_a^b (f + \lambda g)^2 d\alpha = \lambda^2 \left(\int_a^b g d\alpha \right)^2 + 2\lambda \int_a^b f \cdot g d\alpha + \left(\int_a^b f d\alpha \right)^2.$$

We just write that its discriminant is nonpositive.

The proofs of the other claims of the theorem are easy. □

Corollary 1.1. *Suppose f is of bounded variation on $[a, b]$. Then f is Darboux integrable on $[a, b]$.*

Proof. By Theorem 7.3 at page 180, f is a difference of two increasing function. Also by Theorem 1.3 a monotonic function is Darboux integrable and by Theorem 1.7 the difference of two Darboux functions is a Darboux function. Thus we conclude that f is Darboux integrable on $[a, b]$. □

Theorem 1.8. *Consider f a Darboux integrable function on $[a, b]$. Then function g defined on $[a, b]$ by*

$$g(t) = \int_a^t f(s) \, ds$$

is of bounded variation on $[a, b]$.

Proof. Function f is bounded on $[a, b]$; that is, there is a constant $M > 0$ so that $|f(t)| \leq M$, for all $t \in [a, b]$. Then

$$|g(x) - g(y)| = \left| \int_a^x f(s)\,ds - \int_a^y f(s)\,ds \right| = \left| \int_y^x f(s)\,ds \right| \leq \left| \int_y^x |f(s)|\,ds \right|$$
$$\leq M|x - y|.$$

Thus g is a Lipschitz function. By (b) of Remarks 7.1 at page 179 we conclude that g is of bounded variation on $[a, b]$. □

A result occurs allowing us to find the total variation to a function of the integrable derivative.

Theorem 1.9. *Let f be a differentiable function with the Darboux integrable derivative on $[a, b]$. Then*

$$\bigvee_b^a f = \int_a^b |f'(x)|\,dx.$$

Proof. Consider a partition $P = \{x_0, x_1, \ldots, x_n\}$ of $[a, b]$. Then by the Lagrange mean value theorem we have

$$V(f; P) = \sum_{k=0}^{n-1} |f(x_{k+1}) - f(x_k)| = \sum_{k=0}^{n-1} |f'(c_k)|(x_{k+1} - x_k). \qquad (6.15)$$

The derivative of f is Darboux integrable, so it is bounded. Then f is Lipschitz, thus it is of bounded variation. By Theorem 7.2 at page 180, we have that if $\|P\| \to 0$, the left-hand side of (6.15) tends to the total variation of f on $[a, b]$; that is, $\bigvee_b^a f$. Because f' is Darboux integrable, the right hand-side of (6.15) tends to $\int_a^b |f'(x)|\,dx$. □

We now introduce a result on the integrability of the composition of functions.

Theorem 1.10. *Suppose $f \in \mathcal{D}(\alpha)$ on $[a, b]$, $m \leq f \leq M$, g is continuous on $[m, M]$, and $h(x) = g(f(x))$ on $[a, b]$. Then $h \in \mathcal{D}(\alpha)$ on $[a, b]$.*

We introduce the *formula of integration by parts*.

Theorem 1.11. *Suppose f and α are functions of bounded variation on $[a, b]$, and f is also continuous. Then*

$$\int_a^b f\,d\alpha = f(b)\alpha(b) - f(a)\alpha(a) - \int_a^b \alpha\,df.$$

As an immediate application of the previous theorem is the following result. Suppose f is any real polynomial with degree m. If

$$I(t) = \int_0^t e^{t-u} f(u)\,du, \qquad (6.16)$$

where t is any real number, then by repeated integration by parts, Theorem 1.11, we have

$$I(t) = e^t \sum_{j=0}^{m} f^{(j)}(0) - \sum_{j=0}^{m} f^{(j)}(t). \tag{6.17}$$

Furthermore if $\tilde{f}$ denotes the polynomial obtained from f by replacing each coefficient with its absolute value, then

$$|I(t)| \leq \int_0^t e^{t-u} |f(u)| du \leq |t| e^{|t|} \tilde{f}(|t|). \tag{6.18}$$

This result is useful in proving that e is transcendental by Theorem 6.4 at page 282.

Theorem 1.12. (Wallis [2] formula) *There holds*

$$\sqrt{\pi} = \lim_{n \to \infty} \frac{1}{\sqrt{n}} \frac{2^{2n}(n!)^2}{(2n)!}. \tag{6.19}$$

Proof. Consider

$$I_n = \int_0^{\pi/2} \sin^n x \, dx$$

for $n = 2, 3, \ldots$. The right-hand side is well defined. We integrate it by parts

$$I_n = -\sin^{n-1} x \cos x \Big|_0^{\pi/2} + (n-1) \int_0^{\pi/2} \sin^{n-2} x \cos^2 x \, dx$$

$$= (n-1)I_{n-2} - (n-1)I_n.$$

Thus we have $I_n = ((n-1)/n)I_{n-2}$. Then

$$I_{2k} = \frac{2k-1}{2k} I_{2k-2} = \cdots = \frac{2k-1}{2k} \frac{2k-3}{2k-2} \cdots \frac{1}{2} \int_0^{\pi/2} dx$$

$$= \frac{\pi}{2} \frac{(2k-1)(2k-3)\cdots 1}{(2k)(2k-2)\cdots 2},$$

$$I_{2k+1} = \frac{2k}{2k+1} I_{2k-1} = \cdots = \frac{2k}{2k+1} \frac{2k-2}{2k-1} \cdots \frac{2}{3} I_1 = \frac{(2k)(2k-2)\cdots 2}{(2k+1)(2k-1)\cdots 3}.$$

Thus

$$\alpha_k = \frac{I_{2k}}{I_{2k+1}} = \frac{\pi}{2}(2k+1) \left(\frac{(2k-1)(2k-3)\cdots 1}{(2k)(2k-2)\cdots 2} \right)^2.$$

Because $0 \leq x \leq \pi/2$, one has

---

[2] John Wallis, 1616–1703.

$$\sin^{2k+1} x \leq \sin^{2k} x \leq \sin^{2k-1} x \text{ and } I_{2k+1} \leq I_{2k} \leq I_{2k-1}.$$

Thus

$$1 \leq \alpha_k = I_{2k}/I_{2k+1} \leq I_{2k-1}/I_{2k+1} = 1 + 1/2k,$$

and

$$\alpha_k \xrightarrow{k \to \infty} 1,$$

$$\sqrt{\pi} = \lim_{k \to \infty} \frac{1}{\sqrt{k}} \frac{(2k)(2k-2)\cdots 2}{(2k+1)(2k-1)\cdots 1} = \lim_{k \to \infty} \frac{1}{\sqrt{k}} \frac{((2k)(2k-2)\cdots 2)^2}{(2k)!}$$

$$= \lim_{k \to \infty} \frac{1}{\sqrt{k}} \frac{2^{2k}(k!)^2}{(2k)!}. \quad \square$$

6.2 Integrability of sequences and series of functions

Theorem 2.1. *Let α be a monotonically increasing function on $[a, b]$, $f_n \in \mathcal{D}(\alpha)$ on $[a, b]$, $n \in \mathbb{N}$. Suppose that the sequence (f_n) converges uniformly to f on $[a, b]$. Then $f \in \mathcal{D}(\alpha)$ on $[a, b]$ and*

$$\int_a^b f \, d\alpha = \lim_{n \to \infty} \int_a^b f_n \, d\alpha. \tag{6.20}$$

Proof. First we prove that $f \in \mathcal{D}(\alpha)$ on $[a, b]$ and so the left-hand side of (6.20) is meaningful.

Choose $\varepsilon > 0$. Then choose $\eta > 0$ so that

$$\eta(\alpha(a) - \alpha(b)) \leq \varepsilon/3. \tag{6.21}$$

By the uniform convergence of (f_n) there exists an $m \in \mathbb{N}$ such that

$$|f_m(x) - f(x)| \leq \eta, \quad \forall x \in [a, b]. \tag{6.22}$$

Because each $f_n \in \mathcal{D}(\alpha)$ on $[a, b]$, we choose a partition P of $[a, b]$ such that

$$U(P, f_m, \alpha) - L(P, f_m, \alpha) \leq \varepsilon/3. \tag{6.23}$$

By (6.22) $f(x) \leq f_m(x) + \eta$, and by (6.21),

$$U(P, f, \alpha) \leq U(P, f_m, \alpha) + \varepsilon/3. \tag{6.24}$$

Similarly, by $f(x) \geq f_m(x) - \eta$, it follows that

$$L(P, f, \alpha) \geq L(P, f_m, \alpha) - \varepsilon/3. \tag{6.25}$$

Combining (6.23), (6.24), and (6.25), we obtain

$$U(P, f, \alpha) - L(P, f, \alpha) \leq \varepsilon,$$

which, by Theorem 1.2, implies that $f \in \mathcal{D}(\alpha)$ on $[a, b]$.

Because (f_n) converges uniformly on $[a, b]$ to f, there exists n_ε so that $n > n_\varepsilon$ implies that

$$|f_n(x) - f(x)| \leq \varepsilon, \quad \forall\, x \in [a, b].$$

Then, for $n > n_\varepsilon$,

$$\left| \int_a^b f d\alpha - \int_a^b f_n d\alpha \right| = \left| \int_a^b (f - f_n) d\alpha \right| \leq \int_a^b |f - f_n| d\alpha \leq \varepsilon(\alpha(b) - \alpha(a)).$$

Because ε is arbitrary, (6.20) follows.   □

Corollary 2.2. *If* $f_n \in \mathcal{D}(\alpha)$ *on* $[a, b]$ *for all* $n \in \mathbb{N}$, $f(x) = \sum f_n(x)$, *for all* $x \in [a, b]$, *and the series converging uniformly on* $[a, b]$, *then*

$$\int_a^b f d\alpha = \sum \int_a^b f_n d\alpha.$$

6.3 Improper integrals

Two key assumptions appeared during the precess of defining the Darboux integral: the compactness of the set where the integrand is defined (namely [a,b]) and the boundedness of the integrand. Now we try to make sense of a notion of the integral supposing that one or both assumptions are denied.

We are concerned with functions defined on one of the following noncompact intervals: $I = [a, b[$ for $a \in \mathbb{R}$, $b \in \overline{\mathbb{R}}$; $I =]a, b]$ for $a \in \overline{\mathbb{R}}$, $b \in \mathbb{R}$; $I =]a, b[$ for $a \in \overline{\mathbb{R}}$, $b \in \overline{\mathbb{R}}$.

An integral defined on one of the above Is is said to be an *improper integral* or *generalized Darboux integral*. There is no restriction on the boundedness of the integrand.

Let f be a function defined on $I = [a, +\infty[$ so that for every $b \in I$ there exists the Darboux integral $\int_a^b f(x)dx$. Define

$$\int_a^{+\infty} f(x)dx = \lim_{b \to \infty} \int_a^b f(x)dx,$$

provided the limit exists. In this case we say that the *improper integral* $\int_a^{+\infty} f(x)dx$ *converges;* otherwise it *diverges*.

Similarly, let f be a function defined on $I =]-\infty, b]$ so that for every $a \in I$ there exists $\int_a^b f(x)dx$. Define

$$\int_{-\infty}^b f(x)dx = \lim_{a \to -\infty} \int_a^b f(x)dx,$$

provided the limit exists. In this case we say that the *improper integral* $\int_{-\infty}^{b} f(t)dt$ *converges*; otherwise it *diverges*. We have

$$\int_{-\infty}^{b} f(x)dx = \int_{-b}^{+\infty} f(-x)dx.$$

We define

$$\int_{-\infty}^{+\infty} f(x)dx = \int_{\mathbb{R}} f(x)dx = \lim_{\substack{a \to -\infty \\ b \to +\infty}} \int_{a}^{b} f(x)dx.$$

The *improper integral* $\int_{-\infty}^{+\infty} f(x)dx$ *converges* provided the limit exists; otherwise, we say that the *improper integral diverges*.

Often, we use the relation

$$\int_{-\infty}^{+\infty} f(x)dx = \int_{-\infty}^{a} f(x)dx + \int_{a}^{+\infty} f(x)dx, \quad a \in \mathbb{R}.$$

Examples 3.1. (a) We study the convergence of the improper integral

$$\int_{a}^{+\infty} \frac{dx}{x^p}, \quad a > 0. \tag{6.26}$$

Suppose $p = 1$. For any $b > a$, we have $\int_{a}^{b} dx/x = \ln(b/a)$. Then

$$\int_{a}^{b} \frac{dx}{x} \xrightarrow{b \to \infty} +\infty.$$

Suppose $p \neq 1$. For any $b > a$, we have

$$\int_{a}^{b} \frac{dx}{x^p} = \frac{1}{1-p}(b^{1-p} - a^{1-p}) \xrightarrow{b \to \infty} \begin{cases} a^{1-p}/(p-1), & p > 1 \\ \infty, & p < 1. \end{cases}$$

We conclude that the improper integral (6.26) converges if and only if $p > 1$.

(b) We study the convergence of the improper integral

$$\int_{a}^{+\infty} e^{-px}dx, \quad p \in \mathbb{R}. \tag{6.27}$$

Suppose $p = 0$. Then $\int_{a}^{b} e^{-px}dx = \int_{a}^{b} dx = b - a \xrightarrow{b \to \infty} \infty.$
Suppose $p \neq 0$. Then

$$\int_{a}^{b} e^{-px}dx = \frac{e^{-ap} - e^{-bp}}{p} \xrightarrow{b \to \infty} \begin{cases} (1/p)e^{-ap}, & p > 0 \\ -\infty, & p < 0. \end{cases}$$

We conclude that the improper integral (6.27) converges if and only if $p > 0$.

(c) Consider the improper integral

$$\int_0^1 \frac{dx}{x^p}.$$ (6.28)

For $0 < a < 1$, we have

$$\int_a^1 \frac{dx}{x^p} = \begin{cases} \int_a^1 dx/x = -\ln a & \xrightarrow{a \to 0,\ p=1} \infty, \\ 1/(1-p) - a^{1-p}/(1-p) & \begin{cases} \xrightarrow{a \to 0,\ p<1} 1/(1-p), \\ \xrightarrow{a \to 0,\ p>1} \infty. \end{cases} \end{cases}$$

We conclude (6.28) converges if and only if $p < 1$.   △

Lemma 6.2. *Suppose $f : [a, \infty[\to [0, \infty[$ is Darboux integrable on every $[a, b]$, with $b > a$. Then the improper integral $\int_a^\infty f(t)dt$ converges if and only if there exists $M > 0$ so that*

$$\int_a^b f(x)dx \leq M, \quad \forall b > a.$$ (6.29)

Proof. Suppose $\int_a^\infty f(x)dx$ converges. Then $\lim_{b \to \infty} \int_a^b f(x)dx$ exists. Denote this limit by M. Because f has nonnegative values, $\int_a^b f(x)dx$ is increasing in respect to b. Now the conclusion follows.

Suppose (6.29) holds. Because f has nonnegative values, $\int_a^b f(x)dx$ is increasing in respect to b. So, there exists $\lim_{b \to \infty} \int_a^b f(x)dx$.   □

Theorem 3.1. (Cauchy) *Suppose $f : [a, \infty[\to \mathbb{R}$ is Darboux integrable on every $[a, b]$, with $b > a$. Then the improper integral $\int_a^\infty f(x)\,dx$ converges if and only if for every $\varepsilon > 0$ there exists $p > a$ so that for every $d > c > p$ we have*

$$\left| \int_c^d f(x)dx \right| < \varepsilon.$$

Proof. We define $F(t) = \int_a^t f(x)dx$ and apply to F Theorem 1.2 at page 148. Then F has a limit at $+\infty$ if and only if $\varepsilon > 0$ there exists $p > a$ so that for every $d > c > p$ we have

$$|F(d) - F(c)| = \left| \int_c^d f(x)dx \right| < \varepsilon.$$ □

Consider $f : [a, \infty] \to \mathbb{R}$. Suppose $|f|$ is Darboux integrable on every $[a, b]$, with $b > a$. If the improper integral $\int_a^\infty |f(x)|dx$ converges, then we say that $\int_a^\infty f(x)dx$ *converges absolutely*.

Lemma 6.3. *Consider* $f : [a, \infty] \to \mathbb{R}$. *Suppose* $|f|$ *is Darboux integrable on every* $[a, b]$, *with* $b > a$. *Then* $\int_a^\infty f(x)dx$ *is absolutely convergent if and only if there exists a positive* K *so that*

$$\int_a^b |f(x)|dx < K, \quad \forall b > a.$$

Theorem 3.2. *Consider* $f : [a, \infty] \to \mathbb{R}$. *Suppose* f *is Darboux integrable on every* $[a, b]$, *with* $b > a$ *and* $\int_a^\infty f(x)dx$ *is absolutely convergent. Then it is convergent.*

Proof. By Theorem 3.1 applied to $|f|$, we have that for every $\varepsilon > 0$ there exists $p > a$ so that for every $d > c > p$,

$$\int_c^d |f(x)|dx < \varepsilon.$$

By (g) of Theorem 1.7 we have

$$\left| \int_c^d f(x)dx \right| < \int_c^d |f(x)|dx < \varepsilon, \quad \forall d > c > p.$$

Again by Theorem 3.1, the conclusion follows. $\square$

Consider $f : [a, \infty] \to \mathbb{R}$. If $\int_a^\infty f(x)dx$ is convergent whereas $\int_a^\infty f(x)dx$ is not absolutely convergent, then we say $\int_a^\infty f(x)dx$ is *semiconvergent*.

Remark. Consider $f : [0, \infty] \to \mathbb{R}$ defined by

$$f(x) = \frac{(-1)^n}{n+1}, \quad x \in [n, n+1[, \quad n \in \mathbb{N}.$$

Then

$$\int_0^\infty f(x)dx = \sum_{n=0}^\infty (-1)^n/n + 1 = \ln 2$$

and

$$\int_0^\infty |f(x)|dx = \lim_{b \to \infty} \int_0^b |f(x)|dx = \infty.$$

Thus $\int_0^\infty f(x)dx$ is semiconvergent. $\triangle$

Lemma 6.4. *Consider* $f : [a, \infty[\to \mathbb{R}$. *Suppose* f *is Darboux integrable on* $[a, \infty[$ *and there exists* $\lim_{x \to \infty} f(x)$. *Then* $\lim_{x \to \infty} f(x) = 0$.

Proof. Suppose $\lim_{x \to \infty} f(x) = l > 0$. Then there exists $b > a$ so that $f(x) > l/2$ for every $x \geq b$. For $d > c > b$ we have

$$\int_c^d f(x)dx > l(d-c)/2,$$

so the general test of convergence Theorem 3.1 does not hold.

The case $\lim_{t\to\infty} f(x) = l < 0$ can be treated similarly. $\Box$

Remark 3.1. The condition in Lemma 6.4 is only necessary. For it consider $f : [a, \infty[\to \mathbb{R}$ defined by $f(x) = \sin x^2$. Then $\int_0^\infty \sin x^2 dx = \sqrt{2\pi}/4$, [85, p. 443]. More generally, for $\theta \in [0, \pi/4]$

$$\int_0^\infty e^{-x^2 \cos(2\theta)} \cos(x^2 \sin(2\theta))dx = \frac{\sqrt{\pi}}{2}\cos\theta \xrightarrow{\theta=\pi/4} \int_0^\infty \cos x^2 dx = \frac{\sqrt{2\pi}}{4},$$

$$\int_0^\infty e^{-x^2 \cos(2\theta)} \sin(x^2 \sin(2\theta))dx = \frac{\sqrt{\pi}}{2}\sin\theta \xrightarrow{\theta=\pi/4} \int_0^\infty \sin x^2 dx = \frac{\sqrt{2\pi}}{4}.$$

Other general relations on the above integrals appear in [126, Chapter III].

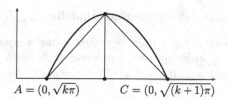

$$A = (0, \sqrt{k\pi}) \qquad C = (0, \sqrt{(k+1)\pi})$$

Fig. 6.1. Figure for Remark 3.1

We estimate the integral $\int_0^\infty |\sin x^2| dx$. Consider a positive a large enough. We interpret $\int_0^a |\sin x^2| dx$ as the area of the domain bounded by the horizontal axis $0x$ and the graph of the function $g(x) = |\sin x^2|$, for $x \in [0, a]$. This area is bounded below by the sum of areas to certain triangles ($\triangle ABC$ s), as shown in Figure 6.1. This is always possible because the function involved is concave between any two consecutive roots. Let m be the largest positive integer so that $\sqrt{m\pi} \le a$. Obviously, $a \to \infty$ implies $m \to \infty$. Then

$$\int_0^a |\sin x^2| dx \ge \sum_{k=0}^{m-1} \frac{\sqrt{(k+1)\pi} - \sqrt{k\pi}}{2} \ge \frac{\sqrt{m\pi}}{2} \xrightarrow{a\to\infty} \infty.$$

We conclude that the integral $\int_0^\infty \sin x^2 dx$ is semiconvergent. $\triangle$

A sufficient condition assuring the conclusion of Lemma 6.4 is supplied below.

Theorem 3.3. (Barbălat[3]) *Consider* $f : [a, \infty[\to \mathbb{R}$ *a uniformly continuous and Darboux integrable function on* $[a, \infty[$. *Then* $\lim_{x \to \infty} f(x) = 0$.

Lemma 6.5. *Consider* $f \in C^1[0, \infty[$ *so that*

$$\int_0^\infty |f'(x)| \, dx < \infty.$$

Then $\lim_{x \to \infty} f(x)$ *exists.*

Proof. For $0 \leq x < y$, we have

$$|f(y) - f(x)| = \left| \int_x^y f'(t) \, dt \right| \leq \int_x^y |f'(t)| \, dt. \tag{6.30}$$

By Theorem 3.1 of Cauchy we have that the last integral of (6.30) tends to zero whenever $x, y \to \infty$. Then

$$\lim_{x, y \to \infty} |f(y) - f(x)| = 0.$$

By Theorem 1.2 at page 148, we conclude that $\lim_{x \to \infty} f(x)$ exists. $\square$

Theorem 3.4. *Consider* $f, g : [a, \infty[\to [0, \infty[$ *and suppose that both functions are Darboux integrable on every interval* $[a, b]$ *with* $b > a$. *Moreover, there exists a constant* $c > 0$ *so that* $f(x) \leq cg(x)$, *for every* $x \in [a, \infty[$. *Then*

(a) *If* $\int_a^\infty g(x)dx$ *is convergent, then* $\int_a^\infty f(x)dx$ *is convergent.*
(b) *If* $\int_a^\infty f(x)dx$ *diverges, then* $\int_a^\infty g(x)dx$ *diverges.*

Proof. (a) Because $\int_a^\infty g(x)dx$ is convergent, by Theorem 3.1 for every $\varepsilon > 0$ there exists $p > a$ so that for every $v > u > p$ we have

$$\left| \int_u^v g(x)dx \right| < \varepsilon/c.$$

Then

$$\left| \int_u^v f(x)dx \right| < c \left| \int_u^v g(x)dx \right| < \varepsilon.$$

(b) It follows that

$$\int_a^b f(x)dx \xrightarrow{b \to \infty} \infty.$$

Then

$$\int_a^b g(x)dx \xrightarrow{b \to \infty} \infty,$$

thus $\int_a^\infty g(x)dx$ diverges. $\square$

---

[3] Ioan Barbălat, 1907–1988.

Corollary 3.3. *Consider $f, g : [a, \infty[\to [0, \infty[$ and suppose that both functions are Darboux integrable on every interval $[a, b]$ with $b > a$. Moreover, suppose that $\lim_{x \to \infty} f(x)/g(x) = l$. Then*

(a) *If $l < \infty$ and $\int_a^\infty g(x)dx$ converges, $\int_a^\infty f(x)dx$ converges.*
(b) *If $l > 0$ and $\int_a^\infty g(x)dx$ diverges, $\int_a^\infty f(x)dx$ diverges.*
(c) *If $l \in]0, \infty[$, then $\int_a^\infty f(x)dx$ converges if and only if $\int_a^\infty g(x)dx$ converges.*

Theorem 3.5. *Consider a decreasing function $f : [0, \infty[\to [0, \infty[$. Then $\int_0^\infty f(x)\,dx$ converges if and only if $\sum f(n)$ converges.*

Proof. For every natural k and $x \in [k, k+1]$ we have $f(k+1) \le f(x) \le f(k)$. Then

$$f(k+1) \le \int_k^{k+1} f(x)\,dx \le f(k)$$

and summing up

$$\sum_{k=1}^{n+1} f(k) \le \int_0^{n+1} f(x)\,dx \le \sum_{k=0}^n f(k). \tag{6.31}$$

Suppose $\sum f(n)$ converges. Then there exists a positive M so that for every $b \le n + 1$ by (6.31), we have

$$\int_0^b f(x)dx \le \int_0^{n+1} f(x)dx \le \sum_{k=0}^n f(k) \le M.$$

Hence by Lemma 6.2, $\int_0^\infty f(x)\,dx$ converges.

Suppose $\int_0^\infty f(x)\,dx$ converges. Then there exists a positive M such that by (6.31) for all n, $\sum_{k=1}^n f(k) \le M$. Then $\sum_{k=0}^\infty f(k)$ converges. $\square$

Theorem 3.6. *Consider $f \in C^1[0, \infty[$ so that*

$$\int_0^\infty |f'(x)|\,dx < \infty.$$

Then $\int_0^\infty f(x)\,dx$ converges if and only if $\sum f(n)$ converges.

Proof. By Lemma (6.5) we have that $\lim_{x \to \infty} f(x)$ exists. Suppose $\lim_{x \to \infty} f(x) = 0$ because otherwise both $\int_0^\infty f(x)\,dx$ and $\sum f(n)$ diverge. But

$$\int_0^\infty f(x)\,dx \text{ exists if and only if } \lim_{b \to \infty} \int_0^b f(x)\,dx \text{ exists.}$$

We remark that

$$\lim_{b \to \infty} \int_0^b f(x)\,dx \text{ exists if and only if } \lim_{b \to \infty} \int_0^{\lfloor b \rfloor} f(x)\,dx \text{ exists.}$$

Indeed, because $\lim_{x \to \infty} f(x) = 0$,

$$\int_0^b f(x)\,dx - \int_0^{\lfloor b \rfloor} f(x)\,dx = \int_{\lfloor b \rfloor}^b f(x)\,dx \xrightarrow{b \to \infty} 0.$$

Denote

$$I_n = \int_0^n f(x)\,dx \text{ and } s_n = \sum_{k=0}^n f(k).$$

Then

$$I_n - s_n = \sum_{k=0}^{n-1} \int_k^{k+1} f(x)\,dx - \sum_{k=0}^{n-1} f(k) - f(n)$$

$$= \sum_{k=0}^{n-1} \int_k^{k+1} (f(x) - f(k))\,dx - f(n) = \sum_{k=0}^{n-1} \int_k^{k+1} \left(\int_k^x f'(y)\,dy \right) dx - f(n).$$

Denote $a_k = \int_k^{k+1} \left(\int_k^x f'(y)\,dy \right) dx$. Then

$$|a_k| \le \int_k^{k+1} \left(\int_k^x |f'(y)|\,dy \right) dx = \int_k^{k+1} |f'(y)|\,dy.$$

From the absolute convergence of f' we have that $\sum_0^\infty |a_k|$ converges. Then $\sum_0^\infty a_k$ converges as well. Necessarily we have $\lim a_n = 0$. But this precisely means that

$$\lim_{n \to \infty} (I_n - s_n) = 0. \quad \square$$

Theorem 3.7. (Abel) *Suppose $f, g : [a, b[\to [0, \infty[$, $b \in \overline{\mathbb{R}}$, $b > a$, both are integrable on any compact interval contained in $[a, b[$, and*

(i) *f is monotonic and $\lim_{x \to b} f(x) = 0$.*
(ii) *There exists $M > 0$ so that $|\int_a^t g| < M$, for each $t \in [a, b[$.*

Then $f \cdot g$ is Darboux integrable on $[a, b[$.

Theorem 3.8. (Dirichlet) *Suppose $f, g : [a, b[\to \mathbb{R}$, $b \in \overline{\mathbb{R}}$, $b > a$, both are integrable on any compact interval contained in $[a, b[$, and*

(i) *f is monotonic and bounded on $[a, b[$.*
(ii) *g is Darboux integrable on $[a, b[$.*

Then $f \cdot g$ is Darboux integrable on $[a, b[$.

We introduce the formula of integration by parts.

Theorem 3.9. *Let $-\infty < a < b \le \infty$. Consider two functions $f, g : [a, b[\to \mathbb{R}$ of class C^2 such that there exists $\lim_{x \uparrow b} f(x)g(x)$ and the integral $\int_a^b f'g$ converges. Then $\int_a^b fg'$ converges and*

$$\int_a^b fg' = \lim_{x \uparrow b} f(x)g(x) - f(a)g(a) - \int_a^b f'g. \tag{6.32}$$

Remark. In many parts of mathematics one can see the following improper integral

$$\int_0^\infty e^{-x^2}\,dx. \tag{6.33}$$

It is clear that the integral is finite and well defined. We already supplied three proofs: one by Example 1.1 at page 333, the other two by (a) and (b) in Examples 5.1 at page 351.   △

6.4 Euler integrals

6.4.1 Gamma function

Euler introduced a C^∞ function that interpolates the factorial whenever the argument of the function is a positive integer. This function is the *Euler integral of the second kind* or the *gamma function* and it is defined by

$$\Gamma(x) = \int_0^\infty t^{x-1} e^{-t}\,dt. \tag{6.34}$$

Theorem 4.1. *The gamma function is defined for $x > 0$.*

Proof. We write

$$\Gamma(x) = \int_0^1 t^{x-1} e^{-t}\,dt + \int_1^\infty t^{x-1} e^{-t}\,dt = I_1(x) + I_2(x),$$

where $I_1(x) = \int_0^1 t^{x-1} e^{-t}\,dt$ and $I_2(x) = \int_1^\infty t^{x-1} e^{-t}\,dt$. Because $|e^{-t} t^{x-1}| \le t^{x-1}$ for $x > 0$, by the comparison test theorem 3.4, we conclude $I_1(x)$ converges. $I_2(x)$ converges because $\lim_{t \to \infty} t^{x-1} e^{-t} \le \lim_{t \to \infty} e^{-t/2} = 0$.   □

Theorem 4.2. (Euler, 1730) *For $x > 0$,*

$$\Gamma(x) = \int_0^1 [-\ln(u)]^{x-1}\,du, \qquad \Gamma(x) = 2\int_0^\infty u^{2x-1} e^{-u^2}\,du.$$

Proof. Change the variables $t = -\ln u$, respectively, $t = u^2$ in (6.34).   □

Theorem 4.3. *The gamma function is of class C^∞ on $]0, \infty[$.*

Corollary 4.4. *One has*

$$\Gamma'(x) = \int_0^\infty t^{x-1} e^{-t} \ln t\,dt \quad \text{and} \quad \Gamma^{(n)}(x) = \int_0^\infty t^{x-1} e^{-t} \ln^n t\,dt.$$

Theorem 4.4. (Reduction formula) *There hold*

$$\Gamma(1) = 1, \tag{6.35}$$
$$\Gamma(x+1) = x\,\Gamma(x), \quad x > 0. \tag{6.36}$$

Proof. We have $\Gamma(1) = \int_0^\infty e^{-t}dt = 1$ and

$$\Gamma(x+1) = \int_0^\infty t^x e^{-t}dt = -t^x e^{-t}\big|_0^\infty + x \int_0^\infty t^{x-1}e^{-t}dt = x\Gamma(x). \quad \square$$

The reduction formula (6.36) is said to be the *functional equation* of the gamma function Γ.

The functional equation (6.36) does not have a unique solution. Other solutions satisfying both (6.35) and (6.36) are supplied by the functions $\cos(2m\pi x)\Gamma(x)$, where $m \in \mathbb{Z} \setminus \{0\}$. Adding extra assumptions, the gamma function is the unique solution of (6.35) and (6.36) as follows from the next result given in [19].

Theorem 4.5. *There is a unique function* $f :]0, +\infty[\to]0, +\infty[$ *such that* $\ln(f(x))$ *is convex and*

$$f(1) = 1 \quad and \quad f(x+1) = xf(x).$$

The gamma function can be extended on the whole real axis except on the nonpositive integers $\{0, -1, -2, \ldots\}$ by considering for $x \in]0, 1[$ the functional equation of it under the form

$$\Gamma(x) = \Gamma(x+1)/x.$$

By the previous relation it follows

$$\Gamma(x) = \frac{\Gamma(x+n)}{x(x+1)\cdots(x+n-1)}, \quad x + n > 0.$$

Euler introduced another gamma function for $x > 0$ and $p \in \mathbb{N}^*$:

$$\Gamma_p(x) = \frac{p!p^x}{x(x+1)\cdots(x+p)} = \frac{p^x}{x(1+x/1)\cdots(1+x/p)}.$$

Then

$$\Gamma(x) = \lim_{p \to \infty} \Gamma_p(x).$$

Immediately we have

$$\Gamma_p(1) = p/(p+1) \quad and \quad \Gamma_p(x+1) = (p/(x+p+1)) \cdot x \cdot \Gamma_p(x),$$

and (6.35) and (6.36) follow.

Remark. We note that the domain of Γ_p can be extended for real but nonpositive integer numbers. $\triangle$

Because

$$p^x = e^{x \ln p} = e^{x(\ln p - 1 - 1/2 - \cdots - 1/p)}e^{x+x/2+\cdots+x/p},$$

we get

$$\Gamma_p(x) = \frac{1}{x}\frac{1}{x+1}\frac{2}{x+2}\cdots\frac{p}{x+p}p^x = \frac{e^{x(\ln p-1-1/2-\cdots-1/p)}e^{x+x/2+\cdots+x/p}}{x(1+x/1)\cdots(1+x/p)}$$

or

$$\Gamma_p(x) = e^{x(\ln p-1-1/2-\cdots-1/p)}\frac{1}{x}\frac{e^x}{1+x}\frac{e^{x/2}}{1+x/2}\cdots\frac{e^{x/p}}{1+x/p}. \tag{6.37}$$

By (6.37) we introduce the *Weierstrass formula* for $\Gamma_p(x)$.

Theorem 4.6. *For every* $x \in \mathbb{R} \setminus \{0, -1, -2, \dots\}$ *we have*

$$\frac{1}{\Gamma(x)} = xe^{\gamma x}\prod_{n=1}^{\infty}\left(1+\frac{x}{n}\right)e^{-x/n}, \tag{6.38}$$

where γ *is the Euler–Mascheroni constant.*

For certain values of the argument, the Γ function has closed values.

Theorem 4.7. *For* $n \in \mathbb{N}$, *one has*

$$\Gamma(n+1) = n!, \tag{6.39}$$

$$\Gamma\left(\frac{1}{2}\right) = \sqrt{\pi}, \tag{6.40}$$

$$\Gamma\left(n+\frac{1}{2}\right) = \frac{1\cdot 3\cdots(2n-1)}{2^n}\sqrt{\pi}, \quad \Gamma\left(n+\frac{1}{3}\right) = \frac{1\cdot 4\cdots(3n-2)}{3^n}\Gamma\left(\frac{1}{3}\right),$$

$$\Gamma\left(n+\frac{1}{4}\right) = \frac{1\cdot 5\cdots(4n-3)}{4^n}\Gamma\left(\frac{1}{4}\right), \quad \Gamma\left(\frac{1-2n}{2}\right) = \frac{(-1)^n 2^n}{1\cdot 3\cdots(2n-1)}\sqrt{\pi}.$$

Proof. (6.39) follows from (6.35) and (6.36).

Taking $t = u^2$, we have

$$\Gamma(1/2) = \int_0^\infty \frac{e^{-t}dt}{\sqrt{t}} = 2\int_0^\infty e^{-u^2}\,du = \sqrt{\pi}.$$

The last equality follows by Example 1.1 at page 333.

The next three equalities follow by (6.36).   □

Remarks. No closed form to $\Gamma(1/3)$ or $\Gamma(1/4)$ is known. The last equality in Theorem 4.7 allows computation of the Γ function for some negative values of the argument.   △

We now introduce the *complement formula* of the Γ function.

Theorem 4.8. *Suppose* x *and* $-x$ *are neither negative or null integers. Then*

$$\Gamma(x)\Gamma(1-x) = \frac{\pi}{\sin\pi x}. \tag{6.41}$$

Proof. By (6.38) we have

$$\frac{1}{\Gamma(x)}\frac{1}{\Gamma(-x)} = -x^2 e^{\gamma x} e^{-\gamma x} \prod_{n=1}^{\infty} \left(\left(1+\frac{x}{n}\right) e^{-x/n} \left(1-\frac{x}{n}\right) e^{x/n}\right).$$

Because $\Gamma(-x) = -\Gamma(1-x)/x$,

$$\frac{1}{\Gamma(x)}\frac{1}{\Gamma(1-x)} = x \prod_{n=1}^{\infty} \left(1-\frac{x^2}{n^2}\right).$$

By

$$\sin \pi x = \pi x \prod_{n=1}^{\infty} \left(1 - x^2/n^2\right)$$

(this identity may be found in [126]), we get (6.41).   □

Corollary 4.5. *One has*

$$\Gamma\left(\frac{1}{2}\right) = \sqrt{\pi}, \quad \Gamma\left(\frac{1}{3}\right)\Gamma\left(\frac{2}{3}\right) = \frac{2\pi\sqrt{3}}{3}, \quad \text{and } \Gamma\left(\frac{1}{4}\right)\Gamma\left(\frac{3}{4}\right) = \pi\sqrt{2}.$$

Proof. Substitute $x = 1/2$, $x = 1/3$, respectively $x = 1/4$ in (6.41).   □

Theorem 4.9. *Suppose x and $x + 1/2$ are neither negative or null integers. Then*

$$\Gamma(x)\Gamma\left(x+\frac{1}{2}\right) = \frac{\sqrt{\pi}}{2^{2x-1}}\Gamma(2x). \tag{6.42}$$

Corollary 4.6. *One has*

$$\Gamma(1/6) = 2^{-1/3}\sqrt{3/\pi}\,\Gamma^2(1/3).$$

Proof. Substitute $x = 1/6$ in (6.42).   □

Theorem 4.10. *Suppose $x, x + 1/n, \ldots, x + (n-1)1/n$ are neither negative or null integers. Then*

$$\Gamma(x)\Gamma\left(x+\frac{1}{n}\right)\Gamma\left(x+\frac{2}{n}\right)\cdots\Gamma\left(x+\frac{n-1}{n}\right) = (2\pi)^{(n-1)/2}n^{1/2-nx}\,\Gamma(nx).$$

Remark. Theorem 4.9 follows from Theorem 4.10 for $n = 2$.   △

Corollary 4.7. *One has*

$$\Gamma\left(\frac{1}{n}\right)\Gamma\left(\frac{2}{n}\right)\cdots\Gamma\left(\frac{n-1}{n}\right) = \frac{(2\pi)^{(n-1)/2}}{\sqrt{n}}.$$

Proof. Substitute $x = 1/n$ in the relation in Theorem 4.10.   □

The gamma function admits an asymptotic expansion as shown in [70, part 2, p. 293].

Theorem 4.11. *For $x > 0$ one has*

$$\Gamma(x+1) = \sqrt{2\pi x}\, x^x\, e^{-x} \left(1 + \frac{1}{12x} + \frac{1}{288x^2} - \frac{139}{51840x^3} + \frac{O(1)}{x^4} \right).$$

The Stirling[4] formula follows at once.

Theorem 4.12. *For $n \in \mathbb{N}^*$ one has*

$$n! = \sqrt{2\pi n}\, n^n\, e^{-n}\, (1 + \alpha_n), \tag{6.43}$$

where $\alpha_n \to 0$ as $n \to \infty$.

Lemma 6.6. *The largest binomial coefficient tends asymptotically to $2^n \sqrt{2/(n\pi)}$ as n tends to ∞.*

Proof. By Theorem 1.25 at page 95 we have for a given n, that the sequence of binomial coefficients $\left(\binom{n}{k}\right)_{k=0,1,\ldots,n}$ is unimodal and the biggest term is given for $k = \lfloor n/2 \rfloor$. We consider only the case n where is even. The other case runs similarly. For n, by formula (6.43) of Stirling, we have

$$\binom{n}{n/2} = \frac{n!}{((n/2)!)^2} \sim \frac{(n/e)^n \sqrt{2n\pi}}{((n/(2e))^{n/2}\sqrt{n\pi})^2} = 2^n \sqrt{\frac{2}{n\pi}}. \quad □$$

6.4.2 Beta function

The *Euler integral of the first kind* or the *beta function* is the integral defined as

$$B(x,y) = \int_0^1 t^{x-1}(1-t)^{y-1}dt, \quad x,y > 0. \tag{6.44}$$

x and y are parameters.

Theorem 4.13. *The beta function is defined for all $x, y > 0$.*

Proof. If $x, y \geq 1$, the integrand is a continuous function in t, so the integral in the right-hand side is proper.

Suppose now that at least one of the parameters belongs to the open $]0, 1[$ interval. In this case the integrand is not defined on $t = 0$ or $t = 1$ or both. Then we write

$$B(x,y) = \int_0^{1/2} t^{x-1}(1-t)^{y-1}dt + \int_{1/2}^1 t^{x-1}(1-t)^{y-1}dt = I_1(x,y) + I_2(x,y),$$

---

[4] James Stirling, 1710–1761.

where $I_1(x,y) = \int_0^{1/2} t^{x-1}(1-t)^{y-1}dt$ and $I_2(x,y) = \int_{1/2}^1 t^{x-1}(1-t)^{y-1}dt$. Obviously, each of the integrands of $I_1(x,y)$ and $I_2(x,y)$ have at most one point on which they are not defined.

If $t \in [0,1/2]$ and $x \in]0,1[$, then t^{x-1} is unbounded whereas $(1-t)^{y-1}$ is continuous in t, and therefore is bounded by a certain constant C. Thus the integrand of I_1 is bounded by $C \cdot t^{x-1}$. Now invoking (c) in Examples 3.1, we conclude that the integral I_1 is defined for all $x,y > 0$.

In a similar way we discuss the existence of I_2. $\square$

Theorem 4.14. *The beta function is continuous on* $]0,\infty[\times]0,\infty[$.

Proof. It is sufficient to show that the integrand of $B(x,y)$ is uniformly convergent in respect to the parameters x and y, for $x \geq x_0 > 0$ and $y \geq y_0 > 0$, where x_0 and y_0 are fixed and positive. Because $x_0 - 1 \leq x - 1$ and $y_0 - 1 \leq y - 1$ for $(x,y) \in]0,\infty[\times]0,\infty[$, we have $t^{x-1}(1-t)^{y-1} \leq t^{x_0-1}(1-t)^{y_0-1}$. Because $\int_0^1 t^{x_0-1}(1-t)^{y_0-1}dt$ is defined (Theorem 4.14), it follows by the Weierstrass criterion that the integral in (6.44) is uniformly convergent for $x \geq x_0 > 0$ and $y \geq y_0 > 0$. Thus the continuity of the beta function follows. $\square$

Theorem 4.15. *One has*

(a) $B(x,y) = B(y,x)$.
(b) $B(x,1) = B(1,x) = 1/x$.
(c) $B(x+1,y) + B(x,1+y) = B(x,y)$.
(d) $B(x,y) = ((x+y)/y)B(x,y+1)$.

Proof. (a) Change the variable $t = 1 - u$.
(b) It follows from (a) and from (c) in Examples 3.1.
(c) We have

$$B(x+1,y) + B(x,1+y) = \int_0^1 t^{x-1}(1-t)^{y-1}(t+1-t)dt = B(x,y).$$

(d) By integration by parts we have $B(x+1,y) = (x/y)B(x,y+1)$. Then by (c) we have

$$B(x,y) = B(x+1,y) + B(x,1+y)$$
$$= (x/y)B(x,y+1) + B(x,y+1)((x+y)/y)B(x,y+1).   \square$$

Remark. The equality

$$B(x,y) = ((x+y)/y)\,B(x,y+1)$$

in (d) is said to be the *functional equation* of the beta function. $\triangle$

The two Euler functions are tied by the next result.

Theorem 4.16. *For $x, y > 0$ one has*

$$B(x, y) = \frac{\Gamma(x)\,\Gamma(y)}{\Gamma(x+y)}. \tag{6.45}$$

Proof. It is given at (c) in Examples 5.1 at page 351.   □

Corollary 4.8. *Given $m, n, p \in \mathbb{N}$, one has*

$$\frac{1}{\binom{mn}{pn}} = (mn+1)\int_0^1 x^{pn}(1-x)^{(m-p)n}dx. \tag{6.46}$$

Proof. Apply (6.45).   □

Proposition 4.1. *There hold*

$$B\left(1/2, 1/2\right) = \pi. \tag{6.47}$$
$$B\left(1/3, 2/3\right) = 2\sqrt{3}\pi/3. \tag{6.48}$$
$$B\left(1/4, 3/4\right) = \pi\sqrt{2}. \tag{6.49}$$
$$B\left(x, 1-x\right) = \frac{\pi}{\sin \pi x}. \tag{6.50}$$
$$B\left(x, 1\right) = 1/x. \tag{6.51}$$
$$B\left(x, n\right) = \frac{(n-1)!}{x(x+1)\ldots(x+n-1)}. \tag{6.52}$$
$$B\left(m, n\right) = \frac{(m-1)!(n-1)!}{(m+n-1)!}. \tag{6.53}$$

Proof. Following the definition of the beta function one has

$$B\left(\frac{1}{2}, \frac{1}{2}\right) = \int_0^1 \frac{dt}{\sqrt{t}\,\sqrt{1-t}} = 2\int_0^1 \frac{du}{\sqrt{1-u^2}} = 2\arcsin u\big|_0^1 = \pi.$$

A different approach is based on (6.45) and (6.40) which yield

$$B\left(1/2, 1/2\right) = \Gamma^2\left(1/2\right)/\Gamma(1) = \pi.$$

Thus (6.47) is proved.

By (6.45) and (6.41) we have

$$B\left(\frac{1}{3}, \frac{2}{3}\right) = \Gamma\left(\frac{1}{3}\right)\Gamma\left(\frac{2}{3}\right) = \Gamma\left(\frac{1}{3}\right)\Gamma\left(1-\frac{1}{3}\right) = \frac{\pi}{\sin \pi/3},$$

and thus (6.48) is proved.

Equality (6.49) is shown similarly to (6.48).

By (6.45) and (6.41), (6.50) is clear.

Equality (6.51) follows from the definition of the beta function

$$B(x,1) = \int_0^1 t^{x-1}\, dt = 1/x,$$

or by (6.45) and (6.36),

$$B(x,1) = \Gamma(x)\Gamma(1)/\Gamma(x+1) = \Gamma(x)/(x\Gamma(x)) = 1/x.$$

By (6.45) and (6.36) we have

$$B(x,n) = \frac{\Gamma(x)\Gamma(n)}{\Gamma(x+n)} = \frac{(n-1)!\Gamma(x)}{(x+n-1)\cdots(x+1)x\Gamma(x)},$$

and thus (6.52) is proved.

Equality (6.53) follows from (6.52). $\square$

6.5 Polylogarithms

For $x \in \mathbb{R}$, $|x| < 1$, and $n \in \mathbb{N}$ we define the *polylogarithm of order* n by

$$L_n(x) = \sum_{k=1}^{\infty} \frac{x^k}{k^n}. \tag{6.54}$$

Note

$$L_0(x) = x/(1-x) \quad \text{and} \quad L_1(x) = -\ln(1-x).$$

For $n \geq 2$, the polylogarithm of order n converges on $[-1,1]$. Suppose $n \geq 2$. Then

$$L_n(1) = \zeta(n), \quad L_n(-1) = \left(1/2^{n-1} - 1\right)\zeta(n).$$

Therefore $L_2(1) = \pi^2/6$ and $L_2(-1) = -\pi^2/12$.

Theorem 5.1. *One has*

$$\int_0^1 \frac{\ln^n t}{t - \frac{1}{x}}\, dt = (-1)^{n+1}\, n! \cdot L_{n+1}(x), \tag{6.55}$$

$$\frac{dL_n(x)}{dx} = \frac{L_{n-1}(x)}{x}, \quad n \geq 1, \tag{6.56}$$

$$L_n(x) = \int_0^x \frac{L_{n-1}(t)}{t}\, dt, \quad n \geq 1, \tag{6.57}$$

$$L_2(x) + L_2(1-x) = \zeta(2) - \ln x \cdot \ln(1-x), \quad x \in \,]0,1[\,, \tag{6.58}$$

$$L_2(x) + L_2\left(\frac{x}{x-1}\right) = -\frac{1}{2}\ln^2(1-x), \quad x \in \left[-1, \frac{1}{2}\right], \tag{6.59}$$

$$L_n(x) + L_n(-x) = \frac{1}{2^{n-1}} L_n(x^2), \qquad (6.60)$$

$$L_3(x) + L_3\left(\frac{x}{x-1}\right) + L_3(1-x)$$

$$= \zeta(3) + \zeta(2)\ln(1-x) - \frac{1}{2}\ln(x)^2\ln(1-x) + \frac{1}{2}\ln^3(1-x), \quad x \in \left[-1, \frac{1}{2}\right].$$

Proof. Equality (6.55) can be proved easily by induction. For $n = 0$, we have

$$\int_0^1 \frac{dt}{t - 1/x} = -x\int_0^1 \frac{dt}{1 - tx} = -x\int_0^1 \left(\sum_{k=0}^\infty x^k t^k\right) dt = -\sum_{k=0}^\infty x^{k+1}\int_0^1 t^k\,dt$$

$$= -\sum_{k=1}^\infty \frac{x^k}{k} = -L_1(x).$$

For $n = 1$, we have

$$\int_0^1 \frac{\ln t\,dt}{t - 1/x} = -x\int_0^1 \frac{\ln t\,dt}{1 - tx} = -x\int_0^1 \left(1 + \sum_{k=1}^\infty\right)\ln t\,dt$$

$$= -x\int_0^1 \ln t\,dt + \sum_{k=1}^\infty x^{k+1}\int_0^1 t^k\ln t\,dt = x + \sum_{k=2}^\infty x^k/k^2 = L_2(x).$$

For $n = 2$, we have

$$\int_0^1 \frac{\ln^2 t\,dt}{t - 1/x} = -x\int_0^1 \frac{\ln^2 t\,dt}{1 - tx} = -x\left(\int_0^1 \ln^2 t\,dt + \sum_{k=1}^\infty x^k\int_0^1 t^k\ln^2 t\,dt\right).$$

But

$$\int_0^1 \ln^2 t\,dt = -2\int_0^1 \ln t\,dt = 2$$

and

$$\int_0^1 t^k\ln^2 t\,dt = 2/(k+1)^3.$$

Therefore

$$\int_0^1 \frac{\ln^2 t\,dt}{t - 1/x} = -2\sum_{k=0}^\infty \frac{x^{k+1}}{(k+1)^3} = (-1)^{2+1}2!L_3(x).$$

It is now clear how the general case is obtained following the same path.
    (6.56) is clear.
    (6.57) follows from (6.56).
    Consider $f(x) = L_2(x) + L_2(1-x) + \ln x \cdot \ln(1-x)$ and note that $f'(x) = 0$ for all $x \in \,]0,1[$. Then f is a constant on $x \in \,]0,1[$. We may consider $\lim_{x\downarrow 0} f(x) = \zeta(2)$ and thus (6.58) is proved. From here immediately follows

$$L_2\left(\frac{1}{2}\right) = \sum_{n=1}^{\infty} \frac{1}{2^n n^2} = \frac{\pi^2}{12} - \frac{1}{2}\ln^2 2. \tag{6.61}$$

The derivatives of the both sides in (6.59) coincide. By (6.61), the equality is established.

For (6.60) successively we have

$$\int_0^1 \frac{\ln^n t}{t - 1/x}\, dt + \int_0^1 \frac{\ln^n t}{t + 1/x}\, dt = \int_0^1 \left(\frac{1}{t - 1/x} + \frac{1}{t + 1/x}\right)\ln^n t\, dt$$

$$= \int_0^1 \left(\frac{2t}{t^2 - 1/x^2}\right)\ln^n t\, dt = \frac{1}{2^n}\int_0^1 \frac{\ln^n u}{u - 1/x^2}\, du. \quad \square$$

6.6 e and π are transcendental

We saw at page 101 that e is irrational. Now we show that some well-known constants are irrational. This approach is based on the Niven[5] polynomials. We show that π^2, $\ln 2$, and e^a, where $a \in \mathbb{Q}\setminus\{0\}$, are irrational.

The Niven polynomials are defined by

$$P_n(x) = \frac{x^n(1-x)^n}{n!} = \frac{1}{n!}\sum_{k=n}^{2n} c_k x^k,$$

where c_k are integers. The Niven polynomials enjoy the following properties

$$0 < P_n(x) < \frac{1}{n!}, \quad 0 < x < 1,$$

$$P_n(0) = 0,$$

$$P_n^{(m)}(0) = 0, \quad m < n \text{ or } m > 2n,$$

$$P_n^{(m)}(0) = \frac{m!}{n!}c_m, \quad n \le m \le 2n$$

$$P_n(x) = P_n(1-x), \quad 0 < x < 1.$$

Hence P_n and its derivatives take integral values at $x = 0$ and $x = 1$.

Lambert[6] established in 1761 that e and $\ln 2$ are irrational. A general result holds.

Theorem 6.1. *Suppose $a > 0$ is an integer. Then e^a is an irrational number.*

Proof. Suppose $e^a = p/q$, where p, q are positive integers. Consider the polynomial

$$U(x) = a^{2n}P_n(x) - a^{2n-1}P_n'(x) + \cdots - aP_n^{(2n-1)}(x) + P_n^{(2n)}(x).$$

---

[5] Ivan Morton Niven, 1915–1999.
[6] Johann Heinrich Lambert, 1728–1777.

From the properties of the Niven polynomials we have that $U(0)$ and $U(1)$ are integers. On the other side we have

$$q(e^{ax}U(x))' = qe^{ax}(aU(x) + U(x)') = qe^{ax}a^{2n+1}P_n(x),$$

and

$$qa^{2n+1}\int_0^1 e^{ax}P_n(x)dx = q(e^{ax}U(x))|_0^1 = pU(1) - qU(0).$$

The integral is strictly positive and bounded; more precisely we have the bounds

$$0 < qa^{2n+1}\int_0^1 e^{ax}P_n(x)dx < \frac{qa^{2n}(e^a - 1)}{n!} < 1,$$

for n sufficiently large. This contradicts the fact that the integral is an integer number. Hence e^a is irrational.   □

Corollary 6.9. *Suppose p/q is a nonnull rational number. Then $e^{p/q}$ is irrational.*

Proof. If $e^{p/q}$ is rational, so is $[e^{p/q}]^q = e^p$. This is impossible based on Theorem 6.1. If $p < 0$, we apply the theorem for $1/e^p$.   □

Corollary 6.10. e^2 *and* $\sqrt{e}$ *are irrational numbers.*

Theorem 6.2. *Suppose p/q is a positive rational number not equal to 1. Then $\ln(p/q)$ is an irrational number.*

Proof. Suppose $\ln(p/q) = a/b$, $a, b \in \mathbb{Z}$, $b \neq 0$. We take the exponential of both sides and it results in $p/q = e^{a/b}$. In the left-hand side we have a rational number and in the right-hand side based on Corollary 6.9 we have an irrational number. This is a contradiction and we conclude that $\ln(p/q)$ is an irrational number.   □

Corollary 6.11. $\ln 2$ *is an irrational number.*

Theorem 6.3. π^2 *is an irrational number.*

Proof. Suppose π^2 is rational. Then there exist p, q natural numbers with $\pi^2 = p/q$, $q \neq 0$. Define the polynomial

$$Q(x) = q^n(\pi^{2n}P(x) - \pi^{2n-2}P_n^{(2)}(x) + \pi^{2n-4}P_n^{(4)}(x) - \cdots + (-1)^n P_n^{(2n)}(x)).$$

From the properties of the Niven polynomials we get that $Q(0)$ and $Q(1)$ are integers. But

$$\frac{d}{dx}\left(\frac{dQ(x)}{dx}\sin \pi x - Q(x)\pi \cos \pi x\right)$$
$$= (Q''(x) + \pi^2 Q(x))\sin \pi x = \pi^2 p^n P_n(x) \sim \pi x.$$

Then

$$\pi p^n \int_0^1 P(x) \sin \pi x \, dx = Q(1) + Q(0).$$

In this equality the right-hand side is an integer. The left-hand side is strictly positive and is bounded as appears below.

$$0 < \pi p^n \int_0^1 P(x) \sin \pi x \, dx < \frac{\pi p^n}{n!} < 1.$$

Thus we get a contradiction. Hence π^2 is an irrational number.   □

Corollary 6.12. π *is an irrational number.*

Some famous constants are not known to be irrational, such as the Euler–Mascheroni constant γ, π^e, 2^e, ...
We saw that the real number set is uncountable, Corollary 2.14 at page 24. At the same time we saw that the algebraic number set is countable, Theorem 2.19 at page 25. So, there has to be at least one nonalgebraic real number. A good candidate is number e.

Theorem 6.4. (Hermite) e *is transcendental.*

Proof. Suppose e is algebraic, so that

$$a_0 + a_1 e + a_2 e^2 + \cdots + a_n e^n = 0 \tag{6.62}$$

for some integers $n > 0$, $a_0 \neq 0$, $a_1, a_2, \ldots, a_n$. We compare estimates for

$$J = a_0 I(0) + a_1 I(1) + a_2 I(2) + \cdots + a_n I(n),$$

where $I(t)$ is defined by (6.16) with

$$f(x) = x^{p-1}(x-1)^p \cdots (x-n)^p,$$

p denoting a large prime. From (6.17) and (6.62) we have

$$J = -\sum_{j=0}^m \sum_{k=0}^m a_k f^{(j)}(k),$$

where $m = (n+1)p - 1$. Now, $f^{(j)}(k) = 0$ if $j < p$, $k > 0$ and if $j < p-1$, $k = 0$, and thus for all j, k other than $j = p-1$, $k = 0$, $f^{(j)}(k)$ is an integer divisible by $p!$. Furthermore,

$$f^{(p-1)}(0) = (p-1)!(-1)^{np}(n!)^p.$$

Then, if $p > n$, $f^{(p-1)}(0)$ is an integer divisible by $(p-1)!$ but not by $p!$. Thus, if also, $p > |a_0|$, then J is a nonzero integer divisible by $(p-1)!$. Thus $|J| \geq (p-1)!$.

The estimate $\tilde{f}(k) \leq (2n)^m$ together with (6.18) gives

$$|J| \leq |a_1|e\tilde{f}(1) + \cdots + |a_n|ne^n\tilde{f}(n) \leq c^p$$

for some c independent of p. The estimates are inconsistent if p is sufficiently large and the contradiction proves the theorem. $\square$

Theorem 6.5. (Lindemann[7]) π is transcendental.

A nice and deep result regarding the transcendental numbers is the following.

Theorem 6.6. (Gelfond[8]) *If*

(i) a *is algebraic, nonzero, and not equal to 1;*
(ii) b *is algebraic and irrational,*

then a^b is transcendental.

6.7 The Grönwall inequality

A basic tool in many results connected to differential equations and inclusions is the following one called the Grönwall[9] inequality.

Lemma 6.7. (Grönwall) *Suppose a continuous function $x : [a, b] \to \mathbb{R}$ satisfies*

$$0 \leq x(t) \leq c + \int_a^t h(s)x(s)ds, \quad t \in [a, b] \tag{6.63}$$

for some nonnegative constant c and some nonnegative integrable function $h : [a, b] \to \mathbb{R}$. Then

$$0 \leq x(t) \leq c + c\int_a^t h(r)\exp\left(\int_r^t h(s)ds\right) dr, \quad t \in [a, b]. \tag{6.64}$$

Proof. Case 1. Suppose $c > 0$.
  Denote

$$w(t) = c + \int_a^t h(s)x(s)ds.$$

Then $w'(t) = h(t)x(t)$, $w(t) > 0$, and $w(a) = c$. From (6.63) it follows that $x(t) \leq w(t)$, for every $t \in [a, b]$. Thus the following sequence of implications holds

---

[7] Carl Louis Ferdinand von Lindemann, 1852–1939.
[8] Alexander Osipovich Gelfond, (Александр Осипович Гельфонд), 1906–1968.
[9] Thomas Hakon Grönwall, 1877–1932.

$$w'(t) = h(t)x(t) \le h(t)w(t) \implies \frac{w'(t)}{w(t)} \le h(t) \implies$$

$$\int_a^t \frac{w'(s)}{w(s)} ds \le \int_a^t h(s)ds \implies \ln w(t) \le \int_a^t h(s)ds + \ln c \implies$$

$$w(t) \le c \cdot \exp\left(\int_a^t h(s)ds\right) \implies x(t) \le c \cdot \exp\left(\int_a^t h(s)ds\right).$$

Substituting in (6.63), we get (6.64).

Case 2. Suppose that (6.63) holds for $c = 0$. Based on case 1, the lemma is true for every $c > 0$. Then passing to the limit $c \downarrow 0$, the lemma is true in this case as well. $\square$

Another form of the Grönwall lemma is the following.

Lemma 6.8. *Let x, f, h, and g be nonnegative continuous real-valued functions defined on the interval $[a, b]$. Suppose that h' exists and is a nonnegative continuous function. Suppose, moreover, that following inequality holds*

$$x(t) \le c + \int_a^t f(s)x(s)ds + \int_a^t f(s)h(s)\left(\int_a^s g(u)x(u)du\right)ds,$$

for $a \le u \le s \le t \le b$ and some nonnegative constant c. Then for every $t \in [a, b]$ it holds

$$x(t) \le c\left(1 + \int_a^t f(s)\exp\left(\int_a^s (f(u) + g(u)h(u) + h'(u)\int_a^u g(v)dv)du\right)ds\right).$$

6.8 Exercises

6.1. Find

$$\lim_{n \to \infty} \sum_{k=1}^n \frac{n}{n^2 + k^2}.$$

6.2. For $ab \ne 0$, find

$$I(a, b) = \int_0^{\pi/2} \frac{dx}{a^2 \cos^2 x + b^2 \sin^2 x}.$$

6.3. Let f be a continuous function on $[0, 1]$ and a be a positive number satisfying

$$\int_0^1 f(x)\, dx = a, \quad 0 < f(x) \le a^{2/3}, \quad \forall x \in [0, 1].$$

Show that $\int_0^1 \sqrt{f(x)}\, dx \ge a^{2/3}$.

6.4. Let $f_1, f_2, \ldots, f_n$ be positive continuous functions on $[0, 1]$ and denote

$$\int_0^1 f_k(x)\, dx = a_k, \quad k = 1, 2, \ldots, n.$$

Show there exists $\xi \in [0, 1]$ such that

$$f_1(\xi) f_2(\xi) \cdots f_n(\xi) \leq a_1 a_2 \cdots a_n.$$

6.5. Find

$$\lim_{n \to \infty} \int_0^\pi \cos(x^n)\, dx.$$

6.6. Let f be a continuous function on $[0, 1]$ and g be a continuous and periodic function of period T. Show that

$$\lim_{n \to \infty} \int_0^1 f(x) g(nx)\, dx = \frac{1}{T} \int_0^T g(t)\, dt \int_0^1 f(x)\, dx.$$

6.7. Let $f : [a, b] \to \mathbb{R}$ be a differentiable function satisfying $|f'(x)| \leq 1$ on $[a, b]$. Consider a partition $a = x_0 < x_1 < \cdots < x_{n-1} < x_n = b$ with

$$\max_{k=1,2,\ldots,n} (x_k - x_{k-1}) \leq 1/(b - a),$$

and $\xi_k \in [x_{k-1}, x_k[$. Show

$$\left| \int_a^b f(x)\, dx - \sum_{k-1}^n f(\xi_k)(x_k - x_{k-1}) \right| \leq \frac{1}{2},$$

and the estimate is sharp.

6.8. Let $f : [a, b] \to [0, 1]$ be a Darboux integrable function. Show that for every positive ε there exists a step function $g : [a, b] \to \{0, 1\}$ such that for every interval $[\alpha, \beta] \subset [a, b]$ it holds

$$\left| \int_\alpha^\beta (f(x) - g(x)) dx \right| < \varepsilon.$$

6.9. Let $f : \mathbb{R} \to \mathbb{R}$ be a continuously differentiable function up to the order p. Denote

$$M_0 = \sup_{x \in \mathbb{R}} |f(x)|, \quad M_k = \sup_{x \in \mathbb{R}} |f^{(k)}(x)|, \quad k = 1, 2, \ldots, p.$$

Show that

$$M_k \leq 2^{k(p-k)/2} M_0^{1-k/p} M_p^{k/p}, \quad k = 0, 1, \ldots, p.$$

6.10. Find all indefinite differentiable functions $f : \mathbb{R} \to \mathbb{R}$ satisfying the relation $f(x + y) = f(x) + f(y) + 2xy$, $x, y \in \mathbb{R}$, $x \neq y$.

6.11. Prove that for any real numbers $a_1, a_2, \ldots, a_n$ the following inequality holds.

$$\left(\sum_{k=1}^{n} a_k\right)^2 \leq \sum_{k,l=1}^{n} \frac{kl}{k+l-1} a_k a_l.$$

6.12. Let f be a continuous periodic function of period 2 such that f is increasing on $[0,1]$, decreasing on $[1,2]$, and $f(x) = f(2-x)$, for all $x \in [0,1]$. Show that the minimum of

$$\int_0^2 f(x)f(x+\alpha)\,dx$$

is attained at $\alpha = 1$.

6.13. Suppose $u \in C^1]0, \infty[$ so that u is positive and $u' > 0$. Show that if

$$\int_1^\infty \frac{dx}{u(x) + u'(x)} < \infty, \qquad (6.65)$$

then

$$\int_1^\infty \frac{dx}{u(x)} < \infty.$$

6.14. Consider the series

$$\sum_{n=1}^{\infty} 2x\left(n^2 e^{-n^2 x^2} - (n-1)^2 e^{-(n-1)^2 x^2}\right), \quad x \in [0,1].$$

May we integrate it term by term?

6.15. Expand the function $f(x) = 3x/(x^2 + 5x + 6)$, $x \in \mathbb{R} \setminus \{-2, -3\}$, in power series about 0.

6.16. Expand the function $f(x) = (\arctan x)^2$, $x \in \mathbb{R}$, in power series about 0.

6.17. Find $\int_0^\infty x^n e^{-x} dx$, $n \in \mathbb{N}^*$.

6.18. Find $\int_0^\infty x \ln x\, dx/(1-x^2)^2$.

6.19. Evaluate $\int_0^1 (\ln(1/(1-x)))dx$.

6.20. Evaluate $\int_0^{\pi/2} \sin^{a-1} u \cos^{b-1} u\, du$, $a, b > 0$.

6.21. Consider the function $f(x,y) = (x/y^2)e^{-x^2/y^2}$, for $(x,y) \in [0,1] \times]0, \infty[$. Show that

$$\lim_{y \to 0} \int_0^1 f(x,y)dx \neq \int_0^1 \lim_{y \to 0} f(x,y)dx.$$

6.9 References and comments

Many parts of the present chapter follow [115] and [87]. We indicate some sources more precisely.

The proof of Theorem 1.10 may be found in [115, p. 110], and the proof of Theorem 1.10 may be found in [115, p. 111].

For the proof of the integration formula by parts we suggest [115, p. 122].

Theorem 3.3 of Barbălat is a powerful tool to deduce asymptotic stability of nonlinear systems, specially time-varying systems. This theorem has been generalized to finite-dimensional spaces and then applied to study stability results to robust model predictive control, that is, systems subject to bounded disturbances in [51].

An elementary proof of Theorem 4.5 is given in [6].

Theorem 6.5 is from [81], and Theorem 6.4 of Hermite is from [65].

The Gelfond theorem 6.6 supplies a partial answer to the seventh Hilbert's exercise.[10] However, the general case where β is irrational has not been solved. The Gelfond theorem establishes the transcendentality of $2^{\sqrt{2}}$, let's say.

It is not known whether, for example, $\pi + e$ or the Euler-Mascheroni constant γ is transcendental.

A comprehensive work dedicated to inequalities involving differentials and integrals of functions is [89]. Among others, it contains a survey on Grönwall type inequalities.

Exercises 6.3–6.9, 6.12, and 6.13 are taken from [117], and exercises 6.10 and 6.11 are from [96].

Exercise 6.14 is from [52].

---

[10] Hilbert introduced a list of 23 exercises at the Second International Congress in Paris on August 8, 1900. The seventh reads as: let $\alpha \neq 0$, $\alpha \neq 1$ be algebraic and β irrational. Is α^β then transcendental?

7

Differential Calculus on $\mathbb{R}^n$

This chapter is devoted to presenting some basic results on differential calculus of functions defined and/or having values mainly in finite-dimensional Euclidean spaces.

7.1 Linear and bounded mappings

In this section we give a very short introduction to the theory of linear and bounded mappings. These results are necessary for a comprehensive and brief presentation of the main results regarding the theory of differential calculus to a vector function of several variables.

We recall the definition of linear mapping from page 17. Let X and Y be two vector spaces over the same field $\mathbb{K}$ and let $f : X \rightarrow Y$ be a mapping. f is said to be additive provided

$$f(x + y) = f(x) + f(y), \quad \forall x, y \in X.$$

f is said to be homogeneous provided

$$f(\lambda x) = \lambda f(x), \quad \forall \lambda \in \mathbb{K}, \ x \in X.$$

f is said to be linear if it is additive and homogeneous; that is, f satisfies

$$f(\alpha x + \beta y) = \alpha f(x) + \beta f(y), \quad \forall \alpha, \beta \in \mathbb{K}, \ x, y \in X.$$

Remark. Every additive mapping satisfies the relation $f(0) = 0$; that is, it transforms the null element in X into the null element in Y. $\quad \triangle$

The surprising fact is that for the case of normed spaces (we are not interested in a more general case) in order to establish the continuity of a linear mapping on the whole space it is enough to prove that it is continuous at the origin. More precisely,

Theorem 1.1. *Let X and Y be two normed spaces over the same field $\mathbb{K}$. For a linear mapping $f : X \to Y$ the following statements are equivalent.*

(a) *f is continuous on X.*
(b) *f is continuous at $0 \in X$.*
(c) *$\|f(x)\|$ is bounded on the closed unit ball $B[0,1]$.*

Proof. The case $(a) \implies (b)$ is obvious.

The case $(b) \implies (c)$. Consider the open unit ball $B_Y(0,1)$ in Y. Its counterimage $f^{-1}(B_Y(0,1))$ is an open neighborhood of the origin in X (Theorem 2.3, page 153). Then there exists a positive r such that

$$\|x\| \le r \implies \|f(x)\| \le 1.$$

We show that $\|x\| \le 1$ implies $\|f(x)\| \le 1/r$. Set $y = rx$, $\|x\| \le 1$. Then $\|f(y)\| = r\|f(x)\|$ and because $\|f(y)\| \le 1$, it follows that $\|f(x)\| \le 1/r$. Thus $\|f(x)\|$ is bounded on the closed unit ball in X.

The case $(c) \implies (a)$. The assumption says that there exists an $M > 0$ such that $\|f(x)\| \le M$ for every $\|x\| \le 1$. Then for every $x \in X$ we have that

$$\|f(x)\| \le M\|x\|.$$

Take an $a \in X$ and an $\varepsilon > 0$. Choose any $x \in X$ with $\|x - a\| \le \varepsilon/M$. Then

$$\|f(x) - f(a)\| = \|f(x - a)\| \le M(\varepsilon/M).$$

Thus function f is continuous at $a \in X$. Because a has been chosen arbitrarily, it follows that function f is continuous on the whole space X. $\quad \square$

From the previous result we have the following corollary.

Corollary 1.1. *Let X and Y be normed spaces over the same field $\mathbb{K}$. A linear mapping $f : X \to Y$ is continuous if and only if there exists a nonnegative real M satisfying*

$$\|f(x)\| \le M\|x\|, \quad \forall x \in X.$$

Remarks. (a) Hereafter the notions of a linear and continuous operator and a linear and bounded operator are synonymous.

(b) Maybe the simplest example of a linear and bounded operator from a normed space X into a normed space Y is the *null operator;* that is, it satisfies

$$f(x) = 0, \quad \forall x \in X. \quad \triangle$$

For the finite-dimensional case one has a deeper result.

Theorem 1.2. *Let Y be a normed space over the field $\mathbb{K}$ and $f : \mathbb{K}^n \to Y$ be a linear mapping. Then f is continuous.*

Proof. Consider $\{e_1, \ldots, e_n\}$ the canonical basis in $\mathbb{K}^n$. Every $x = (x_1, x_2, \ldots, x_n)$ in $\mathbb{K}^n$ can be written as $x = \sum_{k=1}^n x_k e_k$. Because f is linear and $|x_k| \leq \|x\|$, we further write

$$\|f(x)\| = \|f\left(\sum x_k e_k\right)\| = \|\sum x_k f(e_k)\|$$
$$\leq \sum |x_k| \cdot \|f(e_k)\| \leq \|x\| \sum \|f(e_k)\|.$$

Taking $M = \sum \|f(e_i)\|$, we get that for every $x \in \mathbb{K}^n$, $\|f(x)\| \leq M\|x\|$, and M does not depend upon x. Thus by Theorem 1.1, the proof is complete. $\square$

The set of all linear and bounded mappings from a normed space X into a normed space Y is denoted by $L(X, Y)$; the two normed spaces are defined over the same field $\mathbb{K}$. This set is structured as a vector space defining the sum of two linear and bounded mappings in $L(X, Y)$ and the multiplication of a linear and bounded mapping in $L(X, Y)$ by a scalar in $\mathbb{K}$ accordingly to the following rules.

$$f, g \in L(X, Y), \ (f + g)(x) = f(x) + g(x), \quad \forall x \in X,$$
$$\alpha \in \mathbb{K}, \ f \in L(X, Y), \ (\alpha f)(x) = \alpha f(x), \quad \forall x \in X.$$

To each $f \in L(X, Y)$ we denote its *norm* by

$$\|f\| = \sup_{\|x\| \leq 1} \|f(x)\|. \tag{7.1}$$

Note that the right-hand side is a finite nonnegative number, by (c) of Theorem 1.1.

Proposition 1.1. *Consider X and Y two normed spaces over the same field $\mathbb{K}$. The function defined by (7.1) is a norm on $L(X, Y)$.*

Proof. Take 0 the null linear and bounded mapping. Obviously, $\|0\| = \sup_{\|x\| \leq 1} \|0(x)\| = 0$. Conversely, suppose that $\|f\| = 0$, for an $f \in L(X, Y)$. Then $f(x) = 0$ for every x in an open ball B of the origin. But every element outside this ball can be compressed by a positive scalar until it belongs to B. Taking into account the linearity of f, we conclude that f vanishes on all elements outside of B as well.

Suppose an arbitrary scalar α has been selected. Then

$$\|\alpha f\| = \sup_{\|x\| \leq 1} \|\alpha f(x)\| = |\alpha| \sup_{\|x\| \leq 1} \|f(x)\|.$$

We check the triangle inequality. Choose two arbitrary elements $f, g \in L(X, Y)$. Then

$$\|f + g\| = \sup_{\|x\| \leq 1} \|(f + g)(x)\| = \sup_{\|x\| \leq 1} \|f(x) + g(x)\|$$
$$\leq \sup_{\|x\| \leq 1} (\|f(x)\| + \|g(x)\|) \leq \sup_{\|x\| \leq 1} \|f(x)\| + \sup_{\|x\| \leq 1} \|g(x)\| = \|f\| + \|g\|. \quad \square$$

Theorem 1.3. *Consider X and Y two normed spaces over the same field $\mathbb{K}$ and $f : X \to Y$ a linear and continuous mapping. Then $\|f\| = \inf M$, where M is from Corollary 1.1. Also, $\|f\| = \sup_{\|x\|=1} \|f(x)\|$.*

Theorem 1.4. *Let X, Y, and Z be normed spaces over the same field $\mathbb{K}$ and two linear and continuous operators $A : X \to Y$ and $B : Y \to Z$. Then $\|BA\| \le \|A\| \cdot \|B\|$.*

Proposition 1.2. *Consider X and Y two finite-dimensional normed spaces over the same field and a bijective mapping $A \in L(X,Y)$. Then its inverse mapping $A^{-1} \in L(Y,X)$ and $\|A^{-1}\| \ge 1/\|A\|$.*

Theorem 1.5. *Let Ω be the set of linear and invertible mappings in $L(\mathbb{R}^n, \mathbb{R}^n)$.*

(a) *If $A \in \Omega$, $B \in L(\mathbb{R}^n, \mathbb{R}^n)$, and $\|B - A\| \cdot \|A^{-1}\| < 1$, then $B \in \Omega$.*
(b) *Ω is an open subset of $L(\mathbb{R}^n, \mathbb{R}^n)$ and the mapping $A \to A^{-1}$ is continuous on Ω.*

A *functional* is a mapping having scalar values. Below we introduce a particular case of a representation theorem to linear and bounded functionals defined on Hilbert spaces.

Theorem 1.6. (Riesz[1]) *For every linear mapping $f : \mathbb{K}^n \to \mathbb{K}$ one can find a constant $\chi \in \mathbb{K}^n$ fulfilling*

(a) *$f(x) = \langle x, \chi \rangle$, for every $x \in \mathbb{K}^n$.*
(b) *$\|f\| = \|\chi\|$.*
(c) *χ is the unique element in $\mathbb{K}^n$ satisfying (a).*

Proof. Consider $\{e_1, \ldots, e_n\}$ the canonical basis in $\mathbb{K}^n$. Every $x = (x_1, \ldots, x_n)$ in $\mathbb{K}^n$ is written as $x = \sum_{i=1}^{n} x_i e_i$. Because f is linear,

$$f(x) = f \left(\sum x_i e_i \right) = \sum x_i f(e_i).$$

We take $\chi = (\overline{f(e_1)}, \ldots, \overline{f(e_n)})$, where the bar denotes the complex conjugation. Then (a) is fulfilled.

Furthermore we distinguish two cases.

If $\chi = 0$, then $\|\chi\| = 0$. Also, $f(x) = 0$, for all $x \in \mathbb{K}^n$. Hence $\|f\| = 0$. Suppose $\chi \ne 0$. Then

$$\|f\| = \sup_{\|x\| \le 1} \|f(x)\| \ge |f(\chi/\|\chi\|)| = \langle \chi/\|\chi\|, \chi \rangle = \|\chi\|,$$

and we have one inequality for (b). For the reverse inequality we use the Cauchy–Buniakovski–Schwarz inequality from page 50. Then

---

[1] Frigyes Riesz, 1880–1956.

$$\|f\| = \sup_{\|x\| \le 1} \|f(x)\| = \sup_{\|x\| \le 1} \|\langle x, \chi \rangle\| \le \sup_{\|x\| \le 1} \|x\| \cdot \|\chi\| = \|\chi\|.$$

Thus (b) is proved.

To prove the uniqueness of χ, suppose that there is a $\psi \in \mathbb{K}^n$ such that $f(x) = \langle x, \psi \rangle$, for all $x \in \mathbb{K}^n$. Then $\langle x, \chi - \psi \rangle = 0$, for all $x \in \mathbb{K}^n$. Putting $x = \chi - \psi$, we get that

$$0 = \langle \chi - \psi, \chi - \psi \rangle = \|\chi - \psi\|^2 \implies \psi = \chi.$$

Thus the proof is complete. $\square$

7.1.1 Multilinear mappings

Let X_1, X_2, ..., X_n, and Y be vector spaces over the same field. The mapping

$$f : X_1 \times X_2 \times \cdots \times X_n \to Y$$

is said to be *multilinear* (*bilinear* for $n = 2$) if for every $k \in \{1, 2, \ldots, n\}$ and for every element $a_i \in X_i$ the mappings

$$X_k \ni x_k \mapsto f(a_1, \ldots, a_{k-1}, x_k, a_{k+1}, \ldots, a_n)$$

are linear. It follows immediately that a multilinear function f vanishes whenever at least one of the variables x_i is zero. So, a multilinear function f is zero on the origin $(0, 0, \ldots, 0)$. It follows from the definition that

$$f(\lambda_1 x_1, \ldots, \lambda_n x_n) = (\lambda_1 \cdots \lambda_n) \cdot f(x_1, \ldots, x_n).$$

Example. Suppose $n = 2$. A mapping $f : X_1 \times X_2 \to Y$ is bilinear whenever for any $x, y \in X_1$, $u, v \in X_2$, and $\alpha \in \mathbb{K}$, the following relations hold

$$f(x + y, u) = f(x, u) + f(y, u), \qquad f(\alpha x, u) = \alpha f(x, u),$$
$$f(x, u + v) = f(x, u) + f(x, v), \qquad f(x, \alpha u) = \alpha f(x, u). \quad \triangle$$

Similar to Theorem 1.1 the following holds.

Theorem 1.7. *Let X_1, X_2, ..., X_n and Y be normed spaces over $\mathbb{K}$. For a multilinear mapping $f : X_1 \times X_2 \times \cdots \times X_n \to Y$ the following statements are equivalent.*

(a) *f is continuous on $X_1 \times X_2 \times \cdots \times X_n$.*
(b) *f is continuous at $(0, 0, \ldots, 0) \in X_1 \times X_2 \times \cdots \times X_n$.*
(c) *$\|f(x_1, \ldots, x_n)\|$ is bounded on the closed cube $\|x_1\| \le 1, \ldots, \|x_n\| \le 1$.*

The set of multilinear and bounded mappings from $X_1 \times \cdots \times X_n$ into Y is denoted by $L(X_1 \times \cdots \times X_n; Y)$. It is a vector subspace of the space of functions defined on $X_1 \times \cdots \times X_n$ with values in Y, thus it is itself a vector space. The space $L(X_1 \times \cdots \times X_n; Y)$ can be endowed with the norm

$$\|f\| = \sup\{\|f(x_1, \ldots, x_n)\| \mid \|x_k\| \le 1, \ k = 1, \ldots, n\}.$$

In the sequel we consider that the space $L(X_1 \times \cdots \times X_n; Y)$ is endowed with this norm. If $n = 1$, we denote this space as $L(X, Y)$, as before.

A multilinear mapping f is said to be *symmetric* if $X_1 = \cdots = X_n$ and $f(x_1, \ldots, x_n)$ remains invariant under all permutations of the variables $x_1, \ldots, x_n$.

Example. Here is an example of a bilinear and continuous mapping. Let X, Y, and Z be three normed spaces over the same field $\mathbb{K}$. Consider the mapping

$$h : L(Y, Z) \times L(X, Y) \to L(X, Z) \text{ defined as } h(g, f) = g \circ f,$$

the composition of two linear and continuous mappings $g : Y \to Z$ and $f : X \to Y$. It is trivial to check that h is bilinear. Moreover, h is continuous because

$$\|h\| = \|g \circ f\| \le \|g\| \cdot \|f\|. \quad \triangle$$

Theorem 1.8. *Let X, Y, and Z be three normed spaces over the same field $\mathbb{K}$. Then there exists a linear isometry between $L(X, Y; Z)$ and $L(X, L(Y, Z))$.*

We have defined the concept of isometric metric spaces at page 64. Hence, we may identify the spaces $L(X, Y; Z)$ and $L(X, L(Y, Z))$.

7.1.2 Quadratic mappings

A real *quadratic mapping* with respect to a matrix A with real entries and a variable $h = (h_1, \ldots, h_n) \in \mathbb{R}^n$ is a mapping $\Phi : \mathbb{R}^n \to \mathbb{R}$ defined as

$$\Phi(h) = h \, A \, h^t, \tag{7.2}$$

where h^t is the transpose of h and

$$A = \begin{bmatrix} a_{11} & a_{12} & \cdots & a_{1\,n-1} & a_{1\,n} \\ a_{21} & a_{22} & \cdots & a_{2\,n-1} & a_{2\,n} \\ \cdots & \cdots & \cdots & \cdots & \cdots \\ a_{n\,1} & a_{n\,2} & \cdots & a_{n\,n-1} & a_{n\,n} \end{bmatrix}.$$

We write

$$\Phi(h) = \sum_{i=1}^{n} \sum_{j=1}^{n} a_{i,j} h_i h_j. \tag{7.3}$$

Matrix A is said to be *symmetric* if $a_{ij} = a_{ji}$, for all $i, j \in \mathbb{N}_n^*$.

The quadratic mapping is said to be *positive definite* (*negative definite*) if for every nonzero h we have $\Phi(h) > 0$ ($\Phi(h) < 0$). Φ is said to be *definite* if it is either positive definite or negative definite. The quadratic mapping is said to be *alternating* if it assumes both positive and negative values. The quadratic mapping is said to be *semi-definite* if it assumes either nonpositive values or nonnegative values.

The following determinants are said to be the *principal minors*.

$$A_1 = a_{11}, \quad A_2 = \begin{vmatrix} a_{11} & a_{12} \\ a_{21} & a_{22} \end{vmatrix}, \quad A_3 = \begin{vmatrix} a_{11} & a_{12} & a_{13} \\ a_{21} & a_{22} & a_{23} \\ a_{31} & a_{32} & a_{33} \end{vmatrix}, \dots,$$

$$A_n = \begin{vmatrix} a_{11} & a_{12} & \cdots & a_{1n} \\ a_{21} & a_{22} & \cdots & a_{2n} \\ \cdots & \cdots & \cdots & \cdots \\ a_{n1} & a_{n2} & \cdots & a_{nn} \end{vmatrix}.$$

Theorem 1.9. (Sylvester[2]) *Consider a symmetric matrix A and the quadratic form* (7.2). *The quadratic form is positive definite if and only if*

$$A_1 > 0, \quad A_2 > 0, \dots, \quad A_n > 0.$$

The quadratic form is negative definite if and only if

$$A_1 < 0, \quad A_2 > 0, \quad A_3 < 0, \dots, (-1)^n A_n > 0.$$

One can offer more information about the behavior of a positive definite quadratic mapping.

Theorem 1.10. *The quadratic form* (7.3) *is positive definite if and only if there exists a positive λ such that*

$$\Phi(x) \geq \lambda \|x\|^2, \quad \forall x \in \mathbb{R}^n. \tag{7.4}$$

Proof. If (7.4) holds, the quadratic mapping is positive definite.

Suppose the quadratic mapping is positive definite. Consider the unit sphere S; that is, $S = \{x \mid x \in \mathbb{R}^n, \|x\| = 1\}$. Then S is bounded and closed (the mapping $\|\cdot\| \to \mathbb{R}$ is continuous and S is the counterimage of the number 1 by this mapping). Thus S is a compact subset of $\mathbb{R}^n$. Φ being continuous, it attains its minimum value on the compact set S. This minimum value is positive because Φ is positive definite and S does not contain the origin of the space. Denote by λ the minimum value of Φ on S. Then

$$\Phi(y) \geq \lambda, \quad \forall y \in S. \tag{7.5}$$

Consider now a nonzero $x \in \mathbb{R}^n$. Then $y = x/\|x\| \in S$. Substituting y in (7.5), we get (7.4). $\square$

---

[2] James Joseph Sylvester, 1814–1897.

Theorem 1.11. *The quadratic form* (7.3) *is negative definite if and only if there exists a positive* λ *such that*

$$\Phi(x) \leq -\lambda \|x\|^2, \quad \forall\, x \in \mathbb{R}^n.$$

7.2 Differentiable functions

In this section we introduce the concepts of Gâteaux[3] and Fréchet[4] differentiability of a function. Below we consider only the case $\mathbb{K} = \mathbb{R}$.

7.2.1 Variations

Let Y be a normed space and let I be a nonempty and open interval of the real line. The *first derivative* $f'(t_0)$ of $f : I \to Y$, at $t_0 \in I$, is defined by

$$f'(t_0) = \lim_{t \to t_0} \frac{f(t) - f(t_0)}{t - t_0}$$

if the limit exists, where the limit is taken in the sense of the norm of Y. We note that this derivative is unique and that if f has a first derivative at a point t_0, then f is continuous at t_0. The proof is similar to the proof of Theorem 1.1 at page 191. Higher-order derivatives are defined inductively as we already did at page 208.

We now define the Gâteaux variation of a function at a point.

Let f be a mapping from an open subset A of X into Y, where X and Y are normed spaces. Let x_0 be a point in A and h an arbitrary nonzero fixed element in X. Then $x_0 + th \in A$ for all real t with $|t| \leq \varepsilon(x_0, h)$. Let

$$\tau = \sup\{\varepsilon \mid |t| \leq \varepsilon \implies x_0 + th \in A\}.$$

Then $f(x_0 + th)$ is defined for $|t| < \tau$. If

$$\frac{d}{dt} f(x_0 + th)\bigg|_{t=0}$$

exists, it is called the *Gâteaux variation* or the *weak differential* of f at x_0 with increment h and is denoted by

$$\delta f(x_0; h). \tag{7.6}$$

If f has a Gâteaux variation, called the *G-variation*, at every point $x \in A$, then f has a first variation on A.

Similarly, f has an nth variation $\delta^n f(x_0; h)$ at a point x_0 if the function $f(x_0 + th)$ has an nth order derivative with respect to t at $t = 0$.

---

[3] M. R. Gâteaux, –1914.
[4] René Maurice Fréchet, 1878–1973.

It follows from the definition that the nth variation is homogeneous in h of degree n; that is, if $\delta^n f(x_0; h)$ exists and λ is a scalar, then $\delta^n f(x_0; \lambda h) = \lambda^n \delta f(x_0; h)$.

We emphasize that the weak differential is neither necessarily linear nor continuous in h.

Remarks. (i) Consider

$$f(x, y) = \begin{cases} \dfrac{xy^2}{x^2 + y^2}, & (x, y) \neq (0, 0) \\ 0, & (x, y) = (0, 0). \end{cases}$$

For each $h = (h_1, h_2)$, the G-variation exists and is equal to $h_1 h_2^2 (h_1^2 + h_2^2)^{-1}$, but the mapping $h = (h_1, h_2) \to h_1 h_2^2 (h_1^2 + h_2^2)^{-1}$ is not linear in h.

(ii) If f has a Gâteaux variation at x_0, then f is continuous in the direction h; that is,

$$\lim_{t \to 0} \| f(x_0 + th) - f(x_0) \| = 0,$$

h is fixed, but is not necessarily continuous at x_0. $\triangle$

7.2.2 Gâteaux differential

We define the Gâteaux differential of a function at a point.

If $\delta f(x_0, h)$ in (7.6) is linear and bounded in h, then it is called the *Gâteaux differential* or the *G-differential* of f at x_0 with increment h and is denoted by $D f(x_0; h)$.

The G-differential provides in some sense a local approximation property.

Theorem 2.1. *Let A be an open set in X and let f be a mapping from A to Y. Then f is G-differentiable at $x \in A$ if and only if f can be represented as*

$$f(x + h) - f(x) = L(x, h) + r(x, h) \tag{7.7}$$

for every $h \in X$ for which $x + h \in A$, where $L(x, h)$ is linear and continuous in h and

$$\lim_{t \to 0} \| r(x, th) \| / t = 0, \quad \text{for each } h. \tag{7.8}$$

Proof. We note that if such a representation exists, it is unique. Indeed, suppose there exists another representation with L' and r'. Then

$$L(x, h) - L'(x, h) = \lim_{t \to 0} \frac{L(x, th) - L'(x, th)}{t} = \lim_{t \to 0} \frac{r'(x, th) - r(x, th)}{t} = 0.$$

If the representation (7.7) holds, then

$$\frac{d}{dt} f(x_0 + th) \Big|_{t=0} = \lim_{t \to 0} \frac{f(x + th) - f(x)}{t} = L(x, h) + \lim_{t \to 0} \frac{r(x, th)}{t} = L(x, h).$$

Thus the G-variation exists and is linear and continuous in h.

Conversely, if the G-differential exists, then

$$\frac{f(x+tk) - f(x)}{t} = D f(x,k) + \varepsilon(x,tk),$$

where $\varepsilon(x,tk) \to 0$ as $t \to 0$. Letting $tk = h$, we get (7.7), where $r(x,h) = t\varepsilon(x,h)$ and thus (7.8) holds. □

Another condition for a G-variation to be a G-differential is the next one.

Theorem 2.2. *Let the G-variation of a mapping f exists in some neighborhood of the point x_0 and let $\delta f(x,h)$ be continuous in x at x_0. Moreover, assume that $\delta f(x,h)$ is continuous in h at 0. Then $\delta f(x_0,h)$ is a G-differential.*

7.2.3 Fréchet differential

In this subsection we define the concept of a Fréchét differentiable function.

Let $(X, \|\cdot\|_X)$ and $(Y, \|\cdot\|_Y)$ be two normed vector spaces. Consider A an open and nonempty subset of X. A function $f : A \to Y$ is said to be a *Fréchet differentiable* (and we write F-differentiable) function at a point $x \in A$ if representation (7.7) holds, where $L(x,h)$ is linear and continuous in h and moreover

$$\lim_{h \to 0} \|r(x,h)\|/\|h\| = 0. \tag{7.9}$$

Remark. Obviously, (7.9) implies (7.8). Hence each F-differentiable function is G-differentiable. From here it follows that representation (7.7) holds also for F-differentiable functions.   △

We write $L(x_0,h) = d f(x_0,h) = f'(x_0)h$ and call it the *Fréchét differential* or the *F-differential* of f at x_0 with increment h. The mapping $d f(x_0,\cdot) = f'(x_0)(\cdot)$, which is a bounded linear operator, is said to be the *Fréchet derivative* or the *F-derivative* of f at x_0. We emphasize that $d f(x_0,h)$ is an element in Y whereas $f'(x_0)$ is an element in $L(X,Y)$.

The implication relationship between F-differentiability and G-differentiability follows from the next theorem.

Theorem 2.3. *A function f is F-differentiable at x_0 if and only if representation (7.7) holds, where $L(x,h)$ is linear and continuous in h and*

$$\lim_{t \to 0} \|r(x,th)\|/t = 0, \tag{7.10}$$

uniformly with respect to h on each set $\|h\| = constant$.

Proof. Without any loss of generality, we may prove it for the set $\|h\| = 1$.

If f is Fréchét differentiable at x_0, then

$$\lim_{\|h\| \to 0} \|r(x, h)\|/\|h\| = 0.$$

Letting $h = tk$, where $\|k\| = 1$, we get (7.10) uniformly on $\|k\| = 1$.

Conversely, if (7.10) holds uniformly on each bounded set, then, by Theorem 2.1, f has a G-differential $D f(x_0, h)$ at x_0. Thus for any $\varepsilon > 0$, there exists $\eta > 0$ such that

$$\|(f(x_0 + th) - f(x_0))/t - D f(x_0, h)\| < \varepsilon,$$

whenever $|t| < \eta$. That is,

$$f(x_0 + th) - f(x_0) = D f(x_0, th) + r(x_0, th),$$

where $|t| < \eta$, and

$$\|r(x_0, th)\|/\|th\| < \varepsilon, \quad \forall h \text{ with } \|h\| = 1.$$

Letting $k = th$, we get $f(x_0 + k) - f(x_0) = D f(x_0, k) + r(x_0, k)$, where

$$\lim_{k \to 0} \|r(x_0, k)\|/\|k\| = 0.$$

Hence, $D f(x_0, h) = d f(x_0, h)$. □

Remarks. (i) Thus if f is F-differentiable at x_0, then f is G-differentiable at x_0. Furthermore,

$$d f(x_0, h) = D f(x_0, h) = \delta f(x_0; h).$$

The converse holds if f is a function of one variable, but does not necessarily hold in higher dimensions. To show the last assertion we introduce the function $f : \mathbb{R}^2 \to \mathbb{R}$ by

$$f(x, y) = \begin{cases} \dfrac{x(x^2 + y^2)}{y}, & y \neq 0 \\ 0, & y = 0. \end{cases}$$

Then f has a G-variation at $(0,0)$, which is (trivially) continuous and linear in h. In this case

$$r(0, h) = \begin{cases} \dfrac{h_1(h_1^2 + h_2^2)}{h_2}, & h_2 \neq 0 \\ 0, & h_2 = 0, \end{cases}$$

and hence (7.10) holds. However, f is not F-differentiable at $(0,0)$. For if we let $h_n - (n^{-1/2}, n^{-1})$, then $h_n \to 0$ and

$$\|r(0, h_n)\|/\|h_n\| = \sqrt{1 + 1/n} \xrightarrow{n \to \infty} 1.$$

(ii) The uniqueness of the F-differential of a function follows from the uniqueness of the G-differential of it, Theorem 2.1.

(iii) If f is continuous at x_0, then the requirement of continuity of $d\,f(x_0, h)$ in h is redundant. It follows from the inequality

$$\|d\,f(x_0, h)\| \le \|f(x_0 + h) - f(x_0) - d\,f(x_0, h)\| + \|f(x_0 + h) - f(x_0)\|,$$

which shows that $d\,f(x_0, h)$ is continuous at $h = 0$ and hence continuous everywhere.

(iv) Often one denotes $h = d\,x$ and then the common notation is used

$$d\,f(x, d\,x) = f'(x)d\,x. \quad \triangle$$

Suppose $A \subset X$ is open and nonempty and $f : A \to Y$ is Fréchet differentiable on each point in A. Then function f is said to be *Fréchet differentiable on A*. Suppose $f : A \to Y$ is Fréchet differentiable on A. Then function f is said to be *continuously differentiable on A* if f' maps continuously A in $L(\mathbb{R}^n, \mathbb{R}^m)$; that is, for every $x \in A$ and $\varepsilon > 0$ one can find a positive δ such that for all $y \in A$ with $\|x - y\| < \delta$ it holds $\|f'(x) - f'(y)\| < \varepsilon$. In such a case we write $f \in C^1(A)$.

Examples. Consider X and Y two normed spaces.
(i) Let A be a nonempty and open subset of X. Choose an arbitrary, but fixed, element in Y; let it be x_0. Define the constant function $f : A \to Y$ by $f(x) = x_0$, for all $x \in A$. Then

$$f(x + h) - f(x) = 0 = L(x, h) + r(x, h)$$

Taking $L(x, \cdot) = 0 = r(x, h)$, all assumptions are satisfied. Hence the F-derivative of a constant function is the null linear and bounded operator.
(ii) Suppose that a linear and bounded operator $f : X \to Y$ is given. Then $f(x + h) = f(x) + f(h)$ and

$$f(x + h) - f(x) = f(h) + 0.$$

So we may consider $f'(x) = f$, $x \in X$, and $r(x, h) = 0$, for all $x, h \in X$. $\quad \triangle$

7.2.4 Properties of the Fréchet differentiable functions

Theorem 2.4. *Consider X, Y normed spaces, A a nonempty and open subset of X, and a Fréchet differentiable function $f : A \to Y$ at $x \in A$. Then f is continuous at x.*

Proof. We write

$$f(x + h) - f(x) = L(x, h) + r(x, h),$$

for $\|h\|$ sufficiently small, where $L(x,h)$ is linear and bounded in h and

$$\lim_{h \to 0} \frac{\|r(x,h)\|}{\|h\|} = 0.$$

Because $L(x,h)$ is linear and bounded in h, by Corollary 1.1, there exists an $M \geq 0$ so that

$$\|L(x,h)\| \leq M \cdot \|h\|, \quad \forall h \in X.$$

Because $\lim_{h \to 0} \|r(x,h)\|/\|h\| = 0$, for every positive ε there exists $\delta > 0$ such that $\|h\| < \delta$ implies $\|r(x,h)\| \leq \varepsilon \|h\|$. Then for $\|h\| < \delta$,

$$\|f(x+h) - f(x)\| \leq (M + \varepsilon)\|h\|.$$

Thus the theorem is proved. $\square$

Theorem 2.5. *Consider X, Y normed spaces, A a nonempty and open subset of X, and $f : A \to Y$ and $g : A \to Y$ Fréchet differentiable functions at $x \in A$. Then $f + g$ is Fréchet differentiable at x. Let λ be a real. Then λf is Fréchet differentiable at x.*

Proof. By the hypotheses we have

$$\lim_{h \to 0} \frac{\|f(x+h) - f(x) - f'(x)h\|}{\|h\|} = 0,$$

$$\lim_{h \to 0} \frac{\|g(x+h) - g(x) - g'(x)h\|}{\|h\|} = 0.$$

Then

$$0 \leq \lim_{h \to 0} \frac{\|(f(x+h) - g(x+h)) - (f(x) - g(x)) - (f'(x) - g'(x))(h)\|}{\|h\|}$$

$$\leq \lim_{h \to 0} \frac{\|f(x+h) - f(x) - f'(x)h\|}{\|h\|} + \lim_{h \to 0} \frac{\|g(x+h) - g(x) - g'(x)h\|}{\|h\|} = 0.$$

The second statement can be proved similarly. $\square$

Remark. From the proof of the previous theorem it follows that

$$(f+g)'(x) = f'(x) + g'(x) \text{ and } (\lambda f)'(x) = \lambda f'(x). \quad \triangle$$

Theorem 2.6. (The chain rule) *Consider X, Y, Z three normed spaces, A a nonempty and open subset of X, B an open subset of Y, a Fréchét differentiable function $f : A \to Y$ at $x \in A$ with $f(x) \in B$, and another Fréchet differentiable function $g : B \to Z$ at $f(x)$.*

Then $\varphi = g \circ f$ is a Fréchét differentiable function at x and

$$\varphi'(x) = (g \circ f)'(x) = g'(f(x)) \circ f'(x), \tag{7.11}$$

where $\circ$ denotes the composition of the two linear and bounded operators.

We note that the simplest case of the above result has been introduced by Theorem 1.4 at page 195.

Proof. Denote $y = f(x)$, $L = f'(x)$, $M = g'(y)$, and

$$u(h) = f(x + h) - f(x) - Lh, \quad v(k) = g(y + k) - g(y) - Mk,$$

for any h and k for which $f(x + h)$, and $g(y + k)$ are defined. Then

$$\|u(h)\| = \varepsilon(h)\|h\|, \quad \|v(k)\| = \eta(k)\|k\|,$$

such that $\varepsilon(h) \to 0$ as $\|h\| \to 0$ and $\eta(k) \to 0$ as $\|k\| \to 0$.

Fix an h such that $x + h \in A$ and take $k = f(x + h) - f(x)$. Then

$$\|k\| = \|Lh + u(h)\| \le (\|L\| + \varepsilon(h))\|h\|,$$

and

$$\varphi(x + h) - \varphi(x) - MLh = g(y + k) - g(y) - MLh = Mk + v(k) - MLh$$
$$= M(k - Lh) + v(k) = Mu(k) + v(k).$$

Then for any nonzero h such that $x + h \in A$ it follows that

$$\frac{\|\varphi(x + h) - \varphi(x) - MLh\|}{\|h\|} \le \|M\|\varepsilon(h) + (\|L\| + \varepsilon(h))\eta(k).$$

Suppose now that h tends to 0. Then $\varepsilon(h) \to 0$ and $k \to 0$. Therefore $\eta(k) \to 0$. Hence $\varphi'(x) = ML$. □

Now comes a generalization of the Lagrange mean value theorem, Theorem 2.3 at page 199. The generalization is twofold. From one side the space considered here is more general. From the other side the differentiability is required for a part of the space only.

Theorem 2.7. (Denjoy–Bourbaki) *Consider a Banach space X, $a, b \in \mathbb{R}$, $a < b$, a sequence (s_n), with $s_n \in [a, b]$, for all $n \in \mathbb{N}^*$, $f : [a, b] \to X$, and $g : [a, b] \to \mathbb{R}$. Assume that*

(a) *Functions f and g are continuous on $[a, b]$.*
(b) *f and g are differentiable on $[a, b] \setminus S$, where $S = \{s_n \mid n \in \mathbb{N}^*\}$.*
(c) *$\|f'(t)\| \le g'(t)$ for all $t \in [a, b] \setminus S$.*

Then

$$\|f(b) - f(a)\| \le g(b) - g(a).$$

Remark. If in Theorem 2.7 the set S is empty, assumption (a) is useless. △

Corollary 2.2. *Consider X a Banach space, a function $f : [a, b] \to X$ continuous on $[a, b]$ and differentiable on $[a, b] \setminus S$, where S is an at most countable set, and $\|f'(t)\| \le M$ for every $t \in [a, b] \setminus S$. Then*

$$\|f(b) - f(a)\| \le M\|b - a\|.$$

Proof. We take $g(t) = Mt$ and apply Theorem 2.7. $\quad\square$

Corollary 2.3. *Consider* X *a Banach space, a function* $f : [a, b] \to X \cdot$ *continuous on* $[a, b]$ *and differentiable on* $[a, b] \setminus S$, *where* S *is an at most countable set, and* $f'(t) = 0$ *for every* $t \in [a, b] \setminus S$. *Then* f *is constant on* $[a, b]$.

Proof. Take $x, y \in [a, b]$, $x < y$. Obviously,

$$\|f'(t)\| < 1/n, \quad \forall t \in [a, b] \setminus S, \quad n \in \mathbb{N}^*.$$

Invoking Corollary 2.2, we get

$$\|f(x) - f(y)\| \leq |x - y|/n, \quad \forall n \in \mathbb{N}^*.$$

Passing $n \to \infty$, we conclude that $f(x) = f(y)$. $\quad\square$

Theorem 2.8. *Consider* A *a nonempty, open, and convex subset of a Banach space* X *and a function* $f : A \to Y$ *differentiable on* A, *where* Y *is a Banach space. Then*

$$\|f(b) - f(a)\| \leq \|b - a\| \sup_{x \in [a,b]} \|f'(x)\|, \quad \forall a, b \in A,$$

where $[a, b]$ *is the interval generated by* a *and* b; *that is,* $[a, b] = \{z \mid z = pa + (1 - p)b, \ p \in [0, 1]\}$.

Proof. Consider two points a, b in A and the function $g : [0, 1] \to A$ defined by

$$g(t) = tu + (1 - t)v, \quad t \in [0, 1].$$

g is differentiable on $[0, 1]$ and $g'(t) = v - u$, for every $t \in [0, 1]$. Because A is a convex set, $g([0, 1]) \subset A$.

Define $h = f \circ g$. Then

$$h'(t) = f'(g(t)) \circ g'(t) = f'(tu + (1 - t)v)(v - u), \quad \forall t \in [0, 1].$$

Thus

$$\|h'(t)\| \leq \|f'(tu + (1 - t)v)\| \cdot \|u - v\| \leq M, \quad \forall t \in [0, 1],$$

where

$$M = \|u - v\| \sup_{t \in [0,1]} \|f'(tu + (1 - t)v)\| = \|u - v\| \sup_{x \in [a,b]} \|f'(x)\|.$$

Applying Corollary 2.2 for h, we get

$$\|f(v) - f(u)\| = \|h(1) - h(0)\| \leq M(1 - 0) = \|u - v\| \sup_{x \in [a,b]} \|f'(x)\|. \quad\square$$

Corollary 2.4. *Consider* X *and* Y *Banach spaces,* A *a nonempty, open, and convex subset of* X, *and a differentiable function* $f : A \to Y$. *Suppose there exists* $M \geq 0$ *such that* $\|f'(x)\| \leq M$, *for every* $x \in A$. *Then*

$$\|f(b) - f(a)\| \leq M\|b - a\|, \quad \forall a, b \in A.$$

7.3 Partial derivatives

Let A be a nonempty and open subset of $\mathbb{R}^n$ and $f : A \to \mathbb{R}^m$ be a mapping. Consider $\{e_1, e_2, \ldots, e_n\}$ the canonical basis of $\mathbb{R}^n$ and $\{u_1, u_2, \ldots, u_m\}$ the canonical basis of $\mathbb{R}^m$. If $f = (f_1, f_2, \ldots, f_m)$, we may write $f(x) = \sum_{i=1}^{m} f_i(x)u_i$, for every $x \in A$ or $f_i(x) = \langle f(x), u_i \rangle$, $i = 1, 2, \ldots, m$.

The *partial derivative* of f in respect to the jth variable at $x \in A$ is denoted as $\partial f(x)/\partial x_j$ and defined by

$$\frac{\partial f(x)}{\partial x_j} = \lim_{t \to 0} \frac{f(x + te_j) - f(x)}{t} = \lim_{t \to 0} \sum_{i=1}^{m} \frac{f_i(x + te_j) - f_i(x)}{t} u_i$$

$$= \left(\lim_{t \to 0} \frac{f_1(x + te_j) - f_1(x)}{t}, \ldots, \lim_{t \to 0} \frac{f_m(x + te_j) - f_m(x)}{t} \right)$$

$$= \left(\frac{\partial f_1(x)}{\partial x_j}, \ldots, \frac{\partial f_m(x)}{\partial x_j} \right)$$

provided the limits exist.

Theorem 3.1. *Let A be a nonempty and open subset of $\mathbb{R}^n$, $a \in A$, and $f : A \to \mathbb{R}^m$ be a Fréchet differentiable function at a. Then there exist all partial derivatives $\partial f_i(a)/\partial x_j$, $i = 1, 2, \ldots, m$, $j = 1, \ldots, n$ and the following relation holds between the derivative of f and its partial derivatives*

$$f'(a)e_j = \sum_{i=1}^{m} \frac{\partial f_i(a)}{\partial x_j} u_i, \quad 1 \le j \le n. \tag{7.12}$$

Proof. Fix an index j. Because f is differentiable at a we write

$$f(a + te_j) - f(a) = f'(a)(te_j) + r(te_j), \quad \text{with} \quad \frac{\|r(te_j)\|}{t} \xrightarrow{t \to 0} 0.$$

By the linearity of $f'(a)$, we further write

$$\lim_{t \to 0} \frac{f(a + te_j) - f(a)}{t} = \lim_{t \to 0} \sum_{i=1}^{m} \frac{f_i(a + te_j) - f_i(a)}{t} u_i$$

$$= \sum_{i=1}^{m} \lim_{t \to 0} \frac{f_i(a + te_j) - f_i(a)}{t} u_i = f'(a)e_j. \quad \square$$

Remarks. (i) The converse of the previous theorem is false. Consider $f : \mathbb{R}^2 \to \mathbb{R}$ defined by

$$f(x, y) = \begin{cases} \dfrac{xy}{x^2 + y^2}, & (x, y) \ne (0, 0) \\ 0, & (x, y) = (0, 0). \end{cases} \tag{7.13}$$

This function has partial derivatives at $(0,0)$ and, at the same time, it is not continuous at this point. Hence it is not differentiable at the origin.

The existence of the partial derivatives follows by

$$\frac{\partial f(0,0)}{\partial x} = \lim_{x \to 0} \frac{f(x,0) - f(0,0)}{x} = 0 = \lim_{y \to 0} \frac{f(0,y) - f(0,0)}{y} = \frac{\partial f(0,0)}{\partial y}.$$

The function is discontinuous at the origin because if we consider a sequence $(x_n, y_n)_n$ tending to the origin along the straight line $y = ax$, $a \neq 0$, then

$$\lim_{n \to \infty} \frac{x_n y_n}{x_n^2 + y_n^2} = \frac{a}{1 + a^2}.$$

(ii) We are considering the function given by Exercise 4.2 in Section 4.10, namely the function $f : \mathbb{R}^2 \to \mathbb{R}$ defined by

$$f(x,y) = \begin{cases} \dfrac{x^3 + y^3}{x^2 + y^2}, & (x,y) \neq (0,0) \\ 0, & (x,y) = (0,0). \end{cases}$$

We saw that function f is continuous on $\mathbb{R}^2$. Moreover it has partial derivatives at $(0,0)$ because

$$\frac{\partial f(0,0)}{\partial x} = \lim_{x \to 0} \frac{f(x,0) - f(0,0)}{x} = 1,$$
$$\frac{\partial f(0,0)}{\partial y} = \lim_{y \to 0} \frac{f(0,y) - f(0,0)}{y} = 1.$$

But function f is not differentiable at $(0,0)$. Suppose it is so. Then there exist scalars a, b such that

$$\frac{\left| f(h,k) - f(0,0) - (a,b) \begin{pmatrix} h \\ k \end{pmatrix} \right|}{\sqrt{h^2 + k^2}} = \frac{\left| \dfrac{h^3 + k^3}{h^2 + k^2} - (ah + bk) \right|}{\sqrt{h^2 + k^2}}$$
$$= \begin{cases} |9/5 - (2a + b)|/\sqrt{5}, & h = 2k, \\ |1 - a|, & k = 0, \\ |1 - b|, & h = 0. \end{cases}$$

The right-hand sides tend to zero, thus from the last two cases we have $a = b = 1$. Obviously, for these values the first case is far from zero. Hence function f is not differentiable at zero.

(iii) We introduce a differentiable function at zero having discontinuous partial derivatives at zero. Let

$$f(x,y) = \begin{cases} (x^2 + y^2) \sin \dfrac{1}{\sqrt{x^2 + y^2}}, & (x,y) \neq (0,0) \\ 0, & (x,y) = (0,0). \end{cases}$$

Then

$$\frac{\partial f(x,y)}{\partial x} = \begin{cases} 2x \sin \dfrac{1}{\sqrt{x^2+y^2}} - \dfrac{x}{\sqrt{x^2+y^2}} \cos \dfrac{1}{\sqrt{x^2+y^2}}, & (x,y) \neq (0,0) \\ 0, & (x,y) = (0,0). \end{cases}$$

We can check immediately that this partial derivative is discontinuous at zero. By symmetry, we conclude the same thing regarding the other partial derivative.

Function f is differentiable at zero.

(iv) However, a kind of converse of Theorem 3.1 is given by Theorem 3.2 below. $\triangle$

Corollary 3.5. *Under the assumptions of Theorem 3.1 it follows that*

$$f'(a) = \begin{bmatrix} \dfrac{\partial f_1(a)}{\partial x_1} & \dfrac{\partial f_1(a)}{\partial x_2} & \cdots & \dfrac{\partial f_1(a)}{\partial x_{n-1}} & \dfrac{\partial f_1(a)}{\partial x_n} \\ \dfrac{\partial f_2(a)}{\partial x_1} & \dfrac{\partial f_2(a)}{\partial x_2} & \cdots & \dfrac{\partial f_2(a)}{\partial x_{n-1}} & \dfrac{\partial f_2(a)}{\partial x_n} \\ \cdots & \cdots & \cdots & \cdots & \cdots \\ \dfrac{\partial f_{m-1}(a)}{\partial x_1} & \dfrac{\partial f_{m-1}(a)}{\partial x_2} & \cdots & \dfrac{\partial f_{m-1}(a)}{\partial x_{n-1}} & \dfrac{\partial f_{m-1}(a)}{\partial x_n} \\ \dfrac{\partial f_m(a)}{\partial x_1} & \dfrac{\partial f_m(a)}{\partial x_2} & \cdots & \dfrac{\partial f_m(a)}{\partial x_{n-1}} & \dfrac{\partial f_m(a)}{\partial x_n} \end{bmatrix}. \qquad (7.14)$$

Corollary 3.6. *Under the assumptions of Theorem 3.1 it follows that for all* $h = (h_1, \ldots, h_n) \in \mathbb{R}^n$,

$$f'(a)h = f'(a)(h_1, \ldots, h_n) = \left(\sum_{i=1}^n \frac{\partial f_1(a)}{\partial x_i} h_i, \ldots, \sum_{i=1}^n \frac{\partial f_m(a)}{\partial x_i} h_i \right). \qquad (7.15)$$

Theorem 3.2. *Suppose A is a nonempty and open subset of $\mathbb{R}^n$ and $f : A \to \mathbb{R}^m$. Then f is a continuously differentiable function on A if and only if it has continuous partial derivatives on A.*

Proof. Suppose $f \in C^1$. From (7.12) it follows that

$$\frac{\partial f_i(x)}{\partial x_j} = (f'(x)e_j)u_i, \quad 1 \leq j \leq n, \ 1 \leq i \leq m.$$

and

$$\frac{\partial f_i(x)}{\partial x_j} - \frac{\partial f_i(y)}{\partial x_j} = \{[f'(x) - f'(y)]e_j\}u_i.$$

Because $|u_i| = |e_j| = 1$, we conclude

$$\left| \frac{\partial f_i(x)}{\partial x_j} - \frac{\partial f_i(y)}{\partial x_j} \right| \leq |(f'(x) - f'(y))e_j| \leq \|f'(x) - f'(y)\|.$$

Thus the partial derivative $\partial f_i(x)/\partial x_j$ is continuous on A.

The proof of the converse implication is omitted.   □

From Theorem 2.6 and Corollary 3.5 we get the formula of partial derivatives to compound functions.

Theorem 3.3. *Suppose all assumptions of Theorem 2.6 are satisfied. Then*

$$\frac{\partial g_i(x)}{\partial x_j} = \sum_{l=1}^{m} \frac{\partial g_i(y)}{\partial y_l} \cdot \frac{\partial f_l(x)}{\partial x_j}, \quad \forall i = 1, \ldots, k, \quad j = 1, \ldots, n. \tag{7.16}$$

Proof. It follows from (7.14) and (7.11).   □

7.3.1 The inverse function theorem and the implicit function theorem

Consider the following example. Given the function $f : [1, 2] \to [1, \infty]$ by $f(x) = x^2$. It is obvious that this function has a positive derivative, thus it is strictly increasing. Hence it is invertible. We immediately conclude that if a real-valued function of a real variable has a derivative of constant sign (either positive or negative) on an interval, then the function is invertible on that interval. See also Theorem 1.5 at page 196.

Fortunately, a similar result holds in the higher-dimension case. It is called the *inverse function theorem*.

Theorem 3.4. *Suppose f is a C^1 mapping on a nonempty open set $M \subset \mathbb{R}^n$ into $\mathbb{R}^n$, $f'(a)$ is invertible for some $a \in M$, and $b = f(a)$. Then*

(a) *There exist two open sets U and V in $\mathbb{R}^n$ such that $a \in U$, $b \in V$, f is one-to-one on U, and $f(U) = V$.*

(b) *If g is the inverse of f (its existence follows from (a)), defined on V by $g(f(x)) = x$, $x \in U$, then g is a C^1 mapping on V.*

For a proof see [115, Theorem 9.4] or [61, p. 59]. Both proofs are based on the Banach fixed point theorem 5.1 at page 167.

Proof. Denote $A = f'(a)$ and consider a positive λ satisfying

$$2\lambda\|A^{-1}\| = 1. \tag{7.17}$$

Because f' is continuous on a there exists an open ball centered at a, say U, so that

$$\|f'(x) - A\| < \lambda, \quad \forall x \in U. \tag{7.18}$$

To each $y \in \mathbb{R}^n$ we assign the following function

$$\varphi(x) = x + A^{-1}(y - f(x)), \quad \forall x \in M. \tag{7.19}$$

Note that $f(x) = y$ if and only if $\varphi(x) = x$; that is, x is a fixed point of φ. Inasmuch as

$$\varphi'(x) = I - A^{-1}f'(x) = A^{-1}(A - f'(x)), \quad \forall\, x \in M,$$

by Theorem 1.4, (7.17), and (7.18), we get that

$$\|\varphi'(x)\| < 1/2, \quad \forall\, x \in U. \tag{7.20}$$

Recall Corollary 2.4 to get

$$\|\varphi(w) - \varphi(z)\| \le (1/2)\|w - z\|, \quad \forall\, w, z \in U. \tag{7.21}$$

It follows that φ has at most one fixed point in U; equivalently there is at most one $x \in U$ satisfying $f(x) = y$. Thus f is one-to-one on U. Denote $V = f(U)$. So f is a continuous bijection from U onto V.

We show that V is open. Consider a $q \in V$ and its corresponding $p \in U$ so that $f(p) = q$. Consider an open ball B centered in p of radius $r > 0$ so that $\mathrm{cl}\, B \subset U$. We show that $y \in V$ whenever $\|y - q\| < \lambda r$. Fix such a point y. By (7.19) one has

$$\|\varphi(p) - p\| = \|A^{-1}(y - q)\| \le \|A^{-1}\| \cdot \|y - q\| < \|A^{-1}\|\lambda r = r/2.$$

If $p \in \mathrm{cl}\, B$, then by (7.21) we have

$$\|\varphi(x) - p\| \le \|\varphi(x) - \varphi(p)\| + \|\varphi(p) - p\| \le \|x - p\|/2 + r/2 \le r.$$

Thus $\varphi(x) \in \mathrm{cl}\, B$. We note that φ is a contraction from $\mathrm{cl}\, B$ to itself. So φ has a unique fixed point $x \in \mathrm{cl}\, B$. For this x one has $f(x) = y \in f(\mathrm{cl}\, B) \subset f(U) = V$. Thus (a) is proved.

Consider $y, y + k \in V$. Then one can find $x, x + h \in U$ so that $y = f(x)$ and $y + k = f(x + h)$. By φ in (7.19) it holds

$$\varphi(x + h) - \varphi(x) = h + A^{-1}(f(x) - f(x + h)) = h - A^{-1}k.$$

By (7.21) one has that $\|h - A^{-1}k\| \le \|h\|/2$. Therefore $\|A^{-1}k\| \ge \|h\|/2$ and

$$\|h\| \le 2\|A^{-1}\| \cdot \|k\| = \|k\|/\lambda. \tag{7.22}$$

By (7.17), (7.18), and Proposition 1.2, it follows that $f'(x)$ is invertible for each $x \in U$. For it, if $f'(x) = 0$ for some $x \in U$, $\|A\| < \lambda$. Then

$$1/\lambda < 1/\|A\| \le \|A^{-1}\| = 1/(2\lambda),$$

a contradiction.

Denote by T the inverse of f'. Because

$$g(y + k) - g(y) - Tk = h - Tk = -T(f(x + h) - f(x) - f'(x)h),$$

by (7.22) it follows that

$$\frac{\|g(y + k) - g(y) - Tk\|}{\|k\|} \le \frac{\|T\|}{\lambda} \cdot \frac{\|f(x + h) - f(x) - f'(x)h\|}{\|h\|}. \tag{7.23}$$

By (7.22) we have that $h \to 0$ whenever $k \to 0$. So if $h \to 0$, the left-hand side of (7.23) tends to 0. Thus $g'(y) = T$. By the construction of T, $f'(x) = f'(g(y))$. Hence

$$g'(y) = (f'(g(y)))^{-1}. \tag{7.24}$$

We note that g is a continuous function from V onto U and, moreover, f' is a continuous application from U in the set Ω of all invertible mappings in $L(\mathbb{R}^n, \mathbb{R}^n)$. Taking the inverse, it is continuous as it follows from Theorem 1.5. Invoking (7.24) we conclude that g is a C^1 mapping on V. $\square$

We turn now to the implicit function theorem.

Consider the straight line $X = \{(x,y) \in \mathbb{R}^2 \mid ax + by + c = 0\}$ with $a^2 + b^2 > 0$. If $a \neq 0$, we say that x is defined implicitly by the function $ax + by + c = 0$, meaning that for any real y there exists a unique real x satisfying $ax + by + c = 0$. This is a function defining x in respect y. Consider now the unit circle $X = \{(x,y) \in \mathbb{R}^2 \mid x^2 + y^2 = 1\}$. It is obvious that for every y with

- $|y| > 1$ there is no x so that $(x,y) \in X$.
- $y = \pm 1$ there exists a unique x so that $(x,y) \in X$.
- $|y| < 1$ there are two xs so that $(x,y) \in X$.

Therefore, not for every real y does there exist a unique x on the unit circle. Hence one needs additional hypotheses in order to get a unique x for a given y or to define a function supplying x in respect y. This is an existence problem.

Another question is if one can find a formula to get x for a given y. The answer is negative. But the situation is not so bad, because under some assumptions one can get the derivative of the function supplying x in respect to y.

Fortunately, a similar result holds in the higher-dimension case. It is called the *implicit function theorem*. We present some preparatory notation for it. Consider $x = (x_1, \ldots, x_n) \in \mathbb{R}^n$ and $y = (y_1, \ldots, y_m) \in \mathbb{R}^m$. We write $(x,y) = (x_1, \ldots, x_n, y_1, \ldots, y_m) \in \mathbb{R}^{n+m}$. Suppose we are given a nonempty set $S \subset \mathbb{R}^{n+m}$ and a function $f : S \to \mathbb{R}^n$. Consider the equation

$$f(x,y) = 0; \tag{7.25}$$

we want to introduce a sufficient condition guaranteeing that (7.25) defines a unique function $x = g(y)$ so that $f(g(y), y) = 0$ on a properly chosen set.

Theorem 3.5. *Suppose f is a C^1 mapping on an open set $S \subset \mathbb{R}^{n+m}$ into $\mathbb{R}^n$. Suppose $(a,b) \in S$, $(a \in \mathbb{R}^n, b \in \mathbb{R}^m)$ $f(a,b) = 0$, $A = f'(a,b)$ satisfying that A_x is nonsingular, where*

$$A_x = \begin{bmatrix} \dfrac{\partial f_1(a,b)}{\partial x_1} & \dfrac{\partial f_1(a,b)}{\partial x_2} & \cdots & \dfrac{\partial f_1(a,b)}{\partial x_{n-1}} & \dfrac{\partial f_1(a,b)}{\partial x_n} \\ \dfrac{\partial f_2(a,b)}{\partial x_1} & \dfrac{\partial f_2(a,b)}{\partial x_2} & \cdots & \dfrac{\partial f_2(a,b)}{\partial x_{n-1}} & \dfrac{\partial f_2(a,b)}{\partial x_n} \\ \cdots & \cdots & \cdots & \cdots & \cdots \\ \dfrac{\partial f_{n-1}(a,b)}{\partial x_1} & \dfrac{\partial f_{n-1}(a,b)}{\partial x_2} & \cdots & \dfrac{\partial f_{n-1}(a,b)}{\partial x_{n-1}} & \dfrac{\partial f_{n-1}(a,b)}{\partial x_n} \\ \dfrac{\partial f_n(a,b)}{\partial x_1} & \dfrac{\partial f_n(a,b)}{\partial x_2} & \cdots & \dfrac{\partial f_n(a,b)}{\partial x_{n-1}} & \dfrac{\partial f_n(a,b)}{\partial x_n} \end{bmatrix}.$$

Then there are open sets $U \subset \mathbb{R}^{n+m}$ with $(a,b) \in U$ and $W \subset \mathbb{R}^m$ with $b \in W$ such that for every $y \in W$ there corresponds a unique x so that

$$(x,y) \in U \text{ and } f(x,y) = 0.$$

Denote this function by g, $g : W \to \mathbb{R}^n$, $x = g(y)$. Then g is a C^1 mapping on W such that $g(b) = a$,

$$f(g(y),y) = 0, \quad \forall y \in W \tag{7.26}$$

and

$$g'(b) = -(A_x)^{-1} A_y, \tag{7.27}$$

where

$$A_y = \begin{bmatrix} \dfrac{\partial f_1(a,b)}{\partial y_1} & \dfrac{\partial f_1(a,b)}{\partial y_2} & \cdots & \dfrac{\partial f_1(a,b)}{\partial y_{m-1}} & \dfrac{\partial f_1(a,b)}{\partial y_m} \\ \dfrac{\partial f_2(a,b)}{\partial y_1} & \dfrac{\partial f_2(a,b)}{\partial y_2} & \cdots & \dfrac{\partial f_2(a,b)}{\partial y_{m-1}} & \dfrac{\partial f_2(a,b)}{\partial y_m} \\ \cdots & \cdots & \cdots & \cdots & \cdots \\ \dfrac{\partial f_{n-1}(a,b)}{\partial y_1} & \dfrac{\partial f_{n-1}(a,b)}{\partial y_2} & \cdots & \dfrac{\partial f_{n-1}(a,b)}{\partial y_{m-1}} & \dfrac{\partial f_{n-1}(a,b)}{\partial y_m} \\ \dfrac{\partial f_n(a,b)}{\partial y_1} & \dfrac{\partial f_n(a,b)}{\partial y_2} & \cdots & \dfrac{\partial f_n(a,b)}{\partial y_{m-1}} & \dfrac{\partial f_n(a,b)}{\partial y_m} \end{bmatrix}.$$

Under the assumptions of the implicit function theorem we say that g is *defined implicitly* by (7.26) on W.

Proof. Define

$$h(x,y) = (f(x,y),y), \quad \forall (x,y) \in S. \tag{7.28}$$

Then h is of class C^1 on S. We claim that $h'(a,b)$ is an invertible element in $L(\mathbb{R}^{n+m})$. Because $f(a,b) = 0$, by the definition of the Fréchet differentiability of f, we have

$$f(a+p, b+q) = A(p,q) + r(p,q).$$

Because

$$h(a+p, b+q) - h(a,b) = (f(a+p, b+q), k) = (A(p,q), q) + (r(p,q), 0),$$

$h'(a, b)$ is a linear operator from $\mathbb{R}^{n+m}$ to $\mathbb{R}^{n+m}$ and assigns to (p, q) the point $(A(p, q), q)$. If the image is $0 \in \mathbb{R}^{n+m}$, then $A(p, q) = 0$ and $q = 0$. This implies that $A_x p = 0 \in \mathbb{R}^n$. Then $p = 0$, because A_x is nonsingular. This implies that $h'(a, b)$ is one-to-one. Thus it is onto as well and thus invertible. Apply the inverse function Theorem 3.4 to get open sets U and V so that $(a, b) \in U$, $(0, b) \in V$, and h is a bijective mapping from U onto V. Define W as the set of all $y \in \mathbb{R}^m$ such that $(0, y) \in V$. Because V is open, W is open as well.

If $y \in W$, $(0, y) = h(x, y)$ for some $(x, y) \in U$ and by (7.28), $f(x, y) = 0$. Suppose that for the same y, $(z, y) \in U$ and $f(z, y) = 0$. Then

$$h(z, y) = (f(z, y), y) = (f(x, y), y) = h(x, y).$$

Because h is one-to-one on U, $z = x$. Thus to every $y \in W$ we assign a unique x with $(x, y) \in U$ and $f(x, y) = 0$. Denote this function by g. Thus for $y \in W$, $(g(y), y) \in U$ and (7.26) holds. Then

$$h(g(y), y) = (0, y), \quad y \in W. \tag{7.29}$$

Let $h^{-1} : V \to U$ be the inverse of h. Then by the inverse function theorem 3.4, h^{-1} is of class C^1 and by (7.29)

$$(g(y), y) = h^{-1}(0, y), \quad y \in W. \tag{7.30}$$

Because h^{-1} is of class C^1, h is of class C^1 as well. Now we prove (7.27). Denote $\varphi(y) = (g(y), y)$. Then

$$\varphi'(y)k = (g'(y)k, k), \quad y \in W, \ k \in \mathbb{R}^m. \tag{7.31}$$

By (7.26), $f(\varphi(y)) = 0$ on W. By the chain rule theorem 2.6, we have

$$f'(\varphi(y)) \circ \varphi'(y) = 0.$$

If $y = b$, then $\varphi(y) = (a, b)$ and $f'(\varphi(y)) = A$. Thus

$$A\varphi'(b) = 0. \tag{7.32}$$

By (7.32), (7.31), and $A(p, q) = A_x p + A_y q$, we have

$$A_x g'(b)q + A_x q = A(g'(b)q, q) = A\varphi'(b)q = 0, \quad \forall q \in \mathbb{R}^m.$$

Thus

$$A_x g'(b) + A_y = 0,$$

which is equivalent to (7.27). $\quad\square$

7.3.2 Directional derivatives and gradients

Consider A a nonempty open subset in $\mathbb{R}^n$, $M_0 = (x_1^0, x_2^0, \ldots, x_n^0) \in A$, a function $f : A \to \mathbb{R}$, and a vector $v = (v_1, v_2, \ldots, v_n)$ of unitary norm; that is, $\|v\| = (\sum v_i^2)^{1/2} = 1$. The equations of the straight line l through M_0 and parallel to v are

$$x_k = x_i^0 + t v_k, \quad t \in \mathbb{R}, \ k \in \mathbb{N}_n^*.$$

On the straight line l, f is a composite function of one real variable; let it be t. If this function possesses for $t = 0$ a derivative in respect to the variable t, then this derivative is said to be the *directional derivative* of f in the direction v and it is denoted by $\dfrac{\partial f(M_0)}{\partial t}$. Then admitting the assumptions of Theorem 3.3, we can write

$$\frac{\partial f(M_0)}{\partial t} = \sum_{i=1}^n \frac{\partial f(M_0)}{\partial x_i} \frac{\partial x_i}{\partial t} = \sum_{i=1}^n \frac{\partial f(M_0)}{\partial x_i} v_i.$$

As before, consider $f : A \to \mathbb{R}$ and suppose it has partial derivatives at $M_0 \in A$. Then the vector

$$\mathrm{grad} f|_{M_0} = \left(\frac{\partial f(M_0)}{\partial x_1}, \frac{\partial f(M_0)}{\partial x_2}, \ldots, \frac{\partial f(M_0)}{\partial x_n} \right) \tag{7.33}$$

is said to be the *gradient* of f at M_0.

7.4 Higher-order differentials and partial derivatives

In this section we give a rather general definition of what a higher-order differentiable function is for functions of several variables. First we introduce the concept of second-order differentiability.

Let $(X, \| \cdot \|_X)$ and $(Y, \| \cdot \|_Y)$ be normed vector spaces. Consider A an open and nonempty subset of X and a differentiable function $f : A \to Y$. Then there exists the derivative mapping

$$f' : A \to L(X, Y).$$

We ask if this function is differentiable as well.

Function f is said to be *twice (Frechét) differentiable at a point* $a \in A$ if the mapping f' is differentiable at a. In this very case we denote by $f''(a)$ or by $f^{(2)}(a)$ the (Fréchet) derivative of f' at point a. Then

$$f''(a) \in L(X, L(X, Y)). \tag{7.34}$$

Function f is said to be *twice (Frechét) differentiable on* A provided it is twice (Frechét) differentiable at every point $a \in A$. Then the relation $x \mapsto f''(x)$ defines the mapping

$$f'' : A \to L(X, L(X, Y)).$$

Based on Theorem 1.8, we conclude that f'' is a bilinear and continuous mapping from $X \times X$ into Y of the form

$$X \times X \ni (x, y) \mapsto (f'' \cdot x) \cdot y = (f''x)y. \tag{7.35}$$

The previous relation means that, because $x \in X$ is a vector and f'' is bilinear and continuous from X into $L(X, Y)$, $f'' \cdot x \in L(X, Y)$, and applying it to $y \in X$, one gets $(f'' \cdot x) \cdot y \in Y$.

Function f *belongs to class* C^2 on A provided function f is twice differentiable on A and f'' is continuous. Equivalently, f' belongs to C^1 on A.

Theorem 4.1. (Schwarz) *Suppose $f : A \to Y$ is twice differentiable at $a \in A$. Then the mapping $f''(a) \in L(X, Y)$ is a symmetrical mapping; that is,*

$$(f''(a)x)y = (f''(a)y)x, \quad \forall x, y \in X. \tag{7.36}$$

Proof. Because f is twice differentiable at $a \in A$, there exist $f''(a) \in L(X, L(X, Y))$ so that

$$\lim_{x \to a} \frac{\|f'(x) - f'(a) - f''(a)(x - a)\|}{\|x - a\|} = 0;$$

that is, for every $\varepsilon > 0$ there exists an open ball $B(a, r) \subset A$ such that

$$\|f'(x) - f'(a) - f''(a)(x - a)\| < \varepsilon \|x - a\|, \quad \forall x \in B(a, r). \tag{7.37}$$

Set $u, v \in B(a, \varepsilon/2)$ and define the function $g : [0, 1] \to Y$ by

$$g(t) = f(a + tu + v) - f(a + tu). \tag{7.38}$$

Then

$$g(1) - g(0) = f(a + u + v) - f(a + u) - f(a + v) - f(a), \tag{7.39}$$
$$g'(t) = f'(a + tv + u)u - f'(a + tu)u.$$

We note the right-hand side of (7.39) is symmetrical in x and y. Both points $a + tv + u$ and $a + tu$ belong to $B(a, r)$. Substituting $x = a + tv + u$, respectively, $x = a + tu$ in (7.37), we find

$$\|f'(a + tv + u) - f'(a) - f''(a)(tv + u)\| \le \varepsilon \|tv + u\| \le \varepsilon(\|v\| + \|u\|), \tag{7.40}$$
$$\|f'(a + tu) - f'(a) - f''(a)(tu)\| \le \varepsilon \|tu\| \le \varepsilon(\|v\| + \|u\|). \tag{7.41}$$

Then by (7.38), (7.40), and (7.41) we have

$$\|(f''(a)v)u - g'(t)\| \le \|f'(a + tu + v)u - f'(a + tu)u - (f''(a)v)u)\|$$
$$\le \|(f'(a + tu + v) - f'(a) - f''(a)(tu + v)) \cdot \|u\| \tag{7.42}$$
$$+ \|(f'(a + tu) - f'(a) - (f''(a)(tu)\| \cdot \|u\| \le 2(\|u\| + \|v\|) \cdot \|u\|.$$

By Theorem 2.8, we write

$$\|g(1) - g(0) - g'(0)\| \leq \sup_{t \in [0,1]} \|g'(t) - g'(0)\|. \tag{7.43}$$

Using (7.42) and (7.43), we get the estimate

$$\|g(1) - g(0) - (f''(a)v)u\| \leq \|g(1) - g(0) - g'(0)\| + \|g'(0) - (f''(a)v)u\|$$

$$\leq \sup_{t \in [0,1]} \|g'(t) - g'(0)\| + 2(\|u\| + \|v\|) \cdot \|u\|$$

$$\leq \sup_{t \in [0,1]} \|g'(t) - (f''(a)v)u\| + \|(f''(a)v)u - g'(0)\| + 2\varepsilon\|u\|(\|u\| + \|v\|)$$

$$< 6\varepsilon\|u\|(\|u\| + \|v\|). \tag{7.44}$$

The difference $g(1) - g(0)$ is symmetrical in respect to u and v, therefore we get

$$\|g(1) - g(0) - (f''(a)u)v\| < 6\varepsilon\|v\|(\|u\| + \|v\|). \tag{7.45}$$

Thus by (7.44) and (7.45) we have

$$\|(f''(a)v)u - (f''(a)u)v\|$$

$$\leq \|(f''(a)v)u - g(1) + g(0)\| + \|g(1) - g(0) - (f''(a)u)v\|$$

$$\leq 6\varepsilon\|u\|(\|u\| + \|v\|) + 6\varepsilon\|v\|(\|u\| + \|v\|) < 6\varepsilon(\|u\| + \|v\|)^2.$$

For every $x, y \in X$ one can find $t > 0$ and $u, v \in B(a, r/2)$ with $x = tu$ and $y = tv$. Thus $\|(f''(a)x)y - (f''(a)y)x\| < 6\varepsilon(\|x\| + \|y\|)^2$. Hence

$$(f''(a)x)y = (f''(a)y)x. \quad \square$$

7.4.1 The case $X = \mathbb{R}^n$

Consider A a nonempty and open subset of $\mathbb{R}^n$ and $f : \mathbb{R}^n \to \mathbb{R}^m$ a twice differentiable function at $a \in A$. Then f is differentiable on a neighborhood of a. From (7.15), it follows that for every $h = (h_1, \ldots, h_n) \in \mathbb{R}^n$,

$$f'(a)h = \sum_{i=1}^{n} \frac{\partial f(a)}{\partial x_i} \cdot h_i = \left(\sum_{i=1}^{n} \frac{\partial f_1(a)}{\partial x_i} \cdot h_i, \ldots, \sum_{i=1}^{n} \frac{\partial f_m(a)}{\partial x_i} \cdot h_i \right). \tag{7.46}$$

Substituting f by f' we get

$$f''(a)(k_1, \ldots, k_n) = \sum_{i=1}^{n} \frac{\partial f'(a)}{\partial x_i} \cdot k_i, \quad \forall (k_1, \ldots, k_n) \in \mathbb{R}^n.$$

It follows at once that

$$(f''(a)(k_1, \ldots, k_n))(h_1, \ldots, h_n) = \sum_{i=1}^{n} \left(\frac{\partial f'(a)}{\partial x_i} \cdot k_i \right) (h_1, \ldots, h_n). \tag{7.47}$$

We note that

$$\frac{\partial f'(a)}{\partial x_i} \in L(\mathbb{R}^n, L(\mathbb{R}^n, \mathbb{R}^m)),$$

thus

$$\sum_{i=1}^{n} \frac{\partial f'(a)}{\partial x_i} \cdot k_i \in L(\mathbb{R}^n, \mathbb{R}^m);$$

that is, the image of a vector $(h_1, \ldots, h_n) \in \mathbb{R}^n$ is a vector in $\mathbb{R}^m$.

To find $\partial f'(a)/\partial x_i$ we use (7.46), getting

$$\left(\frac{\partial f'(a)}{\partial x_i} \cdot k_i \right)(h_1, \ldots, h_n) = \sum_{j=1}^{n} \left(\frac{\partial}{\partial x_i} \left(\frac{\partial f}{\partial x_j} \right)(a) \cdot k_i \right) \cdot h_j. \qquad (7.48)$$

The derivative $\partial/\partial x_i \, (\partial f/\partial x_j)$ at a is denoted as

$$\frac{\partial}{\partial x_i} \left(\frac{\partial f}{\partial x_j} \right)(a) = \frac{\partial^2 f}{\partial x_i \, \partial x_j}(a) = \frac{\partial^2 f(a)}{\partial x_i \, \partial x_j}.$$

From (7.47) and (7.48) it follows

$$(f''(a)(k_1, \ldots, k_n))(h_1, \ldots, h_n) = \sum_{i,j}^{n} \left(\frac{\partial^2 f(a)}{\partial x_i \, \partial x_j} \cdot k_i \right) \cdot h_j.$$

Proposition 4.3. (Schwarz) *Suppose A is a nonempty and open subset of $\mathbb{R}^n$ and $f : A \to \mathbb{R}^m$ is a twice differentiable function on A. Then for every $a \in A$*

$$\frac{\partial^2 f(a)}{\partial x_i \, \partial x_j} = \frac{\partial^2 f(a)}{\partial x_j \, \partial x_i}, \quad i, j = 1, \ldots, n.$$

Let $f : A \to Y$ be a twice differentiable function. Then there exists the *second-order derivative* mapping

$$f'' : A \to L(X, X; Y).$$

For short we denote by $L_2(X, Y)$ the Banach space $L(X, X; Y)$ of bilinear and continuous mappings from $X \times X$ into Y. Similarly, $L_n(X, Y)$ denotes the Banach space of nth-linear and continuous mappings from the n-times Cartesian product $X \times \cdots \times X$ into Y.

We can ask about the differentiability of f''. If it is differentiable at $a \in A$ ($A \subset X$ is nonempty and open), then we denote by $f'''(a)$ or $f^{(3)}(a)$ the derivative of f'' at a (i.e., the *third-order derivative of f at a*). It is an element of $L(X, L_2(X, Y))$ (isometric to $L_3(X, Y)$). By induction, we define the *nth-order differentiability of f* at $a \in A$ and its *nth-order derivative* $f^{(n)}(a) \in L_n(X, Y)$. We say that f is *n times differentiable at a* if

- There exists an open neighborhood V of a such that f is $n - 1$ times differentiable at every $x \in V$.

- The mapping $x \mapsto f^{(n-1)}(x)$ from V into $L_{n-1}(X, Y)$ is differentiable on a.

Then the derivative of $f^{(n-1)}(x)$ at a is denoted as $f^{(n)}(a)$ and is said to be the nth-order derivative of f at a.

We denote by $f^{(n)}(a)(h_1, \ldots, h_n)$ the image of the vector $(h_1, \ldots, h_n) \in X \times \cdots \times X$ through the mapping $f^{(n)}(a)$.

Function f is said to *belong to the class C^n* on A whenever f is n times differentiable on every point in A and the mapping

$$f^{(n)} : A \to L_n(X, Y)$$

is continuous. By definition

$$f^{(0)} = f.$$

Thus C^0 on A contains precisely the continuous functions on A.

Function f is said to *belong to the class C^∞* on A whenever $f \in C^n(A)$ for every $n \in \mathbb{N}$.

Similar to Theorem 4.1 the following holds.

Theorem 4.2. *Suppose $f : A \to Y$ is n times differentiable at $a \in A$. Then the derivative $f^{(n)}(a) \in L_n(X, Y)$ is a multilinear symmetrical mapping from $X \times \cdots \times X$ into Y. Formally,*

$$f^{(n)}(a)(h_1, \ldots, h_n) = f^{(n)}(a)(h_{\sigma(1)}, \ldots, h_{\sigma(n)}),$$

where $\sigma : \{1, \ldots, n\} \to \{1, \ldots, n\}$ is a permutation.

7.5 Taylor formula

Similar to the scalar case, (5.38) at page 211, we have the vector-valued Taylor formula of vector argument. Let A be a nonempty open set in $\mathbb{R}^n$.

Theorem 5.1. (Taylor) *Suppose $f : A \to \mathbb{R}^m$ belongs to the class C^{p+1} on A and $[a, a+h] \subset A$. Then*

$$f(a + h) = f(a) + f'(a)h + \frac{1}{2}f''(a)(h, h) + \cdots + \frac{1}{p!}f^{(p)}(a)(h)^p$$

$$+ \int_0^1 \frac{(1 - t)^p}{p!}f^{(p+1)}(a + th)(h)^{p+1}\, dt.$$

Theorem 5.2. *Suppose $f : A \to \mathbb{R}^m$ is $p+1$ times differentiable. Moreover, admit*

$$\|f^{(p+1)}(a)\| \le M, \quad \forall a \in A.$$

Then

$$\left\| f(a + h) - f(a) - f'(a)h - \cdots - \frac{1}{p!}f^{(p)}(a)(h)^p \right\| \le M \frac{\|h\|^{p+1}}{(p+1)!}.$$

Theorem 5.3. *Under the assumptions in Theorem 5.1, one has*

$$f(a + h) = f(a) + f'(a)h + \cdots + \frac{1}{p!} f^{(p)}(a)(h)^p + R_p(h), \qquad (7.49)$$

where

$$\lim_{h \to 0} \frac{\|R_p(h)\|}{\|h\|^p} = 0.$$

Corollary 5.7. *Suppose $m = 1$. Then (7.49) becomes*

$$f(a_1 + h_1, \ldots, a_n + h_n) = f(a_1, \ldots, a_n) + \frac{1}{1!} \sum_{k=1}^{n} \frac{\partial f(a_1, \ldots, a_n)}{\partial x_k} h_k$$

$$+ \frac{1}{2!} \sum_{\substack{k_1 + \cdots + k_n = 2 \\ k_1, \ldots, k_n \in \mathbb{N}}} \frac{\partial^2 f(a_1, \ldots, a_n)}{\partial x_1^{k_1} \cdots \partial x_n^{k_n}} h_1^{k_1} \ldots h_n^{k_n} + \cdots$$

$$+ \frac{1}{p!} \sum_{\substack{k_1 + \cdots + k_n = p \\ k_1, \ldots, k_n \in \mathbb{N}}} \frac{\partial^p f(a_1, \ldots, a_n)}{\partial x_1^{k_1} \cdots \partial x_n^{k_n}} h_1^{k_1} \ldots h_n^{k_n} + R_p(h).$$

7.6 Problems of local extremes

7.6.1 First-order conditions

We recall some definitions from page 197. Let f be a real-valued function defined on a metric space X. We say that f has a *local maximum* at a point $p \in X$ if there exists $\delta > 0$ such that $f(q) \leq f(p)$ for all $q \in B(p, \delta)$. We say that f has a *strict local maximum* at a point $p \in X$ if it is a local maximum point and there exists $\delta > 0$ such that $f(q) < f(p)$ for all $q \in B(p, \delta) \setminus \{p\}$. Analogously, we say that f has a *local minimum* at a point $p \in X$ if there exists $\delta > 0$ such that $f(q) \geq f(p)$ for all $q \in B(p, \delta)$. f has a *strict local minimum* at a point $p \in X$ if it is a local minimum point and there exists $\delta > 0$ such that $f(q) > f(p)$ for all $q \in B(p, \delta) \setminus \{p\}$. A local minimum or maximum point is called a local extreme point.

Similar to Theorem 2.1 at page 197 one has the following first-order necessary condition.

Theorem 6.1. (Fermat) *Suppose A is a nonempty and open subset of a Banach space X, $f : A \to \mathbb{R}$ has an local extreme point at a, and f is differentiable at a. Then $f'(a) = 0$.*

Proof. Choose an arbitrary vector $x \in X$. Consider the function $g(t) = f(a + tx)$ defined in a sufficiently small neighborhood of $0 \in \mathbb{R}$. Then function g has a local extreme point at $t = 0$. Based on Theorem 2.1 at page 197, we infer that $g'(0) = 0$. Because

$$g'(t) = f'(a+tx)x,$$

it follows that $g'(0) = f'(a)x$, for every $x \in X$. Hence $f'(a) = 0$.   □

The set of points $a \in A$ fulfilling $f'(a) = 0$ is said to be the *set of stationary points*

The converse of the previous result fails. A stationary point is not necessarily a local extreme point. It is enough to consider the following function, $\mathbb{R} \times \mathbb{R} \ni (x, y) \mapsto x^2 - y^2$. A stationary point that is not a local extreme point is said to be a *saddle point*. For a finer study of local extreme points we need more sophisticated tools, that is, higher-order derivatives.

7.6.2 Second-order conditions

Theorem 6.2. *Suppose A is a nonempty and open subset of $\mathbb{R}^n$ and $f : A \to \mathbb{R}$ is a twice differentiable function at $a \in A$. Furthermore, we suppose f has a local minimum point at a. Then $f''(a) \geq 0$.*

Proof. $f''(a) \geq 0$, means that the bilinear and symmetric form $f''(a)$ is nonnegative; that is, $f''(a)(h, h) \geq 0$, for all $h \in \mathbb{R}^n$.

Because $f'(a) = 0$, by the Taylor formula, Corollary 5.7, we have for h small enough

$$f(a+h) - f(a) = (1/2)f''(a)(h, h) + R_2(h),$$

where $|R_2(h)| = o(\|h\|^2)$. Because a is a local minimum point we still have

$$f''(a)(h, h) + 2R_2(h) \geq 0.$$

We settle down h and for real t such that $|t|$ is small enough it follows

$$f''(a)(th, th) + 2R_2(th) \geq 0.$$

For fixed h we have

$$f''(a)(th, th) = t^2 f''(a)(h, h), \quad 2R_2(th) = t^2 \varepsilon(t, h), \quad \lim_{t \to 0} \varepsilon(t, h) = 0.$$

Then for t small enough

$$f''(a)(h, h) + \varepsilon(t, h) \geq 0.$$

Thus, if $\varepsilon(t, h)$ tends to 0 by t, it results at the limit

$$f''(a)(h, h) \geq 0. \quad □$$

Similarly, one can prove the next result.

Theorem 6.3. *Suppose A is a nonempty and open subset of $\mathbb{R}^n$ and $f : A \to \mathbb{R}$ is a twice differentiable function at $a \in A$. Furthermore, we suppose f has a local maximum point at a. Then $f''(a) \leq 0$.*

We introduce a sufficient condition for the existence of a local extreme point.

Theorem 6.4. *Suppose A is a nonempty and open subset of $\mathbb{R}^n$ and $f : A \to \mathbb{R}$ is a twice differentiable function at $a \in A$. Furthermore, we suppose $f'(a) = 0$ and $f''(a)$ is positively definite. Then a is a local minimum point of f.*

Proof. By the Taylor we have

$$f(a+h) - f(a) = (1/2)f''(a)(h,h) + \varepsilon(h)\|h\|^2,$$

where $\varepsilon(h) \to 0$ as $h \to 0$. By Theorem 1.10 there is a positive λ so that $f''(a)(h,h) \geq \lambda\|h\|^2$. Then

$$f(a+h) - f(a) \geq (\lambda/2 + \varepsilon(h))\|h\|^2.$$

If $\|h\|$ is small enough, $\lambda/2 + \varepsilon(h) > 0$ and thus for nonzero h it follows

$$f(a+h) - f(a) > 0. \quad \square$$

Similarly one has the following theorem.

Theorem 6.5. *Suppose A is a nonempty and open subset of $\mathbb{R}^n$ and $f : A \to \mathbb{R}$ is a twice differentiable function at $a \in A$. Furthermore, we suppose $f'(a) = 0$ and $f''(a)$ is negatively definite. Then a is a local maximum point of f.*

7.6.3 Constraint local extremes

Consider $x = (x_1, \ldots, x_n) \in \mathbb{R}^n$ and $y = (y_1, \ldots, y_m) \in \mathbb{R}^m$. We write $(x,y) = (x_1, \ldots, x_n, y_1, \ldots, y_m) \in \mathbb{R}^{n+m}$. Suppose there are given $f : A \subset \mathbb{R}^{n+m} \to \mathbb{R}$, and $g_i : A \subset \mathbb{R}^{n+m} \to \mathbb{R}$, $i = 1, 2, \ldots, m$, where A is open and nonempty.

Our first task is finding first-order necessary conditions for the problem: find the local extreme points to

$$(x,y) \to f(x,y) \tag{7.50}$$

under the constraints

$$\begin{cases} g_1(x,y) = 0, \\ \ldots \\ g_m(x,y) = 0. \end{cases} \tag{7.51}$$

A point $M_0(x_0^1, \ldots, x_0^n, y_0^1, \ldots, y_0^m)$ is a *local extreme point of (7.50) under the constraints (7.51)* if there exists a neighborhood of M_0 such that, simultaneously, M_0 is a local extreme point of (7.50) and satisfies (7.51).

We introduce a first-order necessary condition for a local extreme point of (7.50) under the constraints (7.51), namely the *Lagrange multipliers method*.

Theorem 6.6. *Suppose A is a nonempty and open subset of $\mathbb{R}^{n+m}$, and function f in (7.50), and the functions g_j, $j = 1, 2, \ldots, m$ in (7.51) are differentiable on A. Moreover, suppose $M_0(x_0^1, \ldots, x_0^n, y_0^1, \ldots, y_0^m) \in A$ is a local extreme point of (7.50) under the constraints (7.51), the partial derivatives of all functions g_j are continuous in respect all $y_1, \ldots, y_m$ in some neighborhood of M_0, and*

$$J = \begin{vmatrix} \dfrac{\partial g_1(M_0)}{\partial y_1} & \cdots & \dfrac{\partial g_1(M_0)}{\partial y_m} \\ \cdots & \cdots & \cdots \\ \dfrac{\partial g_m(M_0)}{\partial y_1} & \cdots & \dfrac{\partial g_m(M_0)}{\partial y_m} \end{vmatrix} \neq 0. \tag{7.52}$$

Then there exist some scalars $\lambda_1, \ldots, \lambda_m$ such that if we define on A the function

$$F(x, y) = f(x, y) + \sum_{j=1}^{m} \lambda_j g_j(x, y),$$

then

$$\frac{\partial F(M_0)}{\partial x_1} = 0, \ldots, \frac{\partial F(M_0)}{\partial x_n} = 0, \quad \frac{\partial F(M_0)}{\partial y_1} = 0, \ldots, \frac{\partial F(M_0)}{\partial y_m} = 0, \tag{7.53}$$

$$g_1(M_0) = 0, \ldots, g_m(M_0) = 0.$$

Function F is said to be a *Lagrangian function*. Scalars $\lambda_1, \ldots, \lambda_m$ are said to be *Lagrangian multipliers*.

Proof. Because M_0 is an extreme point of (7.50) under the constraints (7.51), we have at M_0,

$$\frac{\partial f}{\partial x_1} dx_1 + \cdots + \frac{\partial f}{\partial x_n} dx_n + \frac{\partial f}{\partial y_1} dy_1 + \cdots + \frac{\partial f}{\partial y_m} dy_m = 0 \tag{7.54}$$

and

$$\begin{cases} \dfrac{\partial g_1}{\partial x_1} dx_1 + \cdots + \dfrac{\partial g_1}{\partial x_n} dx_n + \dfrac{\partial g_1}{\partial y_1} dy_1 + \cdots + \dfrac{\partial g_1}{\partial y_m} dy_m = 0, \\ \cdots \\ \dfrac{\partial g_m}{\partial x_1} dx_1 + \cdots + \dfrac{\partial g_m}{\partial x_n} dx_n + \dfrac{\partial g_m}{\partial y_1} dy_1 + \cdots + \dfrac{\partial g_m}{\partial y_m} dy_m = 0. \end{cases} \tag{7.55}$$

Multiply equalities (7.55) by the arbitrary (as yet undetermined) constants $\lambda_1, \ldots, \lambda_m$, respectively. Then we add termwise the equalities thus obtained to (7.54). Denote $F = f + \sum_{j=1}^{m} \lambda_j g_j$. Then at M_0 we get

$$\frac{\partial F}{\partial x_1} dx_1 + \cdots + \frac{\partial F}{\partial x_n} dx_n + \frac{\partial F}{\partial y_1} dy_1 + \cdots + \frac{\partial F}{\partial y_m} dy_m = 0. \tag{7.56}$$

We find some scalars $\lambda_1, \ldots, \lambda_m$ such that

$$\partial F(M_0)/\partial y_j = 0, \quad j = 1, \ldots, m;$$

that is,

$$\frac{\partial f(M_0)}{\partial y_j} + \sum_{k=1}^{m} \lambda_k \frac{\partial g_k(M_0)}{\partial y_j} = 0, \quad j = 1, \ldots, m.$$

By (7.52) this is always possible.

Then in (7.56) remains

$$\frac{\partial F(M_0)}{\partial x_1} d x_1 + \cdots + \frac{\partial F(M_0)}{\partial x_n} d x_n = 0. \tag{7.57}$$

Because under our assumptions the variables $x_1, \ldots, x_n$ are independent, from (7.57) we have

$$\frac{\partial F(M_0)}{\partial x_1} = 0, \ \ldots \ , \frac{\partial F(M_0)}{\partial x_n} = 0.$$

Hence, we have all the conclusions of the theorem.   □

Remark. Theorem 6.6 is a first-order necessary condition for the existence of an extreme point for the constraint problem (7.50) with (7.51). Hence, the set of extreme points of (7.50) with (7.51) is a subset of the set of solutions to (7.53), the unknowns being $x_1, \ldots, x_n, y_1, \ldots, y_m, \lambda_1, \ldots, \lambda_m$.   △

We need a second-order condition to decide more exactly about the nature of solutions to (7.53); that is, which solutions are extreme point and of what kind (maximum or minimum) and which are not. In the latter case we say that the point is a *saddle* one.

We introduce a second-order condition to the constraint problem (7.50) with (7.51). Suppose f and gs are twice differentiable functions on a neighborhood of M_0 and all second-order partial derivatives are continuous at M_0. The construction of the Lagrange function implies that under constraint conditions, the extreme value of f coincides with the extreme value of F. Then it follows from the result obtained in the previous section that we need $d^2 F$ to decide the nature of a point satisfying (7.53). We have

$$d^2 F = \left(\sum_{i=1}^{n} d x_i \frac{\partial}{\partial x_i} + \sum_{j=1}^{m} d y_j \frac{\partial}{\partial y_j} \right)^2 F + \sum_{j=1}^{m} \frac{\partial F}{\partial y_j} d^2 y_j.$$

At a point M_0 of a possible extremum there hold

$$\frac{\partial F}{\partial y_1} = 0, \ \ldots, \ \frac{\partial F}{\partial y_m} = 0.$$

Then $d^2 F$ is given by

$$d^2 F = \left(\sum_{i=1}^{n} d x_i \frac{\partial}{\partial x_i} + \sum_{j=1}^{m} d y_j \frac{\partial}{\partial y_j} \right)^2 F. \tag{7.58}$$

We have to establish whether d^2F is positive or negative definite only under the constraint conditions (7.51). So, we have to replace $dy_1, \ldots, dy_m$ in (7.58) by their values determined from system (7.55).

7.7 Exercises

7.1. Let $a_1, a_2, \ldots, a_n$ be real numbers and $(\mathbb{R}^n, \|\cdot\|_2)$ be the n-dimensional Euclidean space. Consider the mapping $f : \mathbb{R}^n \to \mathbb{R}$, given as $f(x) = a_1x_1 + \cdots + a_nx_n$, $x = (x_1, \ldots, x_n)$. Show that f is a linear mapping and that $\|f\| = \sqrt{a_1^2 + a_2^2 + \cdots + a_n^2}$.

7.2. Consider $\varphi \in C^1(\mathbb{R})$ and $\emptyset \neq D = \text{int}(D) \subset \mathbb{R}^2$. Show that every function $u(x, y) = \varphi(x^2 + y^2)$, $(x, y) \in D$, satisfies the relation

$$y \frac{\partial u}{\partial x} - x \frac{\partial u}{\partial y} = 0.$$

7.3. Suppose $A \subset \mathbb{R}^n$ is open and nonempty and $f : A \to \mathbb{R}$ is a differentiable function on A. Suppose there exists $M > 0$ such that for every $x, y \in A$ one has $\|f(x) - f(y)\| \leq M\|x - y\|$. Show that $\|f'(a)\| \leq M$ for all $a \in A$.

7.4. Consider $f : \mathbb{R}^2 \to \mathbb{R}$ a function of class C^1 so that $f(0, 0) = 0$. Show that

$$f(x, y) = x \int_0^1 \frac{\partial f(tx, ty)}{\partial x} \, dt + y \int_0^1 \frac{\partial f(tx, ty)}{\partial y} \, dt, \quad \forall (x, y) \in \mathbb{R}^2.$$

7.5. Consider a function $f : \mathbb{R}^n \to \mathbb{R}^m$ and $\alpha \geq 0$, $p > 1$. Suppose $\|f(x) - f(y)\| \leq \alpha \|x - y\|^p$, for all $x, y \in \mathbb{R}^n$. Show that f is constant.

7.6. Suppose $f : \mathbb{R}^n \to \mathbb{R}$ is a twice differentiable function satisfying that $f(x) \geq 0$ for all $x \in \mathbb{R}^n$ and $f''(x)(h, h) \leq 0$ for all $x, h \in \mathbb{R}^n$. Show that f is constant.

7.7. Find the extreme values of the function $z(x, y) = x^3 + y^3 - 3xy$, $x, y \in \mathbb{R}$.

7.8. Find the extreme values of the function $f(x, y, z) = xyz$, $x, y, z \in \mathbb{R}$, under the constraints

$$x^2 + y^2 + z^2 = 1 \text{ and } x + y + z = 0. \tag{7.59}$$

7.8 References and comments

Theorem 1.6 may be found in most functional analysis books.
    Theorems 1.7 and 1.8 may be found in [32].
    Theorem 2.7 may be found at [36, p. 230], or [32, Chapter 1, §3].

Theorem 2.8 may be found [36, p. 232].

The notion of variation of a function was introduced in [53]. This paper was introduced at l'Académie des Sciences in Paris on August 4th 1913. It appeared nine years after Gâteaux died in 1914 at the beginning of the First World War.

Theorems 4.1, 4.2, 5.1, and 5.2 may be found in [32].

A survey paper on the genesis of the Lagrange multiplier method is [31].

Double Integrals, Triple Integrals, and Line Integrals

This chapter is devoted to presenting some results on double integrals, triple integrals, n-fold integrals, and line integrals.

8.1 Double integrals

8.1.1 Double integrals on rectangles

Consider a rectangle $R = [a, b] \times [c, d] \subset \mathbb{R}^2$ $(a \leq b, \ c \leq d)$ and a *partition*

$$\begin{aligned} P = &\{a = x_0 \leq x_1 \leq \cdots \leq x_{n-1} \leq x_n = b\} \\ &\times \{c = y_0 \leq y_1 \leq \cdots \leq y_{m-1} \leq y_m = d\} \end{aligned} \tag{8.1}$$

of it. For such a partition P, denote

$$\|P\| = \max_{1 \leq k \leq n, \ 1 \leq l \leq m} \sqrt{(x_k - x_{k-1})^2 + (y_l - y_{l-1})^2}$$

and call it the *norm* of P. For $k = 1, 2, \ldots, n$ and $l = 1, 2, \ldots, m$, denote

$$R_{k,l} = \{(x, y) \mid x_{k-1} \leq x \leq x_k, \ y_{l-1} \leq y \leq y_l\} = [x_{k-1}, x_k] \times [y_{l-1}, y_l].$$

Let $f : R \to \mathbb{R}$ be a bounded function; that is, exist real numbers m and M so that $m \leq f(x, y) \leq M$, for all $(x, y) \in R$. Define

$$m_{k,l} = \inf_{(x,y) \in R_{k,l}} f(x, y), \quad M_{k,l} = \sup_{(x,y) \in R_{k,l}} f(x, y),$$

$$\Delta_{k,l} = (x_k - x_{k-1})(y_l - y_{l-1}),$$

$$L(P, f) = \sum_{k=1}^{n} \sum_{l=1}^{m} m_{k,l} \Delta_{k,l}, \quad U(P, f) = \sum_{k=1}^{n} \sum_{l=1}^{m} M_{k,l} \Delta_{k,l}.$$

$L(P, f)$ is said to be the *lower Darboux sum* of the function f on R and $U(P, f)$ is said to be the *upper Darboux sum* of the function f on R. Because f is bounded, we have

$$m(b-a)(d-c) \leq L(P,f) \leq U(P,f) \leq M(b-a)(d-c).$$

Therefore we put

$$\sup_P L(P,f) = \underline{\int_R} f(x,y)dx\,dy = \underline{\int_R} f(x,y)dy\,dx \tag{8.2}$$

$$\inf_P U(P,f) = \overline{\int_R} f(x,y)dx\,dy = \overline{\int_R} f(x,y)dy\,dx, \tag{8.3}$$

where the sup and the inf are considered over all partitions P of R. The number $\underline{\int_R} f(x,y)dx\,dy$ is said to be the *lower Darboux integral* of the function f on the rectangle R, and $\overline{\int_R} f(x,y)dx\,dy$ is said to be the *upper Darboux integral* of the function f on the rectangle R.

A partition P^* of R is said to be a *refinement* of P if $P \subset P^*$. Suppose two partitions P_1 and P_2 are given on R. Their common refinement P^* is defined by $P^* = P_1 \cup P_2$.

Lemma 8.1. *Let P^* be a refinement of P. Then*

$$L(P,f) \leq L(P^*,f) \tag{8.4}$$

and

$$U(P,f) \geq U(P^*,f). \tag{8.5}$$

Proof. If $P = P^*$, in (8.4) and (8.5) we have equalities. Suppose for the beginning that P^* contains just one extra point. Let (t,s) be this extra point and suppose that $x_{i-1} \leq t \leq x_i$ and $y_{j-1} \leq s \leq y_j$. Set

$$p_1 = \sup_{(x,y)\in[x_{i-1},t]\times[y_{j-1},s]} f(x,y), \quad p_2 = \sup_{(x,y)\in[t,x_i]\times[y_{j-1},s]} f(x,y),$$

$$p_3 = \sup_{(x,y)\in[x_{i-1},t]\times[s,y_j]} f(x,y), \quad p_4 = \sup_{(x,y)\in[t,x_i]\times[s,y_j]} f(x,y).$$

Then $\max\{p_1,p_2,p_3,p_4\} \leq M_{i,j}$. Therefore

$$U(P^*,f) - U(P,f) = p_1(t-x_{i-1})(s-y_{j-1}) + p_2(x_i-t)(s-y_{j-1})$$
$$+ p_3(t-x_{i-1})(y_j-s) + p_4(x_i-t)(y_j-s) - M_{i,j}\Delta_{i,j} \leq 0,$$

and this case is clear. If P^* contains m more points than P, we repeat the above reasoning m times. Thus (8.5) is completely proved.

Inequality (8.4) is proved in a similar way. $\square$

Theorem 1.1. *It holds*

$$\underline{\int_R} f(x,y)dx\,dy \leq \overline{\int_R} f(x,y)dx\,dy.$$

Proof. Choose P_1 and P_2 two arbitrary partitions and let P^* be their common refinement. By Lemma 8.1 we have

$$L(P_1, f) \leq L(P^*, f) \leq U(P^*, f) \leq U(P_2, f).$$

Then

$$L(P_1, f) \leq U(P_2, f).$$

Keeping P_2 fixed and taking the supremum in respect to P_1 in the above inequality, we have

$$\underline{\int_R} f(x, y) dx\, dy \leq U(P_2, f).$$

Taking the infimum in respect to P_2 in the above inequality, we complete the proof. □

If the lower and the upper Darboux integral of f coincide, we say that f is *Darboux integrable* on R and the common value of (8.2) and (8.3) is denoted by

$$\underline{\int_R} f(x, y) dx\, dy = \overline{\int_R} f(x, y) dy\, dx. \tag{8.6}$$

In this case for every partition P we may write

$$L(P, f) \leq \int_R f(x, y) dx\, dy \leq U(P, f). \tag{8.7}$$

The Darboux integrability can be characterized by the next result.

Theorem 1.2. *Function f is Darboux integrable on R if and only if for every $\varepsilon > 0$ there exists a partition P of R such that*

$$U(P, f) - L(P, f) \leq \varepsilon. \tag{8.8}$$

Proof. Condition (8.8) is sufficient because from

$$L(P, f) \leq \underline{\int_R} f(x, y) dx\, dy \leq \overline{\int_R} f(x, y) dx\, dy \leq U(P, f)$$

it follows that

$$0 \leq \overline{\int_R} f(x, y) dx\, dy - \underline{\int_R} f(x, y) dx\, dy \leq \varepsilon;$$

that is, f is Darboux integrable on R.

Suppose that

$$\underline{\int_R} f(x, y) dx\, dy = \overline{\int_R} f(x, y) dx\, dy = \int_R f(x, y) dx\, dy$$

and let $\varepsilon > 0$ be given. Then we can find two partitions P_1 and P_2 such that

$$U(P_2, f) - \int_R f(x, y)dx\, dy \leq \varepsilon/2, \quad \int_R f(x, y)dx\, dy - L(P_1, f) \leq \varepsilon/2.$$

Let P be the common refinement of P_1 and P_2. Then

$$U(P, f) \leq U(P_2, f) \leq \int_R f(x, y)dx\, dy + \frac{\varepsilon}{2} \leq L(P_1, f) + \varepsilon \leq L(P, f) + \varepsilon;$$

that is, inequality (8.8) holds. □

Theorem 1.3. *Suppose f is continuous on R. Then f is Darboux integrable on R. Moreover, to every $\varepsilon > 0$ there corresponds a $\delta > 0$ such that*

$$\left| \sum_{k=1}^{n} \sum_{l=1}^{m} f(t_k, s_l)\Delta_{k,l} - \int_R f(x, y)dx\, dy \right| < \varepsilon \tag{8.9}$$

for every partition P in (8.1) of R with $\|P\| < \delta$, and for every choice of points $(t_k, s_l) \in R_{k,l}$.

Proof. Let $\varepsilon > 0$ be given. Choose $\eta > 0$ such that $(b - a)(d - c)\eta < \varepsilon$. Function f is uniformly continuous on the compact set R, so there exists a $\delta > 0$ such that

$$\|f(x, y) - f(t, s)\| < \eta \tag{8.10}$$

if $\|(x, y) - (t, s)\| < \delta$ and $(x, y), (t, s) \in R$.

Choose a partition P of R such that $\|P\| < \delta$. By (8.10) we have $M_{k,l} - m_{k,l} \leq \eta$, for all $k = 1, 2, \ldots, n$ and $l = 1, 2, \ldots, m$. Thus

$$U(P, f) - L(P, f) = \sum_{k=1}^{n} \sum_{l=1}^{m} (M_{k,l} - m_{k,l})\Delta_{k,l} < \eta \sum_{k=1}^{n} \sum_{l=1}^{m} \Delta_{k,l} < \varepsilon.$$

By the characterization Theorem 1.2, we conclude that f is Darboux integrable on R.

Because the two numbers $\sum_{k=1}^{n} \sum_{l=1}^{m} f(t_k, s_l)\Delta_{k,l}$ and $\int_R f(x, y)dx\, dy$ lie between $U(P, f)$ and $L(P, f)$, inequality (8.9) is proved. □

We now introduce some arithmetic properties of the double Darboux integral.

Theorem 1.4. (a) *Suppose f_1 and f_2 are Darboux integrable on R and p and q are reals. Then $p \cdot f_1 + q \cdot f_2$ is Darboux integrable on R and, moreover,*

$$\int_R (pf_1(x, y) + qf_2(x, y)dx\, dy = p \int_R f_1(x, y)dx\, dy$$

$$+ q \int_R f_2(x, y)dx\, dy. \tag{8.11}$$

(b) *Suppose f_1, f_2 are Darboux integrable on R and $f_1(x, y) \leq f_2(x, y)$, for all $(x, y) \in R$. Then*

$$\int_R f_1(x, y) dx\, dy \leq \int_R f_2(x, y) dx\, dy.$$

(c) *Suppose f is continuous on R. Then exists $(\xi, \eta) \in R$ such that*

$$\int_R f(x, y) dx\, dy = f(\xi, \eta)(b - a)(d - c).$$

(d) *Suppose f and g are continuous on R, $g \geq 0$ on R, and $\int_R g(x, y) dx\, dy > 0$. Then there exists $(\xi, \eta) \in R$ such that*

$$\int_R f(x, y) g(x, y) dx\, dy = f(\xi, \eta) \int_R g(x, y) dx\, dy.$$

(e) *If f and $|f|$ are Darboux integrable on R, then*

$$\left| \int_R f(x, y) dx\, dy \right| \leq \int_R |f(x, y)| dx\, dy.$$

Proof. (a) Pick a positive ε. Then there exists a positive δ_1 so that for every partition P_1 of R with $\|P_1\| \leq \delta_1$ one has

$$U(P_1, f_1) - L(P_1, f_1) < \varepsilon/(2|p| + 1).$$

Similarly, there exists a positive δ_2 so that for every partition P_2 of R with $\|P_2\| \leq \delta_2$ one has

$$U(P_2, f_2) - L(P_2, f_2) < \varepsilon/(2|q| + 1).$$

Let P be the common refinement of P_1 and P_2. Then

$$0 \leq U(P, pf_1 + qf_2) - L(P, pf_1 + qf_2)$$
$$= p(U(P_1, f_1) - L(P_1, f_1)) + q(U(P_2, f_2) - L(P_2, f_2))$$
$$< |p|\varepsilon/(2|p| + 1) + |q|\varepsilon/(2|q| + 1) < \varepsilon.$$

Therefore by Theorem 1.2, we have that $pf_1 + qf_2$ is integrable on R.

We show (8.11). Pick a positive ε. Then exists a positive δ_1 so that for every partition P_1 of R with $\|P_1\| \leq \delta_1$ one has

$$\left| U(P_1, f_1) - \int_R f_1(x, y) dx\, dy \right| < \varepsilon/(2|p| + 1).$$

Similarly, there exists a positive δ_2 so that for every partition P_2 of R with $\|P_2\| \leq \delta_2$ one has

$$\left| U(P_2, f_2) - \int_R f_2(x, y)dx\,dy \right| < \varepsilon/(2|q| + 1).$$

Now, let P be the common refinement of P_1 and P_2. Then

$$\left| U(P, pf_1 + qf_2) - p\int_R f_1(x, y)dx\,dy - q\int_R f_2(x, y)dx\,dy \right|$$

$$= \left| pU(P, f_1) + qU(P, f_2) - p\int_R f_1(x, y)dx\,dy - q\int_R f_2(x, y)dx\,dy \right|$$

$$\leq \left| pU(P, f_1) - p\int_R f_1(x, y)dx\,dy \right| + \left| qU(P, f_2) - q\int_R f_2(x, y)dx\,dy \right|$$

$$< |p|\varepsilon/(2|p| + 1) + |q|\varepsilon/(2|q| + 1) < \varepsilon.$$

Therefore (8.11) is proved.

(b) Because $f_2(x, y) - f_1(x, y) \geq 0$, for every partition P of R we have that $L(P, f_2 - f_1) \geq 0$. By (a) $f_2 - f_1$ is integrable. Now the conclusion follows.

(c) Because (8.7) holds and a continuous function on a compact set attains its extreme values, the result follows.

(d) Successively we have

$$m \cdot g(x, y) \leq f(x, y)g(x, y) \leq M \cdot g(x, y),$$

$$m\int_R g(x, y)dx\,dy \leq \int_R f(x, y)g(x, y)dx\,dy \leq M\int_R g(x, y)dx\,dy,$$

$$m \leq \frac{\int_R f(x, y)g(x, y)dx\,dy}{\int_R g(x, y)dx\,dy} \leq M.$$

The conclusion follows by (c).

(e) Because $-|f(x, y)| \leq f(x, y) \leq |f(x, y)|$, by (a) and (b) we have

$$-\int_R |f(x, y)|dx\,dy \leq \int_R f(x, y)dx\,dy \leq \int_R |f(x, y)|dx\,dy.$$

The conclusion follows.   □

Theorem 1.5. *Consider an integrable function $f : [a, b] \times [c, d] = R \to \mathbb{R}$ and for each x in $[a, b]$ the integral*

$$I(x) = \int_c^d f(x, y)dy$$

exists. Then there exists the iterated integral

$$\int_a^b I(x)dx = \int_a^b \left(\int_c^d f(x, y)dy \right) dx \tag{8.12}$$

and we have

$$\int_R f(x, y)dx\,dy = \int_a^b \left(\int_c^d f(x, y)dy \right) dx. \tag{8.13}$$

Proof. Consider partition (8.1) of R, $\Delta x_k = x_k - x_{k-1}$, and $\Delta y_l = y_l - y_{l-1}$ for $k = 1, 2, \ldots, n$, $l = 1, 2, \ldots, m$. For every $(x, y) \in R_{k,l}$,

$$m_{k,l} \leq f(x, y) \leq M_{k,l}. \tag{8.14}$$

Set $x = \xi_k$ for an arbitrary $\xi_k \in [x_{k-1}, x_k]$ and integrate (8.14) in respect to y on $[y_{l-1}, y_l]$. Then

$$m_{k,l} \Delta y_l \leq \int_{l-1}^{l} f(\xi_k, y) dy \leq M_{k,l} \Delta y_l. \tag{8.15}$$

Summing up (8.15), we get

$$\sum_{l=1}^{m} m_{k,l} \Delta y_l \leq I(\xi_k) \leq \sum_{l=1}^{m} M_{k,l} \Delta y_l. \tag{8.16}$$

We multiply (8.16) by Δx_k and sum it up to get

$$\sum_{k=1}^{n} \sum_{l=1}^{m} m_{k,l} \Delta y_l \Delta x_k \leq \sum_{k=1}^{n} I(\xi_k) \Delta x_k \leq \sum_{k=1}^{n} \sum_{l=1}^{m} M_{k,l} \Delta y_l \Delta x_k.$$

or

$$\sum_{k=1}^{n} \sum_{l=1}^{m} m_{k,l} \Delta_{k,l} \leq \sum_{k=1}^{n} I(\xi_k) \Delta_k \leq \sum_{k=1}^{n} \sum_{l=1}^{m} M_{k,l} \Delta_{k,l}. \tag{8.17}$$

Now suppose $\|P\| \to 0$. Because f is integrable on R, the extreme left-hand side and the extreme right-hand side in (8.17) tend to $\int_R f(x, y) dx \, dy$. Then (8.12) is proved. Now (8.13) follows. $\square$

8.1.2 Double integrals on simple domains

We discuss now the existence of the Darboux integral on a planar set that is not necessarily a rectangle.

A subset D of $\mathbb{R}^2$ is said to be a *simple domain in respect to the horizontal axis* if it can be written as $D = \{(x, y) \mid a \leq x \leq b, \ g(x) \leq y \leq h(x),$ $g, h : [a, b] \to \mathbb{R}\}$, g, h are continuous on $[a, b]$ and of class C^1 on $]a, b[$.

Similarly, subset D of $\mathbb{R}^2$ is said to be a *simple domain in respect to the vertical axis* if it can be written as $D = \{(x, y) \mid c \leq y \leq d, \ g(y) \leq x \leq h(y),$ $g, h : [c, d] \to \mathbb{R}\}$, g, h are continuous on $[c, d]$ and of class C^1 on $]c, d[$.

A subset D of $\mathbb{R}^2$ is said to be a *simple domain* if it is either a simple domain in respect to the horizontal axis or a simple domain in respect to the vertical axis.

Suppose D is a simple domain and $f : D \to \mathbb{R}$. Suppose R is a bounded rectangle containing D. Define $\overline{f} : R \to \mathbb{R}$ by

$$\overline{f}(x, y) = \begin{cases} f(x, y), & (x, y) \in D \\ 0, & (x, y) \in R \setminus D. \end{cases}$$

Function f is said to be Darboux integrable on D if and only if $\overline{f}$ is Darboux integrable on R. If f is integrable on D we write

$$\int_D f(x,y)dx\,dy = \int_R \overline{f}(x,y)dx\,dy.$$

Remark. The previous definition is acceptable if it does not depend upon rectangle R. A sufficient condition for a function f to be integrable on a simple domain D is that f is continuous on R and the boundary of R is formed by a finite number of images of functions of class C^1. $\triangle$

Theorem 1.6. *Consider D a simple domain in respect to the horizontal axis and an integrable function $f : D \to \mathbb{R}$. Suppose that for every x the integral*

$$\int_{y_1(x)}^{y_2(x)} f(x,y)dy$$

exists. Then there exists the iterated integral

$$\int_a^b \left(\int_{y_1(x)}^{y_2(x)} f(x,y)dy \right) dx$$

(a and b are the smallest, respectively, largest, abscissae of the points of D) and we have

$$\int_D f(x,y)dx\,dy = \int_a^b \left(\int_{y_1(x)}^{y_2(x)} f(x,y)dy \right) dx.$$

Proof. Similar to the proof of Theorem 1.5. $\square$

Theorem 1.7. *Consider D a simple domain in respect to the vertical axis and an integrable function $f : D \to \mathbb{R}$. Suppose that for every x the integral*

$$\int_{x_1(y)}^{x_2(y)} f(x,y)dx$$

exists. Then there exists the iterated integral

$$\int_c^d \left(\int_{x_1(y)}^{x_2(y)} f(x,y)dx \right) dy$$

(c and d are the smallest, respectively, largest, ordinates of the points of D) and we have

$$\int_D f(x,y)dx\,dy = \int_c^d \left(\int_{x_1(y)}^{x_2(y)} f(x,y)dx \right) dy.$$

Theorem 1.8. *Consider D a simple domain in $\mathbb{R}^2$ and an integrable function $f : D \to \mathbb{R}$. Suppose $D = D_1 \cup D_2$, such that $D_1 \cap D_2 = \emptyset$ and D_1, D_2 are simple domains. Then*

$$\int_D f(x,y)dx\,dy = \int_{D_1} f(x,y)dx\,dy + \int_{D_2} f(x,y)dx\,dy.$$

Theorem 1.9. *Consider D a simple domain in $\mathbb{R}^2$, and $T : D \to \mathbb{R}^2$ a one-to-one mapping of class C^1 so that $\det T'$ is nonzero on D, $T(x,y) = (u,v)$. Suppose $f : T(D) \to \mathbb{R}$ is integrable. Then*

$$\int_{T(D)} f\,du\,dv = \int_D (f \circ T)|\det T'|\,dx\,dy. \tag{8.18}$$

Example 1.1. We introduce a proof of the improper integral (6.33) at page 271. Denote

$$I_a = \int_0^a e^{-x^2}\,dx.$$

Then we have

$$I_a^2 = \int_0^a e^{-x^2}dx \cdot \int_0^a e^{-y^2}dy = \int_0^a \int_0^a e^{-(x^2+y^2)}dx\,dy = \int_0^{\pi/2} d\theta \int_0^a \rho e^{-\rho^2}\,d\rho$$

$$= \frac{\pi}{2}\left(-\frac{1}{2}e^{-r^2}\right)_0^a = \frac{\pi}{4}\left(1 - e^{-a^2}\right).$$

Thus

$$I^2 = \lim_{a\to\infty} I_a^2 = \pi/4 \implies I = \sqrt{\pi}/2. \quad \triangle$$

8.2 Triple integrals

8.2.1 Triple integrals on parallelepipeds

Consider a parallelepiped $R = [a_1,b_1] \times [a_2,b_2] \times [a_3,b_3] \subset \mathbb{R}^3$ ($a_1 \leq b_1$, $a_2 \leq b_2$, and $a_3 \leq b_3$) and a *partition*

$$P = \{a_1 = x_0 \leq x_1 \leq \cdots \leq x_{n-1} \leq x_n = b_1\} \tag{8.19}$$
$$\times \{a_2 = y_0 \leq y_1 \leq \cdots \leq y_{m-1} \leq y_m = b_2\}$$
$$\times \{a_3 = z_0 \leq z_1 \leq \cdots \leq z_{\mu-1} \leq z_\mu = b_3\}$$

of it. For a partition P, we denote

$$\|P\| = \max_{1\leq i\leq n,\, 1\leq j\leq m,\, 1\leq k\leq \mu} \sqrt{(x_i - x_{i-1})^2 + (y_j - y_{j-1})^2 + (z_k - z_{k-1})^2}$$

and call it the *norm* of P. For $i = 1, 2, \ldots, n$ $j = 1, 2, \ldots, m$, $k = 1, 2, \ldots, \mu$, denote

$$R_{i,j,k} = \{(x,y,z) \mid x_{i-1} \leq x \leq x_i,\ y_{j-1} \leq y \leq y_j,\ z_{k-1} \leq z \leq z_k\}$$
$$= [x_{i-1}, x_i] \times [y_{j-1}, y_j] \times [z_{k-1}, z_k].$$

Let $f : R \to \mathbb{R}$ be a bounded function; that is, there exist real numbers m and M so that $m \leq f(x,y,z) \leq M$, for all $(x,y,z) \in R$. For $i = 1, 2, \ldots, n$ $j = 1, 2, \ldots, m,\ k = 1, 2, \ldots, \mu$, define

$$m_{i,j,k} = \inf_{(x,y,z) \in R_{i,j,k}} f(x,y,z),\quad M_{i,j,k} = \sup_{(x,y,z) \in R_{i,j,k}} f(x,y,z),$$

$$\Delta_{i,j,k} = (x_i - x_{i-1})(y_j - y_{j-1})(z_k - z_{k-1}),$$

and

$$L(P,f) = \sum_{i=1}^{n}\sum_{j=1}^{m}\sum_{k=1}^{\mu} m_{i,j,k}\Delta_{i,j,k},\quad U(P,f) = \sum_{i=1}^{n}\sum_{j=1}^{m}\sum_{k=1}^{\mu} M_{i,j,k}\Delta_{i,j,k}.$$

$L(P,f)$ is said to be the *lower Darboux sum* of the function f on R and $U(P,f)$ is said to be the *upper Darboux sum* of the function f on R. Because f is bounded, we have

$$m(b_1-a_1)(b_2-a_2)(b_3-a_3) \leq L(P,f) \leq U(P,f) \leq M(b_1-a_1)(b_2-a_2)(b_3-a_3).$$

Therefore we put

$$\sup_{P} L(P,f) = \underline{\int_R} f(x,y,z)\, dx\, dy\, dz = \underline{\int_R} f(x,y,z)\, dy\, dx\, dz \tag{8.20}$$

$$= \cdots = \underline{\int_R} f(x,y,z)\, dz\, dy\, dx,$$

$$\inf_{P} U(P,f) = \overline{\int_R} f(x,y,z)\, dx\, dy\, dz = \overline{\int_R} f(x,y,z)\, dy\, dx\, dz \tag{8.21}$$

$$= \cdots = \overline{\int_R} f(x,y,dz)\, dz\, dy\, dx$$

where the sup and the inf are considered over all partitions P of R. The number $\underline{\int_R} f(x,y,z)\, dx\, dy\, dz$ is said to be the *lower Darboux integral* of the function f on the rectangle R, and $\overline{\int_R} f(x,y,z)\, dx\, dy\, dz$ is said to be the *upper Darboux integral* of the function f on the rectangle R.

A partition P^* of R is said to be a *refinement* of P if $P \subset P^*$. Suppose two partitions P_1 and P_2 are given on R. Their common refinement P^* is defined by $P^* = P_1 \cup P_2$.

Lemma 8.2. *Let P^* be a refinement of P. Then*

$$L(P,f) \leq L(P^*,f) \tag{8.22}$$

and

$$U(P, f) \geq U(P^*, f). \tag{8.23}$$

Proof. If $P = P^*$, in (8.22) and (8.23) we have equalities. Suppose for the beginning that P^* contains just one extra point. Let (t, s, u) be this extra point and suppose that $x_{i-1} \leq t \leq x_i$, $y_{j-1} \leq s \leq y_j$, and $z_{k-1} \leq u \leq z_k$. Following a similar approach as in Lemma 8.1 we find that

$$U(P^*, f) - U(P, f) \leq 0.$$

If P^* contains m points more than P, we repeat the above reasoning m times. Thus (8.23) is proved.

Inequality (8.22) is proved in a similar way.   □

Theorem 2.1. *It holds that*

$$\underline{\int_R} f(x, y, z)\, dx\, dy\, dz \leq \overline{\int_R} f(x, y, z)\, dx\, dy\, dz.$$

Proof. Choose P_1 and P_2 two arbitrary partitions and let P^* be their common refinement. By Lemma 8.3 we have

$$L(P_1, f) \leq L(P^*, f) \leq U(P^*, f) \leq U(P_2, f).$$

Then

$$L(P_1, f) \leq U(P_2, f).$$

Keeping P_2 fixed and taking the supremum in respect to P_1 in the above inequality, we have

$$\underline{\int_R} f(x, y, z)\, dx\, dy\, dz \leq U(P_2, f).$$

Taking the infimum in respect to P_2 in the above inequality, we complete the proof.   □

If the lower and the upper Darboux integral of f are equal, we say that f is *Darboux integrable* on R and the common value of (8.20) and (8.21) is denoted by

$$\int_R f(x, y, z)\, dx\, dy\, dz = \cdots = \int_R f(x, y, z)\, dz\, dy\, dx.$$

In this case we may write

$$m(b_1 - a_1)(b_2 - a_2)(b_3 - a_3) \leq L(P, f) \leq \int_R f(x, y, z)\, dx\, dy\, dz \tag{8.24}$$

$$\leq U(P, f) \leq M(b_1 - a_1)(b_2 - a_2)(b_3 - a_3).$$

The Darboux integrability can be characterized by the next result.

Theorem 2.2. *Function f is Darboux integrable on R if and only if for every $\varepsilon > 0$ there exists a partition P of R such that*

$$U(P, f) - L(P, f) \leq \varepsilon. \tag{8.25}$$

Proof. Condition (8.25) is sufficient because from

$$L(P, f) \leq \underline{\int_R} f(x, y, z) \, dx \, dy \, dz \leq \overline{\int_R} f(x, y, z) \, dx \, dy \, dz \leq U(P, f)$$

it follows that

$$0 \leq \overline{\int_R} f(x, y, z) \, dx \, dy \, dz - \underline{\int_R} f(x, y, z) \, dx \, dy \, dz \leq \varepsilon;$$

that is, f is Darboux integrable on R.

Suppose that

$$\overline{\int_R} f(x, y, z) \, dx \, dy \, dz = \underline{\int_R} f(x, y, z) \, dx \, dy \, dz = \int_R f(x, y, z) \, dx \, dy \, dz$$

and let $\varepsilon > 0$ be given. Then we can find two partitions P_1 and P_2 such that

$$U(P_2, f) - \int_R f(x, y, z) \, dx \, dy \, dz \leq \frac{\varepsilon}{2}, \quad \int_R f(x, y, z) \, dx \, dy \, dz - L(P_1, f) \leq \frac{\varepsilon}{2}.$$

Let P be the common refinement of P_1 and P_2. Then

$$U(P, f) \leq U(P_2, f) \leq \int_R f(x, y, z) \, dx \, dy \, dz + \varepsilon/2$$
$$\leq L(P_1, f) + \varepsilon \leq L(P, f) + \varepsilon;$$

that is, inequality (8.25) holds.   □

Theorem 2.3. *Suppose f is continuous on R. Then f is Darboux integrable on R. Moreover, to every $\varepsilon > 0$ there corresponds a $\delta > 0$ such that*

$$\left| \sum_{i=1}^{n} \sum_{j=1}^{m} \sum_{k=1}^{\mu} f(t_i, s_j, u_k) \Delta_{i,j,k} - \int_R f(x, y, z) \, dx \, dy \, dz \right| \leq \varepsilon \tag{8.26}$$

for every partition P in (8.19) of R with $\|P\| < \delta$, and for every choice of points $(t_i, s_j, u_k) \in R_{i,j,k}$.

Proof. Let $\varepsilon > 0$ be given. Choose $\eta > 0$ such that $(b_1 - a_1)(b_2 - a_2)(b_3 - a_3)\eta < \varepsilon$. Function f is uniformly continuous on the compact set R, so there exists a $\delta > 0$ such that

$$\|f(x, y, z) - f(t, s, u)\| < \eta \tag{8.27}$$

if $\|(x, y, z) - (t, s, u)\| < \delta$ and $(x, y, z), (t, s, u) \in R$.

Now, choose a partition P of R such that $\|P\| < \delta$. By (8.27) we have $M_{i,j,k} - m_{i,j,k} \leq \eta$, for all $i = 1, 2, \ldots, n$, $j = 1, 2, \ldots, m$, and $k = 1, 2, \ldots, \mu$. Thus

$$U(P, f) - L(P, f) = \sum_{i=1}^{n} \sum_{j=1}^{m} \sum_{k=1}^{\mu} (M_{i,j,k} - m_{i,j,k}) \Delta_{i,j,k}$$

$$< \eta \sum_{i=1}^{n} \sum_{j=1}^{m} \sum_{l=1}^{\mu} \Delta_{i,j,k} < \varepsilon.$$

By the characterization theorem 1.2, we conclude that f is Darboux integrable on R.

The two numbers

$$\sum_{i=1}^{n} \sum_{j=1}^{m} \sum_{k=1}^{\mu} f(t_i, s_j, u_k) \Delta_{i,j,k} \quad \text{and} \quad \int_R f(x, y, z) \, dx \, dy \, dz$$

lie between $U(P, f)$ and $L(P, f)$, therefore inequality (8.26) is proved.   □

We introduce some arithmetic properties of the triple Darboux integral.

Theorem 2.4. (a) *Suppose f_1 and f_2 are Darboux integrable on R and p and q are reals. Then $p \cdot f_1 + q \cdot f_2$ is Darboux integrable on R and, moreover,*

$$\int_R (pf_1(x, y, z) + qf_2(x, y, z) \, dx \, dy \, dz = p \int_R f_1(x, y, z) \, dx \, dy \, dz \tag{8.28}$$

$$+ q \int_R f_2(x, y, z) \, dx \, dy \, dz.$$

(b) *Suppose f_1, f_2 are Darboux integrable on R and $f_1(x, y, z) \leq f_2(x, y, z)$, for all $(x, y, z) \in R$. Then*

$$\int_R f_1(x, y, z) \, dx \, dy \, dz \leq \int_R f_2(x, y, z) \, dx \, dy \, dz.$$

(c) *Suppose f is continuous on R. Then there exists $(\xi, \eta, \zeta) \in R$ such that*

$$\int_R f(x, y, z) \, dx \, dy \, dz = f(\xi, \eta, \zeta)(b_1 - a_1)(b_2 - a_2)(b_3 - a_3).$$

(d) *Suppose f and g are continuous on R, $g \geq 0$ on R, and $\int_R g(x, y, z) \, dx \, dy \, dz > 0$. Then there exists $(\xi, \eta, \zeta) \in R$ such that*

$$\int_R f(x,y,z)g(x,y,z)\, dx\, dy\, dz = f(\xi,\eta,\zeta)\int_R g(x,y,z)\, dx\, dy\, dz.$$

(e) *If f and $|f|$ are Darboux integrable on R, then*

$$\left|\int_R f(x,y,z)\, dx\, dy\, dz\right| \le \int_R |f(x,y,dz)|\, dx\, dy\, dz.$$

Proof. (a) Pick a positive ε. Then there exists a positive δ_1 so that for every partition P_1 of R with $\|P_1\| \le \delta_1$ one has

$$U(P_1,f_1) - L(P_1,f_1) < \varepsilon/(2|p|+1).$$

Similarly, there exists a positive δ_2 so that for every partition P_2 of R with $\|P_2\| \le \delta_2$ one has

$$U(P_2,f_2) - L(P_2,f_2) < \varepsilon/(2|q|+1).$$

Let P be the common refinement of P_1 and P_2. Then

$$0 \le U(P,pf_1+qf_2) - L(P,pf_1+qf_2)$$
$$= p(U(P_1,f_1) - L(P_1,f_1)) + q(U(P_2,f_2) - L(P_2,f_2))$$
$$< |p|\varepsilon/(2|p|+1) + |q|\varepsilon/(2|q|+1) < \varepsilon.$$

So by Theorem 2.2, we have that $pf_1 + qf_2$ is integrable on R.

We show equality (8.28). Pick a positive ε. Then there exists a positive δ_1 so that for every partition P_1 of R with $\|P_1\| \le \delta_1$ one has

$$\left|U(P_1,f_1) - \int_R f_1(x,y,z)dx\, dy\, dz\right| < \varepsilon/(2|p|+1).$$

Similarly, exists a positive δ_2 so that for every partition P_2 of R with $\|P_2\| \le \delta_2$ one has

$$\left|U(P_2,f_2) - \int_R f_2(x,y,z)dx\, dy\, dz\right| < \varepsilon/(2|q|+1).$$

Let P be the common refinement of P_1 and P_2. Then

$$\left|U(P,pf_1+qf_2) - p\int_R f_1(x,y,z)\, dx\, dy\, dz - q\int_R f_2(x,y,z)\, dx\, dy\, dz\right|$$
$$= \left|pU(P,f_1) + qU(P,f_2) - p\int_R f_1(x,y,z)\, dx\, dy\, dz - q\int_R f_2(x,y,z)\, dx\, dy\, dz\right|$$
$$\le \left|pU(P,f_1) - p\int_R f_1(x,y,z)\, dx\, dy\, dz\right|$$
$$+ \left|qU(P,f_2) - q\int_R f_2(x,y,z)\, dx\, dy, dz\right| < |p|\varepsilon/(2|p|+1) + |q|\varepsilon/(2|q|+1) < \varepsilon.$$

So (8.28) is proved.

(b) Because $f_2(x, y, z) - f_1(x, y, z) \geq 0$, for every partition P of R we have that $L(P, f_2 - f_1) \geq 0$. By (a), $f_2 - f_1$ is integrable. The conclusion follows.

(c) Because (8.24) holds and a continuous function on a compact set attains its extreme values, the result follows.

(d) Successively we have

$$m \cdot g(x, y, z) \leq f(x, y, z)g(x, y, z) \leq M \cdot g(x, y, z),$$

$$m \int_R g(x, y, z) dx \, dy \, dz \leq \int_R f(x, y, z)g(x, y, z) \, dx \, dy \, dz$$

$$\leq M \int_R g(x, y, z) \, dx \, dy \, dz,$$

$$m \leq \frac{\int_R f(x, y, z)g(x, y, z) \, dx \, dy \, dz}{\int_R g(x, y, dz) \, dx \, dy \, dz} \leq M.$$

The conclusion follows by (c).

(e) Because $-|f(x, y, z)| \leq f(x, y, z) \leq |f(x, y, z)|$, by (a) and (b) we have

$$- \int_R |f(x, y, z)| \, dx \, dy \, dz \leq \int_R f(x, y, z) \, dx \, dy \, dz \leq \int_R |f(x, y, z)| \, dx \, dy \, dz.$$

The conclusion follows.   □

Theorem 2.5. *Consider an integrable function $f : [a_1, b_1] \times [a_2, b_2] \times [a_3, b_3] = R \rightarrow \mathbb{R}$ and for each x in $[a_1, b_1]$ the integral*

$$I(x) = \int_{[a_2, b_2] \times [a_3, b_3]} f(x, y, z) \, dy \, dz$$

exists. Then there exists the iterated integral

$$\int_{[a_1, b_1]} I(x) \, dx = \int_{[a_1, b_1]} \left(\int_{[a_2, b_2] \times [a_3, b_3]} f(x, y, z) \, dy \, dz \right) dx \qquad (8.29)$$

and we have

$$\int_R f(x, y, z) \, dx \, dy \, dz = \int_{[a_1, b_1]} \left(\int_{[a_2, b_2] \times [a_3, b_3]} f(x, y, z) \, dy \, dz \right) dx. \qquad (8.30)$$

Proof. Consider partition (8.19) of R, $\Delta x_i = x_i - x_{i-1}$, $\Delta y_j = y_j - y_{j-1}$, and $\Delta z_k = z_k - z_{k-1}$ for $i = 1, 2, \ldots, n$, $j = 1, 2, \ldots, m$, $k = 1, 2, \ldots, \mu$. For every $(x, y, z) \in R_{i,j,k}$,

$$m_{i,j,k} \leq f(x, y, z) \leq M_{i,j,k}. \qquad (8.31)$$

Set $x = \xi_i$ for an arbitrary $\xi_i \in [x_{i-1}, x_i]$ and integrate (8.31) in respect to (y, z) on $[y_{j-1}, y_j] \times [z_{k-1}, z_k]$. Then

$$m_{i,j,k}\Delta y_j \Delta z_k \leq \int_{[y_{j-1},y_j]\times[z_{k-1},z_k]} f(\xi_i,y,z)dy\,dz \leq M_{i,j,k}\Delta y_j \Delta z_k. \quad (8.32)$$

Summing up (8.32), we get

$$\sum_{j=1}^{m}\sum_{k=1}^{\mu} m_{i,j,k}\Delta y_j \Delta z_k \leq I(\xi_i) \leq \sum_{j=1}^{m}\sum_{k=1}^{\mu} M_{i,j,k}\Delta y_j \Delta z_k. \quad (8.33)$$

We multiply (8.33) by Δx_i and sum it up to get

$$\sum_{i=1}^{n}\sum_{j=1}^{m}\sum_{k=1}^{\mu} m_{i,j,k}\Delta x_i \Delta y_j \Delta z_k \leq \sum_{i=1}^{n} I(\xi_k)\Delta x_i \leq \sum_{i=1}^{n}\sum_{j=1}^{m}\sum_{k=1}^{\mu} M_{i,j,k}\Delta x_i \Delta y_j \Delta z_k.$$

or

$$\sum_{i=1}^{n}\sum_{j=1}^{m}\sum_{k=1}^{\mu} m_{i,j,k}\Delta_{i,j,k} \leq \sum_{i=1}^{n} I(\xi_k)\Delta x_i \leq \sum_{i=1}^{n}\sum_{j=1}^{m}\sum_{k=1}^{\mu} M_{i,j,k}\Delta_{i,j,k}. \quad (8.34)$$

Suppose $\|P\| \to 0$. Because f is integrable on R, the extreme left-hand side and the extreme right-hand side in (8.34) tend to $\int_R f(x,y,z)\,dx\,dy\,dz$. Then (8.29) is proved. Now (8.30) follows. $\square$

8.2.2 Triple integrals on simple domains

We discuss the existence of the Darboux integral on a set in $\mathbb{R}^3$ that is not necessarily a parallelepiped.

A subset D_3 of $\mathbb{R}^3$ is said to be a *simple domain in respect to the xOy plane* if it can be written as $D_3 = \{(x,y,z) \mid (x,y) \in D_2, g(x,y) \leq z \leq h(x,y), g,h : D_2 \to \mathbb{R}\}$, g,h are continuous on D_2 and of class C^1 on int D_2, and D_2 is compact and its boundary is formed by arcs of functions of class C^1.

Similarly one can define a simple domain in respect to the yOz or to zOx planes.

A subset D of $\mathbb{R}^3$ is said to be a *simple domain* if it is a simple domain in respect to the one of the planes xOy, yOz, or zOx.

Suppose D is a simple domain in respect to the xOy plane and $f : D \to \mathbb{R}$. Suppose R is a bounded parallelepiped containing D. Define $\overline{f} : R \to \mathbb{R}$ by

$$\overline{f}(x,y,z) = \begin{cases} f(x,y,z), & (x,y,z) \in D \\ 0, & (x,y,z) \in R \setminus D. \end{cases}$$

Function f is said to be Darboux integrable on D if and only if $\overline{f}$ is integrable on R. If f is integrable on D we write

$$\int_D f(x,y,z)\,dx\,dy\,dz = \int_R \overline{f}(x,y,z)\,dx\,dy\,dz.$$

Remark. The previous definition is acceptable if it does not depend upon rectangle R. A sufficient condition for a function f to be integrable on a simple domain D is that f is continuous on R and the boundary of R is formed by a finite number of images of functions of class C^1. $\triangle$

Theorem 2.6. *Consider D_3 a simple domain in respect to the xOy plane and an integrable function $f : D_3 \to \mathbb{R}$. Suppose that for every (x, y) in the projection of D_3 on xOy the integral*

$$\int_{g(x,y)}^{h(x,y)} f(x, y, z)\, dz$$

exists. Then, if D_2 is the Cartesian projection of D_3 on the xOy plane, there exists the iterated integral

$$\int_{D_2} \left(\int_{g(x,y)}^{h(x,y)} f(x, y, z)\, dz \right) dx\, dy$$

and we have

$$\int_{D_3} f(x, y, z)\, dx\, dy\, dz = \int_{D_2} \left(\int_{g(x,y)}^{h(x,y)} f(x, y, z)\, dz \right) dx\, dy.$$

Proof. Similar to the proof of Theorem 2.5. $\square$

Theorem 2.7. *Consider D a simple domain in $\mathbb{R}^3$ and an integrable function $f : D \to \mathbb{R}$. Suppose $D = D_1 \cup D_2$, such that $D_1 \cap D_2 = \emptyset$ and D_1, D_2 are simple domains. Then*

$$\int_D f(x, y, z)\, dx\, dy\, dz = \int_{D_1} f(x, y, z)\, dx\, dy\, dz + \int_{D_2} f(x, y, z)\, dx\, dy\, dz.$$

Theorem 2.8. *Consider D a simple domain in $\mathbb{R}^3$, $T : D \to \mathbb{R}^3$ a one-to-one mapping of class C^1 so that $\det T'$ is nonzero on D, and $T(x, y, z) = (u, v, w)$. Suppose $f : T(D) \to \mathbb{R}$ is integrable. Then*

$$\int_{T(D)} f\, du\, dv\, dw = \int_D (f \circ T)|\det T'|\, dx\, dy\, dz. \tag{8.35}$$

8.3 *n*-fold integrals

8.3.1 *n*-fold integrals on hyperrectangles

Similar to the cases of double and triple integrals, one can define the n-fold integral, $n \in \mathbb{N}^*$, $n \geq 4$. Let $R = \prod_{i=1}^{n} [a_i, b_i]$ be a hyperrectangle, where $a_i \leq b_i$, $i = 1, 2, \ldots, n$. Take a partition

$$P = \prod_{i=1}^{n} \{a_i = x_{i0} \le x_{i1} \le \cdots \le x_{in_i} = b_i\} \tag{8.36}$$

of R. For a partition P, we denote

$$\|P\| = \max_{1 \le i_1 \le n_1, \ldots, 1 \le i_n \le n_n} \sqrt{\sum_{k=1}^{n}(x_{ki_k} - x_{ki_k-1})^2}$$

and call it the *norm* of P. For $i_1 = 1, 2, \ldots, n_1, \ldots, i_n = 1, 2, \ldots, n_n$, denote

$$R_{i_1,\ldots,i_n} = [x_{1,i_1-1}, x_{1,i_1}] \times \cdots \times [x_{n,i_n-1}, x_{n,i_n}].$$

Let $f : R \to \mathbb{R}$ be a bounded function; that is, there exist real numbers m and M so that $m \le f(x_1, \ldots x_n) \le M$, for all $(x_1, \ldots, x_n) \in R$. Define

$$m_{i_1,\ldots,i_n} = \inf\{f(x_1, \ldots, x_n) \mid (x_1, \ldots, x_n) \in R_{i_1,\ldots,i_n}\},$$
$$M_{i_1,\ldots,i_n} = \sup\{f(x_1, \ldots, x_n) \mid (x_1, \ldots, x_n) \in R_{i_1,\ldots,i_n}\},$$
$$\Delta_{i_1,\ldots,i_n} = \prod_{k=1}^{n}(x_{ki_k} - x_{ki_k-1}), \quad 1 \le i_1 \le n_1, \ldots, 1 \le i_n \le n_n$$

and

$$L(P,f) = \sum_{i_1=1}^{n_1} \cdots \sum_{i_n=1}^{n_n} m_{i_1,\ldots,i_n} \Delta_{i_1,\ldots,i_n},$$

$$U(P,f) = \sum_{i_1=1}^{n_1} \cdots \sum_{i_n=1}^{n_n} M_{i_1,\ldots,i_n} \Delta_{i_1,\ldots,i_n}.$$

$L(P,f)$ is said to be the *lower Darboux sum* of the function f on R and $U(P,f)$ is said to be the *upper Darboux sum* of the function f on R. Because f is bounded, we have

$$m \prod_{k=1}^{n}(b_k - a_k) \le L(P,f) \le U(P,f) \le M \prod_{k=1}^{n}(b_k - a_k).$$

Therefore we put

$$\sup_P L(P,f) = \underline{\int}_R f(x_1, \ldots, x_n)\, dx_1 \ldots dx_n = \underline{\int}_R f(x_1, \ldots, x_n)\, dx_2\, dx_1 \ldots dx_n$$

$$= \cdots = \underline{\int}_R f(x_1, \ldots, x_n)\, dx_n \ldots dx_1, \tag{8.37}$$

$$\inf_P U(P,f) = \overline{\int}_R f(x_1, \ldots, x_n)\, dx_1 \ldots dx_n = \overline{\int}_R f(x_1, \ldots, x_n)\, dx_2\, dx_1 \ldots dx_n$$

$$= \cdots = \overline{\int}_R f(x_1, \ldots, x_n)\, dx_n \ldots dx_1. \tag{8.38}$$

where the sup and the inf are considered over all partitions P of R. The number $\int_R f(x_1,\ldots,x_n)\,dx_1\ldots dx_n$ is said to be the *lower Darboux integral* of the function f on the rectangle R, whereas $\overline{\int}_R f(x_1,\ldots,x_n)\,dx_1\ldots dx_n$ is said to be the *upper Darboux integral* of the function f on the rectangle R.

A partition P^* of R is said to be a *refinement* of P if $P \subset P^*$. Suppose two partitions P_1 and P_2 are given on R. Their common refinement P^* is defined by $P^* = P_1 \cup P_2$.

For the sake of completeness we introduce the main statements that are similar to the cases $n = 2$ and $n = 3$.

Lemma 8.3. *Let P^* be a refinement of P. Then*

$$L(P, f) \leq L(P^*, f) \text{ and } U(P, f) \geq U(P^*, f).$$

Theorem 3.1. *It holds*

$$\int_R f(x_1,\ldots,x_n)\,dx_1\ldots dx_n \leq \overline{\int}_R f(x_1,\ldots,x_n)\,dx_1\ldots dx_n.$$

If the lower and the upper Darboux integrals of f are equal, we say that f is *Darboux integrable* on R and the common value of (8.37) and (8.38) is denoted by

$$\int_R f(x_1,\ldots,x_n)\,dx_1\ldots dx_n = \cdots = \int_R f(x_1,\ldots,x_n)\,dx_n\ldots dx_1.$$

In this case we may write

$$m \prod_{k=1}^{n} (b_k - a_k) \leq L(P, f) \leq \int_R f(x_1,\ldots,x_n)\,dx_1\ldots dx_n$$

$$\leq U(P, f) \leq M \prod_{k=1}^{n} (b_k - a_k).$$

The Darboux integrability can be characterized by the next result.

Theorem 3.2. *Function f is Darboux integrable on R if and only if for every $\varepsilon > 0$ there exists a partition P of R such that*

$$U(P, f) - L(P, f) \leq \varepsilon.$$

Theorem 3.3. *Suppose f is continuous on R. Then f is Darboux integrable on R. Moreover, to every $\varepsilon > 0$ there corresponds a $\delta > 0$ such that*

$$\left| \sum_{i_1=1}^{n_1} \cdots \sum_{i_n=1}^{n_n} f(t_{i_1},\ldots,t_{i_n})\Delta_{i_1,\ldots,i_n} - \int_R f(x_1,\ldots,x_n)\,dx_1\ldots dx_n \right| \leq \varepsilon$$

for every partition P in (8.36) of R with $\|P\| < \delta$, and for every choice of points $(t_{i_1},\ldots,t_{i_n}) \in R_{i_1,\ldots,i_n}$.

We introduce some arithmetic properties of the n-fold Darboux integral.

Theorem 3.4. (a) *Suppose f_1 and f_2 are Darboux integrable on R and p and q are reals. Then $p \cdot f_1 + q \cdot f_2$ is Darboux integrable on R and, moreover,*

$$\int_R (p f_1(x_1, \ldots, x_n)\, dx_1 \ldots dx_n + q f_2(x_1, \ldots, x_n))\, dx_1 \ldots dx_n$$

$$= p \int_R f_1(x_1, \ldots, x_n)\, dx_1 \ldots dx_n + q \int_R f_2(x_1, \ldots, x_n)\, dx_1 \ldots dx_n.$$

(b) *Suppose f_1, f_2 are Darboux integrable on R and $f_1(x_1, \ldots, x_n) \le f_2(x_1, \ldots, x_n)$, for all $(x_1, \ldots, x_n) \in R$. Then*

$$\int_R f_1(x_1, \ldots, x_n)\, dx_1 \ldots dx_n \le \int_R f_2(x_1, \ldots, x_n)\, dx_1 \ldots dx_n.$$

(c) *Suppose f is continuous on R. Then there exists $(\xi_1, \ldots, \xi_n) \in R$ such that*

$$\int_R f_1(x_1, \ldots, x_n)\, dx_1 \ldots dx_n = f(\xi_1, \ldots, \xi_n) \prod_{k=1}^{n} (b_k - a_k).$$

(d) *Suppose f and g are continuous on R, $g \ge 0$ on R, and $\int_R g(x_1, \ldots, x_n)\, dx_1 \ldots dx_n > 0$. Then there exists $(\xi_1, \ldots, \xi_n) \in R$ such that*

$$\int_R f(x_1, \ldots, x_n) g(x_1, \ldots, x_n)\, dx_1 \ldots dx_n$$

$$= f(\xi_1, \ldots, \xi_n) \int_R g(x_1, \ldots, x_n)\, dx_1 \ldots dx_n.$$

(e) *If f and $|f|$ are Darboux integrable on R, then*

$$\left| \int_R f_1(x_1, \ldots, x_n)\, dx_1 \ldots dx_n \right| \le \int_R |f_1(x_1, \ldots, x_n)|\, dx_1 \ldots dx_n.$$

Theorem 3.5. *Consider an integrable function $f : R \to \mathbb{R}$ and for each $x_1 \in [a_1, b_1]$ the integral*

$$I(x_1) = \int_{\prod_{k=2}^{n} [a_k, b_k]} f(x_1, x_2, x_3, \ldots, x_n)\, dx_2 \ldots dx_n$$

exists. Then there exists the iterated integral

$$\int_{[a_1, b_1]} I(x_1)\, dx_1 = \int_{[a_1, b_1]} \left(\int_{\prod_{k=2}^{n} [a_k, b_k]} f(x_1, x_2, x_3, \ldots, x_n)\, dx_2 \ldots dx_n \right) dx_1$$

and we have

$$\int_R f(x_1, \ldots, x_n)\, dx_1 \ldots, dx_n$$

$$= \int_{[a_1, b_1]} \left(\int_{\prod_{k=2}^{n} [a_k, b_k]} f(x_1, x_2, x_3, \ldots, x_n)\, dx_2 \ldots dx_n \right) dx_1.$$

8.3.2 n-fold integrals on simple domains

Here we briefly discuss the existence of the Darboux integral on a set in $\mathbb{R}^n$ which is not necessarily a hyperrectangle.

A subset D_n of $\mathbb{R}^n$ is said to be a *simple domain in respect to the* $Ox_1 \ldots x_{n-1}$ *plane* if it can be written as $D_n = \{(x_1, \ldots, x_n) \mid (x_1, \ldots x_{n-1}) \in D_{n-1}, \ g(x_1, \ldots, x_{n-1}) \leq x_n \leq h(x_1, \ldots, x_{n-1}), \ g, h : D_{n-1} \to \mathbb{R}\}$, g, h are continuous on D_{n-1} and of class C^1 on $\operatorname{int} D_{n-1}$, and D_{n-1} is compact and its boundary is formed by images of functions of class C^1.

Similarly one can define a simple domain in respect to other planes, that is, $Ox_2 \ldots x_n$ or $Ox_1 x_3 \ldots x_n$ or others.

A subset D of $\mathbb{R}^n$ is said to be a *simple domain* if it is a simple domain in respect to the one of the planes $Ox_{i_1} x_{i_2} \ldots x_{i_{n-1}}$, where $i_1, i_2, \ldots, i_{n-1}$ are pairwise distinct numbers in $\{1, 2, \ldots, n\}$.

Suppose D in $\mathbb{R}^n$ is a simple domain in respect to the $Ox_1 \ldots x_{n-1}$ plane and $f : D \to \mathbb{R}$. Suppose R is a bounded hyperrectangle containing D. Define $\overline{f} : R \to \mathbb{R}$ by

$$\overline{f}(x_1, \ldots, x_n) = \begin{cases} f(x_1, \ldots, x_n), & (x_1, \ldots, x_n) \in D \\ 0, & (x_1, \ldots, x_n) \in R \setminus D. \end{cases}$$

Function f is said to be Darboux integrable on D if and only if $\overline{f}$ is integrable on R. If f is integrable on D we write

$$\int_D f(x_1, \ldots, x_n) \, dx_1 \ldots dx_n = \int_R \overline{f}(x_1, \ldots, x_n) \, dx_1 \ldots dx_n.$$

Remark. The previous definition is acceptable if it does not depend upon hyperrectangle R. A sufficient condition for a function f to be integrable on a simple domain D in $\mathbb{R}^n$ is that f is continuous on R and the boundary of R is formed by a finite number of images of functions of class C^1. $\triangle$

Theorem 3.6. *Consider $D_n \subset \mathbb{R}^n$ a simple domain in respect to the $Ox_1 x_2 \ldots x_{n-1}$ plane and an integrable function $f : D_n \to \mathbb{R}$. Suppose that for every $(x_1 x_2 \ldots x_{n-1}, x_n)$ in the projection of D_n on $Ox_1 x_2 \ldots x_{n-1}$ the integral*

$$\int_{g(x_1, \ldots, x_{n-1})}^{h(x_1, \ldots, x_{n-1})} f(x_1 x_2 \ldots x_{n-1}, x_n) \, dx_n$$

exists. Then, if D_{n-1} is the Cartesian projection of D_n on $Ox_1 x_2 \ldots x_{n-1}$, there exists the iterated integral

$$\int_{D_{n-1}} \left(\int_{g(x_1, \ldots, x_{n-1})}^{h(x_1, \ldots, x_{n-1})} f(x_1, \ldots, x_n) \, dx_n \right) dx_1 \ldots dx_{n-1}$$

and we have

$$\int_{D_n} f(x_1 x_2 \ldots x_{n-1}, x_n) \, dx_1 \ldots dx_n$$

$$= \int_{D_{n-1}} \left(\int_{g(x_1,\ldots,x_{n-1})}^{h(x_1,\ldots,x_{n-1})} f(x_1 x_2 \ldots x_{n-1}, x_n) \, dx_n \right) dx_1 \ldots dx_{n-1}.$$

Theorem 3.7. *Consider D a simple domain in $\mathbb{R}^n$ and an integrable function $f : D \to \mathbb{R}$. Suppose $D = D_1 \cup D_2$, such that $D_1 \cap D_2 = \emptyset$ and D_1, D_2 are simple domains in $\mathbb{R}^n$. Then*

$$\int_D f(x_1, \ldots, x_n) \, dx_1 \ldots dx_n$$

$$= \int_{D_1} f(x_1, \ldots, x_n) \, dx_1 \ldots dx_n + \int_{D_2} f(x_1, \ldots, x_n) \, dx_1 \ldots dx_n.$$

Theorem 3.8. *Denote $R = \prod_{i=1}^m [a_i, b_i]$ and let (f_n) be a sequence of functions such that every $f_n : R \to \mathbb{R}$ is integrable. Suppose $f_n \xrightarrow{u} f$, where $f : R \to \mathbb{R}$. Then f is integrable and*

$$\lim_{n \to \infty} \int_R f_n = \int_R f.$$

8.4 Line integrals

An *arc* in $\mathbb{R}^n$ is a continuous function $\gamma : I \to \mathbb{R}^n$, where I is a nonempty interval in $\mathbb{R}$. The image of γ (the set $\gamma(I)$) is the *trajectory* of the arc γ. The Cartesian projections of the arc γ on the planes of coordinates are said to be the parametric equations of γ; that is, $\gamma = (\gamma_1, \ldots, \gamma_n)$. A parameterized arc $\gamma : I \to \mathbb{R}^n$ is said to be *smooth* if γ is differentiable on I, and γ' is continuous and nonnull on I.

8.4.1 Line integrals with respect to arc length

Consider γ a smooth arc in $\mathbb{R}^3$. Suppose $\gamma : I \to \mathbb{R}^3$ and $\gamma = (\varphi, \psi, \chi)$. Then the *length* of the arc γ is given by

$$\int_I \sqrt{(\varphi'(t))^2 + (\psi'(t))^2 + (\chi'(t))^2} \, dt. \tag{8.39}$$

If the arc belongs to the xOy plane, then χ is null on I.

Suppose $\varphi : [a, b] \to \mathbb{R}$, $\psi : [a, b] \to \mathbb{R}$, and $\chi : [a, b] \to \mathbb{R}$ define a smooth arc γ in $\mathbb{R}^3$, and f is a continuous function on the image of γ. Then the *line integral with respect to the arc length* of f on γ is

$$\int_a^b f(\varphi(t), \psi(t), \chi(t)) \sqrt{(\varphi'(t))^2 + (\psi'(t))^2 + (\chi'(t))^2} \, dt$$

or $\int_\gamma f ds$, or even $\int_\gamma f(x, y, z) ds$.

Two arcs (φ, ψ, χ) and $(\tilde{\varphi}, \tilde{\psi}, \tilde{\chi})$, $\varphi, \psi, \chi, \tilde{\varphi}, \tilde{\psi}, \tilde{\chi} : [a, b] \to \mathbb{R}$, are said to be *equivalent* if there exists a bijective and differentiable function $\eta : [a, b] \to [\tilde{a}, \tilde{b}]$ with η' nonzero on $[a, b]$ such that $\varphi(t) = \tilde{\varphi}(\eta(t))$, $\psi(t) = \tilde{\psi}(\eta(t))$, and $\chi(t) = \tilde{\chi}(\eta(t))$.

Remarks. (a) This definition can be extended to the case when γ is a finite union of arcs of class C^1 and f is bounded having a finite number of points of discontinuity.

(b) $\int_\gamma f ds$ remains unchanged if an arc is substituted by another equivalent to it. $\triangle$

8.4.2 Line integrals with respect to axis

Suppose $\varphi : [a, b] \to \mathbb{R}$, $\psi : [a, b] \to \mathbb{R}$, and $\chi : [a, b] \to \mathbb{R}$ define a smooth arc γ in $\mathbb{R}^3$ and P, Q, and R are continuous functions defined on the image of γ. Then the *line integral with respect to the axis* of the functions P, Q, and R on γ is

$$\int_a^b (P\left(\varphi(t), \psi(t), \chi(t)\right)\varphi'(t) + Q(\varphi(t), \psi(t), \chi(t))\psi'(t)$$
$$+ R(\varphi(t), \psi(t), \chi(t))\chi'(t))\, dt$$

or $\int_\gamma P(x, y, z)dx + Q(x, y, z)dy + R(x, y, z)dz$.

If the arc belongs to the xOy plane, then χ is zero on $[a, b]$.

Remarks. (a) This definition can be extended to the case when γ is a finite union of arcs of class C^1 and P, Q, and R are bounded having a finite number of points of discontinuity.

(b) Consider (φ, ψ, χ) and $(\tilde{\varphi}, \tilde{\psi}, \tilde{\chi})$, $\varphi, \psi, \chi, \tilde{\varphi}, \tilde{\psi}, \tilde{\chi} : [a, b] \to \mathbb{R}$, two equivalent and class C^1 arcs and let P, Q, R be continuous on the image of these arcs. Then $\int_\gamma Pdx + Qdy + Rdz$ and $\int_{\tilde{\gamma}} Pdx + Qdy + Rdz$ coincide if $\eta' > 0$; otherwise they have opposite signs.

(c) If a planar curve is given, then z is null. $\triangle$

8.4.3 Green formula

Let D be a connected, closed, bounded, planar domain in $\mathbb{R}^2$ whose boundary is smooth and such that a parallel to the horizontal or vertical axis intersects it in at most two points. Denote $\Gamma = \text{fr}\, D$. The sense on Γ is counter-clockwise.

Theorem 4.1. (Green[1])*Let f and g be two continuous functions on D having continuous partial derivatives on some open set containing D. Then*

---

[1] George Green, 1793–1841.

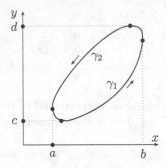

Fig. 8.1. Green formula

$$\int_\Gamma f(x,y)dx + g(x,y)dy = \int_D \left(\frac{\partial g}{\partial x} - \frac{\partial f}{\partial y}\right) dx\,dy. \qquad (8.40)$$

The integral in the left-hand side of (8.40) is a line integral with respect to the axis on the smooth arc Γ, and the integral in the right-hand side of (8.40) is a double integral on a connected, closed, bounded, planar domain in $\mathbb{R}^2$. The idea is suggested by Figure 8.1.

Proof. We decompose the arc Γ in two arcs γ_1 and γ_2 such that γ_1 is defined by $y = t$, $x = \psi(t)$, and $c \le t \le d$, whereas γ_2 is defined by $y = c + d - t$, $x = \varphi(c + d - t)$, and $c \le t \le d$. We suppose $\varphi(c) = \psi(c)$ and $\varphi(d) = \psi(d)$. Then we have

$$\int_D \frac{\partial g(x,y)}{\partial x}\,dxdy = \int_c^d dy \int_{\varphi(y)}^{\psi(y)} \frac{\partial g(x,y)}{\partial x}\,dx = \int_c^d (g(\psi(y),y) - g(\varphi(y),y))\,dy$$

$$= \int_\Gamma g dy.$$

Similarly, we get

$$\int_D \frac{\partial f(x,y)}{\partial y}\,dxdy = -\int_\Gamma f dx.$$

From these two results, the conclusion follows.   □

Remark. Theorem 4.1 of Green can be generalized in several respects. We mention here just one. We may consider that D is a finite union of connected, closed, bounded, planar domains in $\mathbb{R}^2$ whose boundaries are smooth and such that a parallel to the horizontal or vertical axis intersects each component in at most two points.   △

Corollary 4.1. *Assume the hypotheses of Theorem 4.1. Then the area of D is given by*

$$\frac{1}{2}\int_\Gamma x\,dy - y\,dx.$$

8.5 Integrals depending on parameters

Consider a hyperrectangle $R = A \times B$, where $A = \prod_{i=1}^{n}[a_i, b_i]$ and $B = \prod_{j=1}^{m}[c_j, d_j]$ $(a_i \leq b_i, \ i = 1, \ldots, n, \ c_j \leq d_j), \ j = 1, \ldots, m$, and a function $f : R \to \mathbb{R}$ integrable over x on A for each $y \in B$. Then a function $B \ni y \to I(y) \in \mathbb{R}$ is defined by

$$I(y) = \int_A f(x, y) \, dx. \tag{8.41}$$

A function of this sort is said to be an *integral depending on a parameter*.

Similarly, if $f : R \to \mathbb{R}$ is integrable over y on B for each $x \in A$, then a function $A \ni x \to J(x) \in \mathbb{R}$ is defined by

$$J(x) = \int_B f(x, y) \, dy. \tag{8.42}$$

Theorem 5.1. *Suppose $f : R \to \mathbb{R}$ integrable over y on B for each $x \in A$ and $f(x, y) \xrightarrow{u} h(y)$ on B as $x \to x_0 \in A$; then h is integrable on B and*

$$\lim_{x \to x_0} J(x) = \int_B h(y) \, dy.$$

Proof. Consider a sequence $(x_n)_{n \in \mathbb{N}^*}$ in A so that $x_n \to x_0$. Then $f_n(y) = f(x_n, y)$ is integrable on $[c, d]$ and $f_n(y) \xrightarrow{u} h(y)$ on B. By Theorem 3.8 we have that h is integrable on B and

$$\lim_{n \to \infty} \int_B f_n(y) dy = \int_B h(y) \, dy.$$

Because the sequence $(x_n)_{n \in \mathbb{N}^*}$ in A is arbitrary, it follows that

$$\lim_{x \to x_0} J(x) = \int_B h(y) \, dy. \quad \square$$

Theorem 5.2. *Suppose $f : R \to \mathbb{R}$ integrable over y on B for each $x \in A$ and f is continuous at $x_0 \in A$ uniformly in respect to $y \in B$. Then J is continuous at x_0.*

Proof. Because f is continuous at $x_0 \in A$ uniformly in respect to $y \in B$ for every $\varepsilon > 0$ there exists $\delta > 0$ so that for each $x \in A$, $\|x - x_0\| < \delta$ and each $y \in B$ we have

$$|f(x, y) - f(x_0, y)| < \frac{\varepsilon}{\prod_{j=1}^{m}(d_j - c_j)}.$$

Then

$$\|J(x) - J(x_0)\| \leq \int_B |f(x, y) - f(x_0, y)| \, dy < \int_B \frac{\varepsilon \, dy}{\prod_{j=1}^{m}(d_j - c_j)} = \varepsilon,$$

for all $x \in A$ with $\|x - x_0\| < \delta$. Thus J is continuous at x_0. $\quad \square$

Theorem 5.3. *Suppose $f : R \to \mathbb{R}$ is continuous and $\dfrac{\partial f(x,y)}{\partial x_i}$ exists for each $i = 1, \ldots, n$ on R. Then $\dfrac{\partial f(x,y)}{\partial x_i}$ is continuous on R for each $i = 1, \ldots, n$, J in (8.42) has continuous derivative on A, and*

$$\frac{\partial J(x)}{\partial x_i} = \int_B \frac{\partial f(x,y)\, dy}{\partial x_i}, \quad \forall i = 1, 2, \ldots, n, \quad x \in A.$$

Proof. It is enough to suppose that $n = 1$. Then by the Lagrange mean value theorem

$$\left| \frac{J(x) - J(x_0)}{x - x_0} - \int_B \frac{\partial f(x_0,y)\, dy}{\partial x} \right| \leq \int_B \left| \frac{f(x,y) - f(x_0,y)}{x - x_0} - \frac{\partial f(x_0,y)}{\partial x} \right| dy$$

$$= \int_B \left| \frac{\partial f(c,y)}{\partial x} - \frac{\partial f(x_0,y)}{\partial x} \right| dy,$$

where c is between x and x_0. Because $\partial f(x,y)/\partial x_i$ is uniformly continuous on R, for every $\varepsilon > 0$ there exists $\delta > 0$ so that for each $x \in A$ with $\|x - x_0\| < \delta$ and each $y \in B$ it holds

$$\left| \frac{\partial f(x,y)}{\partial x} - \frac{\partial f(x_0,y)}{\partial x} \right| < \frac{\varepsilon}{\prod_{j=1}^{m}(d_j - c_j)}.$$

It follows that J is differentiable at x_0 and

$$J'(x_0) = \int_B \frac{\partial f(x_0,y)\, dy}{\partial x}.$$

By Theorem 5.2 it follows that J' is continuous an A. $\square$

Corollary 5.2. *Suppose $a \in \mathbb{R} \setminus \{0\}$. Then*

$$\int_0^1 \frac{dx}{(a^2 + x^2)^2} = \frac{1}{2a^3} \arctan \frac{1}{a} + \frac{1}{2a^2(1 + a^2)} \implies \int_0^1 \frac{dx}{(1 + x^2)^2} = \frac{\pi}{8} + \frac{1}{4};$$

$$\int_0^1 \frac{dx}{(a^2 + x^2)^3} = \frac{3}{8a^5} \arctan \frac{1}{a} + \frac{3}{8a^4(1 + a^2)} + \frac{1}{4a^2(1 + a^2)^2}$$

$$\implies \int_0^1 \frac{dx}{(1 + x^2)^3} = \frac{3\pi}{32} + \frac{1}{4}.$$

Theorem 5.4. *Consider the functions $u, v : A \to B$ of class C^1 on A with $u \leq v$ componentwise and a continuous function $f : R \to \mathbb{R}$. Suppose $\dfrac{\partial f(x,y)}{\partial x_i}$ are continuous for each $i = 1, \ldots, n$. Then $J : A \to \mathbb{R}$ defined by*

$$J(x) = \int_{u(x)}^{v(x)} f(x,y)\, dy$$

is of class C^1 on A and

$$\frac{\partial J(x)}{\partial x_i} = \int_{u(x)}^{v(x)} \frac{\partial f(x,y)\,dy}{\partial x_i} + \frac{\partial v(x)}{\partial x_i} f(x,v(x)) - \frac{\partial u(x)}{\partial x_i} f(x,u(x)),$$

for all $i = 1, 2, \ldots, n$ and $x \in A$.

Proof. Denote $g(x) = G(x, u(x), v(x))$, where

$$G(x, u, v) = \int_u^v f(x, y)\,dy.$$

By Theorem 5.3 and Theorem 3.3 page 307 we get that function g has partial derivatives on A and

$$\frac{\partial g(x)}{\partial x_i} = \frac{\partial G(x, u(x), v(x))}{\partial x_i} + \frac{\partial G(x, u(x), v(x))}{\partial u} \cdot \frac{\partial u(x)}{\partial x_i}$$

$$+ \frac{\partial G(x, u(x), v(x))}{\partial v} \cdot \frac{\partial v(x)}{\partial x_i}$$

$$= \int_{u(x)}^{v(x)} \frac{\partial f(x,y)}{\partial x_i}\,dy - f(x, u(x), v(x))\frac{\partial u(x)}{\partial x_i} + f(x, u(x), v(x))\frac{\partial v(x)}{\partial x_i}.$$

By Theorem 5.2 and Theorem 2.2 at page 153 we get that $\partial g(x)/\partial x_i$ is continuous on A and thus J is of class C^1 on A. □

Theorem 5.5. (Fubini [2]) *Consider an integrable function $f : R \to \mathbb{R}$ which is integrable over x on A for each $y \in B$ and integrable over y on B for each $x \in A$. Then $J : A \to \mathbb{R}$ defined by (8.42) is integrable on A, the function $I : B \to \mathbb{R}$ defined by (8.41) is integrable on B, and*

$$\iint_{A \times B} f(x, y)\,dxdy = \int_A J(x)\,dx = \int_B I(y)\,dy;$$

that is,

$$\iint_{A \times B} f(x, y)\,dxdy = \int_A \left(\int_B f(x, y)\,dy \right) dx = \int_B \left(\int_A f(x, y)\,dx \right) dy.$$

Examples 5.1. (a) The first proof of the improper integral (6.33) at page 271 follows [24, p. 27]. Denote

$$f(x) = \left(\int_0^x e^{-t^2}\,dt \right)^2 \quad \text{and} \quad g(x) = \int_0^1 \frac{e^{-x^2(t^2+1)}}{t^2 + 1}\,dt, \quad x \geq 0.$$

Then for every nonnegative x,

---

[2] Guido Fubini, 1879–1943.

$$f'(x) + g'(x) = 2 \int_0^x e^{-x^2} e^{-t^2} dt - 2 \int_0^1 x e^{-x^2(t^2+1)} dt$$

$$= 2 \int_0^x e^{-x^2-t^2} dt - 2 \int_0^1 x e^{-x^2(t^2+1)} dt$$

$$= 2 \int_0^1 x e^{-x^2-u^2 x^2} du - 2 \int_0^1 x e^{-x^2(t^2+1)} dt = 0.$$

Therefore there exists a real number c so that $f + g = c$ on the whole nonnegative half line. So

$$c = f(0) + g(0) = \int_0^1 dt/(t^2 + 1) = \pi/4.$$

Hence

$$\frac{\pi}{4} = \lim_{x \to \infty} (f(x) + g(x)) = \lim_{x \to \infty} f(x) = \left(\int_0^\infty e^{-t^2} dt \right)^2,$$

and thus the result is proved.
(b) We introduce another proof[3] of the same improper integral (6.33) at page 271. Denote

$$I = \int_0^\infty e^{-x^2} dx.$$

Take $x = \alpha y$, $\alpha > 0$. Then

$$I = \int_0^\infty e^{-\alpha^2 y^2} \alpha \, dy.$$

Multiply the last relation by $e^{-\alpha^2}$ and integrate it in respect to α from 0 to ∞; that is,

$$I \cdot \int_0^\infty e^{-\alpha^2} d\alpha = \int_0^\infty \int_0^\infty e^{-\alpha^2(1+y^2)} \alpha \, dy d\alpha$$

or

$$I^2 = - \int_0^\infty \left(\frac{e^{-\alpha^2(1+y^2)}}{2(1+y^2)} \right)_{\alpha=0}^{\alpha \to \infty} dy = \int_0^\infty \frac{dy}{2(1+y^2)} = \frac{1}{2} \arctan y \Big|_0^\infty = \frac{\pi}{4}.$$

Thus $I = \sqrt{\pi}/2$.
(c) We show the functional equation of the beta function; that is,

$$B(x, y) = \frac{x+y}{y} B(x, y+1), \quad x, y > 0.$$

We have

---

[3] This proof was introduced by Camille Jordan in the academic year 1892–1983 at the analysis lectures at the Polytechnique in Paris.

$$\Gamma(x)\Gamma(y) = 4 \int_0^\infty u^{2x-1} e^{-u^2} du \int_0^\infty v^{2y-1} e^{-v^2} du$$

$$= 4 \int_0^\infty e^{-(u^2+v^2)} u^{2x-1} v^{2y-1} dudv.$$

Change the variables $u = \rho \cos\theta$, $v = \rho \sin\theta$, getting

$$\Gamma(x)\Gamma(y) = 4 \int_0^\infty \int_0^{\pi/2} e^{-r^2} r^{2(x+y)-1} \cos^{2x-1}\theta \cdot \sin^{2y-1}\theta \, dr \, d\theta$$

$$= 2 \int_0^\infty e^{-r^2} r^{2(x+y)-1} dr \cdot 2 \int_0^{\pi/2} \cos^{2x-1}\theta \cdot \sin^{2y-1}\theta \, d\theta$$

$$= \Gamma(x+y)B(y,x). \quad \triangle$$

8.6 Exercises

8.1. Evaluate $\int_D x^2 y dx\, dy$, $\quad D = \{(x,y) \mid 0 \le x \le 1, \ 0 \le y \le 2\}$.

8.2. Evaluate $\int_D (x^2 + y)dx\, dy$, $\quad D = \{(x,y) \mid 0 \le x \le 1, \ 0 \le y \le 2\}$.

8.3. Evaluate $\int_D (x+y)dx\, dy$, where D is the compact set between the curves $y = x^2$ and $x = y^2$.

8.4. Evaluate $\int_D (x^2 + y^2)dx\, dy$, $\quad D = \{(x,y) \mid x^2 + y^2 \le 1\}$.

8.5. Evaluate $\int_D (x^2 + y^2)dx\, dy$, where D is the bounded domain bounded by the curves $x^2 + y^2 = 1$, $\quad y = \sqrt{3}x$, $\quad x = \sqrt{3}y$.

8.6. Evaluate

$$\int_D e^{(x^3+y^3)/(xy)} dx\, dy, \tag{8.43}$$

where D is the bounded domain bounded by the curves $y^2 = 2px$, $\quad x^2 = 2py$, where $p > 0$.

8.7. Evaluate $\int_D x^2 yz dx\, dy\, dz$, $D = \{(x,y,z) \mid 0 \le x \le 1, \ 0 \le y \le 2, \ 0 \le z \le 1\}$.

8.8. Evaluate

$$\int_D (x^2 + y + z)dx\, dy\, dz, \tag{8.44}$$

where $D = \{(x,y,z) \mid 0 \le x \le 1, \ 0 \le y \le 2, \ 0 \le z \le 1\}$.

8.9. Evaluate $\int_D (x \cdot y \cdot z)dx\, dy\, dz$, where D is the compact set given by $x^2 + y^2 \le z \le 1$.

8.10. Evaluate $\int_D (x^2 + y^2 + z^2)dx\, dy\, dz$, $D = \{(x,y,z) \mid x^2 + y^2 + z^2 \le 1\}$.

8.11. Find the volume of the closed unit ball (in the Euclidean metric) in $\mathbb{R}^3$.

8.12. Consider n a positive integer, $r > 0$, and the set $A(r) \subset \mathbb{R}^n$ given as $A(r) = \{(x_1, \ldots, x_n) \mid x_k \geq 0, \ k = 1, \ldots, n, \ x_1 + \cdots + x_n \leq r\}$. Consider $f : \mathbb{R}^n \to \mathbb{R}$ a differentiable function in a neighborhood of $0_n \in \mathbb{R}^n$. Show that

$$\lim_{r \downarrow 0} \frac{1}{r^{n+1}} \int_{A(r)} (f(x_1, \ldots, x_n) - f(0_n)) dx_1 \ldots dx_n = \frac{1}{(n+1)!} \sum_{k=1}^{n} \frac{f(0_n)}{\partial x_k}.$$

8.13. Find the line integral of $f(x, y) = \sqrt{2(2-y)}$ along the curve $x(t) = t - \sin t$, $y(t) = 1 - \cos t$, $t \in [0, \pi/2]$.

8.14. Find the line integral of $f(x, y, z) = z(x^2 + y^2)$ along the spiral $x(t) = t \cos t$, $y(t) = t \sin t$, $z(t) = t$ for $t \in [0, 1]$.

8.15. Suppose $a > 0$. Evaluate $I = \int_\gamma dx/(x^3 + y^3)$, where γ is the curve given by $x = a \cos t$, $y = a \sin t$, and $t \in [0, \pi/2]$.

8.16. Let γ be the segments generated by the points $A_1 = (1, 0, 0)$, $A_2 = (0, 1, 0)$, $A_3 = (0, 0, 1)$ and suppose that the sense of movement on the triangle $A_1 A_2 A_3$ is $A_1 \to A_2 \to A_3 \to A_1$. Find the integral $I = \int_\gamma y \, dx + z \, dy + x \, dz$.

8.17. Find the area of the ellipse $x^2/a^2 + y^2/b^2 = 1$, $a, b > 0$.

8.18. For $ab \neq 0$, find $F(a, b) = \int_0^{\pi/2} dx/(a^2 \cos^2 x + b^2 \sin^2 x)^2$.

8.19. For $ab \neq 0$, find $F(a, b) = \int_0^{\pi/2} \ln(a^2 \sin^2 x + b^2 \cos^2 x) dx$.

8.20. Suppose $a, b > 0$. Find $\int_0^1 f(x) \, dx$ if $f(0) = 0$, $f(1) = b - a$, and $f(x) = (x^b - x^a)/\ln x$, whenever $0 < x < 1$.

8.21. For $a > 0$, evaluate

$$\int_0^\infty e^{-ax} \frac{\sin(bx) - \sin(cx)}{x} \, dx.$$

8.7 References and comments

The theoretical results of this chapter can be found in many textbooks. We mention only two of them, namely [70] and [88].

9
Constants

The aim of this chapter is to introduce several constants, various ways to define them, and their utility. Many more scholarly excellent expositions concerning them are available. Our treatment of these constants is necessarily incomplete.

9.1 Pythagoras's constant

Sometimes it is considered that Pythagoras's constant is $\sqrt{2} = 1.4142135623\ldots$. We introduced it at page 11. More exactly, we proved that no rational number p satisfies the relation $p^2 = 2$, Subsection 1.2.1.

9.1.1 Sequences approaching $\sqrt{2}$

We already saw that the sequence (x_n) defined by (3.82) at page 140, with $a = 2$, converges to $\sqrt{2}$.

Another elementary and ancient algorithm to define it consists in using the double sequence

$$\begin{cases} a_{n+1} = a_n + 2b_n, \\ b_{n+1} = a_n + b_n, \end{cases} \tag{9.1}$$

which verifies

$$\frac{a_{n+1}}{b_{n+1}} = \frac{a_n/b_n + 2}{a_n/b_n + 1}.$$

Thus

$$\lim x_n = \lim \frac{a_n}{b_n} = \sqrt{2}.$$

This simple algorithm uses only additions of integers and eventually a final division. Denoting the error by $\varepsilon_n = |\sqrt{2} - a_n/b_n|$, one easily can check that $\varepsilon_{n+1} < \varepsilon_n/5$. This involves a geometrical convergence.

It may be convenient to use the matrix representation of the sequence. We write the sequence as

$$\begin{pmatrix} a_{n+1} \\ b_{n+1} \end{pmatrix} = \begin{pmatrix} 1 & 2 \\ 1 & 1 \end{pmatrix} \begin{pmatrix} a_n \\ b_n \end{pmatrix}.$$

Denote

$$A = \begin{pmatrix} 1 & 2 \\ 1 & 1 \end{pmatrix}.$$

Thus we get

$$\begin{pmatrix} a_{n+1} \\ b_{n+1} \end{pmatrix} = A^n \begin{pmatrix} a_1 \\ b_1 \end{pmatrix}.$$

It is possible to improve the *order of convergence* of $\sqrt{2}$.

The matrix representation may be used with $n = 2^p$ and due to the relation

$$A^n = \begin{pmatrix} a_n & 2b_n \\ 1 & 1 \end{pmatrix}.$$

We note that taking $a = 2$ in Exercise 3.4 at page 140, we get another algorithm of finding $\sqrt{2}$.

9.2 Archimedes' constant

Archimedes' constant is π.

Archimedes considered inscribed and circumscribed regular polygons of 96 sides and deduced that

$$3 + \frac{10}{71} < \pi < 3 + \frac{1}{7}.$$

9.2.1 Recurrence relation

Proposition 2.1. *Let a_n and b_n be the perimeters of the circumscribed and inscribed regular n-gon and a_{2n} and b_{2n} be the perimeters of the circumscribed and inscribed regular $2n$-gon. Then*

$$a_{2n} = \frac{2a_n b_n}{a_n + b_n}, \tag{9.2}$$

$$b_{2n} = \sqrt{a_{2n} b_n}. \tag{9.3}$$

Proof. Consider a circle of radius $r = 1$ and denote by s_c and s_i the side lengths of circumscribed, respectively, inscribed regular n-gon. If $\alpha_n = \pi/n$, then

$$s_c = 2 \tan \alpha_n, \quad s_i = 2 \sin \alpha_n$$

and

$$a_n = 2n \tan \alpha_n, \quad b_n = 2n \sin \alpha_n.$$

Applying

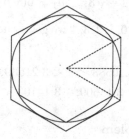

Fig. 9.1. Archimedes' algorithm for $n = 6$

$$\tan \frac{x}{2} = \frac{\tan x \sin x}{\tan x + \sin x}$$

one successively has

$$\frac{2a_n b_n}{a_n + b_n} = \frac{2 \cdot 2n \tan \alpha_n \sin \alpha_n}{2n \tan \alpha_n + \sin \alpha_n} = 4n \frac{\tan \alpha_n \sin \alpha_n}{\tan \alpha_n + \sin \alpha_n} = 4n \tan \alpha_{2n} = a_{2n}.$$

Thus (9.2) holds. Also

$$\sqrt{a_{2n} b_n} = 2n \sqrt{2 \tan \alpha_{2n} 2 \sin \alpha_{2n} \cos \alpha_{2n}} = 4n \sqrt{\sin^2 \alpha_{2n}} = 4n \sin \alpha_{2n} = b_{2n},$$

and (9.3) holds as well.   □

Sometimes the Archimedes' algorithm is called the Borchardt–Pfaff algorithm, page 140.

Remark. We note that for every $n \in \mathbb{N}^*$

$$b_n < 2\pi < a_n.$$

As we argued at page 140, the sequences (a_n) and (b_n) converge to the same limit. Hence

$$\lim a_n = \lim b_n = 2\pi.$$

Thus we supplied an answer to Exercise 3.11 at page 140, which was based on geometrical arguments.   △

Remark. Suppose we start from a hexagon, take as initial values

$$c_0 = a_6 = 4\sqrt{3}, \quad d_0 = b_6 = 6,$$

and consider the sequences

$$c_n = a_{6 \cdot 2^n}, \quad d_n = b_{6 \cdot 2^n}$$

These are subsequences of (a_n), respectively (b_n). Thus their limit is 2π. Then

$$c_1 = 24(2 - \sqrt{3}), \quad d_1 = 6(\sqrt{6} - \sqrt{2}).$$

From (c_n) and (d_n) for $n = 0, 1, 2, 3, 4$ we get some numerical approximations of π which are listed below.

$$3.00000 < \pi < 3.46410, \quad 3.10583 < \pi < 3.21539, \quad 3.13263 < \pi < 3.15966,$$
$$3.13935 < \pi < 3.14609, \quad 3.14103 < \pi < 3.14271.$$

9.2.2 Buffon needle problem

This problem introduces π in a natural way by probabilistic arguments.

The Buffon[1] needle problem reads as follows, according to [55, p. 36]. A plane is partitioned by parallel straight lines separated by a distance of $2a$. A needle of length $2l$ $(l < a)$ is thrown at random onto the plane. Find the probability p of the needle touching one of the lines.

The answer to this problem is

$$p = \frac{2l}{a\pi}.$$

Therefore repeating this experiment of a large number of times, results in a number of frequencies approaching p, implicitly π.

9.3 Arithmetic–geometric mean

The arithmetic–geometric mean iteration of Gauss to two nonnegative numbers a and b has been introduced by Problem 3.10 at page 140 as the common limit point of the sequences (a_n) and (b_n) defined by

$$a_1 = a \geq 0, \ b_1 = b \geq 0, \quad a_{n+1} = (a_n + b_n)/2, \quad b_{n+1} = \sqrt{a_n b_n}. \tag{9.4}$$

We assume $a \geq b$. From the arithmetic–geometric mean we have $a_n \geq b_n$ and then $a_n \geq (a_n + b_n)/2 = a_{n+1}$ and $b_{n+1} = \sqrt{a_n b_n} \geq b_n$ for every n. Therefore, by induction

$$b \leq b_1 \leq \cdots \leq b_n \leq b_{n+1} \leq \cdots \leq a_{n+1} \leq a_n \leq \cdots \leq a_1 \leq a,$$
$$0 \leq a_{n+1} - b_{n+1} \leq a_{n+1} - b_n = (a_n - b_n)/2 \leq (a_0 - b_0)/2^n,$$
$$a_{n+1} - b_{n+1} = \frac{1}{2}\left((a_n + b_n) - 2\sqrt{a_n b_n}\right) = \frac{1}{2}(\sqrt{a_n} - \sqrt{b_n})^2$$
$$= \frac{(a_n - b_n)^2}{2(\sqrt{a_n} + \sqrt{b_n})^2}.$$

We have denoted the common limit point by $M(a, b)$. Obviously, $M(a, b) = M(b, a)$.

---

[1] Georges Louis Leclerc, Comte de Buffon, 1707–1788.

Note if one of the initial terms is zero, say $a = 0$, then $M(0, b) = 0$, for every $b \geq 0$. Also $M(a, a) = a$. Because $M(a, b) \geq \min\{a, b\}$, hereafter we assume that $a \geq b > 0$.

Denote

$$c_n = \sqrt{a_n^2 - b_n^2}, \quad n \in \mathbb{N}^*.$$

Then

$$c_{n+1} = \frac{a_n - b_n}{2} = \frac{c_n^2}{4a_{n+1}} \leq \frac{c_n^2}{M(a, b)}, \quad n \in \mathbb{N}^*.$$

Thus (c_n) converges quadratically to zero.

Because

$$a_n = a_{n+1} + c_{n+1} \text{ and } b_n = a_{n+1} - c_{n+1}, \tag{9.5}$$

one can define a_n, b_n, and c_n for negative n. Denote $\underline{a}_0 = a_0$ and $\underline{b}_0 = b_0$. Then by (9.5) one defines $(\underline{a}_n)$ and $(\underline{b}_n)$ for negative indices. By induction for any integer n, we have

$$\underline{a}_n = 2^{-n} a_{-n}, \quad \underline{b}_n = 2^{-n} c_{-n}, \text{ and } \underline{c}_n = 2^{-n} b_{-n}. \tag{9.6}$$

Proposition 3.2. *Consider two positive numbers a and b and two sequences (a_n) and (b_n) connected as in (9.4). The arithmetic–geometric mean $M(a, b)$ fulfills the following relations.*

$$M(a, a) = a,$$
$$M(a, b) = M(a_n, b_n), \quad n \geq 0, \tag{9.7}$$
$$M(a, b) = M\left(\frac{a + b}{2}, \sqrt{ab}\right), \tag{9.8}$$
$$\lambda M(a, b) = M(\lambda a, \lambda b), \quad \lambda \geq 0, \tag{9.9}$$
$$M(1, \sqrt{1 - a^2}) = M(1 + a, 1 - a), \quad |a| \leq 1, \tag{9.10}$$
$$M(1, b) = \frac{1 + b}{2} M\left(1, \frac{2\sqrt{b}}{1 + b}\right). \tag{9.11}$$

Proof. The limit of a convergent sequence remains unchanged if we neglect some (finite number of) terms. Thus (9.7) follows. Based on this reasoning, (9.8) and (9.10) hold. The identity (9.9) is obvious. Equality (9.11) follows from (9.8) and (9.9). □

From the third relation we have that the convergence is of order 2 (or quadratic)

$$\left| \frac{a_{n+1} - M(a, b)}{(a_n - M(a, b))^2} \right| = O(1).$$

Theorem 3.1. *Given a and b positive numbers with $a \geq b$. Then*

$$I(a, b) = \int_0^{\pi/2} \frac{d\theta}{\sqrt{a^2 \cos^2 \theta + b^2 \sin^2 \theta}} = \frac{\pi}{2M(a, b)}.$$

$I(a, b)$ denotes the *complete elliptic integral of the first kind.*

Proof. Consider the substitution $y = b \tan \theta$ to get

$$I(a, b) = \frac{1}{2} \int_{-\infty}^{+\infty} \frac{dy}{\sqrt{(a^2 + y^2)(b^2 + y^2)}}.$$

By the substitution $v = (y - ab/y)/2$ one has

$$I(a, b) = \frac{1}{2} \int_{-\infty}^{+\infty} \frac{2 dv}{\sqrt{((a + b)^2 + 4v^2)(ab + v^2)}}.$$

Thus

$$I(a, b) = I\left((a + b)/2, \sqrt{ab}\right)$$
$$= I(a_1, b_1) = \cdots = I(a_n, b_n) = \cdots = I(M(a, b), M(a, b)).$$

Then the last integral is

$$I(a, b) = I(M(a, b), M(a, b)) = \int_0^{\pi/2} \frac{d\theta}{M(a, b)} = \frac{\pi}{2M(a, b)}. \quad \square$$

This theorem allows us to compute the complete elliptic integral of the first kind by means of the arithmetic–geometric mean iteration.

The complete elliptic integral of the second kind is defined as the length of a quarter of an ellipse with major axis a and minor axis b by

$$J(a, b) = \int_0^{\pi/2} \sqrt{a^2 \cos^2\theta + b^2 \sin^2\theta} \, d\theta.$$

Suppose b and b' are nonnegative and connected by $b^2 + b'^2 = 1$.

Theorem 3.2. (Legendre[2]) *One has*

$$L(b) = I(1, b)J(1, b) + I(1, b')J(1, b) - I(1, b)I(1, b') = \frac{\pi}{2}. \tag{9.12}$$

Legendre proved the previous theorem showing that the derivative of the function $L(b)$ with respect to b is constantly null. Therefore the constant value of this function is given by any particular value of b, so we choose $b = 0$.

Theorem 3.3. (Legendre) *One has*

$$2J\left(\frac{a + b}{2}, \sqrt{ab}\right) - J(a, b) = ab \, I\left(\frac{a + b}{2}, \sqrt{ab}\right).$$

---

[2] Adrien-Marie Legendre, 1752–1833.

By the arithmetic–geometric mean iteration we have

$$J(a_n, b_n) = 2J(a_{n+1}, b_{n+1}) - a_n b_n I(a_{n+1}, b_{n+1})$$
$$= 2J(a_{n+1}, b_{n+1}) - a_n b_n I(a_0, b_0).$$

Because

$$4a_{n+1}^2 - 2a_n^2 - 2a_n b_n = -c_n^2,$$

we have

$$2^{n+1}(J(a_{n+1}, b_{n+1}) - a_{n+1}^2 I(a_0, b_0)) - 2^n(J(a_n, b_n) - a_n^2 I(a_0, b_0))$$
$$= 2^{n-1} c_n^2 I(a_0, b_0).$$

From here

$$J(a_0, b_0) = \left(a_0^2 - \frac{1}{2}\sum_{k=0}^{\infty} 2^k(a_k^2 - b_k^2)\right) I(a_0, b_0). \tag{9.13}$$

Thus the two complete elliptic integrals are related by (9.13).

Substitute $a_0 = 1$ and $b_0 = b = b' = 1/\sqrt{2}$ in (9.12) and (9.13). There follow

$$2I(1, 1/\sqrt{2})J(1, 1/\sqrt{2}) - I^2(1, 1/\sqrt{2}) = \frac{\pi}{2},$$

$$\left(1 - \frac{1}{2}\sum_{k=0}^{\infty} 2^k(a_k^2 - b_k^2)\right) I(1, 1/\sqrt{2}) = J(1, 1/\sqrt{2}).$$

Now eliminate $J(1, 1/\sqrt{2})$. We get

$$\left(1 - \sum_{k=0}^{\infty} 2^k(a_k^2 - b_k^2)\right) I^2(1, 1/\sqrt{2}) = \frac{\pi}{2}.$$

From $I(a, b) = \pi/(2M(a, b))$ we obtain

$$\left(1 - \sum_{k=0}^{\infty} 2^k(a_k^2 - b_k^2)\right) \frac{\pi^2}{(2M(1, 1/\sqrt{2}))^2} = \frac{\pi}{2}.$$

Thus we proved a result supplying π by the arithmetic–geometric mean iteration.

Theorem 3.4. *Consider $a_0 = 1$, $b_0 = 1/\sqrt{2}$ and the sequences (a_n) and (b_n) defined by (9.4). Then*

$$\pi = \frac{2M^2(1, 1/\sqrt{2})}{1 - \sum_{k=0}^{\infty} 2^k(a_k^2 - b_k^2)}.$$

Consider an arbitrary $k \in]0, 1[$ and $k' = \sqrt{1 - k^2}$.

Proposition 3.3. *It holds*

$$\lim_{k\downarrow 0}\left(\ln\frac{4}{k} - I(1,k)\right) = 0. \tag{9.14}$$

Proof. Consider

$$A(k) = \int_0^{\pi/2} \frac{k'\sin\theta\,d\theta}{\sqrt{k^2 + (k')^2\cos^2\theta}} \quad \text{and} \quad B(k) = \int_0^{\pi/2}\sqrt{\frac{1 - k'\sin\theta}{1 + k'\sin\theta}}\,d\theta.$$

Because $1 - (k'\sin\theta)^2 = \cos^2 + (k\sin\theta)^2 = (k'\cos\theta)^2 + k^2$, one has

$$I(1,k) = A(k) + B(k).$$

Substitute $u = k'\cos\theta$ into A to get

$$A(k) = \int_0^{k'} \frac{du}{\sqrt{u^2 + k^2}} = \ln\frac{1 + k'}{k}. \tag{9.15}$$

We have

$$\lim_{k\downarrow 0} B(k) = B(0) = \int_0^{\pi/2} \frac{\cos\theta\,d\theta}{1 + \sin\theta} = \ln 2. \tag{9.16}$$

From (9.16) and (9.15), we get (9.14). $\square$

We introduce a quadratic algorithm for π.

Theorem 3.5. (The Brent[3]–Salamin[4] algorithm) *Let* $a_0 = 1$ *and* $b_0 = 1/\sqrt{2}$. *Define*

$$\pi_n = \frac{2a_{n+1}^2}{1 - \sum_{k=0}^n 2^k c_k^2},$$

where $c_n = \sqrt{a_n^2 - b_n^2}$, *and* a_n *and* b_n *are computed by the arithmetic-geometric iteration. Then*

(a) π_n *increases monotonically to* π *and satisfies*

$$\pi - \pi_n \leq \frac{\pi^2 2^{n+4} e^{-\pi 2^{n+1}}}{M^2(1, 1/\sqrt{2})}.$$

(b) *It holds*

$$\pi - \pi_{n+1} \leq \frac{2^{-(n+1)}}{\pi^2}(\pi - \pi_n)^2.$$

---

[3] Richard P. Brent, 1946– .
[4] Eugéne Salamin.

9.4 BBP formulas

The power of advanced computing once more became clear in 1996 when Bailey, Borwein, and Plouffe[5] succeeded in [13] in obtaining a simple formula for π using an algorithm for determining linear relations on $\mathbb{N}$ between constants. Their approach is based on algorithms, not on what can be considered as classical mathematics. Their formula is (9.17). The nice feature of (9.17) consists in obtaining the nth binary or hexadecimal digits of π without knowing its previous binary or hexadecimal digits. Afterwards many other relations for different constants (irrational or even transcendental) have been exhibited, all having the property that for the nth binary or hexadecimal digits of such a number we need not know the previous digits.

9.4.1 Computing the nth binary or hexadecimal digit of π

Until quite recently all methods for computing irrational constants were based on some recurrences. The methods involved in such algorithms required at least huge memory space provided the desired accuracy was high. Thus a question appeared: is it possible to compute directly the n-digit of an irrational number without computing all its previous digits? For certain transcendental constants the answer is positive. Moreover, it can be done in (essentially) linear time $= O(n \log^{O(1)}(n))$ and space $= \log^{O(1)}(n)$.

The classical approach that we now introduce belongs to Bellard.[6]

Theorem 4.1. *It holds*

$$\pi = \sum_{k=0}^{\infty} \frac{1}{16^k} \left(\frac{4}{8k+1} - \frac{2}{8k+4} - \frac{1}{8k+5} - \frac{1}{8k+6} \right). \qquad (9.17)$$

Proof. For each $n \in \mathbb{N}^*$ we have

$$\sum_{k=0}^{\infty} \frac{1}{16^k(8k+n)} = 2^{n/2} \sum_{k=0}^{\infty} \frac{x^{n+8k}}{8k+n} \Bigg|_0^{1/\sqrt{2}} = 2^{n/2} \sum_{k=0}^{\infty} \int_0^{1/\sqrt{2}} x^{n-1+8k}\, dx$$

$$= 2^{n/2} \int_0^{1/\sqrt{2}} \left(\sum_{k=0}^{\infty} x^{n-1+8k} \right) dx = 2^{n/2} \int_0^{1/\sqrt{2}} x^{n-1} \sum_{k=0}^{\infty} \left(x^8 \right)^k dx$$

$$= 2^{n/2} \int_0^{1/\sqrt{2}} \frac{x^{n-1}}{1-x^8}\, dx.$$

Using the previous result to estimate the right-hand side of (9.17) we have

$$\sum_{k=0}^{\infty} \frac{1}{16^k} \left(\frac{4}{8k+1} - \frac{2}{8k+4} - \frac{1}{8k+5} - \frac{1}{8k+6} \right)$$

---

[5] Simon Plouffe, 1956–.
[6] Fabrice Bellard, 1973–.

$$= \int_0^{1/\sqrt{2}} \frac{4\sqrt{2} - 8x^3 - 4\sqrt{2}x^4 - 8x^5}{1 - x^8}\, dx.$$

Now we substitute $y = \sqrt{2}\, x$ to find further

$$16 \int_0^1 \frac{y-1}{y^4 - 2y^3 + 4y - 4}\, dy = 4\int_0^1 \frac{2-y}{y^2 - 2y + 2}\, dy + 4\int_0^1 \frac{y}{y^2 - 2}\, dy$$

$$\int_0^1 \frac{4 - 4y}{y^2 - 2y + 2}\, dy + 4\int_0^1 \frac{1}{1 + (y-1)^2}\, dy + 4\int_0^1 \frac{y}{y^2 - 2}\, dy$$

$$\left(-2\ln(y^2 - 2y + 2) + 4\arctan(y-1) + 2(2 - y^2)\right)\Big|_0^1 = \pi. \qquad \square$$

On the same line the next formulas hold

$$\pi = \sum_{k=0}^\infty \frac{1}{16^k}\left(\frac{4+8r}{8k+1} - \frac{8r}{8k+2} - \frac{4r}{8k+3} - \frac{2+8r}{8k+4} - \frac{1+2r}{8k+5} - \frac{1+2r}{8k+6}\right.$$

$$\left. + \frac{r}{8k+7}\right), \quad \forall r \in \mathbb{R}, \ (\text{Adamchik–Wagon}),$$

$$\pi = \sum_{k=0}^\infty \frac{(-1)^k}{4^k}\left(\frac{2}{4k+1} + \frac{2}{4k+2} + \frac{1}{4k+3}\right),$$

$$\pi^2 = \sum_{k=0}^\infty \frac{1}{16^k}\left(\frac{16}{(8k+1)^2} - \frac{16}{(8k+2)^2} - \frac{8}{(8k+3)^2} - \frac{16}{(8k+4)^2} - \frac{4}{(8k+5)^2}\right.$$

$$\left. - \frac{4}{(8k+6)^2} - \frac{2}{(8k+7)^2}\right),$$

$$\pi\sqrt{2} = \sum_{k=0}^\infty \frac{(-1)^k}{8^k}\left(\frac{4}{6k+1} + \frac{1}{6k+2} + \frac{1}{6k+3}\right).$$

On the line of BBP-formulas, Bellard proved the next result.

Theorem 4.2. *It holds*

$$\pi = \sum_{k=0}^\infty \frac{(-1)^k}{2^{10k}}\left(-\frac{32}{4k+1} - \frac{1}{4k+3} + \frac{256}{10k+1} - \frac{64}{10k+3}\right. \tag{9.18}$$

$$\left. - \frac{4}{10k+5} - \frac{4}{10k+7} + \frac{1}{10k+9}\right).$$

Proof. We need the two equalities, valid for $|a| > \sqrt{2}$,

$$\arctan\frac{1}{a-1} = \sum_{k=1}^\infty \frac{2^{k/2}\sin(k\pi/4)}{ka^k}, \quad \arctan\frac{1}{a+1} = \sum_{k=1}^\infty \frac{2^{k/2}\sin(k3\pi/4)}{ka^k}.$$

We can try to use the Taylor expansion of the arctan function, formula (5.19) at page 226 to get the two equalities. But this way is tremendous.

Another way of approaching the two equalities is based on a bit of rudimentary complex functions, [85], [126]. We note that formula (5.19) at page 226 is still valid for complex numbers of modulus less than 1. Also from complex functions we have that for every nonzero complex number $z = x + iy$, one has

$$\ln z = \ln |z| + i\theta, \tag{9.19}$$

where

$$\theta = \text{Arctan} = \arctan \frac{y}{x} + 2k\pi, \quad k \in \mathbb{Z}.$$

We may take $k = 0$, and we do so. Then from (9.19) we have

$$\ln z = \ln |z| + i \arctan \frac{y}{x}.$$

Therefore

$$\arctan \frac{y}{x} = \text{Im} \ln z.$$

Because

$$\ln\left(1 - \frac{i+1}{a}\right) = \ln\left|1 - \frac{i+1}{a}\right| + i \arctan \frac{1}{1-a}$$

and

$$\ln\left(1 - \frac{i+1}{a}\right) = -\sum_{k=1}^{\infty} \frac{(1+i)^k}{ka^k}$$

$$= -\sum_{k=1}^{\infty} \frac{2^{k/2}(\cos(k\pi/4) + i\sin(k\pi/4))}{ka^k}, \quad |a| > \sqrt{2},$$

we find

$$\arctan \frac{1}{a-1} = \sum_{k=1}^{\infty} \frac{2^{k/2} \sin(k\pi/4)}{ka^k} \tag{9.20}$$

$$= \sum_{k=0}^{\infty} \frac{(-1)^k 2^{2k}}{a^{4k+3}} \left(\frac{a^2}{4k+1} + \frac{2a}{4k+2} + \frac{2}{4k+3} \right).$$

Similarly, from

$$\ln\left(1 - \frac{i-1}{a}\right) = \ln\left|1 - \frac{i-1}{a}\right| + i \arctan \frac{-1}{a+1}$$

we find

$$\arctan \frac{1}{a+1} = \sum_{k=1}^{\infty} \frac{2^{k/2} \sin(k3\pi/4)}{ka^k}$$

$$= \sum_{k=0}^{\infty} \frac{(-1)^k 2^{2k}}{a^{4k+3}} \left(\frac{a^2}{4k+1} - \frac{2a}{4k+2} + \frac{2}{4k+3} \right).$$

Substitute $a = 2$ in (9.20) to get

$$\pi = \sum_{k=0}^{\infty} \frac{(-1)^k}{4^k} \left(\frac{2}{4k+1} + \frac{2}{4k+2} + \frac{1}{4k+3} \right).$$

Now consider the following relations

$$\frac{\pi}{4} = 2 \arctan \frac{1}{2} - \arctan \frac{1}{7}, \qquad (9.21)$$

$$\frac{\pi}{4} = 2 \arctan \frac{1}{3} + \arctan \frac{1}{7}, \quad \frac{\pi}{4} = 2 \arctan \frac{1}{2} + \arctan \frac{1}{3},$$

$$\frac{\pi}{4} = 2 \arctan \frac{1}{2} - \arctan \frac{1}{9} - \arctan \frac{1}{32}.$$

From (9.20) and (9.21), one has

$$\pi = 4 \sum_{k=0}^{\infty} \frac{(-1)^k}{(2k+1)4^k} - \frac{1}{64} \sum_{k=0}^{\infty} \frac{(-1)^k}{1024^k} \left(\frac{32}{4k+1} + \frac{8}{4k+2} + \frac{1}{4k+3} \right).$$

Reordering the terms in the last equality, we get the desired relation (9.18).   □

There are many papers dedicated to the symbolic calculus oriented toward finding different BBP formulas.

In [59] is studied the integral

$$\int_0^1 \frac{a_0 + a_1 x + a_2 x^2}{x^4 - 9} \, dx$$

and it is shown to be equal to

$$\left(\frac{\sqrt{3}}{108} a_0 + \frac{\sqrt{3}}{36} a_2 \right) \pi - \frac{a_1}{12} \ln 2 + \left(\frac{\sqrt{3}}{18} a_0 + \frac{\sqrt{3}}{6} a_2 \right) \operatorname{arctanh} \left(\frac{\sqrt{3}}{3} \right),$$

where arctanh is the inverse function of tanh. Then we take $a_1 = 0$ (and then no logarithm in the result) and $a_0 = -3a_2$ (and then no arctanh in the result). Thus we get the following.

Proposition 4.4. *It holds*

$$\int_0^1 \frac{-3a_2 + a_2 x^2}{x^4 - 9} \, dx = \frac{\sqrt{3}}{18} \pi a_2.$$

But the previous result is obvious because

$$\int_0^1 \frac{dx}{x^2 + 3} = \frac{\sqrt{3}}{18} \pi.$$

BBP formulas exist in any base. An idea is given by the next proposition.

Proposition 4.5. *Suppose $n \in \{3, 4, \ldots\}$, $b \in \mathbb{R} \setminus \{0\}$, and $P(x) = a_0 + a_1 x + \cdots + a_{n-2} x^{n-2}$. Then*

$$\int_0^1 \frac{P(x)}{x^n - b} \, dx = -\frac{1}{b} \sum_{i=0}^{\infty} \sum_{k=0}^{n-1} \frac{1}{b^i} \frac{a_k}{n \cdot i + k + 1}. \tag{9.22}$$

Proposition 4.6. *Suppose $n = 4$, $b = 9$, and $a_2 = -1$. By Propositions 4.4 and 4.5 one has the BBP formula*

$$\pi = \frac{2}{\sqrt{3}} \sum_{n=0}^{\infty} \frac{1}{9^n} \left(\frac{3}{4n + 1} - \frac{1}{4n + 3} \right).$$

From (9.22) by $n = 8$, $b = 16$, and $P(x) = x^5 + x^4 + 2x^3 - 4$ follows (9.17).

We show how the BBP method works. Consider a constant C which in base b is written as

$$C = \sum_{k=0}^{\infty} \frac{1}{b^{ck} \, p(k)},$$

where $b \geq 2$ and c are positive integers and $p(k) \neq 0$ is a polynomial with integer coefficients. To compute the nth digit of C in base b it is sufficient to compute $b^n C$ modulo 1. Thus we have

$$b^n C \mod 1 = \sum_{k=0}^{\infty} \frac{b^{n-ck}}{p(k)} \mod 1$$

$$= \sum_{k=0}^{\lfloor n/c \rfloor} \frac{b^{n-ck} \mod p(k)}{p(k)} \mod 1 + \sum_{k=\lfloor n/c \rfloor + 1}^{\infty} \frac{b^{n-ck}}{p(k)} \mod 1.$$

In each term of the first sum, $b^{n-ck} \mod p(k)$ is computed using the fast exponential algorithm modulo the integer $q(k)$. Division by $q(k)$ and summation are performed using ordinary floating-point arithmetic. Concerning the infinite sum, note that the exponent in the numerator is negative. Thus floating-point arithmetic can again be used to compute its value with sufficient accuracy. The final result, a fraction between 0 and 1, is then converted to the desired base b.

With certain minor modifications, this scheme can be extended to numbers of the form

$$C = \sum_{k=0}^{\infty} \frac{p(k)}{b^{ck} \, q(k)},$$

where $q(k)$ is a polynomial with integer coefficients.

9.4.2 BBP formulas by binomial sums

The first BBP formula by binomial sums appeared in [57]. This is as follows.

Proposition 4.7. *It holds*

$$\pi = \sum_{n=0}^{\infty} \frac{50n - 6}{\binom{3n}{n} 2^n}. \tag{9.23}$$

Proof. By the ratio test, we have that the series in (9.23) converges. By Corollary 4.8 at page 277, we have

$$\frac{1}{\binom{3n}{n}} = (3n+1) \int_0^1 x^{2n}(1-x)^n dx$$

and obviously

$$\frac{1}{\binom{3n}{n} 2^n} = (3n+1) \int_0^1 \frac{x^{2n}(1-x)^n}{2^n} dx = (3n+1) \int_0^1 \left(\frac{x^2(1-x)}{2} \right)^n dx.$$

Invoking Corollary 2.2 at page 263, we can write

$$\sum_{n=0}^{\infty} \frac{50n-6}{\binom{3n}{n} 2^n} = \int_0^1 \sum_{n=0}^{\infty} (50n-6)(3n+1) \left(\frac{x^2(1-x)}{2} \right)^n dx$$

$$= 8 \int_0^1 \frac{28x^6 - 56x^5 + 28x^4 - 97x^3 + 97x^2 - 6}{(x^3 - x^2 + 2)^3} dx$$

$$= \left(\frac{4x(x-1)(x^3 - 28x^2 + 9x + 8)}{(x^3 - x^2 + 2)^3} + 4\arctan(x-1) \right) \Bigg|_0^1 = \pi. \quad \square$$

Example 4.1 *Another formula of the same sort is*

$$\sum_{n=0}^{\infty} \frac{(-1)^n}{2^n \binom{5n}{2n}} \left(\frac{416}{5n+1} + \frac{-44}{5n+2} + \frac{16}{5n+3} + \frac{-14}{5n+4} \right) = 125\pi. \quad \triangle$$

Regarding this sort of BBP formulas, one faces the question if there exists a more general formula such as

$$\pi = \sum_{n=0}^{\infty} \frac{S(n)}{a^n \binom{mn}{pn}}, \quad a, m, n, p \in \mathbb{N}^*, \tag{9.24}$$

where S is a polynomial in n.

We introduce an existence result for a particular case of (9.24), namely

$$\pi = \sum_{n=0}^{\infty} \frac{1}{a^n \binom{mn}{pn}} \left(\frac{b_1}{mn+1} + \frac{b_2}{mn+2} + \cdots + \frac{b_{m-1}}{mn+m-1} \right), \quad b_k \in \mathbb{Q}.$$

Denote

$$T = \sum_{n=0}^{\infty} \frac{1}{a^n \binom{mn}{pn}} \left(\frac{b_1}{mn+1} + \frac{b_2}{mn+2} + \cdots + \frac{b_{m-1}}{mn+m-1} \right), \quad b_k \in \mathbb{Q}.$$

$$(9.25)$$

Theorem 4.3. *There exists a polynomial P of degree $m - 2$ with rational coefficients such that*

$$T = \int_0^1 \frac{aP(x)}{a - x^p(1-x)^{(m-p)}} dx.$$

Proof. By Corollary 4.8 at page 277, we have

$$\sum_{n=0}^{\infty} \frac{1}{a^n \binom{mn}{pn}} \frac{1}{mn+1} = \sum_{n=0}^{\infty} \frac{1}{a^n} \int_0^1 x^{pn}(1-x)^{(m-p)n} dx$$

$$= \int_0^1 \sum_{n=0}^{\infty} \left(\frac{x^p(1-x)^{(m-p)}}{a} \right)^n dx = \int_0^1 \frac{a}{a - x^p(1-x)^{(m-p)}} dx \quad (9.26)$$

To get terms of the form $1/(mn+k)$ in (9.25), we write

$$\int_0^1 x^{pn+k-1}(1-x)^{(m-p)n} dx = \frac{1}{\binom{mn+k}{pn+k-1}}$$

$$= \frac{1}{\binom{mn}{pn}} \frac{(pn+1)(pn+2)\ldots(pn+k-1)}{(mn+1)(mn+2)\ldots(mn+k)}.$$

For a fixed k we decompose the last fraction in (9.26) in simple fractions getting a sequence $(a_{k,j})_{j=1,2,\ldots,k}$ such that

$$\int_0^1 x^{k-1} x^{pn}(1-x)^{(m-p)n} dx = \frac{1}{\binom{mn}{pn}} \left(\frac{a_{k,1}}{mn+1} + \frac{a_{k,2}}{mn+2} + \cdots + \frac{a_{k,k}}{mn+k} \right),$$

$$(9.27)$$

where all $a_{k,j} \in \mathbb{Q}$.

$$\begin{pmatrix} a_{1,1} & a_{2,1} & \cdots & a_{m-1,1} \\ 0 & a_{2,2} & \cdots & a_{m-1,2} \\ \cdots & \cdots & \cdots & \cdots \\ 0 & 0 & 0 & a_{m-1,k} \end{pmatrix} \cdot B = \begin{pmatrix} b_1 \\ b_2 \\ \cdots \\ b_{m-1} \end{pmatrix}.$$

The vector B contains the coefficients of the polynomial

$$P(x) = B^t \cdot \begin{pmatrix} 1 \\ x \\ \cdots \\ x^{m-2} \end{pmatrix}$$

of degree $m - 2$. Using (9.27), it follows

$$\sum_{n=0}^{\infty} \frac{1}{a^n \binom{mn}{pn}} \left(\frac{b_1}{mn+1} + \frac{b_2}{mn+2} + \cdots + \frac{b_{m-1}}{mn+m-1} \right)$$

$$= \sum_{n=0}^{\infty} \int_0^1 P(x) \frac{x^{pn}(1-x)^{(m-p)n}}{a^n} dx = \int_0^1 \frac{aP(x)}{a - x^p(1-x)^{(m-p)}} dx. \quad \square$$

Examples 4.1 (a) On the same path one has

$$\sum_{n=0}^{\infty} \frac{1}{2^n \binom{3n}{n}} \frac{1}{3n+1} = \sum_{n=0}^{\infty} \frac{1}{2^n} \int_0^1 x^{2n}(1-x)^n dx = \int_0^1 \frac{x^{2n}(1-x)^n}{2^n} dx$$

$$= \int_0^1 \frac{2}{2 - x^2(1-x)} dx = \frac{3}{5} \ln 2 + \frac{1}{5}\pi,$$

$$\sum_{n=0}^{\infty} \frac{1}{2^n \binom{3n}{n}} \frac{1}{3n+2} = 3 \int_0^1 \frac{x^{2n+1}(1-x)^n}{2^n} dx - \sum_{n=0}^{\infty} \frac{1}{2^n \binom{3n}{n}} \frac{1}{3n+1}$$

$$= \int_0^1 \frac{2x}{2 - x^2(1-x)} dx - \frac{3}{5} \ln 2 - \frac{1}{5}\pi = -\frac{12}{5} \ln 2 + \frac{7}{10}\pi.$$

Two other formulas can be obtained using a linear combination of the previous two relation.

$$\sum_{n=0}^{\infty} \frac{1}{2^n \binom{3n}{n}} \left(\frac{\alpha}{3n+1} + \frac{\beta}{3n+2} \right) = \left(\frac{3}{5}\alpha - \frac{12}{5}\beta \right) \ln 2 + \left(\frac{1}{5}\alpha + \frac{7}{10}\beta \right) \pi. \tag{9.28}$$

Thus from (9.28) for $\alpha = 4\beta = 4$, we get

$$\frac{1}{3} \sum_{n=0}^{\infty} \frac{1}{2^n \binom{3n}{n}} \left(\frac{8}{3n+1} + \frac{2}{3n+2} \right) = \pi,$$

whereas for $\alpha = (-7/2)\beta = -7$, we get

$$\frac{1}{9} \sum_{n=0}^{\infty} \frac{1}{2^n \binom{3n}{n}} \left(\frac{7}{3n+1} - \frac{2}{3n+2} \right) = \ln 2.$$

(b) In a very similar way one can show

$$\pi = \frac{1}{16807} \sum_{n=0}^{\infty} \frac{1}{2^n \binom{7n}{2n}} \left(\frac{59296}{7n+1} - \frac{10326}{7n+2} - \frac{3200}{7n+3} \right.$$

$$\left. - \frac{1352}{7n+4} - \frac{792}{7n+5} + \frac{552}{7n+6} \right).$$

For it, we start with

$$\int_0^1 x \cdot x^{2n}(1-x)^{5n}dx = B(5n+1, 2n+2) = \frac{1}{\binom{7n}{2n}}\left(\frac{5}{7}\frac{1}{7n+1} - \frac{3}{7}\frac{1}{7n+2}\right),$$

$$\int_0^1 x^2 \cdot x^{2n}(1-x)^{5n}dx = B(5n+1, 2n+3)$$

$$= \frac{1}{\binom{7n}{2n}}\left(\frac{30}{49}\frac{1}{7n+1} - \frac{30}{49}\frac{1}{7n+2} + \frac{4}{49}\frac{1}{7n+3}\right). \quad \triangle$$

Some interesting BBP formulas by binomial sums can be obtained if the parameters of the beta function are not naturals but rationals. Consider now the case

$$\int_0^1 x^{pn+a/b}(1-x)^{mn+c/d}dx = \frac{\Gamma(pn + \frac{a}{b} + 1)\Gamma(mn + \frac{c}{d} + 1)}{\Gamma((p+m)n + \frac{a}{b} + \frac{c}{d} + 2)} \quad (9.29)$$

$$= \frac{(pn + \frac{a}{b})(pn - 1 + \frac{a}{b})\dots(\frac{a}{b})\Gamma(\frac{a}{b})(mn + \frac{c}{d})(mn - 1 + \frac{c}{d})\dots(\frac{c}{d})\Gamma(\frac{c}{d})}{((p+m)n + \Gamma(\frac{a}{b}) + \Gamma(\frac{c}{d}) + 1)((p+m)n + \Gamma(\frac{a}{b}) + \Gamma(\frac{c}{d}))\dots(\Gamma(\frac{a}{b}) + \Gamma(\frac{c}{d}))},$$

where a/b and $c/d \in \mathbb{Q}$. We wish to repeat somehow the previous approaches.
Suppose $a/b = -1/2$ and $c/d = 0$. Then we have

$$\int_0^1 a^{-n}x^{pn-1/2}(1-x)^{mn}dx = a^{-n}\frac{\Gamma(pn + \frac{1}{2})\Gamma(mn + 1)}{\Gamma((p+m)n + \frac{1}{2} + 1)}$$

$$= \int_0^1 \frac{1}{\sqrt{x}}\frac{a}{a - x^p(1-x)}dx.$$

On the other hand, by (9.29), we have

$$\sum_{n=0}^\infty \int_0^1 a^{-n}x^{pn-1/2}(1-x)^{mn}dx = \sum_{n=0}^\infty (-1)^n \frac{(2pn)!(mn+pn)!(mn)!}{a^n(2mn+2pn)!(pn)!}.$$

Consider $p = 3$, $m = 1$, and $a = 2$. We choose a proper polynomial to produce at the end number π. Thus

$$\int_0^1 \frac{1}{\sqrt{x}}(x^3 - 2x^2 + 2x - 2)\sum_{n=0}^\infty \frac{(-1)^n x^{3n}(1-x)^n}{2^n}dx \quad (9.30)$$

$$= \sum_{n=0}^\infty \frac{(-1)^n}{2^n}\int_0^1 (x^3 - 2x^2 + 2x - 2)\frac{x^{3n}(1-x)^n}{\sqrt{x}}dx.$$

On the other side

$$\int_0^1 \frac{1}{\sqrt{x}}(x^3 - 2x^2 + 2x - 2)\sum_{n=0}^\infty \frac{(-1)^n x^{3n}(1-x)^n}{2^n}dx \quad (9.31)$$

$$= \int_0^1 \frac{x^3 - 2x^2 + 2x - 2}{\sqrt{x}}\frac{-2}{x^4 - x^3 - 2}dx = -2\int_0^1 \frac{dx}{\sqrt{x}(x+1)} = -\pi$$

and

$$-\int_0^1 (x^3 - 2x^2 + 2x - 2)\frac{x^{3n}(1-x)^n}{\sqrt{x}}dx$$

$$= -\int_0^1 x^{3n+5/2}(1-x)^n dx + 2\int_0^1 x^{3n+3/2}(1-x)^n dx$$

$$- 2\int_0^1 x^{3n+1/2}(1-x)^n dx + 2\int_0^1 x^{3n-1/2}(1-x)^n dx$$

$$= -\frac{\Gamma(3n+\frac{7}{2})\Gamma(n+1)}{\Gamma(4n+\frac{9}{2})} + 2\frac{\Gamma(3n+\frac{5}{2})\Gamma(n+1)}{\Gamma(4n+\frac{7}{2})}$$

$$- 2\frac{\Gamma(3n+\frac{3}{2})\Gamma(n+1)}{\Gamma(4n+\frac{5}{2})} + 2\frac{\Gamma(3n+\frac{1}{2})\Gamma(n+1)}{\Gamma(4n+\frac{3}{2})}$$

$$= \frac{(6n)!(4n)!n!4^n}{(8n)!(3n)!}\frac{1}{512}\left(\frac{1885}{8n+1} - \frac{965}{8n+3} + \frac{363}{8n+5} - \frac{51}{8n+7}\right).$$

By (9.30) and (9.31) one has

$$\sum_{n=0}^{\infty}(-1)^n 2^n \frac{(6n)!(4n)!n!4^n}{(8n)!(3n)!}\left(\frac{1885}{8n+1} - \frac{965}{8n+3} + \frac{363}{8n+5} - \frac{51}{8n+7}\right) = 2^9 \pi.$$

Hence equality

$$\sum_{n=0}^{\infty}\frac{(-1)^n 2^n \binom{6n}{3n}}{\binom{8n}{4n}\binom{4n}{n}}\left(\frac{1885}{8n+1} - \frac{965}{8n+3} + \frac{363}{8n+5} - \frac{51}{8n+7}\right) = 2^9 \pi$$

follows.

9.5 Ramanujan formulas

Ramanujan proved the next surprising formulas.

Theorem 5.1. *One has*

$$\frac{1}{\pi} = \frac{\sqrt{8}}{9801}\sum_{n=0}^{\infty}\frac{(4n)!}{(n!)^4}\frac{1103 + 26390n}{396^{4n}},$$

$$\frac{1}{\pi} = \sum_{n=0}^{\infty}\binom{2n}{n}^2\frac{42n+5}{2^{12n+4}}.$$

Theorem 5.2.

$$\frac{1}{\pi} = 12\sum_{n=0}^{\infty}(-1)^n\frac{(6n)!}{(n!)^3(3n)!}\frac{13591409 + 545140134n}{(640320^3)^{n+1/2}}.$$

Some Ramanujan formulas were first discovered experimentally. A list of such formulas follows.

Theorem 5.3. *There hold*

$$\frac{2}{\pi} = \sum_{n=0}^{\infty} \frac{(\cos \pi k)(-1)^n}{2^{6n}2^{2k}} \frac{\binom{2n}{n}^3 \binom{2k}{k}^2}{2^{2k}\binom{n-1/2}{k}\binom{n+k}{n}} (4n+1),$$

$$\frac{4}{\pi} = \sum_{n=0}^{\infty} \frac{(\cos \pi k)}{2^{8n}2^{2k}} \frac{\binom{2n}{n}^2 \binom{2k}{k}\binom{2n+2k}{n+k}}{2^{2k}\binom{n-1/2}{k}} (6n+2k+1),$$

$$\frac{2\sqrt{2}}{\pi} = \sum_{n=0}^{\infty} \frac{(\cos \pi k)(-1)^n}{2^{9n}2^{3k}} \frac{\binom{2n}{n}\binom{n+k}{n}\binom{2n+2k}{n+k}^2}{2^{2k}\binom{n-1/2}{k}} (6n+2k+1),$$

$$\frac{16}{\pi} = \sum_{n=0}^{\infty} \frac{(\cos \pi k)(-1)^n}{2^{12n}2^{2k}} \frac{\binom{2n}{n}^2 \binom{2n+2k}{n+k}^2 \binom{n+k}{n}}{2^{2k}\binom{n-1/2}{k}\binom{2n+k}{n}}$$
$$\cdot \frac{84n^2 + 56nk + 52n + 4k^2 + 12k + 5}{2n+k+1},$$

$$\frac{8}{\pi} = \sum_{n=0}^{\infty} \frac{(\cos \pi k)(-1)^n}{2^{10n}2^{2k}} \frac{\binom{2n}{n}^2 \binom{2k}{k}^2 \binom{4n}{2n}}{2^{2k}\binom{2n-1/2}{k}\binom{n+k}{k}} (20n+2k+3),$$

$$\frac{2\sqrt{3}}{\pi} = \sum_{n=0}^{\infty} \frac{(\cos \pi k)3^k}{2^{8n}2^{4k}3^{2n}} \frac{\binom{2n}{n}\binom{4n}{2n}\binom{2k}{k}\binom{2n+2k}{n+k}}{2^{2k}\binom{n-1/2}{k}} (8n+2k+1),$$

$$\frac{16\sqrt{3}}{3\pi} = \sum_{n=0}^{\infty} \frac{(\cos \pi k)(-1)^n 3^k}{2^{12n}2^{4k}3^n} \frac{\binom{2n}{n}\binom{2n+2k}{n+k}\binom{4n+2k}{2n+k}\binom{n+k}{n}}{2^{2k}\binom{n-1/2}{k}}$$
$$\cdot \frac{56n^2 + 36nk + 34n + 4k^2 + 8k + 3}{2n+k+1},$$

$$\frac{8}{\pi^2} = \sum_{n=0}^{\infty} \frac{(\cos \pi k)(-1)^n}{2^{12n}2^{4k}} \frac{\binom{2n}{n}^4 \binom{2n+2k}{n+k}\binom{2k}{k}^2}{2^{2k}\binom{2n-1/2}{k}\binom{n+k}{k}} (20n^2 + 12nk + 8n + 2k + 1),$$

$$\frac{32}{\pi^2} = \sum_{n=0}^{\infty} \frac{(\cos \pi k)}{2^{16n}2^{4k}} \frac{\binom{2n}{n}^4 \binom{4n}{2n}\binom{2k}{k}^3}{2^{2k}\binom{2n-1/2}{k}\binom{n+k}{k}^2} (120n^2 + 84nk + 34k + 10k + 3).$$

A useful approach for proving Ramanujan formulas is based on the $W-Z$ method at page 130.

Theorem 5.4. *The following formula of Ramanujan holds.*

$$\frac{2}{\pi} = \sum_{k=0}^{\infty} (-1)^k (4k+1) \frac{(1/2)_k^3}{k!^3}. \tag{9.32}$$

Proof. At the beginning we prove that

$$\frac{\Gamma(n+3/2)}{\Gamma(3/2)\Gamma(n+1)} = \sum_{k=0}^{\infty}(-1)^k(4k+1)\frac{(1/2)_k^2(-n)_k}{k!^2(n+3/2)_k}. \qquad (9.33)$$

To prove (9.33) for all positive integers n, we define

$$F(n,k) = \frac{\Gamma(3/2)\Gamma(n+1)}{\Gamma(n+3/2)}\sum_{k=0}^{\infty}(-1)^k(4k+1)\frac{(1/2)_k^2(-n)_k}{k!^2(n+3/2)_k},$$

$$G(n,k) = \frac{(2k+1)^2}{(2n+2k+3)(2k+1)}.$$

We note that

$$F(n+1,k) - F(n,k) = G(n,k) - G(n,k-1), \quad n \in \mathbb{N}, \ k \in \mathbb{Z}.$$

Also note that the assumptions of Theorem 3.26 at page 131 are satisfied. Therefore summing the last identity with respect to k, one gets

$$\sum_k F(n,k) = \text{const.}, \quad n \in \mathbb{N}.$$

By plugging in $n = 0$, it follows that $\sum_k F(n,k) = 1$. From here (9.33) follows. Now, by the Carlson theorem at page 228, we have that $\sum_k F(n,k) = 1$ for all real n. Take $n = -1/2$ in (9.33) to get (9.32). $\square$

9.6 Several natural ways to introduce number e

We defined number e as the limit of sequences by Proposition 1.15 at page 99, by (i) in Proposition 1.16 at page 101, and as the value of the exponential function at $x = 1$ by (5.13) at page 225.

By Theorem 6.4 at page 282 we proved that number e is transcendental.

A question arises: does this number occur frequently and naturally in concrete problems? The answer is definitely positive.

Below we introduce only a small number of natural ways to get or to use number e.

The number N of atoms of a *radioactive* disintegration is proportional to the number of atoms present and can be determined from the relation

$$\frac{dN}{dt} = -\lambda N,$$

where λ is the radioactive disintegration constant. Obviously, $\lambda > 0$. Separating the variables and integrating on the $[0,t]$ interval, we successively find

$$\frac{dN}{N} = -\lambda dt, \quad \int_0^t \frac{dN}{N} = -\lambda \int_0^t dt, \quad \ln\frac{N(t)}{N(0)} = -\lambda t.$$

Thus we get

$$N(t) = N(0)e^{-\lambda t},$$

where $N(0)$ is the number of atoms at the moment of time $t = 0$ and $N(t)$ is the number of atoms after the time t has elapsed. Hence, the behavior of a very natural phenomenon is expressed by the number e. The *half-time* T is defined as the moment of time T for which $N(T) = N(0)/2$. This is unique because the function N is strictly decreasing. The half-time and the disintegration constant λ are connected by

$$T = \frac{\ln 2}{\lambda}. \tag{9.34}$$

Consider now, following [72, p. 189], a *cell population* problem. Bacteria are inoculated in a petri dish at a given density. The bacterial density, denoted by x, doubles in 20 hours. Then the model is described by

$$\frac{dx}{dt} = Cx,$$

where C is a positive constant. Reasoning as before we find that, if no other factor is involved, the bacterial density at a moment t is given by

$$x(t) = x(0)e^{Ct}.$$

Hence, number e appears in an extremely simple cell population problem.

The third problem that we introduce comes from probability and it is known under several names: the *hat-check* problem, the *cards* problem, the *party* problem, or the *noncoincidence* problem.

The hat-check problem is as follows. A hat-check girl in a restaurant, having checked n hats, gets them hopelessly scrambled and returns them at random to the n owners. What is the probability p_n that nobody gets his own hat back?

The cards problem is as follows. A deck of cards numbered from 1 to n is shuffled and dealt. What is the probability p_n that no card is in the correct place; that is, the card i is not ith from the top for any i?

The party problem is as follows. At a party there are n pairs of husbands with their wives. Each man dances with just one lady. What is the probability p_n that no husband dances with his wife?

The noncoincidence problem reads as follows. Consider a permutation $\sigma : \mathbb{N}_n^* \to \mathbb{N}_n^*$. Permutation σ has *one coincidence* if there is only one $p \in \mathbb{N}_n^*$ such that $\sigma(p) = p$ and $\sigma(i) \neq i$, for all $i \neq p$. Permutation σ has *two coincidences* if there are distinct $p, q \in \mathbb{N}_n^*$ such that $\sigma(p) = p$, $\sigma(q) = q$, and $\sigma(i) \neq i$, for all $i \in \mathbb{N}_n^* \setminus \{p, q\}$. In a similar way we define a permutation with $k\,(\in \mathbb{N}_n^*)$ coincidences; that is, there are k numbers $p_1, \ldots, p_k \in \mathbb{N}_n^*$ such that $\sigma(p_j) = p_j$, $j = 1, \ldots, k$ and $\sigma(i) \neq i$ for all $i \in \mathbb{N}_n^* \setminus \{p_j \mid j = 1, \ldots, k\}$. Permutation σ has *no coincidence* provided $\sigma(i) \neq i$, for all $i \in \mathbb{N}_n^*$. Consider the set of all permutations $\sigma : \mathbb{N}_n^* \to \mathbb{N}_n^*$. What is the probability p_n that one selects a permutation with no coincidence?

We recall following [55, p. 24] that the probability $P(A)$ of event A is equal to the number of possible trial outcomes favorable to A divided by the number of all possible outcomes.

It is obvious the four problems are equivalent.

We use the permutation language to study our problem. The event A consists in selecting a permutation with no coincidence. The number of all possible outcomes is equal to the number of all permutations $\sigma : \mathbb{N}_n^* \to \mathbb{N}_n^*$, that is, $n!$.

Now we find the number of possible trial outcomes favorable to A. Consider the S of all permutations $\sigma : \mathbb{N}_n^* \to \mathbb{N}_n^*$. Then S can be represented as a union of nonempty and pairwise disjoint subsets; that is, we are looking for a convenable partition. Let S_k be the set of permutations $\sigma : \mathbb{N}_n^* \to \mathbb{N}_n^*$ having precisely k coincidences, $k = 0, 1, 2, \ldots, n-1$. Then $S = \cup_0^{n-1} S_k$ and $S_i \cap S_j = \emptyset$ whenever $i \neq j$. Obviously $|S_{n-1}| = 1$, because there is only one permutation that is the identity function. Then successively we have

$$n! = |S_0| + \binom{n}{1}|S_1| + \binom{n}{2}|S_2| + \cdots + \binom{n}{n-2}|S_{n-2}|,$$

$$1 = \frac{|S_0|}{n!} + \frac{1}{1!}\frac{|S_1|}{(n-1)!} + \frac{1}{2!}\frac{|S_2|}{(n-2)!} + \cdots + \frac{1}{(n-2)!}\frac{|S_{n-2}|}{2!} + \frac{1}{n!},$$

$$p_n = \frac{|S_0|}{n!} = 1 - \frac{1}{1!} + \frac{1}{2!} - \frac{1}{3!} + \cdots + (-1)^n \frac{1}{n!}.$$

Now we get the following surprising result.

Theorem 6.1. *One has*

$$\lim_{n \to \infty} p_n = e^{-1}. \tag{9.35}$$

Thus number e appears due to a limiting process of a simple probability problem.

The card problem has been generalized in the following sense. Consider a deck of kn cards consisting of k suites of n denominations (1 to n) each. We shuffle this deck, deal the cards, and at the same time count from 1 to n, k times. What is the probability $p_{n,k}$ that no card matches the denomination called? The answer is supplied by the next theorem.

Theorem 6.2. *For fixed k, one has*

$$\lim_{n \to \infty} p_{n,k} = e^{-k}.$$

Obviously, for $k = 1$ we recover (9.35).

9.7 Optimal stopping problem

The *optimal stopping problem* or the *secretary problem* or the *marriage problem*, or even the *sultan's dowry problem* is stated in the following way. An unordered sequence of applicants or candidates (distinct real numbers) $a_1, a_2, \ldots, a_n$ are interviewed by an employer one at a time. The employer knows only the number n of applicants but has no prior information about the applicants. The employer has to either accept a_i and end the process or reject a_i and interview a_{i+1}. The decision to accept or reject a_i must be based only on whether $a_i > a_j$, for all $1 \le j < i$. A rejected applicant is not later recalled.

The problem is to select the most highly qualified applicant, that is, the probability of choosing the most highly qualified applicant when the employer looks at $m - 1$ out of n applicants before starting to choose the next. Denote $P(m, n)$ this probability. $P(1, n) = P(n, n) = 1/n$, because picking the first or the last applicant, is just picking one at random.

Suppose we have collected the information from $m - 1$ applicants and we consider the kth in sequence. Obviously,

$$m - 1 \le k \le n.$$

Because the applicants arrive in random order, the chance that this one is the best is $1/n$. But we only consider this applicant if the highest-ranking applicant that we have seen so far was among the first $m - 1$ of the $k - 1$ that we have rejected. The probability of this event is

$$\frac{m-1}{k-1}.$$

Thus the overall chance of achieving the aim of finding the best applicant is

$$\frac{m-1}{n(k-1)}.$$

Because k runs from m to n, we get

$$P(m, n) = \sum_{k=m}^{n} \frac{m-1}{n(k-1)} = \frac{m-1}{n} \sum_{k=m}^{n} \frac{m-1}{k-1}.$$

We determine m so that

$$P(m - 1, n) < P(m, n) > P(m + 1, n). \tag{9.36}$$

From the first inequality in (9.36) we have

$$\frac{m-2}{n} \sum_{k=m-1}^{n} \frac{1}{k-1} < \frac{m-1}{n} \sum_{k=m}^{n} \frac{1}{k-1},$$

$$(m-2)\left(\frac{1}{m-2}+\sum_{k=m}^{n}\frac{1}{k-1}\right) < (m-1)\sum_{k=m}^{n}\frac{1}{k-1},$$

$$1 < \sum_{k=m}^{n}\frac{1}{k-1}.$$

From the second inequality in (9.36) we have

$$\frac{m-1}{n}\sum_{k=m}^{n}\frac{1}{k-1} > \frac{m}{n}\sum_{k=m+1}^{n}\frac{1}{k-1},$$

$$1+(m-1)\sum_{k=m+1}^{n}\frac{1}{k-1} > m\sum_{k=m+1}^{n}\frac{1}{k-1},$$

$$1 > \sum_{k=m+1}^{n}\frac{1}{k-1}.$$

Thus

$$\frac{1}{m}+\frac{1}{m+1}+\cdots+\frac{1}{n-1} < 1 < \frac{1}{m-1}+\frac{1}{m}+\cdots+\frac{1}{n-1}.$$

For large m (implicitly n), by Remark 1.4 at page 104, we have

$$\frac{1}{m}+\frac{1}{m+1}+\cdots+\frac{1}{n-1} \sim \ln\frac{n-1}{m-2} < 1$$
$$< \ln\frac{n-1}{m-1} \sim \frac{1}{m-1}+\frac{1}{m}+\cdots+\frac{1}{n-1}.$$

Hence, m, n has to satisfy

$$1 \sim \ln\frac{n}{m} \implies \frac{m}{n} \sim \frac{1}{e}.$$

The probability under discussion is given by the next theorem.

Theorem 7.1. *One has*

$$P(m,n) \sim \frac{m}{n}\ln\frac{n}{m}.$$

9.8 References and comments

The Pythagoreans proved that $\sqrt{2}$ is irrational. Legend has it that the Pythagorean philosopher Hippassus[7] used geometric methods to prove the irrationality of $\sqrt{2}$ while at sea and, upon notifying his fellows of his discovery, was immediately thrown overboard by the fanatic Pythagoreans.

---

[7] Hippasus of Metapontum ($'I\pi\pi\alpha\sigma\sigma\varsigma$ o $M\epsilon\tau\alpha\pi\acute{o}\nu\tau\iota\sigma\varsigma$), ~ 500 (BC).

The Pythagoras' constant is famous because it's probably the first irrational number discovered. Later, about 2300 years ago, in Book X of the impressive *Elements*, Euclid showed the irrationality of every positive number whose square is a nonsquare positive integer.

The Babylonians gave the impressive approximation

$$\sqrt{2} \approx 1 + \frac{24}{60} + \frac{51}{60^2} + \frac{10}{60^3} = 1.41421296296296\ldots.$$

$17/12$ have been used by Mesopotamians instead of $\sqrt{2}$.

A nice survey on the history until the newest results on computation of the number π may be found in several papers, for example [21] and the references therein.

Proposition 2.1 is from [132].

Other algorithms of computing π may be found at http://numbers.computation.free.fr/Constants/Sqrt2/sqrt2.htlm.

Interesting introductions to the arithmetic–geometric mean iteration and its relation to number π are [39], [23], and [24].

A proof of Theorem 3.3 may be found in [24, p. 13].

Theorems 3.4 and 3.5 may be found in [27], [118], and [24].

Propositions 4.4–4.6 appeared in [59].

Proposition 4.7 may be found in [57] and [4].

Example 4.1 and examples 4.1 are from [58].

Theorem 5.1 has been proved in [112].

Theorem 5.2 has been proved in [35].

We mention several papers regarding the battle for billions of digits of π [10], [1], [25], [11], [111], [9], [20], [104], [16], [4], [12], [26], and [59].

Theorem 5.3 appeared in [62].

Theorem 5.4 appeared in [46].

Theorem 6.2 appeared in [119].

We mention some papers on the optimal stopping problem and the references therein [49], [33], and [68].

Asymptotic and Combinatorial Estimates

The aim of this chapter is to introduce some results on asymptotic and combinatorial estimates.

10.1 Asymptotic estimates

The process of selection between several algorithms regarding the same problem is often based on easy implementation, elegance, time, and space allocation. Often we estimate the time consumption and based on it, we appreciate that one algorithm is faster than another.

Theorem 1.1. (Euler) *Suppose* $f : \mathbb{R} \to \mathbb{R}$ *has a continuous derivative on the interval* $[0, n]$. *Then*

$$\sum_{k=1}^{n-1} f(k) = \int_1^n f(x)\,dx - \frac{1}{2}\left(f(n) - f(1)\right) + \int_1^n \left(\{x\} - \frac{1}{2}\right) f'(x)dx. \quad (10.1)$$

Proof. By integration by parts we find

$$\int_k^{k+1} \left(\{x\} - \frac{1}{2}\right) f'(x)dx = \left(x - k - \frac{1}{2}\right) f(x)\Big|_k^{k+1} - \int_k^{k+1} f(x)\,dx$$

$$= \frac{1}{2}\left(f(k+1) - f(k)\right) - \int_k^{k+1} f(x)\,dx.$$

We add both sides of the previous equality for $1 \le k < n$, and get

$$\int_1^n \left(\{x\} - \frac{1}{2}\right) f'(x)dx = \sum_{k=1}^{n-1} f(k) + \frac{1}{2}\left(f(n) - f(1)\right) - \int_1^n f(x)\,dx.$$

Thus the theorem is proved. $\square$

Corollary 1.1. *Suppose* $f : \mathbb{R} \to \mathbb{R}$ *has a continuous derivative on the interval* $[0, n]$. *Then*

$$\sum_{k=2}^{n} f(k) = \int_{1}^{n} f(x)\, dx + \int_{1}^{n} \{x\} f'(x) dx.$$

Remark. One can write (10.1) as

$$\sum_{k=1}^{n-1} f(k) = \int_{1}^{n} f(x)\, dx - \frac{1}{2}(f(n) - f(1)) + \int_{1}^{n} B_1(\{x\}) f'(x) dx, \qquad (10.2)$$

where $B_1(\cdot)$ is the the Bernoulli polynomial of the first degree. $\triangle$

Theorem 1.2. (Euler summation formula) *Suppose* $f : [1, n] \to \mathbb{R}$ *is of class* C^m, $m, n \in \mathbb{N}^*$. *Then*

$$\sum_{k=1}^{n-1} f(k) = \int_{1}^{n} f(x)\, dx + \sum_{k=1}^{m} \frac{\mathfrak{b}_k}{k!} [f^{(k-1)}(n) - f^{(k-1)}(1)] + R_{nm}, \qquad (10.3)$$

where

$$R_{nm} = \frac{(-1)^{m+1}}{m!} \int_{1}^{n} B_m(\{x\}) f^{(m)}(x)\, dx,$$

B_m *are the Bernoulli polynomials, and* $\mathfrak{b}_k$ *are the Bernoulli numbers.*

Proof. Recall Proposition 8.2 at page 230 saying that

$$B_m'(x) = m B_{m-1}(x), \quad m \in \mathbb{N}^*.$$

By integration by parts and by (5.27) at page 230 we write

$$\frac{1}{m!} \int_{1}^{n} B_m(\{x\}) f^{(m)}(x)\, dx$$

$$= \frac{1}{(m+1)!} \left(B_{m+1}(1) f^{(m)}(n) - B_{m+1}(0) f^{(m)}(1) \right)$$

$$- \frac{1}{(m+1)!} \int_{1}^{n} B_{m+1}(\{x\}) f^{(m+1)}(x)\, dx$$

$$= \frac{(-1)^{m+1} \mathfrak{b}_{m+1}}{(m+1)!} \left(f^{(m)}(n) - f^{(m)}(1) \right) - \frac{1}{(m+1)!} \int_{1}^{n} B_{m+1}(\{x\}) f^{(m+1)}(x)\, dx.$$

Following (10.2) successively, we have

$$\sum_{k=1}^{n-1} f(k) = \int_{1}^{n} f(x)\, dx + \mathfrak{b}_1(f(n) - f(1)) + \int_{1}^{n} B_1(\{x\}) f'(x)\, dx$$

$$= \int_{1}^{n} f(x)\, dx + \mathfrak{b}_1(f(n) - f(1)) + \frac{1}{2}\mathfrak{b}_2(f'(n) - f'(1))$$

$$- \frac{1}{3!} \int_{1}^{n} B_1(\{x\}) f''(x)\, dx$$

$$\cdots$$

$$= \int_1^n f(x)\,dx + \sum_{k=1}^m \frac{b_k}{k!}(f^{(k-1)}(n) - f^{(k-1)}(1))$$

$$+ \frac{(-1)^{m+1}}{m!} \int_1^n B_m(\{x\})f^{(m)}(x)\,dx.$$

Thus (10.3) is proved.   □

Proposition 1.1. *Suppose* $f : [1,n] \to \mathbb{R}$ *is of class* C^m, $m,n \in \mathbb{N}^*$, *and there is a positive constant* c *so that* $|f^{(m)}(x)| < c$. *Then*

$$|R_{nm}| < \frac{4(n-1)c}{(2\pi)^m}.$$

Proof. The estimate follows by (5.24) at page 230.   □

Example 1.1. We show how Theorem 1.2 can be used to get the Stirling formula for $n!$. Set $f(x) = \ln x$, $x \ge 1$. Then

$$f^{(k-1)}(x) = \frac{(-1)^k (k-2)!}{x^{k-1}}.$$

Theorem 1.2 yields

$$\ln(n-1)! = n \ln n - n + 1 - \frac{1}{2}\ln n + \sum_{k=2}^m \frac{(-1)^k B_k}{k(k-1)}\left(\frac{1}{n^{k-1}} - 1\right) + R_{mn}.$$

The limit σ given by

$$\lim_{n\to\infty}\left(\ln n! - n\ln n + n - \frac{1}{2}\ln n\right) = 1 + \sum_{k=2}^m \frac{(-1)^{k+1} B_k}{k(k-1)} + \lim_{n\to\infty} R_{mn}$$

exists because

$$\lim_{n\to\infty} R_{mn} = -\frac{1}{m}\int_1^\infty \frac{B_m(\{x\})}{x^m}\,dx$$

and the improper integral exists because it is bounded by

$$c \cdot \int_1^\infty x^{-m}\,dx,$$

for some constant c. Therefore, we get

$$\ln n! = \left(n + \frac{1}{2}\right)\ln n - n + \sigma + \sum_{k=2}^m \frac{(-1)^k B_k}{k(k-1)n^{k-1}} + O\left(\frac{1}{n^m}\right).$$

In particular, if we choose $m = 5$, we get

$$\ln n! = \left(n + \frac{1}{2}\right)\ln n - n + \sigma + \frac{1}{12n} - \frac{1}{360n^3} + O\left(\frac{1}{n^5}\right).$$

Passing the exponential of the both sides yields

$$n! = e^\sigma \sqrt{n}\left(\frac{n}{e}\right)^n \exp\left(\frac{1}{12n} - \frac{1}{360n^3} + O\left(\frac{1}{n^5}\right)\right).$$

It can be shown that $e^\sigma = \sqrt{2\pi}$. Using

$$\exp(x) = 1 + x + \frac{x^2}{2!} + \frac{x^3}{3!} + \frac{x^4}{4!} + O(x^5),$$

we obtain

$$n! = \sqrt{2\pi n}\left(\frac{n}{e}\right)^n\left(1 + \frac{1}{12n} + \frac{1}{288n^2} - \frac{139}{51840n^3} - \frac{571}{2488320n^4} + O\left(\frac{1}{n^5}\right)\right). \quad \triangle$$

10.2 Algorithm analysis

Divide-and-conquer recurrences appear in many problems tied to analysis of algorithms. Certain sorting methods generate such kind of recurrences. The easier case is covered by the Master Theorem.

Theorem 2.1. (Master Theorem) *Let $a \geq 1$ and $b > 1$ be constants, and $f : \mathbb{N}^* \to \mathbb{R}$ be an asymptotically positive function. Consider the recurrence $T : \mathbb{N}^* \to [0, \infty[$ given by*

$$T(n) = aT(n/b) + f(n),$$

where we interpret n/b as $\lfloor n/b \rfloor$ or $\lceil n/b \rceil$.

(a) *If $f(n) = O(n^{\log_b a - \epsilon})$ for some constant ϵ, then $T(n) = \Theta(n^{\log_b a})$.*
(b) *If $f(n) = \Theta(n^{\log_b a} \log^k n)$ with $k \geq 0$, then $T(n) = \Theta(n^{\log_b a} \log^{k+1} n)$.*
(c) *If $f(n) = \Omega(n^{\log_b a + \epsilon})$ for some constant ϵ, $af(n/b) \leq cf(n)$ for some constant $0 < c < 1$ and for all sufficiently large n, then $T(n) = \Theta(f(n))$.*

Below are some examples of applications of the Master theorem.

$$\begin{aligned}
T(n) &= 3T(n/2) + n^2 &&\Longrightarrow T(n) = \Theta(n^2)\\
T(n) &= 4T(n/2) + n^2 &&\Longrightarrow T(n) = \Theta(n^2 \log n)\\
T(n) &= 2T(n/2) + n &&\Longrightarrow T(n) = \Theta(n \log n)\\
T(n) &= 2^n T(n/2) + n &&a \text{ is not a constant}\\
T(n) &= 1/2\,T(n/2) + n &&a \text{ is not } \geq 1.
\end{aligned}$$

We introduce a more general result than Theorem 2.1. Consider a recurrence of the form

$$T(x) = \begin{cases} \Theta(1), & 1 \le x \le x_0, \\ \sum_{k=1}^{m} a_k T(b_k x) + g(x), & x > x_0 \end{cases} \qquad (10.4)$$

where

(i) $x \ge 1$ is a real number.
(ii) $m \ge 1$ is an integer constant.
(iii) $a_k > 0$ and $b_k \in {]}0, 1{[}$ for all $k = 1, 2, \ldots, m$.
(iv) x_0 is a constant such that

$$x_0 \ge 1/b_k \text{ and } x_0 \ge 1/(1 - b_k), \quad \forall k = 1, 2, \ldots, m.$$

(v) g is a nonnegative function satisfying the *polynomial growth condition*, namely, there exist positive constants c_1, c_2 such that for all $x \ge 1$, $k = 1, 2, \ldots, m$ and $u \in [b_k x, x]$,

$$c_1 g(x) \le g(u) \le c_2 g(x).$$

Theorem 2.2. *Consider the recurrence given in* (10.4) *with the corresponding assumptions. Let p be the unique real number for which*

$$\sum_{k=1}^{m} a_k b_k^p = 1.$$

Then

$$T(x) = \Theta\left(x^p \left(1 + \int_1^x \frac{g(u)}{u^{p+1}} \, du\right)\right). \qquad (10.5)$$

Lemma 10.1. *Suppose g is a nonnegative function satisfying the polynomial growth condition. Then there are positive constants c_3, c_4 such that for $1 \le k \le m$ and all $x \ge 1$,*

$$c_3 g(x) \le x^p \int_{b_k x}^x \frac{g(u)}{u^{p+1}} \, du \le c_4 g(x).$$

Proof. From the polynomial growth condition we have

$$x^p \int_{b_k x}^x \frac{g(u)}{u^{p+1}} \, du \le x^p (x - b_k x) \frac{c_2 g(x)}{\min\{(b_k x)^{p+1}, x^{p+1}\}}$$

$$= \frac{(1 - b_k) c_2}{\min\{1, b_k^{p+1}\}} g(x) \le c_4 g(x)$$

where c_4 is defined so that

$$c_4 \ge \frac{(1 - b_k) c_2}{\min\{1, b_k^{p+1}\}}, \quad \forall k = 1, 2, \ldots, m.$$

Similarly,

$$x^p \int_{b_k x}^x \frac{g(u)}{u^{p+1}}\, du \geq x^p (x - b_k x) \frac{c_1 g(x)}{\max\{(b_k x)^{p+1}, x^{p+1}\}}$$

$$= \frac{(1 - b_k)c_1}{\max\{1, b_k^{p+1}\}}\, g(x) \geq c_3 g(x),$$

where c_3 is defined so that

$$c_3 \leq \frac{(1 - b_k)c_1}{\min\{1, b_k^{p+1}\}}, \quad \forall k = 1, 2, \dots, m. \quad \Box$$

Proof of Theorem 2.2. Consider the intervals $I_0 = [1, x_0]$ and $I_k =]x_0 + k - 1, x_0 + k]$ for $k \geq 1$.

By the definition of x_0, if $x \in I_j$ for some $j \geq 1$, then for $1 \leq k \leq m$, $b_k x \in I_{k'}$ for some $j' < j$ because

$$1 \leq x_0 b_k \leq (x_0 + j - 1)b_k < b_k x \leq b_k(x_0 + j)$$
$$\leq x_0 + j - (1 - b_k)x_0 \leq x_0 + j - 1.$$

Thus the value of T in any interval after $[1, x_0]$ depends only on the values of T in previous intervals.

We first try to find a positive constant c_5 such that for all $x > x_0$

$$T(x) \geq c_5 x^p \left(1 + \int_1^x \frac{g(u)}{u^{p+1}}\, du\right).$$

We prove the previous inequality by induction on j so that $x \in I_j$. For $j = 0$, we have that $T(x) = \Theta(1)$, when $x \in [1, x_0]$, provided c_5 is small enough.

The inductive step is as follows.

$$T(x) = \sum_{k=1}^m a_k T(b_k x) + g(x)$$

$$\geq \sum_{k=1}^m a_k c_5 (b_k x)^p \left(1 + \int_1^{b_k x} \frac{g(u)}{u^{p+1}}\, du\right) + g(x)$$

$$= c_5 x^p \sum_{k=1}^m a_k b_k^p \left(1 + \int_1^x \frac{g(u)}{u^{p+1}}\, du - \int_{b_k x}^x \frac{g(u)}{u^{p+1}}\, du\right) + g(x)$$

$$\geq c_5 x^p \sum_{k=1}^m a_k b_k^p \left(1 + \int_1^x \frac{g(u)}{u^{p+1}}\, du - \frac{c_4}{x^p} g(x)\right) + g(x)$$

$$= c_5 x^p \left(1 + \int_1^x \frac{g(u)}{u^{p+1}}\, du - \frac{c_4}{x^p} g(x)\right) + g(x)$$

$$= c_5 x^p \left(1 + \int_1^x \frac{g(u)}{u^{p+1}}\, du\right) + g(x) - c_5 c_4 g(x)$$

$$\geq c_5 x^p \left(1 + \int_1^x \frac{g(u)}{u^{p+1}}\, du\right)$$

whenever $c_5 < 1/c_4$.

The proof that there is a positive constant c_6 so that for all $x > x_0$,

$$T(x) \leq c_6 x^p \left(1 + \int_1^x \frac{g(u)}{u^{p+1}} \, du \right)$$

is similar. Then (10.5) follows.   □

Remark. If g grows faster than any polynomial, then $T(x) = \Theta(g(x))$. Hence Theorem 2.2 does not necessarily hold if g grows faster than any polynomial.   △

Remark. In analysis of algorithms there often appear recurrences not covered by (10.5). Such an example is supplied by

$$T(x) \leq \sum_{k=1}^m a_k T(\lceil b_k x \rceil) + g(x).   △$$

We now discuss about a larger class of recurrences able to manage floors and ceilings. Namely, we focus on recurrences of the form

$$T(x) = \begin{cases} \Theta(1), & 1 \leq x \leq x_0, \\ \sum_{k=1}^m a_k T(b_k x + h_k(x)) + g(x), & x > x_0 \end{cases} \tag{10.6}$$

where besides assumptions (i)–(v) at page 385 we require

(vi) There is some constant $\epsilon > 0$ for which $|h_i(x)| \leq x/(\log^{1+\epsilon} x)$ for $1 \leq k \leq m$ whenever $x \geq x_0$.

(vii) There exist positive constants c_1, c_2 such that for all $x \geq 1$, $k = 1, 2, \ldots, m$ and $u \in [b_k x + h_k(x), x]$,

$$c_1 g(x) \leq g(u) \leq c_2 g(x);$$

(viii) x_0 is a large enough constant so that for any $k \leq m$ and any $x \geq x_0$,

(viii)$_1$ $$\left(1 - \frac{1}{b_k \log^{1+\epsilon} x} \right)^p \left(1 + 1 \Big/ \log^{\epsilon/2} \left(b_k x + x/\log^{1+\epsilon} x \right) \right)$$
$$\geq 1 + 1/\log^{\epsilon/2} x,$$

(viii)$_2$ $$\left(1 + \frac{1}{b_k \log^{1+\epsilon} x} \right)^p \left(1 - 1 \Big/ \log^{\epsilon/2} \left(b_k x + x/\log^{1+\epsilon} x \right) \right)$$
$$\leq 1 - 1/\log^{\epsilon/2} x,$$

(viii)$_3$ $$\left(1 + 1/\log^{\epsilon/2} x \right) /2 \leq 1,$$

(viii)$_4$ $$2 \left(1 - 1/\log^{\epsilon/2} x \right) \geq 1.$$

Similarly to Lemma 10.1, one can show the following.

Lemma 10.2. *Suppose g is a nonnegative function satisfying* (vii). *Then there are positive constants c_3, c_4 such that for $1 \le k \le m$ and all $x \ge 1$,*

$$c_3 g(x) \le x^p \int_{b_k x + h_k(x)}^{x} \frac{g(u)}{u^{p+1}} \, du \le c_4 g(x).$$

Theorem 2.3. (Akra[1]–Bazzi[2]) *Consider the recurrence given in* (10.6) *with the assumptions* (i)–(iv) *and* (vi)–(viii). *Let p be the unique real number for which*

$$\sum_{k=1}^{m} a_k b_k^p = 1.$$

Then

$$T(x) = \Theta\left(x^p \left(1 + \int_1^x \frac{g(u)}{u^{p+1}} \, du\right)\right). \tag{10.7}$$

Proof. Similar to the proof of Theorem 2.3. We show that there exists a positive constant c_5 such that for all $x > x_0$,

$$T(x) \ge c_5 x^p \left(1 + \frac{1}{\log^{\epsilon/2} x}\right)\left(1 + \int_1^x \frac{g(u)}{u^{p+1}} \, du\right).$$

We show it by induction on j so that $x \in I_j$. For $j = 0$, we have that $T(x) = \Theta(1)$, when $x \in [1, x_0]$, provided c_5 is small enough.

The inductive step is as follows.

$$T(x) = \sum_{k=1}^{m} a_k T(b_k x + h_k(x)) + g(x)$$

$$\ge \sum_{k=1}^{m} a_k c_5 (b_k x + h_k(x))^p \left(1 + 1/\log^{\epsilon/2}(b_k x + h_k(x))\right)$$

$$\cdot \left(1 + \int_1^{b_k x + h_k(x)} \frac{g(u)}{u^{p+1}} \, du\right) + g(x)$$

$$= \sum_{k=1}^{m} a_k b_k^p c_5 x^p \left(1 - \frac{1}{b_k \log^{1+\epsilon} x}\right)\left(1 + 1/\log^{\epsilon/2}\left(b_k x + \frac{x}{\log^{1+\epsilon} x}\right)\right)$$

$$\cdot \left(1 + \int_1^x \frac{g(u)}{u^{p+1}} \, du - \int_{b_k x + h_k(x)}^{x} \frac{g(u)}{u^{p+1}} \, du\right) + g(x)$$

$$\ge \sum_{k=1}^{m} a_k b_k^p c_5 x^p \left(1 + \frac{1}{\log^{\epsilon/2} x}\right)\left(1 + \int_1^x \frac{g(u)}{u^{p+1}} \, du - \frac{c_4}{x^p} g(x)\right) + g(x)$$

$$= c_5 x^p \left(1 + \frac{1}{\log^{\epsilon/2} x}\right)\left(1 + \int_1^x \frac{g(u)}{u^{p+1}} \, du - \frac{c_4}{x^p} g(x)\right) + g(x)$$

---

[1] M. Akra.
[2] L. Bazzi.

$$= c_5 x^p \left(1 + \frac{1}{\log^{\epsilon/2} x}\right)\left(1 + \int_1^x \frac{g(u)}{u^{p+1}}\, du\right)$$

$$+ g(x) - c_5 c_4 \left(1 + \frac{1}{\log^{\epsilon/2} x}\right) g(x)$$

$$\geq c_5 x^p \left(1 + \frac{1}{\log^{\epsilon/2} x}\right)\left(1 + \int_1^x \frac{g(u)}{u^{p+1}}\, du\right)$$

whenever $c_5 < 1/(2c_4)$.

The proof of the upper bound is similar. By induction it is shown there is a positive constant c_6 such that for all $x > x_0$,

$$T(x) \leq c_6 x^p \left(1 - \frac{1}{\log^{\epsilon/2} x}\right)\left(1 + \int_1^x \frac{g(u)}{u^{p+1}}\, du\right).$$

Indeed,

$$T(x) = \sum_{k=1}^m a_k T(b_k x + h_k(x)) + g(x)$$

$$\geq \sum_{k=1}^m a_k c_6 (b_k x + h_k(x))^p \left(1 - 1/\log^{\epsilon/2}(b_k x + h_k(x))\right)$$

$$\cdot \left(1 + \int_1^{b_k x + h_k(x)} \frac{g(u)}{u^{p+1}}\, du\right) + g(x)$$

$$= \sum_{k=1}^m a_k b_k^p c_6 x^p \left(1 + \frac{1}{b_k \log^{1+\epsilon} x}\right)\left(1 - 1/\log^{\epsilon/2}\left(b_k x + x/\log^{1+\epsilon} x\right)\right)$$

$$\cdot \left(1 + \int_1^x \frac{g(u)}{u^{p+1}}\, du - \int_{b_k x + h_k(x)}^x \frac{g(u)}{u^{p+1}}\, du\right) + g(x)$$

$$\geq \sum_{k=1}^m a_k b_k^p c_6 x^p \left(1 - \frac{1}{\log^{\epsilon/2} x}\right)\left(1 + \int_1^x \frac{g(u)}{u^{p+1}}\, du - \frac{c_3}{x^p} g(x)\right) + g(x)$$

$$= c_6 x^p \left(1 - \frac{1}{\log^{\epsilon/2} x}\right)\left(1 + \int_1^x \frac{g(u)}{u^{p+1}}\, du - \frac{c_4}{x^p} g(x)\right) + g(x)$$

$$= c_6 x^p \left(1 - \frac{1}{\log^{\epsilon/2} x}\right)\left(1 + \int_1^x \frac{g(u)}{u^{p+1}}\, du\right)$$

$$+ g(x) - c_6 c_3 \left(1 - \frac{1}{\log^{\epsilon/2} x}\right) g(x)$$

$$\geq c_6 x^p \left(1 - \frac{1}{\log^{\epsilon/2} x}\right)\left(1 + \int_1^x \frac{g(u)}{u^{p+1}}\, du\right)$$

whenever $c_6 < 1/(2c_3)$. Thus (10.7) follows. $\square$

Remark. Intuitively, h_k represents a small perturbation in the argument of T. Because

$$\lfloor b_k x \rfloor = b_k x + (\lfloor b_k x \rfloor - b_k x)$$

and $|\lfloor b_k x \rfloor - b_k x| < 1$, h_k can be used to ignore the floor function. This is also the case of the ceiling function. Therefore by Theorem 2.3,

$$T(n) = T\left(\frac{1}{2}n\right) + n \text{ and } T(n) = T\left(\left\lfloor \frac{1}{2}n \right\rfloor\right) + n$$

have the same asymptotic behavior. $\triangle$

Examples. (a) Consider

$$T(n) = \begin{cases} 1, & n = 0, 1, 2, 3, \\ T\left(\left\lfloor \frac{1}{2}n \right\rfloor\right) + \left(\left\lceil \frac{1}{2}n \right\rceil\right) + n^2, & n = 4, 5, \dots. \end{cases}$$

In this case $p = 1$. Then using Theorem 2.3 we get

$$T(x) = \Theta\left(x^2\left(1 + \int_1^x \frac{u^2}{u^2}\,du\right)\right) = \Theta(x(1+x)) = \Theta(x^2).$$

(b) Consider

$$T(n) = \begin{cases} 0, & n = 0, 1, \\ T\left(\left\lfloor \frac{1}{2}n \right\rfloor\right) + T\left(\left\lceil \frac{1}{2}n \right\rceil\right) + n - 1, & n = 2, 3, \dots. \end{cases}$$

In this case $p = 1$. Then using Theorem 2.3 we get

$$T(x) = \Theta\left(x^2\left(1 + \int_1^x \frac{u-1}{u^2}\,du\right)\right) = \Theta(x(1 + \ln x + 1/x)) = \Theta(x \ln x).   \triangle$$

10.3 Combinatorial estimates

In this section we introduce several combinatorial estimates. These estimates are useful for studying some problems belonging to what is called discrete mathematics.

A finer inequality than the one in Corollary 1.10 at page 100 follows.

Proposition 3.2. *One has*

$$1 < \left(n + \frac{1}{2}\right)\ln\left(1 + \frac{1}{n}\right), \quad \forall n \in \mathbb{N}^*.$$

Proof. Consider the function $f :]0, \infty[\to \infty$ by $f(x) = 1/x$ (Figure 10.1). For $n \in \mathbb{N}^*$, we have

$$\int_n^{n+1} \frac{dx}{x} > \frac{1}{n + 1/2}.$$

Now the result follows. $\square$

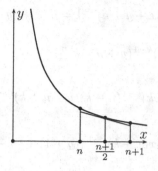

Fig. 10.1. Function $f(x) = 1/x$, $x > 0$

Proposition 3.3. *We have*

$$\left(\frac{n}{e}\right)^n < n! \le 2\left(\frac{n}{2}\right)^n.$$

Proof. The right-hand side inequality has been proved by (1.19) at page 36.

The left-hand side inequality follows immediately by the Stirling formula (6.43) at page 275.

A different approach to the left-hand side inequality is based on induction. For $n = 1$, the inequality holds. Suppose the inequality holds for an integer $n \ge 1$. Then

$$(n+1)! > (n+1)\left(\frac{n}{e}\right)^n = \frac{n+1}{e^n}n^n > \frac{n+1}{e^n}\frac{(n+1)^n}{e} = \left(\frac{n+1}{e}\right)^{n+1}. \quad \square$$

Proposition 3.4. *There hold the following Abel identities.*

$$\sum_{k=0}^{n}\binom{n}{k}x(x+k)^{k-1}(y+n-k)^{n-k} = (x+y+n)^n, \qquad (10.8)$$

$$\sum_{k=0}^{n}\binom{n}{k}(x+k)^{k-1}(y+n-k)^{n-k-1} = \left(\frac{1}{x}+\frac{1}{y}\right)(x+y+n)^{n-1}, \quad (10.9)$$

$$\sum_{k=0}^{n}\binom{n}{k}k^{k-1}(n-k)^{n-k-1} = 2(n-1)(n)^{n-2}. \qquad (10.10)$$

Proof. Identity (10.8) is proven by induction on n. For $n = 0$, it is obvious. Suppose n is a positive integer. Then

$$\frac{\partial (x+y+n)^n}{\partial y} = n(x+y+n)^{n-1} = n\left(x+(y+1)+(n-1)\right)^{n-1},$$

$$\frac{\partial}{\partial y} \sum_{k=0}^{n} \binom{n}{k} x(x+k)^{k-1}(y+n-k)^{n-k}$$

$$= \sum_{k=0}^{n} \binom{n}{k} x(x+k)^{k-1}(n-k)(y+n-k)^{n-k-1}$$

$$= n \sum_{k=0}^{n-1} \binom{n-1}{k} x(x+k)^{k-1}\left((y+1)+(n-1)-k\right)^{(n-1)-k}.$$

By the induction hypothesis, the two right-hand sides are equal. To show (10.8) it is enough to prove it for any y. Choose $y = -x - n$. Then the right-hand side vanishes and the left-hand side is equal to

$$\sum_{k=0}^{n} \binom{n}{k} x(x+k)^{k-1}(-x-k)^{n-k} = x \sum_{k=0}^{n} \binom{n}{k}(-1)^{n-k}(x+k)^{n-1}$$

$$= x \sum_{k=0}^{n} \binom{n}{k}(-1)^{n-k} \sum_{j=0}^{n-1} \binom{n-1}{j} k^j x^{n-j}$$

$$= x \sum_{j=0}^{n-1} \binom{n-1}{j} x^{n-j} \sum_{k=0}^{n} \binom{n}{k}(-1)^{n-k} k^j = x \sum_{j=0}^{n-1} \binom{n-1}{j} x^{n-j} \left\{ \begin{matrix} j \\ n \end{matrix} \right\} = 0,$$

by (3.12). Thus (10.8) is proved.

The left-hand side of (10.8) can be written as

$$\sum_{k=0}^{n} \binom{n}{k} x(x+k)^{k-1}(y+n-k)^{n-k-1}(y+n-k)$$

$$= \sum_{k=0}^{n} \binom{n}{k} x(x+k)^{k-1} y(y+n-k)^{n-k-1}$$

$$+ \sum_{k=0}^{n} \binom{n}{k} x(x+k)^{k-1}(y+n-k)^{n-k-1}(n-k).$$

By (10.8) the last term can be written as

$$\sum_{k=0}^{n-1} n\binom{n-1}{k} x(x+k)^{k-1}\left((y+1)+(n-1)-k\right)^{(n-1)-k}$$

$$= n(x+(y+1)+(n-1))^{n-1} = n(x+y+n)^{n-1}.$$

Hence

$$\sum_{k=0}^{n} \binom{n}{k} x(x+k)^{k-1} y(y+n-k)^{n-k-1}$$

$$= (x+y+n)^n - n(x+y+n)^{n-1} = (x+y)(x+y+n)^{n-1}.$$

Dividing by xy, (10.9) follows.

Subtract

$$\frac{1}{x}(y+n)^{n-1} + \frac{1}{y}(x+n)^{n-1}$$

from the both sides of (10.8). Then

$$\sum_{k=1}^{n-1} \binom{n}{k}(x+k)^{k-1}(y+n-k)^{n-k-1}$$

$$= \frac{1}{x}\left((x+y+n)^{n-1} - (y+n)^{n-1}\right) + \frac{1}{y}\left((x+y+n)^{n-1} - (y+n)^{n-1}\right).$$

Letting $x, y \to 0$, (10.10) follows. □

The Bell numbers have been defined at page 34.

Theorem 3.1. *Let B_n be the nth Bell number. Then*

$$B_n = \frac{1}{e}\sum_{k=0}^{\infty} \frac{k^n}{k!}. \tag{10.11}$$

Let $b(x)$ be the exponential generating function of the sequence (B_n). Then

$$b(x) = \sum_{n=0}^{\infty} \frac{B_n}{n!} x^n = e^{e^x - 1}. \tag{10.12}$$

Proof. Because $\left\{\begin{matrix} n \\ k \end{matrix}\right\} = 0$ for $k > n$, we have

$$B_n = \sum_{k=0}^{\infty} \left\{\begin{matrix} n \\ k \end{matrix}\right\} = \sum_{k=0}^{\infty} \frac{1}{k!} \sum_{m=0}^{k} (-1)^{k-m} \binom{k}{m} m^n$$

$$= \sum_{m=0}^{\infty} \frac{m^n}{m!} \sum_{k=m}^{\infty} \frac{(-1)^{k-m}}{(k-m)!} = \sum_{m=0}^{\infty} \frac{m^n}{m!} \frac{1}{e}.$$

Thus (10.11) is proved.

The exponential generating function of the sequence (B_n) is obtained by (10.11) as follows.

$$b(x) - 1 = \frac{1}{e}\sum_{n=1}^{\infty} \frac{x^n}{n!} \sum_{m=1}^{\infty} \frac{m^{n-1}}{(m-1)!} = \frac{1}{e}\sum_{m=1}^{\infty} \frac{1}{m!}(e^{mx} - 1)$$

$$= \frac{1}{e}\sum_{m=1}^{\infty} \frac{1}{m!} \sum_{n=1}^{\infty} \frac{(mx)^n}{n!} = \frac{1}{e}\left(e^{e^x} - e\right). \quad \square$$

10.3.1 Counting relations, topologies, and partial orders

Let S be a set having n elements, $n \in \mathbb{N}^*$. We have defined the notion of (binary) relation at page 6.

Proposition 3.5. *The number of relations on S is 2^{n^2}.*

Proof. We have n^2 pairs of elements in S. Every subset of pairs is a relation on S.   □

Proposition 3.6. *The number of reflexive relations on S is $2^{n(n-1)}$.*

Proof. We may consider the number of subsets to pairs of elements excluding those of the form (a, a), $a \in S$; that is, there are $n^2 - n$ pairs (a, b) with $a \neq b$.   □

A relation is said to be *irreflexive* if it contains none of the pairs (a, a), $a \in S$.

Proposition 3.7. *The number of irreflexive relations on S is $2^{n(n-1)}$.*

Proof. It is precisely the number of subsets to pairs of elements excluding those of the form (a, a), $a \in S$; that is, there are $n^2 - n$ pairs (a, b) with $a \neq b$.   □

Proposition 3.8. *The number of symmetric relations on S is*

$$2^{\binom{n}{2}+n}.$$

Proof. In a symmetric relation R on S, $a \neq b$ corresponds to the set $\{a, b\}$ so that $\{(a, b), (b, a)\} \subset R$. There are $\binom{n}{2}$ of 2 elements in S. Thus there are $2^{\binom{n}{2}}$ symmetric relations that do not contain any element of the form (a, a), $a \in S$. To each such relation one can add an element of the form (a, a), $a \in S$. Because there are 2^n ways to choose subsets of $\{(a, a) \mid a \in S\}$. Hence the total number of symmetric relations is $2^{\binom{n}{2}}2^n$.   □

A relation is said to be *antisymmetric* if, whenever it contains both (a, b) and (b, a), then $a = b$ (i.e., if a and b are distinct elements in S, then we cannot have (a, b) and (b, a) in the relation at the same time).

Proposition 3.9. *The number of antisymmetric relations on S is*

$$2^n 3^{n(n-1)/2}. \tag{10.13}$$

Proof. We count the possibilities for diagonal ($\{(a, a) \mid a \in S\}$) and off-diagonal elements separately. For diagonal elements there are 2^n possibilities. Suppose we index the elements in S; that is, we can write $S = \{a_1, a_2, \ldots, a_n\}$. There are $n(n-1)/2$ pairs of elements (a_i, a_j) with $i < j$. For each such pair there are three possibilities:

$$(a_i, a_j) \notin S \text{ and } (a_j, a_i) \notin S$$
$$(a_i, a_j) \in S \text{ and } (a_j, a_i) \notin S$$
$$(a_i, a_j) \notin S \text{ and } (a_j, a_i) \in S.$$

Thus we have $3^{n(n-1)/2}$ possibilities. Thus the total number of antisymmetric relations is (10.13).   □

Proposition 3.10. *The number of reflexive and antisymmetric relations on S is*

$$3^{n(n-1)/2}. \tag{10.14}$$

Proof. In this case we can avoid any discussion regarding the elements of the diagonal. Now we invoke the arguments of the second half of the proof to Proposition 3.10 and (10.14) follows.   □

Proposition 3.11. *The number of irreflexive and symmetric relations on S is*

$$2^{n(n-1)/2}. \tag{10.15}$$

Proof. The only pairs involved are under the diagonal, so (10.15) follows.   □

Proposition 3.12. *The number of reflexive and symmetric relations on S is*

$$2^{n(n-1)/2}. \tag{10.16}$$

Proposition 3.13. *Let X be a finite set. There is a bijection between the topologies on X and the quasi-orders on X.*

Proof. The notion of quasi-order is defined at page 8. A quasi-order $\leq$ on X defines a topology, where the open sets are the down-sets $\downarrow A$ for $A \subset X$. This mapping of quasi-orders to topologies is injection which is seen by considering the principal down-sets $\downarrow x$.

Conversely, a topology τ on X defines a quasi-order $\leq_\tau$ by

$$y \leq_\tau x \iff \text{every open set that contains } x \text{ also contains } y.$$

Thus for all open sets $A \in \tau$ we have

$$x \in A \text{ and } y \leq_\tau x \implies y \in A,$$

and hence A is a down-set in the quasi-order.   □

A general formula is not known for the number U_n of topologies on S. A topology is said to be a T_0 *topology* if for any pair of distinct points in S there is an open set containing one point but not the other. A general formula for the number V_n of T_0 topologies on S is not known.

10.3.2 Generalized Fubini numbers

This subsection is dedicated to the introduction and study of generalized Fubini numbers. Their connections to Stirling numbers of the second kind and Eulerian numbers are revealed. The power of numerical series to establish some combinatorial results is made clear.

Choose $n \in \mathbb{N}^*$. The nth *Fubini number* a_n is defined in [37, p. 228] as the number of ways of writing an nth multiple integral. Equivalently, it can be defined as the number of ordered partitions of nonempty subsets of $\mathbb{N}_n^*$. Table 10.1 contains a list of some Fubini numbers.

Table 10.1. Fubini numbers

n=1	1
n=2	3
n=3	13
n=4	75
n=5	541
n=6	4 683
n=7	47 293
n=8	545 835
n=9	7 087 261
n=10	102 247 563
n=11	1 622 632 573
n=12	28 091 567 595
n=13	526 858 348 381
n=14	10 641 342 970 443
n=15	230 283 190 977 853

Regarding a_n, we mention the following results.

(i) Let $\left\{ {n \atop j} \right\}$ be Stirling numbers of the second kind. Then

$$a_n = \sum_{j=1}^{n} j! \left\{ {n \atop j} \right\}. \tag{10.17}$$

(ii) Let σ be a permutation of $\mathbb{N}_n^*$. We say that σ has an increase (decrease) if there there is an i so that $\sigma(i) < \sigma(i+1)$ ($\sigma(i) > \sigma(i+1)$). The Eulerian number $\left\langle {n \atop j} \right\rangle$, as we already defined at page 88, is the number of permutations σ having j increases. Then

$$a_n = \sum_{j=1}^{n} 2^{j-1} \left\langle {n \atop j-1} \right\rangle. \tag{10.18}$$

(iii) It holds

$$a_n = \sum_{j=1}^{\infty} \frac{j^n}{2^{j+1}}. \tag{10.19}$$

(iv) The exponential generating function of the Fubini numbers is

$$\sum_{n=0}^{\infty} a_n \frac{z^n}{n!} = \frac{1}{2 - e^z}, \tag{10.20}$$

where $a_0 = 1$ and $|z| < \ln 2$. We use its equivalent form

$$\sum_{n=1}^{\infty} a_n \frac{z^n}{n!} = \frac{e^z - 1}{2 - e^z}. \tag{10.21}$$

(v) We have the following asymptotic estimate

$$\frac{a_n}{n!} \sim \frac{1}{2(\ln 2)^{n+1}}, \quad n \to \infty. \tag{10.22}$$

The relations (10.17), (10.18), and (10.20) may be found at page 228 in [37]. The relations (10.19), (10.20), and (10.22) are contained in Problem 1.15 in [83]. The relations (10.19) and (10.20) appear in Problem 3.32 in [127]. Number s_n defined by Problem 1.15 in [83] or by Problem 3.32 in [127] is the number of mappings $f : \mathbb{N}_n^* \to \mathbb{N}_n^*$ having the property that if i belongs to the range of f, then j belongs to the range of f, for every $1 \leq j \leq i$. It is easy to remark that $a_n = s_n$, for each $n \in \mathbb{N}^*$, because we can build up a bijection between the set of functions from $\mathbb{N}_n^*$ to $\mathbb{N}_n^*$ with the above-mentioned property and the set of ordered partitions of $\mathbb{N}_n^*$.

In [131] the Fubini numbers are introduced as the numbers of possible combination locks.

Consider a *combination lock* consisting of n buttons that can be pressed in any combination (including multiple buttons at once), but in such a way that each number is pressed exactly once. The number of possible combination locks is given by the number of nonempty disjoint and ordered subsets of the set $\{1, 2, \ldots, n\}$ that contain each number exactly once. By (10.19) we can write

$$a_n = \sum_{j=1}^{\infty} \frac{j^n}{2^{j+1}} = \frac{1}{2} \sum_{j=1}^{\infty} \frac{j^n}{2^j} = \frac{1}{2} \mathrm{Li}_{-n}\left(\frac{1}{2}\right),$$

where $\mathrm{Li}_n(z)$ is the polylogarithm.

We introduce the notion of a k-labeled ordered partition as it appears in [91], [121]. A k-*labeled ordered partition* of $\mathbb{N}_n^*$ is an ordered partition in which each subset is labeled by a positive integer i, where $1 \leq i \leq k$, so that k is given. Then the *generalized Fubini number* $f_{n,k}$ is the cardinal of the set of k-labeled ordered partitions of $\mathbb{N}_n^*$. It is obvious that $f_{n,1} = a_n$, for all $n \in \mathbb{N}^*$. By definition $f_{0,k} = 1$.

We start introducing some properties of $f_{n,k}$s.

Proposition 3.14. *We have that*

$$f_{1,k} = 1, \quad f_{n,k} = 1 + k \sum_{j=1}^{n-1} \binom{n}{j} f_{j,k}, \quad n \geq 2. \tag{10.23}$$

Proof. If $n = 1$, we have only one subset that is labeled 1. Thus the first part of (10.23) is proved.

Suppose $n \geq 2$ and select an integer j, $1 \leq j \leq n - 1$. From $\mathbb{N}_n^*$ we select a subset of cardinal $n - j$ (in $\binom{n}{j}$ ways) and we label it 1. Using the remaining j elements, we can form $k \cdot f_{j,k}$ k-labeled ordered partitions. It results in $k \binom{n}{j} f_{j,k}$ k-labeled ordered partitions. If $j = n$, then we have only one subset which is labeled 1. Hence (10.23) is completely proved.   □

Proposition 3.15. *The exponential generating function of $f_{n,k}$ is*

$$F_k(z) = \sum_{n=1}^{\infty} f_{n,k} \frac{z^n}{n!} = \frac{e^z - 1}{k + 1 - k e^z}. \tag{10.24}$$

Proof. In order to prove (10.24) we use (10.23).

$$F_k(z) = \sum_{n=1}^{\infty} f_{n,k} \frac{z^n}{n!} = z + \sum_{n=2}^{\infty} f_{n,k} \frac{z^n}{n!} = z + \sum_{n=2}^{\infty} \left(1 + k \sum_{j=1}^{n-1} \binom{n}{j} f_{j,k} \right) \frac{z^n}{n!}$$

$$= -1 + e^z + k \sum_{n=2}^{\infty} \sum_{j=1}^{n-1} \binom{n}{j} f_{j,k} \frac{z^n}{n!} = -1 + e^z + k \sum_{n=1}^{\infty} \frac{z^n}{n!} \sum_{i=1}^{\infty} f_{i,k} \frac{z^i}{i!}$$

$$= -1 + e^z + k(e^z - 1) F_k(z).$$

Thus

$$F_k(z) = -1 + e^z + k(e^z - 1) F_k(z).$$

Hence (10.24) follows. For $k = 1$, we get (10.21).   □

Remark. From (10.24), (10.23) immediately follows. Indeed, successively we have

$$(k + 1 - k e^z) F_k(z) = e^z - 1,$$

$$\left(1 - k \sum_{m=1}^{\infty} \frac{z^m}{m!} \right) \left(\sum_{n=1}^{\infty} f_{n,k} \frac{z^n}{n!} \right) = \sum_{n=1}^{\infty} \frac{z^n}{n!}.$$

Identifying the coefficients of z^n in the two sides we get

$$\frac{f_{n,k}}{n!} - k \sum_{\substack{m+j=n \\ m,j \geq 1}} \frac{f_{m,k}}{m! j!} = \frac{1}{n!}, \quad n \geq 1.$$

From here (10.23) follows at once.   △

Proposition 3.16. *It holds*

$$f_{n,k} = \sum_{j=0}^{n-1} \binom{n}{j} f_{j,k} f_{n-j,k-1}, \quad k = 2, 3, \ldots. \tag{10.25}$$

Proof. An analytic and a combinatorial proof are supplied in [91]. We show here only the first one in spite of the perfume of the second one. Consider $k \geq 2$. Then

$$\sum_{n=0}^{\infty} \frac{f_{n,k} z^n}{n!} = 1 + \sum_{n=1}^{\infty} \frac{f_{n,k} z^n}{n!} = \frac{F_k(z)}{F_{k-1}(z)}.$$

From here

$$\sum_{n=1}^{\infty} \frac{f_{n,k} z^n}{n!} = \left(\sum_{n=0}^{\infty} f_{n,k} z^n / n! \right) \left(\sum_{n=1}^{\infty} f_{n,k-1} z^n / n! \right).$$

Identifying the coefficients of z^n in the two sides we get

$$\frac{f_{n,k}}{n!} = \sum_{j=0}^{n-1} \frac{f_{j,k} f_{n-j,k-1}}{j!(n-j)!},$$

that is, (10.25). △

Proposition 3.17. *It holds*

$$f_{n,k} = \sum_{j=1}^{n} k^{j-1} j! \begin{Bmatrix} n \\ j \end{Bmatrix}. \tag{10.26}$$

Proof. The number of ordered partitions of the set $\mathbb{N}_n^*$ such that the number of subsets is j is equal to $j! \begin{Bmatrix} n \\ k \end{Bmatrix}$, [113, p. 91]. Each subset is labeled by a label from the k labels, except the first subset which is labeled 1. □

Remark. The exponential generating function may be obtained by (10.26), as shown in [91]. Taking $k = 1$ in (10.26), we get (10.17). △

Proposition 3.18. *We have*

$$f_{n,k} = \frac{1}{k(k+1)} \sum_{j=1}^{\infty} \left(\frac{k}{k+1} \right)^j j^n. \tag{10.27}$$

For $k = 1$, we get (10.19).

Proof. We use the identity $1/(1 - x) = 1 + x + x^2 + \ldots$, for $|x| < 1$. Successively, we have

$$F_k(z) = -\frac{1}{k} + \frac{1}{k(k+1)} \frac{1}{1 - \frac{k}{k+1}e^z} = -\frac{1}{k} + \frac{1}{k(k+1)} \sum_{j=0}^{\infty} \left(\frac{ke^z}{k+1}\right)^j$$

$$= -\frac{1}{k} + \frac{1}{k(k+1)} \sum_{j=0}^{\infty} \left(\frac{k}{k+1}\right)^j \sum_{n=0}^{\infty} \frac{j^n}{n!} z^n$$

$$= -\frac{1}{k} + \frac{1}{k(k+1)} \sum_{n=0}^{\infty} \sum_{j=0}^{\infty} \left(\frac{k}{k+1}\right)^j j^n \frac{z^n}{n!},$$

from where (10.27) follows. $\square$

Proposition 3.19. *We have*

$$f_{n,k} = \sum_{j=1}^{\infty} k^{n-j}(k+1)^{j-1} \left\langle \begin{matrix} n \\ j \end{matrix} \right\rangle. \tag{10.28}$$

For $k = 1$, we get (10.18).

Theorem 3.2. *Denote $l_k = \ln(1 + 1/k)$, $k \in \mathbb{N}^*$. Then*

$$\frac{f_{n,k}}{n!} = \frac{1}{k(k+1)} \left(\frac{1}{l_k^{n-1}} + \sum_{m=1}^{\infty} \left(\frac{1}{(l_k + 2\pi im)^{n+1}} + \frac{1}{(l_k - 2\pi im)^{n+1}} \right) \right). \tag{10.29}$$

Proof. First we write $F_k(z)$ as a function of cot; then we use the expansion

$$\cot z = \frac{1}{z} + \sum_{k=1}^{\infty} \left(\frac{1}{z - k\pi} + \frac{1}{z + k\pi} \right),$$

where z is not an integer of π.
   Consider $z = 2\pi iu + l_k$. Then

$$F_k(z) = -\frac{1}{k}\left(1 - \frac{1}{k+1-ke^z}\right) = -\frac{1}{k} - \frac{1}{k(k+1)(e^{2\pi iu} - 1)}$$

$$= -\frac{1}{k} - \frac{\cot \pi u - i}{2ik(k+1)} = -\frac{1}{k} + \frac{1}{2k(k+1)} - \frac{\pi \cot \pi u}{2\pi ik(k+1)}$$

$$= -\frac{1}{k} + \frac{1}{2k(k+1)} - \frac{1}{2\pi ik(k+1)}\left(\frac{1}{u} + \sum_{m=1}^{\infty} \left(\frac{1}{u-m} + \frac{1}{u+m}\right)\right)$$

$$= -\frac{1}{k} + \frac{1}{2k(k+1)}$$

$$+ \frac{1}{k(k+1)}\left(\frac{1}{l_k - z} + \sum_{m=1}^{\infty} \left(\frac{1}{l_k - z - 2\pi im} + \frac{1}{l_k - z + 2\pi im}\right)\right),$$

from where (10.29) follows. $\square$
   The expansion (10.29) can be used to get the following to asymptotic estimates as has been done in [91].

Proposition 3.20. *For fixed k we have*

$$\frac{f_{n,k}}{n!} \sim \frac{1}{k(k+1)l_k^{n+1}}, \quad n \to \infty. \tag{10.30}$$

For $k = 1$, we get (10.22).

Proposition 3.21. *For fixed n we have*

$$f_{n,k} \sim \frac{n!}{k(k+1)l_k^{n+1}}, \quad k \to \infty. \tag{10.31}$$

Some values of generalized Fubini numbers are listed in Table 10.2.

Table 10.2. Generalized Fubini numbers

	k=1	k=2	k=3	k=4
n=1	1	1	1	1
n=2	3	5	7	9
n=3	13	37	73	121
n=4	75	365	1 015	2 169
n=5	541	4 501	17 641	48 601
n=6	4 683	66 605	367 927	1 306 809
n=7	47 293	1 149 877	8 952 553	40 994 521
n=8	545 835	22 687 566	248 956 855	1 469 709 369
n=9	7 087 261	503 589 781	7 538 499 561	59 277 466 201
n=10	102 247 563	12 420 052 205	270 733 288 647	2 656 472 295 609

Remark. The problem of finding Fubini numbers is connected to the problem of finding the numbers of logical propositions. More exactly, let $P(x_1, x_2, \ldots, x_n)$ be a propositional function with n free variables. One gets a proposition by binding the variables in some way by the universal or the existential quantifier. This problem is important for one wishing to find the number of nonequivalent propositions from a propositional function by quantification. We mention here just two works on this topic [97] and [114, §6.5].   △

10.3.3 The Catalan numbers and binary trees

We extend the definition of binomial coefficients $\binom{n}{m}$ with $m, n \in \mathbb{N}$ and $m \le n$ to the case of nonnatural entries. Define

$$\binom{r}{k} = \begin{cases} 0, & k < 0 \\ 1, & k = 0 \\ \dfrac{r(r-1)\ldots(r-k+1)}{1 \cdot 2 \cdots k}, & k > 0 \end{cases}$$

for $r \in \mathbb{R}$ and $k \in \mathbb{Z}$. Obviously,

$$\binom{r}{0} = 1, \quad \binom{r}{1} = r, \quad \binom{r}{2} = \frac{r(r-1)}{2},$$

$$\binom{r}{m}\binom{m}{k} = \binom{r}{k}\binom{r-k}{m-k},$$

$$\binom{-r}{k} = (-1)^k \binom{r-k+1}{k}, \quad \forall r \in \mathbb{R}, \quad m, k \in \mathbb{Z}.$$

Then the binomial series (5.18) at page 226 can be written as

$$(1+x)^r = 1 + \sum_{k \geq 1} \binom{r}{k} x^k, \quad |x| < 1. \tag{10.32}$$

The nth Catalan number[3] is defined by

$$C_n = \frac{1}{n+1}\binom{2n}{n}, \quad n \in \mathbb{N}. \tag{10.33}$$

Because

$$C_n = \binom{2n}{n} - \binom{2n}{n-1}, \quad n \in \mathbb{N}^*,$$

we note that C_n is a natural number.

The first Catalan numbers are given in Table 10.3.

Table 10.3. Some Catalan numbers

n	0	1	2	3	4	5	6	7
C_n	1	1	2	5	14	42	132	439

Proposition 3.22. *The Catalan numbers grow asymptotically as*

$$\frac{4^n}{n^{3/2}\sqrt{\pi}}. \tag{10.34}$$

Proof. We use (6.43) in Theorem 4.12 at page 275; that is, for $n \in \mathbb{N}^*$ one has

$$n! = \sqrt{2\pi n}\, n^n\, e^{-n}\,(1 + \alpha_n), \qquad \alpha_n \xrightarrow{n \to \infty} 0,$$

$$(n+1)! = \sqrt{2\pi(n+1)}\,(n+1)^{n+1}\, e^{-(n+1)}\,(1 + \beta_n), \qquad \beta_n \xrightarrow{n \to \infty} 0,$$

$$(2n)! = \sqrt{2\pi(2n)}\,(2n)^{2n}\, e^{-(2n)}\,(1 + \gamma_n), \qquad \gamma_n \xrightarrow{n \to \infty} 0.$$

---

[3] The Catalan numbers were first described in the 18th century by Euler, who was interested in the number of different ways of dividing a polygon into triangles. The numbers are named after Catalan, who discovered the connection to parenthesized expressions.

Then

$$C_n = \frac{1}{n+1}\binom{2n}{n} = \frac{(2n)!}{(n+1)!n!}$$

$$= \frac{1}{\sqrt{\pi}}\frac{e \cdot 4^n}{\sqrt{n+1}}\frac{n^n}{(n+1)(n+1)^n}\frac{1+\gamma_n}{(1+\alpha_n)(1+\beta_n)} \sim \frac{4^n}{n^{3/2}\sqrt{\pi}}. \quad \square$$

The Catalan numbers describe (many other interpretations are contained in [123, pp. 219–229])

- The number of ways a convex polygon with $n+2$ sides can be cut into n triangles by connecting vertices with straight lines. For $n = 3$ we have the cases in Figure 10.2.

Fig. 10.2. Polygon cutting

- The number of different $n+1$ factors can be completely parenthesized (for $n = 3$ we have the following five parenthesizations of four factors: $((ab)c)d$, $(a(bc))d$, $(ab)(cd)$, $a((bc)d)$, and $a(b(cd))$).
- The number of planar binary trees with n nodes (this case is discussed later).
- The number of nondecreasing functions

$$f : \mathbb{N}_n^* \to \mathbb{N}_n^* \quad \text{with} \quad f(x) \leq x, \quad \forall x \in \mathbb{N}^*.$$

For $n = 3$ we have the following functions.

$$\begin{pmatrix} 1 \to 1 \\ 2 \to 1 \\ 3 \to 1 \end{pmatrix}, \quad \begin{pmatrix} 1 \to 1 \\ 2 \to 1 \\ 3 \to 2 \end{pmatrix}, \quad \begin{pmatrix} 1 \to 1 \\ 2 \to 1 \\ 3 \to 3 \end{pmatrix}, \quad \begin{pmatrix} 1 \to 1 \\ 2 \to 2 \\ 3 \to 2 \end{pmatrix}, \quad \begin{pmatrix} 1 \to 1 \\ 2 \to 2 \\ 3 \to 3 \end{pmatrix}.$$

- The number of *Dyck words* of length $2n$. A Dyck word is a string consisting of n as and n bs such that no initial segment of string has more bs than as. For $n = 3$ we have the following Dyck words.

$$aaabbb, \quad abaabb, \quad ababab, \quad aabbab, \quad aababb;$$

- The number of monotonic paths of length $2n$ through an n-by-n grid that that do not rise above the main diagonal. For $n = 3$ we have the cases in Figure 10.3.

Fig. 10.3. Monotonic paths of length $2n$ through an n-by-n grid

Theorem 3.3. *The recurrence of the Catalan numbers is*

$$C_0 = 1, \quad C_{n+1} = \sum_{k=0}^{n} C_k C_{n-k}, \quad n \in \mathbb{N}. \tag{10.35}$$

and its generating function is

$$f(x) = \frac{1 - \sqrt{1 - 4x}}{2x}. \tag{10.36}$$

From here follows (10.33).

Proof. Consider a Dyck word w of length at least $2n + 2$. It can be written in a unique way in the form aw_1bw_2 with (possible empty) Dyck words w_1 and w_2. Then (10.35) follows.

We have $f(x) = 1 + \sum_{n \geq 1} C_n x^n$. By Proposition 3.22 we have that this power series has a positive radius of convergence. Therefore in a certain neighborhood of origin we write $1 + x f^2(x) = f(x)$. From here (10.36) follows.

By (10.36) successively we have

$$f(x) = \frac{1}{2x}\left(1 - \sqrt{1 - 4x}\right) = \frac{1}{2x}\left(1 - (1 - 4x)^{1/2}\right)$$

$$= \frac{1}{2x}\left(1 - \sum_{n \geq 0} \binom{1/2}{n}(-4x)^n\right) = \frac{1}{2x}\left(1 - \sum_{n \geq 0} \binom{1/2}{n}(-1)^n 2^{2n} x^n\right)$$

$$= \sum_{n \geq 1} \frac{1}{n}\binom{2n - 2}{n - 1}x^n = \sum_{n \geq 0} \binom{2n}{n}\frac{x^n}{n + 1}. \qquad \square$$

A *binary tree* is a tree-like structure that is rooted and the vertex of which has at most two children and each child of a vertex is designed as its left or right child.

Let b_n be the number of binary trees having n vertices. Some values of b_n are listed in 10.4.

Figure 10.4 shows the binary trees for $1, 2, 3$, respectively, 4 vertices. We now find the b_ns. By definition we take $b_0 = 1$.

Theorem 3.4. *Suppose a binary tree with n vertices is given. Then*

Table 10.4. Some numbers of binary trees

n	1	2	3	4	5
b_n	1	2	5	14	42

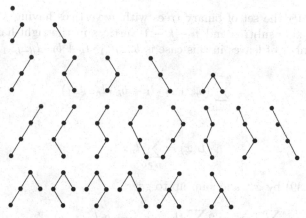

Fig. 10.4. Binary trees

$$b_n = \frac{1}{n+1}\binom{2n}{n} = C_n. \qquad (10.37)$$

Proof. Consider a binary tree with n vertices is given and we mark the root of a binary tree. The remaining $n-1$ vertices are lying either in the left-hand side subtree or in the right-hand side subtree. If the left-hand side subtree contains k vertices, the right-hand side subtree contains $n-k-1$ vertices. Using these n vertices we can form $b_k b_{n-k-1}$ trees, $k=0,1,\ldots,n-1$. Thus

$$b_n = \sum_{k=0}^{n-1} b_k b_{n-k-1}. \quad \square$$

A *leaf* is a terminal vertex (different from a root) of a binary tree. Denote l_n the number of leaves of binary trees with n vertices, $n \geq 2$. By definition we consider $l_0 = 0$ and $l_1 = 1$. Using Figure 10.4, we note in Table 10.5 some numbers of leaves of binary trees.

Table 10.5. Some numbers of leaves

n	2	3	4
l_n	2	6	20

Theorem 3.5. *Consider the binary trees with* n *vertices. Then*

$$l_n = \binom{2n-2}{n-1}. \tag{10.38}$$

Proof. Consider the set of binary trees with n vertices having k vertices in the left-hand side subtree and $n - k - 1$ vertices in the right-hand subtree. Thus the number of leaves in this case is $b_{n-k-1} \cdot l_k + b_k \cdot l_{n-k-1}$. Therefore

$$l_n = \sum_{k=0}^{n-1} (b_{n-k-1} \cdot l_k + b_k \cdot l_{n-k-1}). \tag{10.39}$$

Denote

$$L(x) = \sum_{n \geq 0} l_n x^n.$$

Multiply (10.39) by x^n and sum up to get

$$\sum_{n \geq 2} l_n x^n = 2 \sum_{n \geq 2} (b_{n-k-1} l_k + b_k l_{n-k-1}) x^n.$$

By $l_0 = 0$ and $l_1 = 1$,

$$L(x) - x = 2x L(x) B(x) \implies L(x) = \frac{x}{1 - 2x B(x)} = \frac{x}{\sqrt{1 - 4x}}.$$

Then

$$L(x) = x(1 - 4x)^{-1/2} = x \sum_{n \geq 0} \binom{-1/2}{n} (-4x)^n$$

$$= \sum_{n \geq 0} \binom{2n}{n} x^{n+1} = \sum_{n \geq 1} \binom{2n-2}{n-1} x^n.$$

Hence (10.38) is proved.   □

We look for the number of binary trees with n vertices each one having k leaves. Let $b_n^{(k)}$ be this number. Then

- $b_n^{(k)} = 0$ whenever $k > \left\lfloor \dfrac{n+1}{2} \right\rfloor$.

- $b_n^{(1)} = 2^{n-1}$, for $n \geq 1$ because we can realize a bijection between these binary trees and the ordered $(n-1)$-set of binary elements, say

$$\underbrace{\frac{0}{1}, \frac{0}{1}, \cdots, \frac{0}{1}}_{n-1 \text{ times}}$$

with the convention that 0 means left and 1 means right.

By definition, $b_0^{(0)} = 1$.

Theorem 3.6. *Suppose a binary tree with n vertices is given. Then*

$$b_n^{(k)} = \frac{1}{2k-1}\binom{2k}{k}\binom{n-1}{n-2k+1}2^{n-2k}. \tag{10.40}$$

Proof. If the left-hand side subtree has i vertices and j leaves, then the right-hand side subtree contains $n-i-1$ vertices and $k-j$ leaves. Then the number of these trees is

$$b_i^{(j)} \cdot b_{n-i-1}^{(k-j)}.$$

Summing up in respect to i and j we have

$$b_n^{(k)} = 2b_{n-1}^{(k)} + \sum_{i=1}^{n-2}\sum_{j=1}^{k-1} b_i^{(j)} b_{n-i-1}^{(k-j)}. \tag{10.41}$$

Consider the generating function

$$B^{(k)}(x) = \sum_{n\geq 0} b_n^{(k)} x^n. \tag{10.42}$$

Multiply both sides of (10.41) by x^n and then sum up getting

$$\sum_{n\geq 1} b_n^{(k)} x^n = 2\sum_{n\geq 1} b_{n-1}^{(k)} x^n + \sum_{n\geq 1}\left(\sum_{i=1}^{n-2}\sum_{j=1}^{k-1} b_i^{(j)} b_{n-i-1}^{(k-j)}\right) x^n$$

$$= 2\sum_{n\geq 1} b_{n-1}^{(k)} x^n + \sum_{j=1}^{k-1}\sum_{n\geq 1}\left(\sum_{i=1}^{n-2} b_i^{(j)} b_{n-i-1}^{(k-j)}\right) x^n.$$

From here it follows

$$B^{(k)}(x) = 2xB^{(k)}(x) + x\sum_{j=1}^{k-1} B^{(j)}(x)B^{(k-j)}(x) \implies$$

$$B^{(k)}(x) = \frac{x}{1-2x}\sum_{j=1}^{k-1} B^{(j)}(x)B^{(k-j)}(x). \tag{10.43}$$

Inductively we get

$$B^{(2)}(x) = \frac{x}{1-2x}\left(B^{(1)}(x)\right)^2,$$

$$B^{(3)}(x) = \frac{2x^2}{(1-2x)^2}\left(B^{(1)}(x)\right)^3, \quad B^{(4)}(x) = \frac{5x^3}{(1-2x)^3}\left(B^{(1)}(x)\right)^4.$$

Inspired by the previous equalities, we are looking for the general form of B^k as

$$B^{(k)}(x) = \frac{c_k x^{k-1}}{(1-2x)^{k-1}} \left(B^{(1)}(x) \right)^k, \qquad (10.44)$$

where $c_2 = 1$, $c_3 = 2$, $c_4 = 5$. Substituting (10.44) in (10.43) we get the recurrence for c_ks

$$c_k = \sum_{i=1}^{k-1} c_i \cdot c_{k-i}. \qquad (10.45)$$

If $k = 2$, $c_2 = c_1 c_1$. Thus $c_1 = 1$. By definition, $c_0 = 1$. If $C(x) = \sum_{n \geq 0} c_n x^n$, we have

$$C(x) - 1 - x = (C(x) - 1)^2 \implies C(x) = \frac{3 - \sqrt{1-4x}}{2} \qquad (C(0) = 1).$$

Now

$$C(x) = \frac{3}{2} - \frac{1}{2}(1-4x)^{1/2} = \frac{3}{2} - \frac{1}{2} \sum_{n \geq 0} \frac{-1}{2n-1} \binom{2n}{n} x^n$$

$$= \frac{3}{2} + \sum_{n \geq 0} \frac{1}{2(2n-1)} \binom{2n}{n} x^n = 1 + \sum_{n \geq 1} \frac{1}{2(2n-1)} \binom{2n}{n} x^n.$$

From where immediately

$$c_n = \frac{1}{2(2n-1)} \binom{2n}{n} x^n, \quad n \geq 1.$$

As we noted at the beginning $b_n^{(1)} = 2^{n-1}$, $n \geq 1$. Thus

$$B^{(1)}(x) = \frac{x}{1-2x}$$

and

$$B^{(k)}(x) = \frac{1}{2(2k-1)} \binom{2k}{k} \frac{x^{2k-1}}{(1-2x)^{2k-1}}.$$

Because

$$\frac{1}{(1-x)^m} = \sum_{n \geq 0} \binom{n+m-1}{n} x^n,$$

we have

$$B^{(k)}(x) = \frac{1}{2(2k-1)} \binom{2k}{k} \sum_{n \geq 0} \binom{2k+n-2}{n} 2^n x^{2k+n-1}$$

$$= \frac{1}{2(2k-1)} \binom{2k}{k} \sum_{n \geq 2k-1} \binom{n-1}{n-2k+1} 2^{n-2k+1} x^n.$$

Hence

$$b_n^{(k)} = \frac{1}{2k-1} \binom{2k}{k} \binom{n-1}{n-2k+1} 2^{n-2k},$$

and thus (10.40) is proved. $\square$

10.4 References and comments

The Master Theorem 2.1 may be found in [38] and its generalization Theorems 2.2 and 2.3 may be found in [3] and [80].

Theorem 3.1 may be found in [83, p. 14].

There is no known general formula for the number T_n of transitive relations on S [77], [76], [50].

Equivalent definitions for Fubini numbers are introduced in [122]. Other papers on Fubini numbers are [84] and [98]. The generalized Fubini numbers are introduced in [91].

The Catalan numbers, and Bernoulli, Eulerian, Stirling, and Fibonacci as well are discussed in detail in [60] and [123]. Strictly to Catalan numbers is dedicated [124].

References

1. V. ADAMCHIK and S. WAGON, *π A 2000-year search changes direction*, www.cs.cmu.edu/~adamchik/articles/pi/pi.htm.
2. M. AISSEN, I. J. SCHOENBERG, and A. WHITNEY, *On generating functions of totally positive sequences I*, J. Analyse Math. **2** (1952), 93–103.
3. M. AKRA and L. BAZZI, *On the solutions of linear recurrence equations*, Comput. Optim. Appl. **10** (1998), no. 2, 195–210.
4. G. ALMKVIST, Ch. KRATTENTHALER, and J. PETERSSON, *Some new formulas for π*, Experiment. Math. **12** (2003), no. 4, 441–456.
5. Sz. ANDRÁS and M. MUREŞAN, *Mathematical Analysis and its Applications*, EDP, Bucharest, 2005 (Hungarian).
6. G. E. ANDREWS, R. ASKEY, and R. ROY, *Special Functions*, Cambridge University Press, UK, 1999.
7. L. ARAMĂ and T. MOROZAN, *Problem Book on Mathematical Analysis*, Ed. Universal Pan, Bucharest, 1996 (Romanian).
8. J. ARNDT, *Formulas of the form $k\frac{\pi}{4} = \sum_{i=1}^{n} m_i \arctan(1/x_i)$*, Tech. report, http://www.jjj.de.
9. D. H. BAILEY, *The computation of π to 29,360,000 decimal digits using Borweins' quartically convergent algorithm*, Math. Comp. **50** (1988), no. 181, 283–296.
10. D. H. BAILEY, J. M. BORWEIN, and P. B. BORWEIN, *Ramanujan, modular equations, and approximations to pi or how to compute one billion digits of pi*, Amer. Math. Monthly **96** (1989), no. 3, 201–219.
11. D. H. BAILEY, J. M. BORWEIN, P. B. BORWEIN, and S. PLOUFFE, *The quest for pi*, Math. Intelligencer **19** (1997), no. 1, 50–57, crd.lbl.gov/~dhbailey/dhbpapers/pi-quest.pdf.
12. D. H. BAILEY, J. M. BORWEIN, and R. GIRGENSOHN, *Experimental evaluation of Euler sums*, Experiment. Math. **3** (1994), no. 1, 17–30.
13. D. H. BAILEY, P. B. BORWEIN, and S. PLOUFFE, *On the rapid computation of various polylogarithmic constants*, Math. Comp. **66** (1997), no. 218, 903–913.
14. W. N. BAILEY, *Generalized Hypergeometric Series*, Cambridge Univ. Press, London, 1935, Reprinted by Hafner, New York, 1964.
15. D. M. BĂTINEŢU, *Mathematics Problems for the Second Step of High-school. Sequences*, Albatros, Bucharest, 1979 (Romanian).

16. F. BELLARD, *Computation of the n'th digit of π in any base in $O(n^2)$*, fabrice.bellard.free.fr/pi/, January 10 1997.

17. L. BERGREN, J. BORWEIN, and P. BORWEIN, *Pi: A Source Book*, 3rd ed., Springer-Verlag, New York, 2004.

18. A. A. BLANK, *A simple example of a Weierstrass function*, Amer. Math. Monthly **73** (1966), no. 5, 515–519.

19. H. BOHR and I. MOLLERUP, *Loerbog I Matematisk Analyse*, vol. 3, Kopenhagen, 1922 (Danish).

20. D. BORWEIN, J. M. BORWEIN, and D. M. BRADLEY, *Parametric Euler sum identities*, J. Math. Anal. Appl. **316** (2006), 328–338.

21. J. M. BORWEIN, *The life of π. History and computation*, Australian Colloquia, June 21– July 17 2003, www.cecm.sfu.ca/~jborwein/taks.html.

22. J. M. BORWEIN, D. H. BAILEY, N. J. CALKIN, R. GIRGENSOHN, D. RUSSEL LUKE, and V. H. MOLL, *Experimental Mathematics in Action*, A K Peters, Natick, MA, 2007.

23. J. M. BORWEIN and P. B. BORWEIN, *The arithmetic-geometric mean and fast computation of elementary functions*, SIAM Rev. **26** (1984), no. 3, 351–366.

24. _____, *Pi and the AGM - a Study in Analytic Number Theory and Computational Complexity*, Canadian Mathematical Society Series of Monographs and Advanced Texts, Wiley, New York, 1987.

25. _____, *Challenges in mathematical computing*, (2001), //oldweb.cecm. sfu.ca/preprints/2001pp.html.

26. J. M. BORWEIN, D. J. BROADHURST, and J. KAMNITZER, *Central binomial sums, multiple clausen values, and zeta values*, Experiment. Math. **10** (2001), no. 1, 25–30.

27. R. P. BRENT, *Fast multiple-precision evaluation of elementary functions*, J. Assoc. Comput. Mach. **23** (1976), 242–251.

28. F. BRENTI, *Log-concave and unimodal seqences in algebra, combinatorics, and geometry: an update*, Contemp. Math. **178** (1994), 71–89.

29. J. BROWKIN, J. REMPAŁA, and S. SRASZEWICZ, *25 Lat Olimpiady Matematycznej*, Wydawnictwa Skolne I Pedagogiczne, Warsaw, 1975 (Polish).

30. E. BUSCHE, *Zur Theorie der Function [x]*, J. Reine Angew. Math. **136** (1909), 39 – 81.

31. P. BUSSOTTI, *On the genesis of the Lagrange multipliers*, J. Math. Anal. Appl. **117** (2003), no. 3, 453–459.

32. H. CARTAN, *Calcul Différentiel. Formes Différentielles*, Hermann, Paris, 1967.

33. R. W. CHEN, B. ROSENBERG, and L. A. SHEPP, *A secretary problem with two decision makers*, J. Appl. Probability **34** (1997), no. 4, 1068–1074.

34. W.-S. CHEUNG, *Generalizations of Hölder's inequality*, Internat. J. Math. Math. Sci. **26** (2001), no. 1, 7–10.

35. D. V. CHUDNOVSKY and G. V. CHUDNOVSKY, *The computation of classical constants*, Proc. Nat. Acad. Sci. U.S.A. **86** (1989), 8178–8182.

36. I. COLOJOARĂ, *Mathematical Analysis*, EDP, Bucharest, 1983 (Romanian).

37. L. COMTET, *Advanced Combinatorics*, D. Reidel, Dordrecht, 1974.

38. T. H. CORMEN, C. E. LEISERSON, and R. L. RIVEST, *Introduction to Algorithms*, second ed., MIT Press and McGraw-Hill, Cambridge, MA, 1991.

39. A. COX, *The arithmetic-geometric mean of Gauss*, L'enseignement Mathématique **30** (1984), 275–330.

40. H. DAVENPORT and G. PÓLYA, *On the product of two power series*, Canad. J. Math. **1** (1949), 1–5.

41. B. P. DEMIDOVICH, *Collection of Problems and Exercices on Mathematical Analysis*, 10th ed., Nauka, Moscow, 1990 (Russian).

42. W. G. DOTSON, *Solution to problem 5264*, Amer. Math. Monthly **73** (1966), no. 5, 212.

43. S. B. EKHAD, *Forty "strange" computer-discovered [and computer-proved (of course!)] hypergeometric series evaluations*, www. math.rutgers.edu/~zeilberg/ ekhad/ekhad.html, Oct. 2004.

44. S. B. EKHAD and J. E. MAJEWICZ, *A short computer-generated proof of Abel's identity*, Electron. J. Combin. **3** (1996), no. 2.

45. S. B. EKHAD and M. MOHAMMED, *A WZ proof of a "curious" identity*, Integers **3** (2003), A6.

46. S. B. EKHAD and D. ZEILBERGER, *A WZ proof of Ramanujan's formula for* π, Geometry, Analysis, and Mechanics (J. M. Rassias, ed.), World Scientific, Singapore, 1994, pp. 107–108.

47. R. ENGELKING, *General Topology*, Monografie Matematyczne, PWN, Warszawa, 1977.

48. G. M. FIHTENGOLTZ, *Course on Differential and Integral Calculus*, vol. I-III, Publisher for Mathematics-Physics Literature, Moscow, 1962 (Russian).

49. S. R. FINCH, *Optimal stopping problem*, unpublished note, June 5 2003.

50. _____, *Transitive relations, topologies and partial orders*, unpublished note, June 5 2003.

51. F. A. C. C. FONTES and L. MAGNI, *A generalization of Barbalat's lemma with applications to robust model predictive control*, Proceedings of the MTNS'04 Mathematical Theory of Networks and Systems, Louvain, Belgium, 5-9 July 2004.

52. S. GĂINĂ, E. CĂMPU, and Gh. BUCUR, *Problem Book on Differential and Integral Calculus*, vol. II, Ed. Tehnică, Bucharest, 1966 (Romanian).

53. M. R. GÂTEAUX, *Sur diverses questions de calcul fonctionel*, Bull. Soc. Math. France **50** (1922), no. 1, 1–37, www.numdam.org.

54. R. GAUNTT, *The irrationality of* $\sqrt{2}$, Amer. Math. Monthly **63** (1956), no. 4, 247.

55. B. GNEDENKO, *The Theory of Probability*, Mir, Moscow, 1976.

56. H. GONSKA, P. PIŢUL, and I. RAŞA, *On Peano's form of the Taylor remainder, Voronovskaja's theorem and the commutator of positive linear operators*, Proceedings of the International Conference on Numerical Analysis and Approximation Theory (2006), 55–80, July, 5–8, Cluj-Napoca, Romania.

57. R. W. GOSPER, *Acceleration of series*, Tech. Report AIM-304, MIT, Cambridge, MA, USA, 1974, dspace.mit.edu/handle/1721.1/6088.

58. B. GOURÉVITCH and J. GUILLERA, *Une généralisation des formules BBP aux formules binomiales*, //personal.auna.com/jguillera/, April 2003.

59. M. GOUY, G. HUVENT, and A. LADUREAU, *A la recherche des formules BBP (Bailey, Borwein, Plouffe)*, Stage IREM de LILLE: Calcul formal. Preprint.

60. R. L. GRAHAM, D. E. KNUTH, and O. PATASHNIK, *Concrete Mathematics*, second ed., Addison-Wesley, Reading, MA, 1994, A foundation for computer science.

61. A. GRANAS and J. DUGUNDJI, *Fixed Point Theory*, Springer Monographs in Mathematics, Springer, New York, 2003.

62. J. GUILLERA, *Generators of some Ramanujan's formulas*, Ramanujan J. **11** (2006), no. 1, 41–48.

63. P. R. HALMOS, *Naive Set Theory*, Van Nostrand, Princeton, NJ, 1967.

64. P. HARTMAN, *Ordinary Differential Equations*, Wiley, New York, 1964.

65. Ch. HERMITE, *Sur quelques conséquens arithmétiques des formules de la théorie des fonctions elliptiques*, Acta Math. **5** (1884), no. 1, 297–330.

66. E. HEWITT and K. STROMBERG, *Real and Abstract Analysis*, Springer, New York, 1975, A modern treatment of the theory of functions of a real variable, 3rd printing, Graduate Texts in Mathematics, No. 25.

67. N. R. HOWES, *Modern Analysis and Topology*, Universitext, Springer, New York, 1995.

68. S.-R. HSIAU and J.-R. YANG, *A natural variation of the standard secretary problem*, Statist. Sci. **10** (2000), 639–646.

69. G. HUVENT, *Sur la somme $\sum_{k=0}^{n} 1/\binom{n}{k}$*, //perso.orange.fr/gery.huvent/alternative_index.htm, Nov. 2002.

70. V. A. ILYIN and E. G. POZNYAK, *Fundamentals of Mathematical Analysis*, Mir, Moscow, 1982.

71. T. J. JECH, *Lectures on Set Theory with Particular Emphasis on the Method of Forcing*, Lecture Notes on Mathematics, vol. 217, Springer, Berlin, 1971.

72. D. KAPLAN and L. GLASS, *Understanding Nonlinear Dynamics*, Texts in Applied Mathematics, no. 19, Springer, New York, 1995.

73. J. KARAMATA, *Sur une inégalité relative aux fonctions convexes*, Publ. Math. Univ. Belgrade **1** (1932), 145–158.

74. _____, *Théorémes sur la sommabilité exponentielle et d'autres sommabilités rattachant*, Mathematica (Cluj) **9** (1935), 164–178.

75. M. S. KLAMKIN, *Extension of an inequality*, Univ. Beograd. Publ. Elektrotehn. Fak. Ser. Mat. **7** (1996), 72–73.

76. J. KLAŠKA, *History of the number of finite posets*, Acta Univ. Mathaei Belii Nat. Sci. Ser. Ser. Math. (1997), no. 5, 73–84.

77. _____, *Transitivity and partial order*, Math. Bohem. **122** (1997), no. 1, 75–82.

78. L. F. KLOSINSKI, G. L. ALEXANDERSON, and L. C. LARSON, *The sixty-fourth William Lowell Putnam mathematical competition*, Amer. Math. Monthly **111** (2004), no. 8, 680–690.

79. D. E. KNUTH, *Big omicron and big omega and big theta*, ACM SIGACT News **8** (1976), no. 2, 18–24.

80. T. LEIGHTON, *Notes on better Master theorems for divide-and-conquer recurrences*, Tech. report, MIT, Cambridge, MA, 1996.

81. C. L. F. LINDEMANN, *Die Zahl Pi*, Math. Ann. **20** (1882), 213–225.

82. L. LIU and Y. WANG, *On the log-convexity of combinatorial sequences*, Adv. in Appl. Math. **39** (2007), no. 4, 453–476.

83. L. LOVÁSZ, *Combinatorial Problems and Exercises*, Akadémiai Kiadó and North-Holland, Budapest and Amsterdam, 1979.

84. H. MAASSEN and T. BEZEMBINDER, *Generating random weak orders and the probability of a Condorcet winner*, Soc. Choice Welf. **19** (2002), no. 3, 517–522.

85. O. MAYER, *The Theory of Functions of a Complex Variable*, Acad. R.S.R., Bucharest, 1981 (Romanian).

86. M. MEGAN, *Basis of Mathematical Analysis*, Matuniv, vol. 2, Eurobit, Timişoara, 1997 (Romanian).

87. _____, *Mathematical Analysis*, vol. 2, Mirton, Timişoara, 1999 (Romanian).
88. _____, *Differential and Integral Calculus on* $\mathbb{R}^p$, Faculty of Mathematics, West University Timişoara, Timişoara, 2000 (Romanian).
89. D. S. MITRINOVIČ, J. E. PEČARIČ, and A. M. FINK, *Inequalities Involving Functions and Their Integrals and Derivaties*, Kluwer, Dordrecht, 1991.
90. I. MUNTEAN, *Elementary transcendental functions*, Tech. report, Babeş-Bolyai University, Cluj-Napoca, 1982, (Romanian).
91. M. MUREŞAN, *On the generalized Fubini numbers*, Stud. Cerc. Mat. **37** (1985), no. 1, 70–76 (Romanian).
92. _____, *Introduction to Optimal Control*, Risoprint, Cluj-Napoca, 1999 (Romanian).
93. _____, *Introduction to Set-Valued Analysis*, Cluj University Press, Cluj-Napoca, 1999.
94. _____, *Nonsmooth Analysis and Applications*, Risoprint, Cluj-Napoca, 2001 (Romanian).
95. _____, *Mathematical Analysis and Applications*, Risoprint, Cluj-Napoca, 2005.
96. _____, *Mathematics for Competitions*, Cyprus Math. Soc., Nicosia, 2006.
97. M. MUREŞAN and Gh. TOADER, *A generalization of Fubini numbers*, Studia Univ. Babeş-Bolyai Math. **31** (1986), no. 2, 60–65.
98. Ph. NEMERY and Y. De SMET, *Multicriteria ordered clustering*, Proceedings of the MCDA 63, Porto, Portugal, March 2006.
99. M. NICOLESCU, *Mathematical Analysis*, vol. I, Ed. Tehnică, Bucharest, 1957 (Romanian).
100. S. M. NIKOLSKY, *A Course of Mathematical Analysis*, Mir, Moskow, 1981.
101. I. NIVEN, *Numbers: Rational and Irrational*, The School Mathematics Study Group, Random House, New York, 1961.
102. L. OLSEN, *A new proof of Darboux's theorem*, Amer. Math. Monthly **111** (2004), no. 8, 713 – 715.
103. A. OSTROWSKI, *Sur quelques applications des fonctions convexes et concaves au sens de I. Schur*, J. Math. Pures. Appl. **31** (1952), no. 9, 253–292.
104. P. PAULE, *A proof of a conjecture of Knuth*, Experiment. Math. **5** (1996), no. 2, 83–89.
105. G. PEANO, *Arithmetices Principia, Nova Methodo Exposita*, Boca, Torino, 1898 (Latin).
106. G. PÓLYA and G. SZEGÖ, *Problems and Theorems in Analysis* I, Springer-Verlag, Berlin, 1972.
107. T. POPOVICIU, *Notes sur les fonctions convexes d'ordre supérieur, III*, Mathematica (Cluj) **16** (1940), 74–86.
108. _____, *Notes sur les fonctions convexes d'ordre supérieur, IV*, Disquisit. Math. Phys. **1** (1940), 163–171.
109. _____, *Les fonctiones convexes*, Actualitiés Sci. Ind. no. 992, Hermann et Cie, Paris, 1945.
110. _____, *Numerical Analysis. Basic Notions of Approximative Calculus*, Calculus Theory, Numerical Analysis, and Computer Science, vol. 1, Acad. R. S. R., Bucharest, 1975 (Romanian).
111. S. RABINOWITZ and S. WAGON, *A spigot algorithm for the digits of Pi*, Amer. Math. Monthly **102** (1995), 195–203.
112. S. RAMANUJAN, *Modular equations and approximations to* π, Quart. J. Math. **45** (1914), 350 – 372.

113. J. RIORDAN, *An Introduction to Combinatorial Analysis*, Wiley Publications in Mathematical Statistics, John Wiley & Sons Inc., New York, 1958.

114. N. ROUCHE, P. HABETS, and M. LALOY, *Stability Theory by Liapunov's Direct Method*, Applied Mathematical Sciences, no. 22, Springer, New York, 1977.

115. W. RUDIN, *Principles of Mathematical Analysis*, McGraw-Hill, New York, 1976.

116. I. A. RUS, *Picard operators and applications*, Math. Japon. **58** (2003), no. 1, 191–219.

117. V. A. SADOVNICHIĬ, A. A. GRIGOR'YAN, and S. B. KONYAGIN, *Problems from Student's Mathematics Olympiads*, Ed. Moscow University, 1987 (Russian).

118. E. SALAMIN, *Computation of π using arithmetic-geometric mean*, Math. Comp. **30** (1976), no. 135, 565–570.

119. R. SCOVILLE, *The hat-check problem*, Amer. Math. Monthly **73** (1966), no. 3, 262–265.

120. P. A. SHMELEV, *Series Theory by Problems and Exercises*, Vyshee, Vysshaia shkola, Moskow, 1983 (Russian).

121. N. J. A. SLOANE, *On-line Encyclopedia of Integer Sequences*, www.research. att.com/cgi-bin/access.cgi/as/njas/sequences/eismum.cgi.

122. _____, *On-line Encyclopedia of Integer Sequences*, www.research. att.com/cgi-bin/access.cgi/as/njas/sequences/eisA.cgi?Anum=A0...

123. R. P. STANLEY, *Enumerative Combinatorics. Vol. 2*, Cambridge Studies in Advanced Mathematics, vol. 62, Cambridge University Press, Cambridge, 1999, With a foreword by Gian-Carlo Rota and appendix 1 by Sergey Fomin.

124. _____, *Catalan addendum*, Tech. report, MIT, May 2007, www-math.mit.edu/~rstan/ec/catadd.pdf.

125. Dj. TAKAČI, S. RADENOVIĆ, and A. TAKAČI, *Exercise Book on Series. Numerical Series, Functional Series, and other Problems*, University of Kragujevac, Faculty of Science, Kragujevac, 2000 (Serbian).

126. E. C. TITCHMARSH, *The Theory of Functions*, second ed., Oxford University Press, Oxford, 1939.

127. I. TOMESCU, *Combinatorial Problems and Graph Theory*, EDP, Bucharest, 1981 (Romanian).

128. A. VERNESCU, *Number e and Mathematics of the Exponential*, Ed. Universității din București, Bucharest, 2004 (Romanian).

129. B. VOLK, *Solution to problem 5296*, Amer. Math. Monthly **73** (1966), no. 5, 557.

130. Y. WANG and Y.-N. YEH, *Polynomials with real zeros and Pólya freqency sequences*, J. Combin. Theory Ser. A **109** (2005), no. 1, 63–74.

131. E. W. WEISSTEIN, *Combination Lock*, //mathworld.wolfram.com/CombinationLock.html.

132. _____, *Archimedes' Recurrence Formula*, //mathworld.wolfram.com/ArchimedesRecurrenceFormula.html.

133. H. WILF, *Yet another proof for the enumeration of labelled trees*, www.math.rutgers.edu/~zeilberg/mamarim/mamarimPDF/cayley.pdf.

134. H. WILF and D. ZEILBERGER, *Rational functions certify combinatorial identities*, J. Amer. Math. Soc. **3** (1990), no. 1, 147–158.

135. K. WU and A. LIU, *Rearrangement inequality*, Math. Competitions **8** (1995), no. 1, 53–60.

136. , Gaz. Mat. (Bucharest) (Romanian).
137. _____, Gaz. Mat. Ser. A (Romanian).
138. B.-Q. YUAN, *Refinements of Carleman's inequality*, JIPAM. J. Inequal. Pure Appl. Math. **2** (2001), no. 2, Article 21, 4 p. (electronic).

List of Symbols

Author Index

Subject Index

accumulation point   28, 62
addition   12
application   9
arc   346
  length   346
  smooth   346
arcs
  equivalent   347

ball   58
  closed   58
  open   58
basis   46
  canonical   46

cardinal number   22
ceiling   20
chain rule   195, 301
class   1
collection   1
combination lock   397
compact
  countably   66
  sequentially   66
  set   65
  space   66
complex number set   31
components   154
composition
  of a relation   6
concave   95
congruent to   7
constant
  of Euler–Mascheroni   104

contraction   167
convex   95
convolution
  binomial   96
  ordinary   96
coordinate   41
correspondence   9
covering   65
  open   65
cube root   18

denominator   2
dependent
  linearly   45
derivative
  directional   312
diameter   63
differential
  Gâteaux   297
  weak   296
G-differential   297
dimension   46
  finite   41
directional derivative   312
discontinuity
  first kind   161
  second kind   161
  simple   161
distance   57
  Euclidean   57
  on $\mathbb{C}$ 57
domain
  of a relation   6